Lecture Notes in Computer Science   11965

More information about this series at http://www.springer.com/series/7411

Vladimir M. Vishnevskiy ·
Konstantin E. Samouylov ·
Dmitry V. Kozyrev (Eds.)

Distributed Computer and Communication Networks

22nd International Conference, DCCN 2019
Moscow, Russia, September 23–27, 2019
Revised Selected Papers

 Springer

Editors
Vladimir M. Vishnevskiy
V.A. Trapeznikov Institute of Control
Sciences of Russian Academy of Sciences
Moscow, Russia

Konstantin E. Samouylov ⓘ
Peoples' Friendship University of Russia
Moscow, Russia

Dmitry V. Kozyrev ⓘ
V.A. Trapeznikov Institute of Control
Sciences of Russian Academy of Sciences
Moscow, Russia

Peoples' Friendship University of Russia
Moscow, Russia

ISSN 0302-9743   ISSN 1611-3349 (electronic)
Lecture Notes in Computer Science
ISBN 978-3-030-36613-1   ISBN 978-3-030-36614-8 (eBook)
https://doi.org/10.1007/978-3-030-36614-8

LNCS Sublibrary: SL5 – Computer Communication Networks and Telecommunications

This Springer imprint is published by the registered company Springer Nature Switzerland AG
The registered company address is: Gewerbestrasse 11, 6330 Cham, Switzerland

Preface

This volume contains a collection of revised selected full-text papers presented at the 22nd International Conference on Distributed Computer and Communication Networks (DCCN 2019), held in Moscow, Russia, September 23–27, 2019.

DCCN 2019 is an IEEE (Region 8 + Russia Section) technically cosponsored international conference. It is a continuation of traditional international conferences of the DCCN series, which took place in Sofia, Bulgaria (1995, 2005, 2006, 2008, 2009, 2014); Tel Aviv, Israel (1996, 1997, 1999, 2001); and Moscow, Russia (1998, 2000, 2003, 2007, 2010, 2011, 2013, 2015, 2016, 2017, 2018) in the last 22 years. The main idea of the conference is to provide a platform and forum for researchers and developers from academia and industry from various countries working in the area of theory and applications of distributed computer and communication networks, mathematical modeling, methods of control, and optimization of distributed systems, by offering them a unique opportunity to share their views as well as discuss the perspective developments and pursue collaboration in this area. The content of this volume is related to the following subjects:

1. Communication networks algorithms and protocols
2. Wireless and mobile networks
3. Computer and telecommunication networks control and management
4. Performance analysis, QoS/QoE evaluation, and network efficiency
5. Analytical modeling and simulation of communication systems
6. Evolution of wireless networks toward 5G
7. Centimeter- and millimeter-wave radio technologies
8. Internet of Things and Fog Computing
9. Probabilistic and statistical models in information systems
10. Queuing theory and reliability theory applications
11. High-altitude telecommunications platforms

The DCCN 2019 conference gathered 174 submissions from authors from 26 different countries. From these, 132 high quality papers in English were accepted and presented during the conference. The current volume contains 46 extended papers which were recommended by session chairs and selected by the Program Committee for the Springer post-proceedings.

All the papers selected for the post-proceedings volume are given in the form presented by the authors. These papers are of interest to everyone working in the field of computer and communication networks.

We thank all the authors for their interest in DCCN, the members of the Program Committee for their contributions, and the reviewers for their peer-reviewing efforts.

September 2019

Vladimir Vishnevskiy
Konstantin Samouylov

Organization

DCCN 2019 was jointly organized by the Russian Academy of Sciences (RAS), the V.A. Trapeznikov Institute of Control Sciences of RAS (ICS RAS), the Peoples' Friendship University of Russia (RUDN University), the National Research Tomsk State University, and the Institute of Information and Communication Technologies of Bulgarian Academy of Sciences (IICT BAS).

International Program Committee

V. M. Vishnevskiy (Chair)	ICS RAS, Russia
K. E. Samouylov (Co-chair)	RUDN University, Russia
Ye. A. Koucheryavy (Co-chair)	Tampere University of Technology, Finland
S. M. Abramov	Program Systems Institute of RAS, Russia
S. D. Andreev	Tampere University of Technology, Finland
A. M. Andronov	Riga Technical University, Latvia
N. Balakrishnan	McMaster University, Canada
A. S. Bugaev	Moscow Institute of Physics and Technology, Russia
S. R. Chakravarthy	Kettering University, USA
T. Czachorski	Institute of Computer Science of Polish Academy of Sciences, Poland
A. N. Dudin	Belarusian State University, Belarus
D. Deng	National Changhua University of Education, Taiwan
A. V. Dvorkovich	Moscow Institute of Physics and Technology, Russia
Yu. V. Gaidamaka	RUDN University, Russia
P. Gaj	Silesian University of Technology, Poland
D. Grace	York University, UK
Yu. V. Gulyaev	Kotelnikov Institute of Radio-engineering and Electronics of RAS, Russia
J. Hosek	Brno University of Technology, Czech Republic
V. C. Joshua	CMS College, India
H. Karatza	Aristotle University of Thessaloniki, Greece
N. Kolev	University of São Paulo, Brazil
J. Kolodziej	Cracow University of Technology, Poland
G. Kotsis	Johannes Kepler University Linz, Austria
T. Kozlova Madsen	Aalborg University, Denmark
U. Krieger	University of Bamberg, Germany
A. Krishnamoorthy	Cochin University of Science and Technology, India
A. E. Koucheryavy	Bonch-Bruevich Saint-Petersburg State University of Telecommunications, Russia
Ye. A. Koucheryavy	Tampere University of Technology, Finland

N. A. Kuznetsov	Moscow Institute of Physics and Technology, Russia
L. Lakatos	Budapest University, Hungary
E. Levner	Holon Institute of Technology, Israel
S. D. Margenov	Institute of Information and Communication Technologies of Bulgarian Academy of Sciences, Bulgaria
N. Markovich	ICS RAS, Russia
A. Melikov	Institute of Cybernetics of the Azerbaijan National Academy of Sciences, Azerbaijan
G. K. Miscoi	Academy of Sciences of Moldova, Moldavia
E. V. Morozov	Institute of Applied Mathematical Research of the Karelian Research Centre RAS, Russia
V. A. Naumov	Service Innovation Research Institute (PIKE), Finland
A. A. Nazarov	Tomsk State University, Russia
I. V. Nikiforov	Université de Technologie de Troyes, France
P. Nikitin	University of Washington, USA
S. A. Nikitov	Institute of Radio-engineering and Electronics of RAS, Russia
D. A. Novikov	ICS RAS, Russia
M. Pagano	Pisa University, Italy
E. Petersons	Riga Technical University, Latvia
V. V. Rykov	Gubkin Russian State University of Oil and Gas, Russia
L. A. Sevastianov	RUDN University, Russia
M. A. Sneps-Sneppe	Ventspils University College, Latvia
P. Stanchev	Kettering University, USA
S. N. Stepanov	Moscow Technical University of Communication and Informatics, Russia
S. P. Suschenko	Tomsk State University, Russia
J. Sztrik	University of Debrecen, Hungary
H. Tijms	Vrije Universiteit Amsterdam, The Netherlands
S. N. Vasiliev	ICS RAS, Russia
M. Xie	City University of Hong Kong, Hong Kong, China
Yu. P. Zaychenko	Kyiv Polytechnic Institute, Ukraine

Organizing Committee

V. M. Vishnevskiy (Chair)	ICS RAS, Russia
K. E. Samouylov (Vice Chair)	RUDN University, Russia
D. V. Kozyrev	RUDN University and ICS RAS, Russia
A. A. Larionov	ICS RAS, Russia
S. N. Kupriyakhina	ICS RAS, Russia
S. P. Moiseeva	Tomsk State University, Russia
T. Atanasova	IICT BAS, Bulgaria
I. A. Gudkova	RUDN University, Russia

| S. I. Salpagarov | RUDN University |
| D. Yu. Ostrikova | RUDN University |

Organizers and Partners

Organizers

Russian Academy of Sciences
V.A. Trapeznikov Institute of Control Sciences of RAS
RUDN University
National Research Tomsk State University
Institute of Information and Communication Technologies of Bulgarian Academy of Sciences
Research and Development Company "Information and Networking Technologies"

Support

Information support is provided by the IEEE (Region 8 + Russia Section) and the Russian Academy of Sciences. The conference has been organized with the support of the "RUDN University Program 5-100."

Contents

Analytical Modeling of Distributed Systems

Distributed Systems Applications

Computer and Communication Networks

5G New Radio System Performance Analysis Using Limited Resource Queuing Systems with Varying Requirements

Valeriy Naumov[1] , Vitalii Beschastnyi[2] , Daria Ostrikova[2] ,
and Yuliya Gaidamaka[2,3](✉)

[1] Service Innovations Research Institute, 8 Annankatu Street, Helsinki 00120, Finland
valeriy.naumov@pf.fi
[2] Peoples' Friendship University of Russia (RUDN University),
6 Miklukho-Maklaya Street, Moscow 117198, Russian Federation
{beschastnyy-va,ostrikova-dyu,gaydamaka-yuv}@rudn.ru
[3] Federal Research Center "Computer Science and Control" of the Russian Academy
of Sciences (FRC CSC RAS), 44-2 Vavilov Street, Moscow 119333, Russian Federation

Abstract. Prospective 5G New Radio (NR) systems offer unprecedented capacity boost by the ultradense deployments of small cells operating at mmWave frequencies, with massive available bandwidths. They will facilitate the provisioning of exceptionally demanding mission-critical and resource-hungry applications that are envisaged to utilize the 5G communications infrastructure. In this work, we provide an analytical framework for 5G NR system analysis in terms of queuing theory. We consider a multiservice queuing system with a limited resource with customers that demand varying amount of resource within their service time. Such an approach provides more accurate performance evaluation compared to conventional multiservice models. For the considered model, we propose a method that allows to calculate the stationary probability distribution to the specified accuracy. Our findings are illustrated with a numerical example.

Keywords: 5G NR · Blocking probability · Loss system · Limited resource · Random requirements

1 Introduction

5G NR systems bring a set of new unique challenges to systems designers including much higher propagation losses compared to microwave communications, blockage of propagation paths by small dynamic objects in the channel, the need for efficient electronic beamstreering mechanisms, etc. In real-life outdoor

The publication has been prepared with the support of the "RUDN University Program 5-100" (D.Yu. Ostrikova, visualization). The reported study was partially funded by RFBR, projects Nos. 18-07-00576 (Yu.V. Gaidamaka, methodology and project administration) and 18-37-00380 (V.A. Beschastnyi, numerical analysis).

V. M. Vishnevskiy et al. (Eds.): DCCN 2019, LNCS 11965, pp. 3–14, 2019.
https://doi.org/10.1007/978-3-030-36614-8_1

deployments 5G NR systems mostly suffer from mobile obstacles such as humans and cars, which are often termed "blockers" [1,2].

Depending on the propagation environment and the distance to NR base station (BS), a user equipment (UE) experiencing such type of blockage may either enter outage conditions or lower its modulation and coding scheme such that block error probability at the air interface is satisfied [3]. To target outage situations 3GPP has recently proposed multi-connectivity operation, where several simultaneously active links for adjacent NR BSs are maintained and the connection is transferred between them in case of blockage events [4,5]. When no outage conditions are experiences by UE, the service may continue at the current BS. However, to support the required rate at the air interface more physical resources are needed. When this surplus of resources is not available an ongoing session is dropped or its rate needs to be reduced.

Resource multiserver loss systems are widely used in analysis of communication systems [6,7]. In typical communication scenarios resource is allocated once the session is initiated and the amount of the occupied resource is not changed within the session lifetime. Meanwhile, there are many scenarios that entail resource demand variation, particularly in wireless systems where demand depends on channel quality which is heavily affected by mobility of objects. However, loss systems with varying requirements have not been well studied so far.

In this paper, we analyze a multiserver loss system with varying requirements, derive formulations for the stationary probability distribution, and propose an efficient accurate method that allows to calculate it. Finally, we apply our analytical results to the performance analysis of an 5G NR system. The rest of the paper is organized as follows. First, in Sect. 2 we describe our system model of an NR BS, then in Sect. 3 we introduce the analytical framework can be used as a baseline for building analytical models for 5G NR networks, and the accurate method. Numerical results are presented in Sect. 4. Conclusions are drawn in the last section.

2 System Model

We consider the scenario with 5G NR Base Station (BS) deployment serving point-to-point (unicast) sessions, see Fig. 1. Each BS has a circularly-shaped coverage area of radius R_C estimated using the mmWave propagation model, the set of MCS [8], and network topology. Users are assumed to be randomly distributed according to Poisson Point Process (PPP) with parameter ρ. So the intensity of user requests for service is a Poisson process with parameter $\lambda = \Lambda \rho \pi (d_{LoS}^E)^2$, where Λ is the parameter of exponentially distributed intervals between two consecutive requests from a single UE and ρ is the density of users.

To process a user request BS allocates radio frequency resource of the size that is generally a random variable and determined by the UE location. In accordance with [9], the mmWave path loss L_{dB} for UEs in LoS and nLoS conditions is given by:

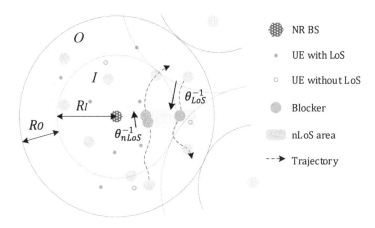

Fig. 1. 5G NR Multiconnectivity deployment.

$$L_{dB}(x) = \begin{cases} 32.4 + 21\log(x) + 20\log f_c, \text{non-blocked}, \\ 47.4 + 21\log(x) + 20\log f_c, \text{blocked}, \end{cases}$$

where f_c is operational frequency measured in MHz, and x is the distance between BS and UE. From these equations we derive maximum distances d_{nLoS}^E and d_{LoS}^E at which a UE can establish a session in LoS and nLoS states respectively by setting value of L_{dB} threshold as the worst possible Signal-to-Noise Ratio (SNR) given by modulation and coding scheme (MCS) for mmWave propagation model [3,10], as shown in (1) and (2):

$$d_{LoS}^E = \left(\frac{P_T \cdot \sqrt[10]{\left(10^{G_T+G_R}\right)}}{\sqrt[10]{10^{N_O W + S}} \cdot 10^{2\log_{10} f_c + 3,24}} \right)^{\frac{1}{\gamma}}, \tag{1}$$

$$d_{nLoS}^E = \left(\frac{P_T \cdot \sqrt[10]{\left(10^{G_T+G_R}\right)}}{\sqrt[10]{10^{N_O W + S}} \cdot 10^{2\log_{10} f_c + 4,74}} \right)^{\frac{1}{\gamma}}, \tag{2}$$

where P_T is the transmit power, G_T - BS antenna gain, G_R - UE antenna gain, N_o - the Johnson-Nyquist noise at one Hz, S - SNR value in dB, W - the bandwidth assigned to the BS, and γ - path loss exponent.

From resource demand point of view, we divide the coverage area into two subareas: the circularly-shaped "inner zone" with radius $R_I = d_{nLoS}^E$; and the annulus-shaped "outer zone" with width $R_O = R_C - R_I$, where R_O is defined by the radius of the coverage area R_C. The coverage area is limited by either d_{nLoS}^E, or Inter-Site Distance (ISD), which is the distance between two adjacent BSs. Thus, $R_C = min(d_{LoS}^E, ISD - d_{nLoS}^E)$, as there is no need to extend the coverage to the inner zone of adjacent BS, where sessions can be maintained regardless of LoS conditions.

LoS towards a NR BS may by blocked by mobile obstacles. In our model we consider the most common type of blockers which is humans that have their blocker radius r_B. Movement of blockers can be modeled with two exponentially distributed random variables with parameters θ_{LoS} and θ_{nLoS} that represent blockage duration (nLoS state intensity) and time intervals between consecutive blockages (LoS state intensity) [6].

Figure 2 illustrates resource allocation process in system time where event a_n^I describes establishment of the n-session from inner zone, a_n^O - establishment of a new n-session from outer zone, d_n^I - end of the n-session service from inner zone, τ_n^{nLoS} - the moment when the UE with n-session enters nLoS conditions, τ_n^{LoS} - the moment when the UE enters LoS conditions, $C = \left\lfloor \frac{W}{s_A} \right\rfloor$ - the number of RBs in the pool of resources, and s_A - the service unit (size of an RB in frequency domain).

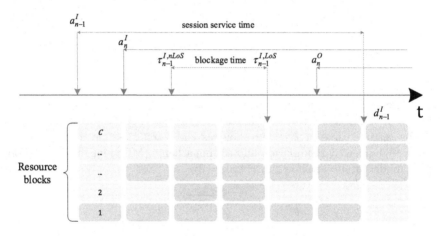

Fig. 2. Resource allocation in system time.

On UE request for new session initiation, NR BS first process the channel quality data and defines a proper MCS. Then, in accordance with selected MCS and requested service type it calculates the required number of RBs r. If r is less or equal to the number of unallocated RBs from the whole pool of resources, the UE request is accepted and the demanded resource is allocated starting from the next time slot. If r is greater than the number of unallocated RBs, BS rejects the request and the session is considered to be dropped.

In our model the resource demands are calculated as mean demands for the corresponding zones and LoS conditions. Particularly, we divide a zone into adjacent annuluses: one for each scheme from the set of MCS. Then, using the spectral efficiently parameter we quantify the demand for each annulus and finally get the weighted demand in accordance with the area of the annulus. Demand of the blocked session from outer zone equals to 0 since the session is handed over to an adjacent NR BS and do not require resource from the origin BS until the end of blockage.

Table 1 describes the mapping between CQIs and spectral efficiency for 3GPP NR systems. This approach follows [6], where authors compute mean demand for the model without multiconnectivity feature.

Table 1. CQI, MCS and Spectral efficiency mapping.

CQI	MCS	Spectral efficiency	SNR in dB	RBs for 50 Mbit/s
1	QPSK, 78/1024	0,15237	−9,478	228
2	QPSK, 120/1024	0,2344	−6,658	149
3	QPSK, 193/1024	0,377	−4,098	93
4	QPSK, 308/1024	0,6016	−1,798	58
5	QPSK, 449/1024	0,877	0,399	40
6	QPSK, 602/1024	1,1758	2,424	30
7	16QAM, 378/1024	1,4766	4,489	24
8	16QAM, 490/1024	1,9141	6,367	19
9	16QAM, 616/1024	2,4063	8,456	15
10	16QAM, 466/1024	2,7305	10,266	13
11	16QAM, 567/1024	3,3223	12,218	11
12	16QAM, 666/1024	3,9023	14,122	9
13	16QAM, 772/1024	4,5234	15,849	8
14	16QAM, 873/1024	5,1152	17,786	7
15	16QAM, 948/1024	5,5547	19,809	7

LoS from a UE towards its NR BS may be blocked while serving an active UE's session. In this case the BS recalculates the new demand for the worsened channel quality. If the number of additionally required RBs is less or equal to the number of unallocated RBs, BS proceeds serving the session. Otherwise, the session is dropped.

In our model we consider the appearance of LoS blockers for users throughout the whole coverage area [11]. On LoS blocker appearance between UE and BS at the distance less then d_{nLoS}^{E} the session can be still maintained by allocating greater amount of resource to compensate the worsened channel quality, thus we define the radius of "inner zone" $R_I = d_{nLoS}^{E}$. The radius of "outer zone" R_O, that defines the distance from BS to the edge of the coverage area, is directly connected to the notion of multiconnectivity. UE from outer zone cannot maintain a session through the closest BS without LoS towards it. However, the session may be handed over to another BS with LoS condition. Therefore, radius of outer zone determines ISD that is $d_{ISD} = R_O + R_I$. If the distance between UE and BS is greater than R_C, user enters outage condition and cannot be provided with a service.

In the outer zone when the LoS to UE is blocked, channel quality decreases to such an extent that it is no longer worth or possible to maintain connection between UE and current BS, and the session is handed over to the adjacent BS. As UE is always served at BS with better channel quality, after handover it will demand greater amount of resource than it was previously allocated by the origin BS. On re-establishment of LoS towards origin BS, the session is handed over back with the former demand. However, at this moment it may appear that origin BS does not have available resource to resume session, in this case it is lost.

3   Analytical Model

3.1   Queuing System Formalization

Consider a multiserver loss system with $N \leq \infty$ servers and a Poisson arrival flow with the rate λ [10]. Assume customer service time be independent of the arrival process, independent of each other, and have a phase-type cumulative distribution function (CDF) $B(x) = 1 - \mathbf{a}e^{x\mathbf{M}}\mathbf{u}$, – where $\mathbf{a}$ is a row vector of length n represents the probability distribution of initial phase, $\mathbf{u}$ - all-ones column vector of length n, and $\mathbf{M} = [\mu_{ij}]$ is non-degenerate square matrix of order n that represents service inter-phase transition rates.

In our model customer demands for resource may vary within the service time and depend on the service phase. On transition to the j-th service phase, the customer demands for r_j resource units. If at the moment of arrival or service phase transition the amount of the demanded resource exceeds the amount of available resource, the customer is considered to be lost. Let us denote $R = \sum_{j=1}^{n} r_j$ the total amount of resource.

The system state at the moment t is described by a Markovian process $\mathbf{X}(t) = (X_1(t), ..., X_n(t))$, where $X_j(t)$ - is the number of customers being served at phase j. The process $\mathbf{X}(t)$ has the finite state space (3).

$$\mathcal{X}(R) = \left\{ (k_1, ..., k_n) \,|\, k_j \geq 0, j = 1, ..., n, \sum_{j=1}^{n} k_j \leq N, \sum_{j=1}^{n} k_j r_j \leq R \right\}. \quad (3)$$

We denote $\mu_{i0} = -\sum_{j=1}^{n} \mu_{ij}$ as service termination rate at i-phase, and $\mu_i = -\mu_{ii}$ - departure rates from i phase. Then, the system of equilibrium equations (SEE) for the process $\mathbf{X}(t)$ can be presented in the form (4) where $I(S)$ is the indicator function of statement S, $\mathbf{r} = (r_1, ..., r_n)$ and $\boldsymbol{\mu} = (\mu_1, ..., \mu_n)$ are column vectors, $\mathbf{e}_i$ is row vector of zeros with figure one at the i-th position.

$$\left(\lambda I\left(\mathbf{ku} < N \right) \sum_{i=1}^{n} I(\mathbf{kr} + r_i \le R)a_i + \mathbf{k\mu} \right) p(\mathbf{k})$$

$$= \lambda \sum_{i=1}^{n} I(k_i > 0)p(\mathbf{k} - \mathbf{e}_i)a_i + I(\mathbf{ku} < N) \sum_{i=1}^{n} p(\mathbf{k} + \mathbf{e}_i)(k_i + 1)\mu_{i0}$$

$$+ I(\mathbf{ku} < N) \sum_{i=1}^{n} \sum_{\substack{j=1 \\ j \ne i}}^{n} I(\mathbf{kr} + r_j > R)p(\mathbf{k} + \mathbf{e}_i)(k_i + 1)\mu_{ij} \qquad (4)$$

$$+ \sum_{i=1}^{n} \sum_{\substack{j=1 \\ j \ne i}}^{n} I(k_j > 0)I(\mathbf{kr} + r_j \le R)p(\mathbf{k} + \mathbf{e}_i - \mathbf{e}_j)(k_i + 1)\mu_{ij}, \mathbf{k} \in \mathcal{X}(R).$$

3.2 Computation Approach

We propose an approach of the loss system analysis with limited resources as an iterative method based on Gauss–Seidel scheme. At the initiation stage we set uniform distribution as the first approximation:

$$p(\mathbf{k}) = \frac{1}{|\mathcal{X}(N, R)|}, \mathbf{k} \in \mathcal{X}(R).$$

Then we calculate consecutive approximations of the probability distribution $\mathbf{p}^m$, where m is the counting number of iteration, until $\left\| \mathbf{p}^m - \mathbf{p}^{m-1} \right\| \le \epsilon$, where ϵ is the required computational accuracy, performing the following steps:

1. Set $C := 0$;
2. Calculate new values of $\mathbf{p}$ using (5);
3. Normalize $p(\mathbf{k})$: $p(\mathbf{k}) = \frac{1}{C}p(\mathbf{k}), \mathbf{k} \in \mathcal{X}(R)$.

$$p(\mathbf{k}) = \frac{1}{\left(\lambda I\left(\mathbf{ku} < N \right) \sum_{i=1}^{n} I(\mathbf{kr} + r_i \le R)a_i + \mathbf{k\mu} \right)}$$

$$\times \left[\lambda \sum_{i=1}^{n} I(k_i > 0)p(\mathbf{k} - \mathbf{e}_i)a_i + I(\mathbf{ku} < N) \sum_{i=1}^{n} p(\mathbf{k} + \mathbf{e}_i)(k_i + 1)\mu_{i0} \right.$$

$$+ I(\mathbf{ku} < N) \sum_{i=1}^{n} \sum_{\substack{j=1 \\ j \ne i}}^{n} I(\mathbf{kr} + r_j > R)p(\mathbf{k} + \mathbf{e}_i)(k_i + 1)\mu_{ij} \qquad (5)$$

$$\left. + \sum_{i=1}^{n} \sum_{\substack{j=1 \\ j \ne i}}^{n} I(k_j > 0)I(\mathbf{kr} + r_j \le R)p(\mathbf{k} + \mathbf{e}_i - \mathbf{e}_j)(k_i + 1)\mu_{ij} \right], \mathbf{k} \in \mathcal{X}(R).$$

4   Numerical Results

In this Section we concentrate on session drop probabilities (6) and mean NR BS resource utilization (7) for a fixed and variable ISD.

$$B = \mathbf{ab}, \mathbf{b} = (b_1, ..., b_n), b_i = \sum_{\substack{\mathbf{k} \in \mathcal{X}(R): \\ \mathbf{kr} + r_i > R}} p(\mathbf{k}) + \sum_{\substack{j \in \{1, ..., n\} \setminus i, \mathbf{k} \in \mathcal{X}(R): \\ \mathbf{kr} - r_j + r_i > R}} p(\mathbf{k}). \qquad (6)$$

$$U = \sum_{\mathbf{k} \in \mathcal{X}} \mathbf{kr} \cdot p(\mathbf{k}). \qquad (7)$$

The default system parameters [6,12] are summarized in Table 2.

Table 2. Input data

Notation	Description	Values
f_c	Operational frequency	28 GHz
W	Bandwidth	1 GHz
s_A	Service unit	1.44 MHz
r_B	Blocker radius	0.4 m
P_T	Transmit power	0.2 W
γ	Path loss exponent	2.1
G_T	BS antenna gain	5.57 dBi
G_R	UE antenna gain	2.58 dBi
$N_o W$	Johnson-Nyquist noise,	−84 dBi
R_O	Outer zone width	200–800 m
R_I	Inner zone radius	200 m
d_{ISD}	Inter site distance	400–1200 m
v	Service Data Rate	50 Mbps
r_1	nLoS session demand in outer zone	16 RBs
r_2	LoS session demand in inner zone	158 RBs
r_3	LoS session demand in outer zone	31 RBs
r_4	nLoS Session demand in inner zone	0 RBs
Λ^{-1}	Session mean inter-arrival time	1000 s
μ^{-1}	Mean service time	30 s
θ_{nLoS}^{-1}	Mean blockage time	2.94 s

For the considered scenario the transition matrix has the form (8) and vector $\mathbf{a} = (\frac{R_I^2}{(R_I+R_O)^2}, 0, 1 - \frac{R_I^2}{(R_I+R_O)^2}, 0)$ which represents initial phase probability distribution. Vector $\mathbf{a}$ has four components, where a_1 and a_3 are equal to the ratio of inner and outer zone areas to the coverage area respectively, while $a_2 = a_4 = 0$

which follows our assumption about impossibility to establish a session in nLoS conditions.

A customer service may be transited between the first and the second phases, as well as between the third and the fourth phases. The first phase corresponds to sessions from inner zone in LoS conditions, while the second - to sessions from inner zone in nLoS conditions. Similarly, the third and the fourth phases correspond to sessions from outer zone in LoS/nLoS conditions.

$$
\mathbf{M} = \begin{pmatrix}
-(\theta^I_{LoS} + \mu) & \theta^I_{LoS} & 0 & 0 \\
\theta^I_{nLoS} & -(\theta^I_{nLoS} + \mu) & 0 & 0 \\
0 & 0 & -(\theta^O_{LoS} + \mu) & \theta^O_{LoS} \\
0 & 0 & \theta^O_{nLoS} & -(\theta^O_{nLoS} + \mu)
\end{pmatrix}. \tag{8}
$$

Figure 3 illustrates the session drop probabilities for two defined zones and the cumulative drop probabilities as a function of density of users ρ. Drop probability in inner zone is significantly lower as the mean demand of I-session is less compared to the outer zone.

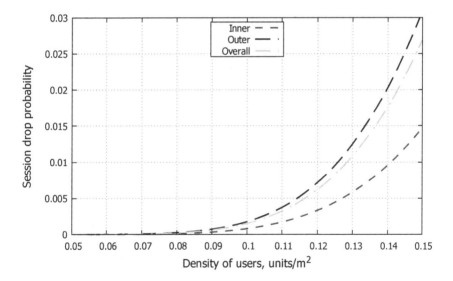

Fig. 3. Session drop probabilities on arrival.

Figure 4 presents the utilization fraction for the two defined zones and the aggregate utilization fraction for LoS and nLoS sessions as a function of density of users ρ. It should be noted that when there is a lack for resources with the load increase, the nLoS session in inner zone suffer most due to their huge demands giving the resource away for sessions with lower demands.

Optimization of ISD is one of the most challenging problems for mmWawe systems [6]. By increasing the density of BSs, i.e. shortening the ISD, it is possible

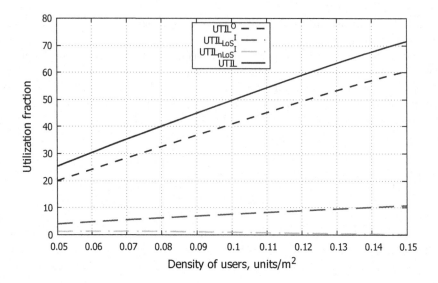

Fig. 4. The mean NR BS resource utilization.

to lower the drop probabilities and improve the quality of experience. However, it requires massive number of expensive equipment to be installed and may not be commercially effective. To address this problem, our modeling approach allows for analyzing session drop probabilities and resource utilization fraction as a function of ISD, as shown in Figs. 5 and 6.

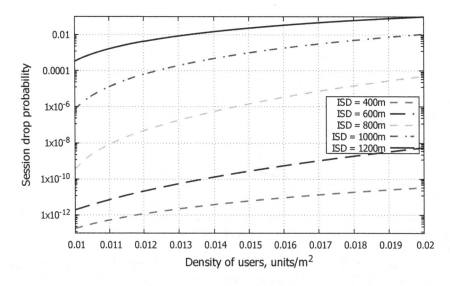

Fig. 5. Session drop probabilities as a function of ISD.

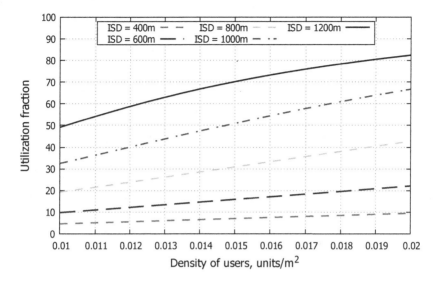

Fig. 6. The mean NR BS resource utilization as a function of ISD.

Both figures show that quality of service and NR BS load is severely undermined in case ISD is greater than 800m. When ISD approaches 1000m, the BS covers the number of users that it is not capable to provide with sufficient resources, resulting in session drop probability higher than 10^{-5}. Thus, knowing the average expected density of users in a specified area, it is possible to predict the cost-effective deployment scheme of NR BSs that will allow to fulfill user and system needs.

5   Conclusion

In this paper, we considered the radio resource allocation process by an 5G NR Base Station. We proposed an analytical model in terms of multiserver queuing system with limited resources and varying resource demands. We also proposed an approximate method for calculation of stationary probability distribution that allows for estimation of user- and system-centric performance metrics that can be further used in performance optimization of practical NR BS deployments, which is highly sought as deploying dense sites increases capital and operating expenditure (CAPEX and OPEX) for operators, thus increasing cost for end-users.

Our future work is to introduce overlapping resource requirements [13] to our model which will allow for modeling multicast services as they are widely used in conventional wired and wireless networks significantly improving resource utilization in presence of users interested in the same content.

References

1. IEEE Standard: 802.11ad-2012: Enhancements for very high throughput in the 60 GHz band. https://ieeexplore.ieee.org/document/6392842/
2. 3GPP: 5G; study on channel model for frequencies from 0.5 to 100 GHz (Release 14), 3GPP TR 38.901, May 2017
3. 3GPP: NR: physical channels and modulation (Release 15), 3GPP TR 38.211, December 2017
4. 3GPP: NR: multi-connectivity; Overall description (Release 15), 3GPP TS 37.34 V15.2.0, June 2018
5. Drago, M., Azzino, T., Polese, M., Stefanović, Č., Zorzi, M.: Reliable video streaming over mmWave with multi connectivity and network coding. In: International Conference on Computing, Networking and Communications (ICNC), pp. 508–512 (2018)
6. Kovalchukov, R., et al.: Improved session continuity in 5G NR with joint use of multi-connectivity and guard bandwidth. In: 2018 IEEE Global Communications Conference (GLOBECOM), pp. 1–7 (2018)
7. Basharin, G., Gaidamaka, Y., Samouylov, K.: Mathematical theory of teletraffic and its application to the analysis of multiservice communication of next generation networks. Autom. Control Comput. Sci. **47**, 62–69 (2013)
8. Petrov, V., et al.: Dynamic multi-connectivity performance in ultra-dense urban mmWave deployments. IEEE J. Sel. Areas Commun. **35**(9), 2038–2055 (2017)
9. Naumov, V., Samouylov, K., Yarkina, N., Sopin, E., Andreev, S., Samuylov, A.: LTE performance analysis using queuing systems with finite resources and random requirements. In: 7th International Congress on Ultra Modern Telecommunications and Control Systems and Workshops (ICUMT), pp. 100–103 (2015)
10. Polese, M., Zorzi, M.: Impact of channel models on the end-to-end performance of Mmwave cellular networks. In: IEEE 19th International Workshop on Signal Processing Advances in Wireless Communications (SPAWC), pp. 1–5 (2018)
11. Venugopal, K., Valenti, M.C., Heath Jr., R.W.: Analysis of millimeter wave networked wearables in crowded environments. In: 2015 49th Asilomar Conference on Signals, Systems and Computers, pp. 872–876 (2015)
12. Mezzavilla, M., et al.: End-to-end simulation of 5G mmWave networks. IEEE Comm. Surv. Tutorials **20**(3), 2237–2263 (2018)
13. Samouylov, K., Gaidamaka, Y.: Analysis of loss systems with overlapping resource requirements. Stat. Pap. **59**(4), 1463–1470 (2018)

On the Performance of LoRaWAN in Smart City: End-Device Design and Communication Coverage

Dmitry Poluektov[1]([⊠]) [iD], Michail Polovov[1], Petr Kharin[1], Martin Stusek[2][iD], Krystof Zeman[2][iD], Pavel Masek[2][iD], Irina Gudkova[1,2,3][iD], Jiri Hosek[2][iD], and Konstantin Samouylov[1,2,3][iD]

[1] Applied Mathematics & Communications Technology Institute,
Peoples' Friendship University of Russia (RUDN University),
Moscow, Russian Federation
poluektov_dmitri@mail.ru
[2] Department of Telecommunication, Brno University of Technology,
Brno, Czech Republic
[3] Institute of Informatics Problems, Federal Research Center Computer Science and Control of the Russian Academy of Sciences, Moscow, Russian Federation

Abstract. Expected communication scenarios within the emerging landscape of Internet of Things (IoT) bring the growth of smart devices connected in the communication network. The communication technologies of greatest interest for IoT are known as Low-Power Wide-Area Networks (LPWANs). Today, there are LPWA technologies (Sigfox, Long-Range Wide Area Network (LoRaWAN), and Narrowband IoT (NB-IoT)) capable to provide energy efficient communication as well as extended communication coverage. This paper provides an analysis of LPWA technologies and describes experimental evaluation of LoRaWAN technology in real conditions. The LoRaWAN technology provides over 150 dB Maximum Coupling Loss (MCL), which together with maximum transmission power (TX) 14 dBm and spreading factor 7 results in theoretical communication distance in units of kilometers. The obtained results from field-deployment in the city of Brno, Czech Republic confirm the initial expectations as it was possible to establish reliable communication between low-end LoRaWAN device and LoRaWAN gateway on the distance up to 6 km.

The described research was supported by the National Sustainability Program under grant LO1401. The publication has been prepared with the support of the "RUDN University Program 5-100" (recipients Irina Gudkova, Dmitry Poluektov, Petr Kharin). The reported study was funded by RFBR, project numbers 18-00-01555(18-00-01685), 19-07-00933 (recipient Konstantin Samouylov). This article is based as well upon support of international mobility project MeMoV, No. CZ.02.2.69/0.0/0.0/16_027/00083710 funded by European Union, Ministry of Education, Youth and Sports, Czech Republic and Brno University of Technology. For the research, infrastructure of the SIX Center (Czech Republic) as well as the 5G Lab RUDN (Russia) was used.

© Springer Nature Switzerland AG 2019
V. M. Vishnevskiy et al. (Eds.): DCCN 2019, LNCS 11965, pp. 15–29, 2019.
https://doi.org/10.1007/978-3-030-36614-8_2

Keywords: LoRaWAN · LPWA · M2M · IoT · Smart cities

1   Introduction

The recent advent of new kind of wireless communication technologies started to form new communication scenarios to enable a broader range of the Internet of Things (IoT) applications [15]. Even though the expected growth of smart devices, published by world's largest telecommunication players in the past, within the IoT may have been overestimated [9], the current forecasts predict the number of smart devices (sensors, actuators etc.) ranging from 20 to 30 billion devices by end of the next year (2020) [6]. When compared with legacy mobile communication technologies (2G/3G/4G) and even with new upcoming heterogeneous networks (5G New Radio (NR)), the communication scenarios for LPWA technologies are not related to high speed data transmissions and low latency. Quite the contrary, the crucial goal of LPWA technologies is to extend the communication coverage and provide possibility to have low hardware complexity of battery powered end-devices [17]. An example of LPWA network is depicted in Fig. 1.

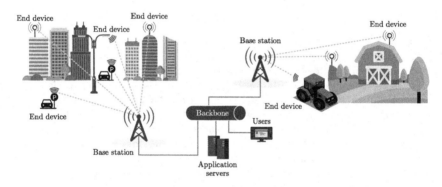

Fig. 1. Typical architecture of an LPWA network.

Although the features of LPWA technologies limit the range of communication scenarios e.g., excluding multimedia services, the number of applications which benefit from the low complexity and low cost of communication modules growths enormous [15]. The reason we can see this trend is that many of LPWA applications are part of massive Machine-Type Communications (mMTC) which belong to the next-generation mobile networks (NGMN) [10]. As the mMTC is about massive deployments of low-cost communication devices, telecommunication players started to define mobile communication technologies to fulfill rising requirements on both end-devices and communication infrastructure [1,8]. These days, two main groups of LPWA technologies exist. The key difference is given to the frequency spectrum in use. First group of technologies (Sigfox, LoRaWAN) uses the license-exempt frequency spectrum i.e., Industrial Scientific and Medical (ISM) bands. On the contrary, the communication technologies defined by the 3rd Generation Partnership Project (3GPP) i.e., Narrowband IoT and LTE Cat-M1 transmit data in the licensed frequency bands. As it was aforementioned, to the contrary of legacy cellular technologies, LPWA technologies have adapted their characteristics to requirements of mMTCs. The summary of key parameters for LPWA technologies in use in Czech Republic, where the field-measurements were conducted, is given in Table 1.

Table 1. Key parameters and characteristics of LPWA technologies (LoRaWAN, Sigfox, NB-IoT, and LTE Cat-M1.)

	LoRaWAN	Sigfox	NB-IoT	LTE Cat-M1
Coverage (MCL)	157 dB	162 dB	164 dB	155 dB
Technology	Proprietary	Proprietary	Open LTE	Open LTE
Spectrum	Unlicensed	Unlicensed	Licensed (LTE/any)	Licensed (LTE/any)
Duty cycle limit	Yes	Yes	No	No
Output power restrictions	Yes (14 dBm = 25 mW)	Yes (14 dBm = 25 mW)	No (23 dBm = 200 mW)	No (23 dBm = 200 mW)
Downlink data rate	0.3–50 kbps	<1 kbps	0.5–27.2 kbps	<300 kbps
Uplink data rate	0.3–50 kbps	<1 kbps	0.3–32.25 kbps	<375 kbps
Battery life/Current consumption	8+ years < 2 uA	10+ years < 2 uA	10+ years < 3 uA	10+ years < 8 uA
Module cost	<$ 10	<$ 10	$ 10 (2019); $ 3 (2020)	<$ 25 (2019)
Security	Medium (AES-128)	Low (AES-128)	Very high (LTE Security)	Very high (LTE Security)

The most popular LPWAN technologies, see Table 1, offer long range data transmissions (up to several tens of kilometers), very low power consumption (years of battery operation), and limited bandwidth (tens of kbps). Another advantage of LPWANs is that they require a much lower investment compared to cellular networks, allowing new telecommunication companies to challenge with current Mobile Network Operators (MNOs). For this reason, many MNOs (e.g., CRa, KPN, Orange, SKTelecom, Bouygues Telecom, Swisscom, and SoftBank) have started to deploy LoRaWAN infrastructure to complement their current cellular networks deployments [13].

1.1 Related Work

Already finished research projects help to understand and define the state of the art related to the LoRa(WAN) deployments in both laboratories and real in-filed scenarios. The parameter which is taken into consideration the most in all of the previous research works focuses on the measurements of coverage distance. Notably, different kind of evaluation techniques were introduced: (i) covered distance [4,5,12,20], (ii) percentage of successfully transmitted data [11], (iii) different payload sizes [2], (iv) different impairments or spreading factors [18,19], and (v) rural vs. urban environments [14].

Exploring the recent research outputs further, in [7], the comparison of a variety of test deployments is provided focusing in detail on selected parameters i.e., Received Signal Strength Indicator (RSSI), Signal to Noise Ratio (SNR), and communication distance. The time needed for data transmissions for different configurations of spreading factor (SF) is given in [16]. In addition, the network throughput as a function of SF for different message sizes is revealed in [3].

1.2 Main Objectives

In this work, we propose the design of a LoRaWAN end-device together with LoRaWAN gateway for supporting data transmissions in Smart City scenario. Since, in the case of LoRaWAN communication technology, the radio equipment is low-cost, we target the created solution to be an example of LoRaWAN end-device in the role of a smart sensor placed in smart houses or utilized in the communication infrastructure of cities.

Even though the research works related to the communication distance in case of LoRaWAN infrastructure do exist, in this work, we cover up the question of reliable communication between the end-device and private LoRaWAN gateway within the infrastructure at Brno University of Technology, Czech Republic. The technology itself was introduced in 2014 and therefore, the aim of this paper is to discuss and share recent findings related to LoRaWAN communication approach. As the LoRaWAN radio module utilizes license-exempt frequency bands, the paper also reveals the spectrum utilization for ISM bands in the Czech Republic and draw the conclusion of the suitability of LoRaWAN technology for mMTC scenarios.

For the aforementioned objectives, the rest of the paper is organized as follows. Section 2 provides the key information of the LoRa(WAN) technology. Further, in Sect. 3, the created testbed is explained as it includes hardware setup and the developed software platform. On top of the testbed, city-specific requirements while transmitting data over LoRaWAN in the city of Brno, Czech Republic are given. The results of the real-life measurements illustrating the practical deployment and utilization of LoRaWAN are presented in Sect. 4. Finally, concluding remarks together with lessons learned are provided in Sect. 5.

2 LoRaWAN Overview

In this section, the physical layer (PHY) is described following the LoRaWAN specification[1]. To cover in nutshell the communication phases of a class A of LoRaWAN device, which belongs to link layer, a scheme of data transmissions is also given in this section, see Fig. 2.

2.1 Physical Layer

As mentioned earlier, the communication between an end-device and a LoRaWAN communication gateway is established utilizing sub-GHz frequency

---

[1] https://lora-alliance.org/resource-hub/lorawanr-specification-v11.

bands which depend on the local frequency regulations[2]. For the purpose of this paper, we address the operation in the EU ISM 868 MHz frequency band. In this configuration, the LoRaWAN enables to use eight PHY options. Six configurations are based on the LoRa modulation with SF ranging between 7 and 12 with dedicated bandwidth 125 kHz. Seventh option uses 250 kHz bandwidth with LoRa modulation and SF 7. The last option implements Gaussian frequency-shift keying (GFSK) which offers up to 50 kbps data rate.

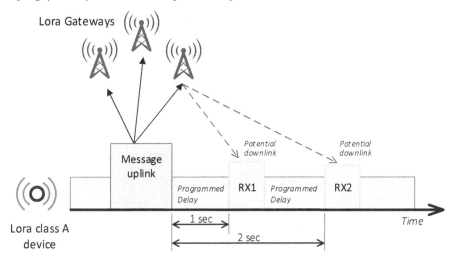

Fig. 2. Communication phases of a class A LoRaWAN device.

The LoRa modulation is based upon chirp spread spectrum (CSS) scheme which uses wide-band linear frequency-modulated pulses. The frequency of those pulses decreases or increases over a specific amount of time based on the encoded information[3]. The utilization of high bandwidth means the radio signals are resistant against in-band and out-of-band interferences, while the use of sufficiently broadband chirps enables to improve robustness against multipath fading while transmitting data. This configuration makes possible to utilize the maximum link budget (MCL) of about 157 dB, which enables to achieve long communication ranges or to significantly reduce the transmit power. Therefore, it is possible to saving the energy of the end-devices as most of them are battery powered. Furthermore, LoRa modulation includes a cyclic error-correcting scheme, which improves the communication robustness by adding redundancy. To improve the spectral efficiency and increase the network capacity, LoRa modulation features six different data rates resulting from orthogonal SF codes. This enables multiple access method on the same channel (frequency) without degrading the communication performance.

---

[2] In the Czech Republic, the regulations are driven by the Czech Telecommunication Office.

[3] https://www.semtech.com/uploads/documents/LoraDesignGuide_STD.pdf.

3   Testbed Proposal

This section describes both the hardware design of end-device and software developed for established LoRaWAN prototype.

3.1   Hardware Setup

The Radio Access Network (RAN) of LoRaWAN network prototype is done utilizing one TTN-GW-868 gateway which enables end-devices to connect to the Internet (towards the remote server i.e., destination node). The end-nodes in use are composed of radio module RN2483 managed by the Arduino Mega micro-controller, see the Fig. 4 where created prototype of LoRaWAN end-device is shown. General infrastructure of intended LoRaWAN network is depicted in Fig. 3.

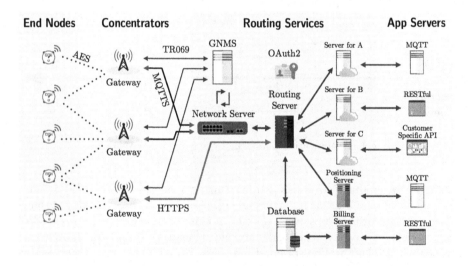

Fig. 3. An illustration scheme of created LoRaWAN infrastructure.

3.2   Software Configuration

To enable the correct data exchange between the RN2483 module and the controlling board, we adopted the existing library RN2483-Arduino-Library[4] targeted for Arduino platform. The library itself contains of 10 functions for setting up the RN2483 module and sending messages. In the process of development of the device for the real-field measurements, we found out that the library lacks functionality to provide implementation of correct data transmissions for

---

[4]   https://github.com/jpmeijers/RN2483-Arduino-Library.

intended LoRaWAN network. As the objectives of our experiment required fine-tuning of all channels i.e., configuration of frequency, cycle, SF, data rate etc., we decided to expand the library with the necessary functions currently missing in the official library. The modification consisted of updating 10 functions and adding 82 additional functions writing them from the scratch. Resulting 92 functions correspond to all the AT commands supported by the RN2483 communication module. On top of the described changes, we developed 6 functions that simplify the processes of setting up the module and sending out data messages, see Table 2.

Table 2. List of selected functions in the modified library for RN2483 module.

	Library functions	Description
1	randomHEXSymbol()	Returns a random hexadecimal number (one digit)
2	getRandomHEXString()	Returns a random hexadecimal string of a given length. To generate "Devaddr", "Deveui" etc.
3	macSetCHPr(String parameter, String channel, String arg0, String arg1 ="")	Sets the specified parameter for the communication channel
4	macSetCH(String channel, String freq, String dcycle, String drrange, String status)	Sets all parameters for a given channel in use
5	macTXHex(String type, String portno, String data)	Sends a string to HEX format. This option enables to send string and not just numbers (valid for LoRaWAN)
6	radioTxHEX(String data)	Sends a string to HEX format. This option enables to send string and not just numbers (valid for LoRa)

Thus, the updated library allows us to use the RN2483 modules with Arduino board as end-device in the architecture which is illustrated in Fig. 3 and can be described as follows:

- **Step 1 [End nodes]:** Configuration of the RN2483 module; generation of messages occurs using the Arduino Mega.
- **Step 2 [End nodes < − > Gateway]:** Transmission of a message by the RN2483 module to the LoRaWAN gateway.
- **Step 3 [Gateway < − > Server LoRaWAN]:** Transmission of the message from the LoRaWAN gateway towards the LoRaWAN server.
- **Step 4 [Web UI]:** Monitoring of all the transmitted messages together with signal quality indicators in the LoRaWAN network.

The Fig. 5 shows the data traffic model. There are four possible options when Arduino generates a data and transmits it to RN2483 module. In a state the RN2483 module receives messages, the module always sends the response back to the Arduino. There exist two responses: (i) the data was sent out or, (ii) the channel is busy. In case of the first state, RN2483 module transmits the message to the LoRaWAN gateway and the module goes into the ACK (Acknowledge) standby mode. In case the channel was busy, the module tries to send out the same message again i.e., the repetition occurs. It is necessary to keep in mind that in case the end-device sent a message and the ACK was not received, then RN2483 module reports this situation to Arduino board by sending NACK (Negative Acknowledge).

The first scenario marked with number 1 in the Fig. 5 corresponds to the ideal case when the message reaches the server and then the ACK returns to the Arduino board. The second scenario stands for the situation where RN2483 module can not send out a message due to the fact that all channels are busy. The third scenario shows the case when the message reaches the remote server via the LoRaWAN gateway, but Arduino does not receive an ACK. The fourth scenario represents the worst case as the transmitted message is not received by the LoRaWAN gateway at all.

The process of communication between the Arduino board and the RN2483 module is illustrated in Algorithm 1.

Algorithm 1. Messaging Process

1: $DataRateIndex = 5$
2: Set Data Rate to $DataRateIndex$
3: $MessageIndex = 1$
4: **while** $MessageIndex \leq 5$ **do**
5:    $Message = $ (message size 20 B)
6:    Send Message $Message$
7:    **while** Channel is busy **do**
8:       Send Message $Message$
9:    Wait 10 seconds
10:   $MessageIndex = MessageIndex + 1$

---

4   Experimental Measurements and Results

In order to assess the practical capabilities and constraints of the LoRaWAN technology, the end-device equipped with RN2483 radio module, see Fig. 4, was configured with SF7, Code Rate (CR) 4/5, bandwidth 125 kHz. Transmission power (TX power) was set to the default TX level i.e., 14 dBm. The message size of transmitted data was 20 B with the transmission interval of 10 s.

For the transmissions, the EU frequency plan was used i.e., three frequencies 863.1 MHz, 868.3 MHz, and 868.5 MHz were utilized. What is important to highlight is the configuration of spreading factor. In this paper, we do use the default SF which equals to 7. This spreading factor does offer the sensitivity of signal level up to -123 dBm. This signal level is almost 15 dB from the sensitivity level in case of SF 12 i.e., -137 dBm. Explanation for this choice is supported by the fact the research team explored the LoRaWAN devices currently available on the market do have the SF 7 configured once the device is used right out of the box.

For characterizing the communication distance and performance of LoRaWAN technology within the private LoRaWAN network, the end-devices were placed in pre-selected points in the city of Brno. At each measurement point, devices were placed approximately at 1,5 m from the ground level for performed on-ground measurements. Figure 6 shows all the measurement points we explored during the measurements. The brown point represents the LoRaWAN gateway located at Brno University of Technology. The measurement points are differentiated by color as the blue points stand for the locations where it was possible to successfully transmit the data between LoRaWAN GW and end-devices. Red points show the locations out of the coverage in case of performed network configuration. Following the above-mentioned, Table 3 presents in detail the obtained data for measurement points where the end-devices were able to communicate with.

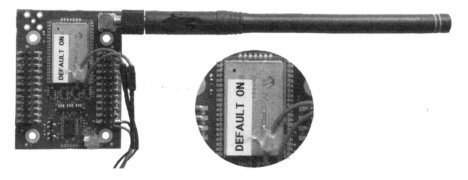

Fig. 4. Microchip RN2483 module based on wireless LoRa technology.

To verify the obtained values for each measurement point during the practical experiments, we decided to use channel attenuation model for communication in urban areas. As the first point, we used the RSSI values from each measurement point to calculate path loss (PL)

$$PL = |RSSI| + SNR + P_{TX} + G_{RX}, \tag{1}$$

where P_{TX} equals to 14 dBm and G_{RX} is 2 dBi. Linear regression was used to derive the path loss exponent n, which equals to 3.8995. Further the EPL value was calculated as

$$EPL = PL_{d0} + 10n\log_{10}\left(\frac{d}{d_0}\right), \tag{2}$$

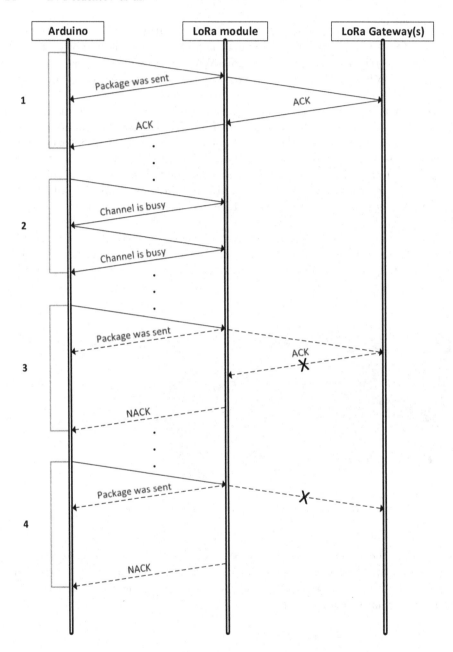

Fig. 5. Simplified visualization of message exchange during data transmission.

where PL_{d0} denotes path loss in reference distance d_0; in our case $173\,\mathrm{m}$ distance was selected.

Table 3. Selected measurement points where the communication between end-device and LoRaWAN gateway was possible to establish, see Fig. 6 for a better understanding of locations of measurement points.

Point no. [-]	Latitude [-]	Longitude [-]	UE to GW distance [km]	Frequency [MHz]	RSSI [dBm]	SNR [dB]
19	49.2205503N	16.6574828E	6.0354	868.3	−108.50	−1.10
36	49.2065614N	16.5957275E	2.6493	868.5	−110.00	−2.15
37	49.2123928N	16.5927236E	2.0022	868.1	−108.75	−3.68
38	49.2158900N	16.5891831E	1.5406	868.5	−107.17	−2.55
39	49.2177753N	16.5878419E	1.3223	868.5	−104.50	2.35
40	49.2218608N	16.5850631E	0.8801	868.1	−102.00	−1.60
41	49.2254764N	16.5826492E	0.5733	868.5	−103.60	4.00
42	49.2281178N	16.5822631E	0.5868	868.1	−94.83	7.70
43	49.2313689N	16.5771239E	0.6117	868.3	−90.17	7.10
44	49.2364686N	16.6275703E	4.0014	868.3	−109.50	−5.20
45	49.2336247N	16.6246092E	3.7122	868.5	−109.00	−8.80
46	49.2319572N	16.6194808E	3.3090	868.1	−102.75	2.93
47	49.2292247N	16.6168414E	3.0722	868.1	−104.80	2.10
48	49.2258478N	16.6179786E	3.1350	868.1	−109.00	−4.90
49	49.2234097N	16.6224633E	3.4734	868.1	−109.17	−2.91
50	49.2228625N	16.6173456E	3.1097	868.5	−103.50	1.81
51	49.2237597N	16.6123031E	2.7350	868.1	−106.67	5.36
52	49.2241097N	16.6071856E	2.3614	868.3	−107.75	0.25
53	49.2259250N	16.6018211E	1.9616	868.1	−108.50	−3.23
54	49.2261347N	16.5973900E	1.6398	868.5	−107.75	0.30
55	49.2240050N	16.5944289E	1.4433	868.3	−109.00	−6.50
56	49.2229122N	16.5913606E	1.2525	868.3	−111.00	−8.50
58	49.2215522N	16.5792797E	0.5990	868.3	−109.50	−5.40
59	49.2237108N	16.5780353E	0.3525	868.3	−97.33	7.25
60	49.2272558N	16.5763831E	0.1737	868.5	−64.20	7.00
61	49.2305631N	16.5718553E	0.5427	868.5	−96.17	4.85
68	49.2525925N	16.5756861E	2.9488	868.5	−114.00	−7.80
69	49.2474100N	16.5791622E	2.3928	868.1	−111.67	−3.40
70	49.2442722N	16.5776389E	2.0334	868.3	−105.67	1.19
71	49.2407558N	16.5741197E	1.6327	868.5	−109.50	−5.80
72	49.2382483N	16.5781108E	1.3742	868.3	−108.00	−1.27
73	49.2372114N	16.5743772E	1.2382	868.1	−108.00	−1.60
74	49.2363008N	16.5702789E	1.1832	868.3	−107.67	−1.38
75	49.2350819N	16.5734544E	1.0059	868.3	−104.00	5.07
76	49.2330642N	16.5757933E	0.7800	868.1	−90.33	8.34
77	49.2325036N	16.5725961E	0.7322	868.3	−92.33	8.26
78	49.2310325N	16.5713947E	0.6040	868.1	−104.50	1.18
79	49.2289447N	16.5722100E	0.3703	868.5	−86.67	9.21
80	49.2260581N	16.5744844E	0.0237	868.1	−93.17	7.73
81	49.2272492N	16.5749994E	0.1308	868.3	−94.17	6.95

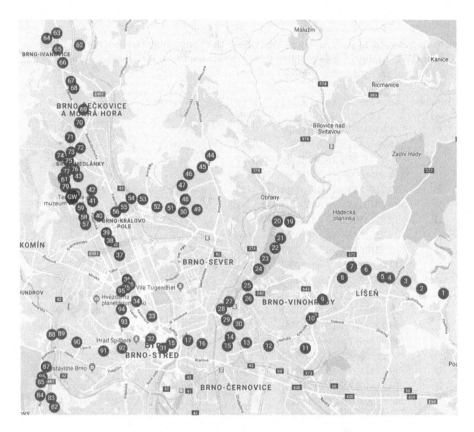

Fig. 6. Map of measurement points in the city of Brno, Czech Republic. (Color figure online)

5   Conclusions

In this paper, we demonstrate the usage of private LoRaWAN network infrastructure for data transmissions in urban area. To evaluate the performance of LoRaWAN technology, a prototype testbed of a LoRaWAN network has been established as an end-devices were constructed from the scratch and the known LoRaWAN gateway was purchased. Table 3 provides the detailed information about communication parameters obtained from the measurement devices capable to communicate with LoRaWAN gateway.

Based on the performed measurements, the signal level of RSSI was ranging from −64 dBm to −111 dBm. As the theoretical sensitivity of LoRa(WAN) module configured with SF 7 is around −123 dBm, the results are in line with theoretical expectations. The Fig. 7 shows the measured path loss (marked by red points) and the expected path loss (solid blue curve) for on-ground scenario based on the measured RSSI.

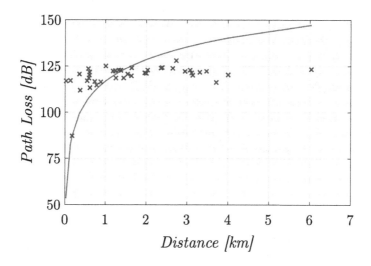

Fig. 7. Path loss for on-ground measurements. (Color figure online)

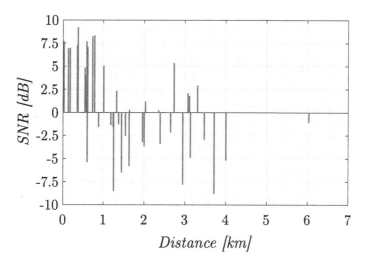

Fig. 8. SNR as a function of communication distance.

Although the presented results indicate the communication possibilities of LoRaWAN technology, the obtained results brought to light the limitations of the technology in question. First concern to point out here is we took into the consideration only measurements using 125 kHz bandwidth with the SF 7. Despite the fact the obtained results provide some degree about the communication capabilities of the technology in use, it would be definitely beneficial to explore in more detail the measurements covering different communication parameters of LoRaWAN technology i.e., bandwidth, spreading factor, code rate, modulations, and environment of use (rural vs. urban areas). In future, the other spreading factors, levels of bandwidth, and modulation schemes will be taken into the account. Also, the LoRaWAN technology is not the only available LPWA technology in Czech Republic at the time of writing this paper (08/2019) (Fig. 8).

For the future evaluation, into the consideration we can take technologies like IEEE 802.15.4k, Narrowband IoT, Sigfox, LTE Cat-M1, or Weightless. Therefore, the evaluation of performance of aforementioned technologies and definition of communication use-cases for them is the rising topic and will be covered in upcoming research works as an output of the long-term cooperation between the Brno University of Technology and the Peoples' Friendship University of Russia (RUDN University).

References

1. LPWAN emerging as fastest growing IoT communication technology - 1.1 billion IoT connections expected by 2023, LoRa and NB-IoT the current market leaders. https://iot-analytics.com/lpwan-market-report-2018-2023-new-report/
2. Aref, M., Sikora, A.: Free space range measurements with Semtech LoRa^{TM} technology. In: 2014 2nd International Symposium on Wireless Systems within the Conferences on Intelligent Data Acquisition and Advanced Computing Systems, pp. 19–23. IEEE (2014)
3. Augustin, A., Yi, J., Clausen, T., Townsley, W.: A study of LoRa: long range & low power networks for the Internet of Things. Sensors **16**(9), 1466 (2016)
4. Cenedese, A., Zanella, A., Vangelista, L., Zorzi, M.: Padova smart city: an Urban Internet of Things experimentation. In: Proceeding of IEEE International Symposium on a World of Wireless, Mobile and Multimedia Networks 2014, pp. 1–6. IEEE (2014)
5. Centenaro, M., Vangelista, L., Zanella, A., Zorzi, M.: Long-range communications in unlicensed bands: the rising stars in the IoT and smart city scenarios. IEEE Wirel. Commun. **23**(5), 60–67 (2016)
6. Cisco Visual Networking Index: Cisco Visual Networking Index: Forecast and Trends, 2017–2022. White Paper (2019)
7. Marais, J.M., Malekian, R., Abu-Mahfouz, A.M.: LoRa and LoRaWAN testbeds: a review. In: 2017 IEEE Africon, pp. 1496–1501. IEEE (2017)
8. Markova, E., et al.: Flexible spectrum management in a Smart City within licensed shared access framework. IEEE Access **5**, 22252–22261 (2017)
9. Nordrum, A., et al.: Popular Internet of Things forecast of 50 billion devices by 2020 is outdated. IEEE Spectr. **18**(3) (2016)
10. Ometov, A., et al.: System-level analysis of IEEE 802.11 ah technology for unsaturated MTC traffic. Int. J. Sens. Netw. **26**(4), 269–282 (2018)
11. Petajajarvi, J., Mikhaylov, K., Roivainen, A., Hanninen, T., Pettissalo, M.: On the coverage of LPWANs: range evaluation and channel attenuation model for LoRa technology. In: 2015 14th International Conference on ITS Telecommunications (ITST), pp. 55–59. IEEE (2015)
12. Petrić, T., Goessens, M., Nuaymi, L., Toutain, L., Pelov, A.: Measurements, performance and analysis of LoRa FABIAN, a real-world implementation of LPWAN. In: 2016 IEEE 27th Annual International Symposium on Personal, Indoor, and Mobile Radio Communications (PIMRC), pp. 1–7. IEEE (2016)

13. Pötsch, A., Haslhofer, F.: Practical limitations for deployment of LoRa gateways. In: 2017 IEEE International Workshop on Measurement and Networking (M&N), pp. 1–6. IEEE (2017)
14. Radcliffe, P.J., Chavez, K.G., Beckett, P., Spangaro, J., Jakob, C.: Usability of LoRaWAN technology in a central business district. In: 2017 IEEE 85th Vehicular Technology Conference (VTC Spring), pp. 1–5. IEEE (2017)
15. Raza, U., Kulkarni, P., Sooriyabandara, M.: Low power wide area networks: an overview. IEEE Commun. Surv. Tutor. **19**(2), 855–873 (2017)
16. Vatcharatiansakul, N., Tuwanut, P., Pornavalai, C.: Experimental performance evaluation of LoRaWAN: a case study in Bangkok. In: 2017 14th International Joint Conference on Computer Science and Software Engineering (JCSSE), pp. 1–4. IEEE (2017)
17. Vitturi, S., Zunino, C., Sauter, T.: Industrial communication systems and their future challenges: next-generation Ethernet, IIoT, and 5G. Proc. IEEE **107**(6), 944–961 (2019)
18. Wendt, T., Volk, F., Mackensen, E.: A benchmark survey of long range (LoRaTM) spread-spectrum-communication at 2.45 GHz for safety applications. In: 2015 IEEE 16th Annual Wireless and Microwave Technology Conference (WAMICON), pp. 1–4. IEEE (2015)
19. Wixted, A.J., Kinnaird, P., Larijani, H., Tait, A., Ahmadinia, A., Strachan, N.: Evaluation of LoRa and LoRaWAN for wireless sensor networks. In: 2016 IEEE Sensors, pp. 1–3. IEEE (2016)
20. Xiong, X., Zheng, K., Xu, R., Xiang, W., Chatzimisios, P.: Low power wide area machine-to-machine networks: key techniques and prototype. IEEE Commun. Mag. **53**(9), 64–71 (2015)

Adaptive Cyclic Polling Systems: Analysis and Application to the Broadband Wireless Networks

V. M. Vishnevsky[1(✉)], O. V. Semenova[1], D. T. Bui[2], and Alexander Sokolov[1]

[1] Institute of Control Sciences of Russian Academy of Sciences,
Profsoyuznaya str. 65, Moscow 117997, Russia
vishn@inbox.ru, olgasmnv@gmail.com, aleksandr.sokolov@phystech.edu
[2] Moscow Institute of Physics and Technology, Institutskiy per. 9, Dolgoprudny,
Moscow Region 141701, Russia

Abstract. In the paper, we consider a stochastic polling system with a cyclic adaptive polling order adequately modelling the behavior of the broadband wireless networks with centralized control. Using the probability generating function method, we obtain the system of the linear algebraic equations for the first and second order moments of the number of packets at an abonent station which allows calculating the mean sojourn time and other performance characteristics. The obtained results are illustrated by numerical estimation of the performance of the IEEE802.11 broadband wireless networks with centralized control.

Keywords: Wireless network · Polling systems · Cyclic adaptive polling · Exhaustive service · Generating function method

1 Introduction

Modern and perspective protocols for broadband wireless networks provide methods and algorithms for solving the "hidden station problem". This problem occurs when the station (hidden station) listens to the wireless environment and does not "see" the transmission/reception of data from other stations in the network due to the distance from the base station or the presence of interference which leads to collisions and the package loss. One of the effective algorithms to fix this problem is the introduction of a polling mechanism which is a cyclic polling of the abonent stations by the base station. An example of the implementation of such algorithms is the PCF (Point Coordination Function) and its development, the HCCA (Controlled Channel Access method) in the IEEE 802.11-2016 standard, as well as the protocols of the perspective millimeter-wave networks, Internet of Things (IoT) and radio frequency identification (RFID) of vehicles [1,21,24]. The base station manages the joint access of all other stations (nodes) of the network to transmit data in the medium based on a specific

Supported by the Russian Foundation for Basic Research (project No. 18-57-00002).

polling algorithm or based on the network priorities. Thus, the base station polls all nodes of the network, keeps the list of stations to be polled and transmit data between all stations of the network accordingly to the polling list.

It is important to note that this approach completely eliminates competitive access to the medium avoiding collisions impossible and it guarantees the priority access to the medium for the time-dependent applications. Despite of the fact that equipment supporting the centralized control is more difficult to design and produce which makes it more expensive, it uses the two most valuable resources of a broadband wireless network much more efficiently: a frequency resource and a bandwidth resource. Centralized control allows to eliminate the problem of "hidden stations" completely, and also allows planning clearly the order the stations get an access to the medium, to manage flexibly the entire operation of the radio cell and change its parameters depending on the specific situation setting up only the base station and not affecting the abonent stations. A significant number of papers (see, for example, [2–14] are devoted to the study of centralized control protocols based on the stochastic polling models. A stochastic polling system (or polling system) is a queuing system with the finite number of queues having a server (or multiple servers) common for all queues. According to a specific rule, a server polls the queues and serves customers that enter the system and wait for their servers in the queues [15]. Models of polling systems are widely used in communication networks, industry, air and rail transportation, health-care systems, maintenance systems, computer communications, traffic light traffic control, etc. The fields of application of polling systems are described in detail in the exhaustive review [16].

The analysis of mathematical models of the stochastic polling systems started its history in the middle of the last century due to the wide practical application of such systems. The theoretical results of the polling system analysis obtained before 1985 are described in detail in the monograph of Takagi [17]. Further development of theoretical results published before 1995 was reflected in the monograph of Borst [18]. The reviews [19,20] covering the period 1995-2017 as well as the monograph [15] are devoted to the generalization and systematization of models and methods for studying stochastic polling systems and their application for the design of broadband wireless networks. They examined polling models that adequately describe the operation of broadband wireless networks based on the IEEE 802.11 protocols with a centralized control. The most recent review [21] on the theory of polling systems was published in 2018. The main purpose of the review was not so much to cover all new publications in the field but to review the papers on the key methods for analyzing the cyclic polling systems with a single server including the heavily loaded systems, and also to discuss a number of complex unsolved problems in the field of polling systems analysis.

The aim of the present paper is to analyze a polling system with an arbitrary finite number of queues and adaptive cyclic polling order. For the first time, such a polling order was introduced in [10] and proposed for the systems where the information about the state of a queue (the number of customers) is available only when the server has finished connecting to the queue and is ready to serve it (the so-called a queue polling moment). In an adaptive polling order, the server polls the queues cyclically but skips (does not poll) in the current cycle those queues that were polled by the server in the previous cycle and were empty at the polling moment. All queues that the server skips in this cycle will be polled in the next cycle. Thus, the polling cycle in such systems is usually shorter than in systems with traditional cyclic polling which improves system performance. The analysis of adaptive polling systems requires keeping the information about the states of queues at the polling moments in the previous cycle which is much more difficult to investigate. In [10], an approximate algorithm was proposed to calculate the performance characteristics including the first two moments of the waiting time. Note that the key parameter of such systems is the probability that the queue will be polled by the server in an arbitrary cycle (it will not be skipped). Knowing these probabilities for all queues, we succeeded to apply the method of probability generating functions (see, e.g., [22]) to the stationary state probability distribution of the number of customers in the queues at the polling moments for the system with adaptive cyclic polling [11]. The method allows to calculate the average waiting time in the queues. Note that in [11], we consider the case of the gated service and the present paper is the further development of the PGF-method for the adaptive cyclic polling in case of the exhaustive service of queues.

2   Mathematical Model

Consider a polling system with a single server attending N queues of $M/G/1$-type with exhaustive service. The ith queue Q_i has a Poisson input of arrivals of intensity λ_i. The service time in the queue has the distribution function $B_i(t)$ with the first and second initial moments $b_i = \int_0^\infty t dB_i(t)$ and $b_i^{(2)} = \int_0^\infty t^2 dB_i(t)$, and the Laplace-Stieltjes transform (LST) $\tilde{B}_i(s) = \int_0^\infty e^{-st} dB_i(t)$. The time the server switches to queue Q_i has the distribution function $S_i(t)$ with the first and second initial moments s_i and $s_i^{(2)}$, respectively, and the LST $\tilde{S}_i(s)$. We assume that in case the server meets N queues empty consecutively at their polling moments (starting from any queue) it stops at the last empty queue and takes a vacation having the distribution function $H(t)$ with the first and second initial moments β and $\beta_i^{(2)}$ and the LST $\tilde{H}(s)$. After a vacation finishes, the server leaves its current position and switches to the next queue. The polling procedure is repeated again. The service of queues is exhaustive, i.e. the server serves the queue until it gets empty, then it departs from the queue and switches to the next queue to be polled.

The stability condition for the polling system considered is $\rho = \sum_{i=1}^N \rho_i < 1$ where $\rho_i = \lambda_i b_i$ is the load of queue Q_i.

The cycle time is supposed to be the time the server spends polling queues from Q_1 and Q_N including a vacation time (if reqiured). The mean cycle time is given by formula

$$C = \frac{\sum_{i=1}^{N} s_i u_i + \beta \prod_{i=1}^{N}(1 - u_i)}{1 - \rho} \tag{1}$$

where u_i is the probability that queue Q_i is polled in the cycle [11]

$$u_i = \frac{1}{1 + e^{-\lambda_i C}}, \quad i = \overline{1, N}.$$

3 Method of Probability Generating Functions

Let X_i^j be the number of customers present in the queue Q_j when server polls the queue Q_i, $i, j = \overline{1, N}$, $A_i(T)$ be the number of Poisson arrivals to the queue Q_i during a random time interval of length T, $\Theta_{i,k}$ be the duration of the server busy period generated by k-th customer in the queue Q_i. Variables $\Theta_{i,k}$ are independent and equally distributed with LST distribution functions $\widetilde{\theta}_i(w)$ which can be found as a solution of the functional equation $\widetilde{\theta}_i(w) = \widetilde{B}_i(w + \lambda_i - \lambda_i \widetilde{\theta}_i(w))$.

The average duration of the busy period of server in queue Q_i is defined as $\theta_i = -\widetilde{\theta}'(0) = \frac{b_i}{1 - \rho_i}$. For the exhaustive service, the evolution of the system state is given by

$$X_{i+1}^j \Big| M_{i+1}^{(0)} = \begin{cases} X_i^j + A_j \left(\sum_{k=1}^{X_i} \Theta_{i,k} + S_{i+1} \right), i \neq j, \\ A_j(S_{i+1}), i = j, \end{cases}$$

$$X_{i+1}^j \Big| M_{i+1}^{(1)} = \begin{cases} X_{i-1}^j + A_j \left(\sum_{k=1}^{X_{i-1}} \Theta_{i-1,k} + S_{i+1} \right), i - 1 \neq j, \\ A_j(S_{i+1}), i - 1 = j, \end{cases}$$

$$X_{i+1}^j \Big| M_{i+1}^{(2)} = \begin{cases} X_{i-2}^j + A_j \left(\sum_{k=1}^{X_{i-2}} \Theta_{i-2,k} + S_{i+1} \right), i - 2 \neq j, \\ A_j(S_{i+1}), i - 2 = j, \end{cases}$$

$$\dots$$

$$X_{i+1}^j \Big| M_{i+1}^{(N-1)} = \begin{cases} X_{i-N+1}^j + A_j \left(\sum_{k=1}^{X_{i-N+1}} \Theta_{i-N+1,k} + S_{i+1} \right), i - N + 1 \neq j, \\ A_j(S_{i+1}), i - N + 1 = j, \end{cases}$$

$$X_{i+1}^j \Big| M_{i+1}^{(N)} = \begin{cases} X_{i-N}^j + A_j \left(\sum_{k=1}^{X_{i-N}} \Theta_{i-N,k} + S_{i+1} + H \right), i - N \neq j, \\ A_j(S_{i+1} + H), i - N = j, \end{cases} \tag{2}$$

for all $n = \overline{0, N}$, where $M_{i+1}^{(n)}$ is the event that the server skipped exactly n queues before polling the current queue Q_{i+1}, i.e. the previously polled queue was Q_{i-n}, if $i > n$, and Q_{i-n+N} otherwise. If $i - n < 0$ we assume that $X_{i-n}^j = X_{i-n+N}^j$.

For the fixed i, the probabilities of $M_{i+1}^{(n)}$, $n = \overline{0, N}$ are calculated as follows:

$$\mathbf{P}\{M_{i+1}^{(0)}\} = u_i, \quad \mathbf{P}\{M_{i+1}^{(1)}\} = (1 - u_i)u_{i-1}, \dots ,$$
$$\mathbf{P}\{M_{i+1}^{(i-1)}\} = (1 - u_i)(1 - u_{i-1}) \cdots (1 - u_2)u_1,$$
$$\mathbf{P}\{M_{i+1}^{(i)}\} = \prod_{k=1}^{i}(1 - u_k)u_N, \tag{3}$$
$$\mathbf{P}\{M_{i+1}^{(i+1)}\} = \prod_{k=1}^{i}(1 - u_k)(1 - u_N)u_{N-1}, \dots ,$$
$$\mathbf{P}\{M_{i+1}^{(N)}\} = \prod_{k=1}^{i}(1 - u_k)(1 - u_N) \cdots (1 - u_{i-1}) = \prod_{k=1}^{N}(1 - u_k).$$

It is easy to see that $\sum_{j=0}^{N}\mathbf{P}\{M_{i+1}^{(j)}\} = 1$.

Let $p_i(r_1, r_2, \dots, r_N)$ be the stationary probability that Q_j has r_j customers at the polling moment of Q_i, $r_j \geq 0$, $i, j = \overline{1, N}$. Consider the probability generating functions (PGFs)

$$F_i(\mathbf{z}) = F_i(z_1, z_2, \cdots, z_N) = \sum_{r_1=0}^{\infty}\sum_{r_2=0}^{\infty} \cdots \sum_{r_N=0}^{\infty} p_i(r_1, r_2, \dots, r_N)z_1^{r_1} \cdots z_N^{r_N}.$$

They can be also presented as

$$F_i(\mathbf{z}) = \mathbf{E}\left[\prod_{j=1}^{N} z_j^{X_i^j}\right], \quad i = \overline{1, N},$$

where $\mathbf{E}$ is the expectation. While using (2), we get

$$F_i(\mathbf{z}) = u_i\mathbf{E}\left[\prod_{j=1}^{N} z_j^{X_{i+1}^j}\,\middle|\,M_{i+1}^{(0)}\right] + (1 - u_i)u_{i-1}\mathbf{E}\left[\prod_{j=1}^{N} z_j^{X_{i+1}^j}\,\middle|\,M_{i+1}^{(1)}\right] + \dots +$$
$$+ (1 - u_1) \cdots (1 - u_N)\mathbf{E}\left[\prod_{j=1}^{N} z_j^{X_{i+1}^j}\,\middle|\,M_{i+1}^{(N)}\right] =$$
$$= u_iM_{i+1}^{(0)}(\mathbf{z}) + (1 - u_i)u_{i-1}M_{i+1}^{(1)}(\mathbf{z}) + \dots + (1 - u_1) \cdots (1 - u_{N-1})u_NM_{i+1}^{(N-1)}(\mathbf{z}) +$$
$$+ (1 - u_1) \cdots (1 - u_N)M_{i+1}^{(N)}(\mathbf{z}) \quad (4)$$

where $u_{i-N} = u_i$, $F_{i-N}(\mathbf{z}) = F_i(\mathbf{z})$, $M_{i+1}^{(l)}(\mathbf{z}) = \mathbf{E}\left[\prod_{j=1}^{N} z_j^{X_{i+1}^j}\,\middle|\,M_{i+1}^{(l)}\right]$, $l = \overline{0, N}$.

Here we have

$$M_{i+1}^{(l)}(\mathbf{z}) = \mathbf{E}\left[\prod_{j=1}^{N} z_j^{X_{i+1}^j} \middle| M_{i+1}^{(l)}\right] = \mathbf{E}_{X_i}\left[\prod_{j=1}^{N} z_j^{X_i^j} \mathbf{E}\left[\prod_{j=1}^{N} z_j^{A_j\left(\sum_{k=1}^{X_i^i} B_{i,k}\right)} \middle| \mathbf{X}_i\right]\right]$$

$$\times \mathbf{E}\left[\prod_{j=1}^{N} z_j^{A_j(S_{i+1})}\right] \quad (5)$$

where $\mathbf{X}_i = (X_i^1, X_i^2, ..., X_i^N)$, $\mathbf{E}[\cdot|X_i]$ is the conditional expectation.

Generating functions $\mathbf{M}_{i+1}^{(l)}(\mathbf{z})$, $l = \overline{0, N}$ determined by equalities [25]

$$\mathbf{M}_{i+1}^{(l)}(\mathbf{z}) = F_{i-l}\left(z_1, z_2, ..., z_{i-l}, \widetilde{\theta}_{i-l}\left(\sum_{j=1}^{N} \lambda_j(1-z_j)\right), z_{i-l+2}, ..., z_N\right) \times \quad (6)$$

$$\times \widetilde{S}_{i-l+1}\left[\sum_{j=1}^{N} \lambda_j(1-z_j)\right], \quad l = \overline{0, N-1},$$

$$\mathbf{M}_{i+1}^{(N)}(\mathbf{z}) = F_{i-N}\left(z_1, z_2, ..., z_{i-N}, \widetilde{\theta}_{i-N}\left(\sum_{j=1}^{N} \lambda_j(1-z_j)\right), z_{i-N+2}, ..., z_N\right) \times$$

$$\times \widetilde{S}_{i-N+1}\left[\sum_{j=1}^{N} \lambda_j(1-z_j)\right] \widetilde{H}\left(\sum_{j=1}^{N} \lambda_j(1-z_j)\right).$$

The mean number of customers $f_i(j) = \mathbf{E}[X_i^j]$ in queue Q_j at a polling moment of Q_i is given by

$$f_i(j) = \mathbf{E}\left[X_i^j\right] = \left.\frac{\partial F_i(\mathbf{z})}{\partial z_j}\right|_{\mathbf{z}=\mathbf{1}},$$

where $\mathbf{1} = (1, 1, ..., 1)$.

Differentiating Eq. (4) by means of (6), we get the following system of linear algebraic equations for $f_i(j)$, $i, j = \overline{1, N}$:

$$f_{i+1}(j) = u_i\left[I_{\{i,j\}}f_i(j) + I_{\{i,j\}}\lambda_j\frac{b_i}{1-\rho_i}f_i(i) + \lambda_j s_{i+1}\right] +$$

$$+ \tau_{1,i}u_{i-1}\left[I_{\{i-1,j\}}f_{i-1}(j) + I_{\{i-1,j\}}\lambda_j\frac{b_{i-1}}{1-\rho_{i-1}}f_{i-1}(i-1) + \lambda_j s_{i+1}\right] + ... +$$

$$+ \tau_{N,i}\left[I_{\{i-N,j\}}f_{i-N}(j) + I_{\{i-N,j\}}\lambda_j\frac{b_{i-N}}{1-\rho_{i-N}}f_{i-N}(i-N) + \lambda_j(s_{i+1}+\beta)\right],$$

$$(7)$$

where $I_{\{i,j\}} = 1$, if $i = j$, and $I_{\{i,j\}} = 0$ otherwise.

The second order moments of the random variables $X_i^j, i, j = \overline{1, N}$ are calculated as the second order derivatives

$$f_i(j, k) = \mathbf{M}\left[X_i^j X_i^k\right] = \frac{\partial^2 F_i(\mathbf{z})}{\partial z_j \partial z_k}\bigg|_{\mathbf{z}=1},$$

$$f_i(i, i) = \mathbf{M}\left[X_i^i(X_i^i - 1)\right] = \frac{\partial^2 F_i(\mathbf{z})}{\partial z_j^2}\bigg|_{\mathbf{z}=1}.$$

By differentiating (7), we get

$$f_{i+1}(j, k) = \left[u_i \frac{\partial^2 \mathbf{M}_{i+1}^{(0)}(\mathbf{z})}{\partial z_j \partial z_k} + \tau_{1,i} u_{i-1} \frac{\partial^2 \mathbf{M}_{i+1}^{(1)}(\mathbf{z})}{\partial z_j \partial z_k} + ... + \tau_{N,i} \frac{\partial^2 \mathbf{M}_{i+1}^{(N)}(\mathbf{z})}{\partial z_j \partial z_k}\right]\bigg|_{\mathbf{z}=1}, \quad (8)$$

here the second order partial derivatives $\frac{\partial^2 \mathbf{M}_{i+1}^{(l)}(\mathbf{z})}{\partial z_j \partial z_k}$, $l = \overline{0, N}$ are calculated as:

$$\frac{\partial^2 \mathbf{M}_{i+1}^{(0)}(\mathbf{z})}{\partial z_j \partial z_k} = \lambda_j \lambda_k s_{i+1}^{(2)} + s_{i+1} \lambda_k f_i(j) + s_{i+1} \lambda_j f_i(k) + f_i(i, j)\frac{b_i \lambda_k}{1 - \rho_i} +$$
$$+ f_i(i)\lambda_j \lambda_k \left[\frac{2b_i s_{i+1}}{1 - \rho_i} + \frac{b_i^{(2)}}{(1 - \rho_i)^3}\right] + f_i(i, k)\frac{b_i \lambda_j}{1 - \rho_i} +$$
$$+ f_i(i, i)\lambda_j \lambda_k \left(\frac{b_i}{1 - \rho_i}\right)^2 + f_i(j, k), i \neq j, i \neq k,$$

$$\frac{\partial^2 \mathbf{M}_{i+1}^{(0)}(\mathbf{z})}{\partial z_j \partial z_k} = \lambda_i \lambda_j s_{i+1}^{(2)} + s_{i+1} \lambda_i f_i(j) + f_i(i)\lambda_i \lambda_j \frac{s_{i+1} b_i}{1 - \rho_i}, i \neq j,$$

$$\frac{\partial^2 \mathbf{M}_{i+1}^{(0)}(\mathbf{z})}{\partial z_j \partial z_k} = \lambda_i^2 s_{i+1}^{(2)}, i, j, k = \overline{1, N},$$

...

$$\frac{\partial^2 \mathbf{M}_{i+1}^{(N)}(\mathbf{z})}{\partial z_j \partial z_k} = \lambda_j \lambda_k(s_{i+1}^{(2)} + \beta^{(2)}) + (s_{i+1} + \beta)\lambda_k f_i(j) + (s_{i+1} + \beta)\lambda_j f_i(k) +$$
$$+ f_i(i)\lambda_j \lambda_k \left[\frac{2b_i(s_{i+1} + \beta)}{1 - \rho_i} + \frac{b_i^{(2)}}{(1 - \rho_i)^3}\right] + f_i(i, j)\frac{b_i \lambda_k}{1 - \rho_i} +$$
$$+ f_i(i, i)\lambda_j \lambda_k \left(\frac{b_i}{1 - \rho_i}\right)^2 + f_i(i, k)\frac{b_i \lambda_j}{1 - \rho_i} + f_i(j, k), i \neq j, i \neq k,$$

$$\frac{\partial^2 \mathbf{M}_{i+1}^{(N)}(\mathbf{z})}{\partial z_j \partial z_k} = \lambda_i \lambda_j(s_{i+1}^{(2)} + \beta^{(2)}) + (s_{i+1} + \beta)\lambda_i f_i(j) + f_i(i)\lambda_i \lambda_j \frac{(s_{i+1} + \beta)b_i}{1 - \rho_i}, i \neq j,$$

$$\frac{\partial^2 \mathbf{M}_{i+1}^{(N)}(\mathbf{z})}{\partial z_j \partial z_k} = \lambda_i^2(s_{i+1}^{(2)} + \beta^{(2)}), i, j, k = \overline{1, N}.$$

The relations (4)–(8) give the system of linear algebraic equations to calculate the second-order moments $f_i(j, k)$, $i, j, k = \overline{1, N}$ which allow to find the mean waiting time W_i in queue Q_i by formula:

$$\mathbf{W_i} = \frac{\lambda_i b_i^{(2)}}{2(1 - \rho_i)} + \frac{f_i(i, i)}{2\lambda_i f_i(i)}, i = \overline{1, N}. \quad (9)$$

4 An Application of Polling Systems

The point coordination function protocol (PCF) and its development, the HCCA method, requires an access point (server) that determines which station is currently has the right to transmit data. In fact, this is a system of cyclic polling with a single server, which serves all queues. The Fig. 1 illustrates the implementation of a PCF polling mode.

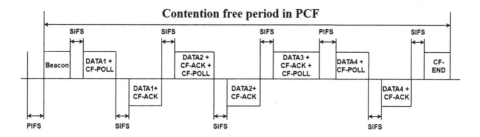

Fig. 1. Contention free repetition interval

An Access Point (AP) transmits data frames (DATA) for destined stations (if available) and also polls all stations to send data that are waiting in their transmission queues, sending them CF-POLL polling service frames (invitation to transmit). The access point polls the stations according to the polling list. The methods for generating and keeping of the polling list are not regulated by the IEEE 802.11-2016 standard, so wireless device developers have no restrictions in how to implement this list. If the method of generating and saving the polling list is a traditional cyclic polling then the following steps are taken:

- Step 0: The AP sends a Beacon (Broadcast Network Identifier Signal) right after the PIFS (PCF Interframe Space - PCF Frame Interval). This message reaches each station (WLAN device) without any risk of collision. Beacon carries information about the possibilities of starting the PCF mode.
- Step 1: After SIFS (Short Interframe Space), the AP starts polling the first station by sending a CF-POLL (polling frame) and transmits data (DATA) to the first station if the AP has data for the first station. After SIFS, the first station transmits its data and CF-ACK (acknowledgment frame) to the AP (if the queue is empty, the station responds with the so-called NULL frame, consisting only of the header and CF-ACK).
- Step 2: After SIFS, the server starts polling the second station, sending CF-POLL + CF-ACK and transmits data to the second station if the AP has data for the second station. After SIFS, the second station transmits its data (if available) and CF-ACK to the AP.
- Step 3: After SIFS, the server starts polling the third station, sending CF-POLL + CF-ACK and transmits data to the third station if the AP has data for the third station. After SIFS, if the third station does not respond, the

AP waits for one time interval (one time interval is Slot time and PIFS = Slot time + SIFS) and starts polling the next station. And so on. After the AP has received data and CF-ACK from the last station, the AP waits for SIFS and sends a CF-END frame to everyone and reports that this is the end of the PCF mode. If the way to create and maintain a polling list is adaptive dynamic polling, then the AP does not poll only those stations that were empty in the previous cycle.

To reduce overhead, the access point can combine a polling frame with data transmission (DATA + CF-POLL frame). Similarly, network stations can combine acknowledgment frames with DATA + CF-ACK data transmission. The following frame types are allowed during PCF mode: DATA – data frame, CF-ACK – confirmation frame, CF-POLL – polling frame, DATA + CF-ACK – combined data and confirmation frame, DATA + CF-POLL – combined data frame and polling, DATA + CF-ACK + CF-POLL – combined data frame, confirmation and polling, CF-ACK + CF-POLL – combined confirmation and polling frame.

- Next, we carry out a comparative numerical analysis of the performance characteristics of a wireless network for two polling mechanisms (adaptive and cyclic polling) with a study of the package of application programs for calculating polling systems developed by the authors of [23].
- The duration period of the SIFS + CF-POLL is the time of switching from one queue to another, and DATA, CF-ACK, DATA + CF-ACK, etc., is the service time for a visit to the queue by the server. We assume that: the number of queues $N = 4$, the intensities of the input Poisson flows $\lambda_1 = \lambda_2 = \frac{1}{600\,\mu s}$, $\lambda_3 = \lambda_4 = \lambda$ take values from 1 to 200 in steps of 10, the queue service time is distributed according to the Poisson law, and has the intensity $\mu_1 = \mu_3 = \mu_4 = \frac{1}{200\,\mu s}$, $\mu_2 = \frac{1}{300\,\mu s}$, the switching time to the queues is distributed exponentially with the parameters mean $s_1 = s_2 = s_3 = s_4 = 600\,\mu s$.
- Denote by

$$V^{(*)} = \sum_{i=1}^{N} \frac{\rho_i}{\rho} V_i^{(*)} \tag{10}$$

- The average sojourn time of customers in the system, obtained as the sum of the average sojourn times in the queues $V_i^{(*)} = W_i^{(*)} + \frac{1}{\mu_i}, i = \overline{1, N}$, with the corresponding shares of the queue loads $\frac{\rho_i}{\rho}$. The symbol in formula (10) indicates the type of survey: c is a cyclic polling and a is an adaptive polling. The dependence of the values $V^{(c)}$ and $V^{(a)}$ for various values of the intensities of the input flows in queues 3 and 4 is shown in Fig. 2, and some of their values are given in Table 1. The calculations were carried out using a software package for evaluating the characteristics of stochastic polling systems [23].

Table 1. Mean sojourn time of customers in polling systems.

λ	Switching time equal to $300\,\mu s$		Switching time equal to $600\,\mu s$	
	Cyclic polling	Adaptive polling	Cyclic polling	Adaptive polling
1	0.00388	0.00344	0.00613	0.00515
10	0.00395	0.00357	0.00627	0.00542
20	0.00406	0.00371	0.00646	0.00569
30	0.00418	0.00387	0.00663	0.00597
40	0.00429	0.00399	0.00679	0.00622
50	0.0044	0.00412	0.00698	0.00648
60	0.00454	0.00428	0.00718	0.00678
70	0.00465	0.00444	0.00742	0.00704
80	0.00481	0.00463	0.00769	0.00734
90	0.00498	0.00479	0.00792	0.00761
100	0.0051	0.00499	0.00816	0.00793
110	0.00529	0.00514	0.00843	0.00819
120	0.00549	0.00535	0.00869	0.00855
130	0.00568	0.00558	0.00911	0.00888
140	0.0059	0.00582	0.00936	0.00926
150	0.00609	0.006	0.00967	0.00965
160	0.00639	0.00621	0.0101	0.01
170	0.0066	0.00652	0.0105	0.0105
180	0.00686	0.00682	0.0109	0.0109
190	0.00725	0.00717	0.0115	0.0115
200	0.00754	0.00747	0.0119	0.0119
210	0.00783	0.00778		
220	0.0083	0.00825		
230	0.00875	0.00871		
240	0.0093	0.00926		
250	0.00976	0.00973		

In this example, the advantage of the average sojourn time of customers in the system $p = \frac{V^{(c)} - V^{(a)}}{V^{(c)}} 100$ when use adaptive polling compared to a cyclic poll is 15.99% in the case when queues 3 and 4 have a small load relative to the first two queues. When the input flow intensity increases then the system load will increase, and the p value gradually decreases to zero, and the adaptive polling system begins to behave like a system with ordinary cyclic polling. Note that in this example, the average switching time between queues is quite large (twice as long as the average service time for one visit), and in this case, the benefit from using adaptive polling is more noticeable then the case, when the average switching time is halved. In the latter case, the p value decreases to 11.34%.

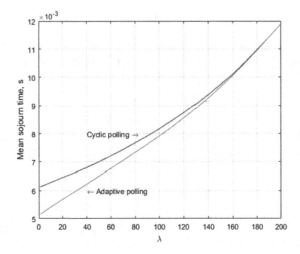

Fig. 2. Mean sojourn time of customers in polling systems with two modes when the switching time is big

Consider another example. Increase the number of system queues to $N = 6$. Let $\mu_i = 10^4, s_i = 1000, i = \overline{1,6}$. The intensities of the input flow in the first three queues are the same and equal to 1000, and $\lambda_4 = \lambda_5 = \lambda_6$ vary from 1 to 1500. In this case, p reaches 27.8% (with $\lambda_4 = \lambda_5 = \lambda_6 = 1$). We can see, that for queues $4 - 6$, the missed rate of polling (equal to the probability $1 - u_i$ of the "non-polling" of queue i in an arbitrary cycle, $i = \overline{4,6}$) is 0.49, that mean, each second cycle of polling these queues are skipped. In this case, also, the main contribution to the benefit p is made by queues $1 - 3$ with a higher load and, accordingly, with large weights $\frac{\rho_i}{\rho}$, and these queues get the advantage of using adaptive polling, while queues 4–6 with a small loading and low weight $\frac{\rho_i}{\rho}$ in this case have a large value $V_i^{(a)}$, and adaptive polling is disadvantage for them. Further, when the intensity of input flows in queues $4 - 6$ increases, the missed rate of polling approaches zero, and adaptive polling allows these queues to decrease the values $V_i^{(a)}, i = \overline{4,6}$, as soon as the intensities of their input flows begin to exceed the intensities of the first three queues.

Thus, in communication systems or in other areas where stochastic polling system models are used, the adaptive cyclic order of queue polling makes sense (at least if the goal of improving the system is to reduce the average waiting time spent in the system) in cases of a relatively large average switching time between queues (if the polling system is asymmetric), as well as in cases of strong asymmetry in loading queues. For symmetrical polling systems, when the system queues are stochastic identical and the distribution of switching times between queues is the same for all queues, adaptive cyclic polling has no advantages over ordinary cyclic polling.

5   Conclusion

The paper proposes a new approach to assessing the performance characteristics of broadband wireless networks with a point coordination function based on stochastic polling models. A method of generating functions has been developed for the analysis of a dynamic adaptive polling system with an exhaustive service. The method allows to obtain a system of equations for the first and second moments of the number of packets in the queues of the system at the polling time, which made it possible to find exact formulas for the average lengths of queues and waiting times in queues. A comparative numerical analysis of the performance characteristics of a wireless network for two polling mechanisms (adaptive and traditional cyclic) was carried out using the software package for the study of polling systems developed by the authors.

References

1. IEEE std 802.11-2016-IEEE standard for information technology telecommunications and information exchange between systems local and metropolitan area networks-specific requirements (2016). https://standards.ieee.org/findstds/standard/802.11-2016.html
2. Bekker, R., Vis, P., Dorsman, J.L., van der Mei, R.D., Winands, E.M.M.: The impact of scheduling policies on the waiting-time distributions in polling systems. Queueing Syst. **79**(2), 145–172 (2015)
3. Shapira, G., Levy, H.: On fairness on polling systems. Ann. Oper. Res. 1–33 (2016) https://doi.org/10.1007/s10479-016-2247-8
4. Saffer, Z., Telek, M., Horváth, G.: Fluid polling system with markov modulated load and gated discipline. In: Takahashi, Y., Phung-Duc, T., Wittevrongel, S., Yue, W. (eds.) QTNA 2018. LNCS, vol. 10932, pp. 86–102. Springer, Cham (2018). https://doi.org/10.1007/978-3-319-93736-6_6
5. Jolles, A., Perel, E., Yechiali, U.: Alternating server with non-zero switchover times and opposite-queue threshold-based switching policy. Perform. Eval. **126**(October), 22–38 (2018)
6. Meyfroyt, T.M.M., Boon, M.A.A., Borst, S.C., Boxma, O.J.: Performance of large-scale polling systems with branching-type and limited service. Perform. Eval. **133**(September), 1–24 (2019)
7. Kim, B., Kim, J.: Analysis of the waiting time distribution for polling systems with retrials and glue periods. Ann. Oper. Res. **277**(2), 197–212 (2019)
8. Gaidamaka, Y.V.: Model with threshold control for analyzing a server with SIP protocol in the overload mode. Autom. Control Comput. Sci. **47**(4), 211–218 (2013)
9. Sonkin, M.A., Moiseev, A.N., Sonkin, D.M., Burtovaya, D.A.: Ob'ektnaya model' prilozheniya dlya imitatsionnogo modelirovaniya tsiklicheskikh sistem massovogo obsluzhivaniya [Object model of application for simulation of cyclic queueing systems]. Vestnik Tomskogo gosudarstvennogo universiteta. Upravlenie, vychislitel'naya tekhnika i informatika, vol. 40, pp. 71–80 (2017)
10. Vishnevsky, V.M., Dudin, A.N., Klimenok, V.I., Semenova, O.V., Shpilev, S.: Approximate method to study M/G/1-type polling system with adaptive polling mechanism. Qual. Technol. Quant. Manag. **2**, 211–228 (2012)

11. Semenova, O.V., Bui, D.T.: Method of generating functions for performance characteristic analysis of the polling systems with adaptive polling and gated service. Commun. Comput. Inf. Sci. **912**, 348–359 (2018)
12. Rykov, V.V.: On analysis of periodic polling systems. Autom. Remote Control **70**(6), 997–1018 (2009)
13. Matveev, A., Feoktistova, V., Bolshakova, K.: On global near optimality of special periodic protocols for fluid polling systems with setups. J. Optim. Theory Appl. **171**(3), 1055–1070 (2016)
14. Zorine, A.V.: On ergodicity conditions in a polling model with Markov modulated input and state-dependent routing. Queueing Syst. **76**(2), 223–241 (2014)
15. Vishnevsky, V.M., Semenova, O.V.: Sistemy pollinga: teoriya i primeneniye v shirokopolosnykh besprovodnykh setyakh [Polling systems: theory and application in broadband wireless networks]. Moscow: Technosfera, p. 309 (2007)
16. Boon, M.A.A., van der Mei, R.D., Winands, E.M.M.: Applications of polling systems. Surv. Oper. Res. Manag. Sci. **16**(2), 67–82 (2011)
17. Takagi, H.: Analysis of Polling Systems. MIT Press, Cambridge (1986)
18. Borst, S.C.: Polling systems. Stichting Mathematisch Centrum, Amsterdam (1996)
19. Vishnevskii, V.M., Semenova, O.V.: Mathematical methods to study the polling systems. Autom. Remote Control **67**(2), 173–220 (2006)
20. Vishnevskii, V.M., Semenova, O.V. : Sistemy adaptivnogo dinamicheskogo pollinga s korrelirovannymi vkhodnymi potokami: preprint. [Adaptive dynamic polling systems with correlated input streams: preprint] Moscow, Institute of Control Science, p. 88 (2017)
21. Borst, S.C., Boxma, O.J.: Polling: past, present, and perspective. TOP **26**(3), 335–369 (2018)
22. Vishnevsky, V.M., Dudin A.N., Klimenok V.I.: Stokhastichesky sistemy s korrelirovannymi vkhodnymi potokami. Theory and primenhenie v telecommunicasionnyx setiax. M.: Reklamno-Izdatensky tsenter "Texnosphera", p. 564 (2018)
23. Vishnevsky, V.M., Semenova, O.V., Bui, D.T.: Programmnyy kompleks otsenki kharakteristik sistem stokhasticheskogo pollinga [Software system for evaluating the characteristics of stochastic polling systems]. Certificate of the State Registration of Computer Programs No. 2019614554 of Russian Federation; registered 08.04.2019 (2019)
24. Vishnevsky, V.M., Larionov, A.A., Ivanov, R.E.: UHF RFID in automatic vehicle identification: analysis and simulation. IEEE J. Radio Freq. Identif. **1**(1), 3–12 (2017). https://doi.org/10.1109/JRFID.2017.2751592
25. Yechiali, U.: Analysis and control of polling systems. In: Donatiello, L., Nelson, R. (eds.) Performance/SIGMETRICS -1993. LNCS, vol. 729, pp. 630–650. Springer, Heidelberg (1993). https://doi.org/10.1007/BFb0013871

Multichannel Diffusion Approximation Models for the Evaluation of Multichannel Communication Networks

Tadeusz Czachórski[1]([⊠]), Godlove Suila Kuaban[1], and Tomasz Nycz[2]

[1] Institute of Theoretical and Applied Informatics Polish Academy of Sciences,
Baltycka 5, 44–100 Gliwice, Poland
{tadek,gskuaban}@iitis.pl
[2] Silesian University of Technology, Akademicka 16, 44–100 Gliwice, Poland
Tomasz.Nycz@polsl.pl

Abstract. The paper proposes steady state and transient state diffusion approximation models for the design and performance evaluation of multichannel communication networks. We concentrate on $G/G/c/c+K$ queueing model and highlight some its applications together with numerical examples to illustrate how this models can be used.

1 Introduction

Multisever queueing models were studied since the very beginning of queueing theory, see Erlang and Engset formulas [7–9] for loss probability or queueing probability in telephone or telegraph connections.

Following much recent Kendal's notation, they concerned $M/M/c/c$, $M/M/c$ and $M/M/c/c/H$, $M/M/c//H$ models, hence having exponentially distributed interarrival times and service times, c parallel service channels, with or without possibility of queueing demands if all channels are busy and having infinite (Erlang) or limited to H (Engset) populations of customers. These models were based on Markov chains. Erlang B and Erlang C formulas are classical results, still used in teletraffic engineering. Such models were also used in the evaluation of modern Internet [1,3,20] or mobile networks [21]. They result in more complex Markov models and application of numerical algorithms, e.g. Kaufman-Roberts recursion [15,19] and convolution algorithms [14].

Queueing models are still extensively used to model computer and telecommunication networks. The rapid growth of telephone switching systems in offices of public switch telephone networks (PSTN), mobile switching centers (MSC) and radio access networks of cellular mobile networks, cloud data center systems and customer service centers has renewed the interest in multiserver queueing models.

Large amounts of multimedia files (such as high definition audio, video, images, etc.) are accessed by an enormous number of mobile users over the Internet [18]. The 4G Long Term Evolution (LTE) and long term evolution advanced

V. M. Vishnevskiy et al. (Eds.): DCCN 2019, LNCS 11965, pp. 43–57, 2019.
https://doi.org/10.1007/978-3-030-36614-8_4

(LTE-A) has been developed to provide high speed cellular radio access networks for mobile network to cope with the high demand for voice, video, image and data traffic over mobile network in order to provide better quality of service and quality of experience to the mobile users. The access network of a radio access network (2G/3G/4G/5G) can be considered as a multichannel access network and modeled using multiserver queueing models. In [11], a queueing model of an LTE access network was proposed where the wireless channel connections are treated as an $M/M/c$ queueing system.

The queueing model representation of the edge node of an optical burst switching network which consist of the packet aggregation buffers that queue up electronic packets into a burst, the burst transmission buffers that store the bursts if all the c optical transmitters are busy. The optical transmission unit that convert the burst from electronic to optical domain and transmitted into the optical channel can be model as a $G/G/c/c + K$ where c is the number of wavelength channels and K is the size of the buffer. This burst transmission buffer or the burst scheduling buffer was studied in [16] using $M/G/c$ queueing model.

HTTP based communication protocols are ubiquitous in modern distributed control/supervision systems, therefore it becomes important to understand the performance of the HTTP protocol working with this type of applications and to develop techniques to investigate the impact of the protocol on the overall performance of these systems. A distributed control/supervision system based on HTTP usually consists of multiple nodes sending control requests or measurements to one or more recipient nodes; these data are used to perform control functions, and are stored or presented to supervising personnel. Typically multiple, concurrent flows of data are carried over HTTP to a given recipient [10].

Cloud and Fog computing paradigms are becoming very popular and many companies are using the rented server infrastructure to execute the computations. In the Infrastructure as a Service (IaaS) concept the virtualized computing resources are used over the Internet. This allows to use the computing infrastructure on demand, renting virtual machines (VMs) when needed [5]. Multiserver $M/M/c$ model was used in [17] to analyze energy consumption in cloud with tasks migration. Approximate analytical model for the performance evaluation of active virtual machines in IaaS clouds using an $M/G/c/c+K$ queueing model where service stations represent nodes, service times represent the time needed to process a task and queues at stations model the queue of virtual machines tasks to be processed at physical server was presented in [2].

2   Diffusion Approximation $G/G/c/c + K$, $G/G/c/c$, $G/G/c$, $G/G/c/c + K/H$, $G/G/c/c/H$, $G/G/c//H$ Models

The diffusion has been developed in the 70-ties [12] to study the performance of wired networks. The steady state diffusion approximation of a $G/G/N/m$ was presented in [22], while the transient analysis of a $G/G/N/N$ was applied to model call center and to study the sliding window mechanism, a popular Call Admission Control (CAC) algorithm in [6]. We develop here and supplement with numerical examples the ideas which preliminary were presented in [5].

2.1 Single Channel Model, a Reminder

Diffusion approximation is a heuristics that replaces the discrete stochastic process $N(t)$ – the number of customers in a queueing system by the continuous diffusion process $X(t)$, e.g. [12,13]. The diffusion Eq. (1) determines the probability density function $f(x,t;x_0)$ of $X(t)$

$$\frac{\partial f(x,t;x_0)}{\partial t} = \frac{\alpha}{2}\frac{\partial^2 f(x,t;x_0)}{\partial x^2} - \beta\frac{\partial f(x,t;x_0)}{\partial x} \tag{1}$$

where βdt and αdt represent the mean and variance of the changes of the diffusion process at dt, and $f(x,t;x_0) = P[x \leq X(t) < x + dx \mid X(0) = x_0]$ approximates the distribution of the number of customers in the service system.

Let $f_A(t)$, $f_B(t)$ denote the interarrival and service time distributions in a queueing system. They are general but not specified, the method requires only their two first moments: means $E[A] = 1/\lambda$, $E[B] = 1/\mu$ and variances $\mathrm{Var}[A] = \sigma_A^2$, $\mathrm{Var}[B] = \sigma_B^2$. Denote also the squared coefficients of variation of the interarrival and service times respectively as $C_A^2 = \sigma_A^2\lambda^2$, $C_B^2 = \sigma_B^2\mu^2$.

In a $G/G/1/N$ system the choice of diffusion parameters is [12]: $\beta = \lambda - \mu$, $\alpha = \sigma_A^2\lambda^3 + \sigma_B^2\mu^3 = C_A^2\lambda + C_B^2\mu$. These values assure that the processes $N(t)$ and $X(t)$ have not only normally distributed changes but also their mean and variance increase in the same way with the observation time.

In case of $G/G/1/N$ queue, the diffusion process should be limited to the interval $[0, N]$ corresponding to possible number of customers inside the system. To ensure it, two barriers are placed at $x = 0$ and $x = N$. In Gelenbe's model when the diffusion process comes to $x = 0$, it remains there for a time exponentially distributed with parameter λ and then jumps instantaneously to $x = 1$. When the diffusion process comes to the barrier at $x = N$ it stays there for a time exponentially distributed with the parameter μ that corresponds to the time for which the queue is saturated and then jumps instantaneously to $x = N - 1$. The diffusion equation supplemented with jumps and probability balance equations for barriers are

$$\frac{\partial f(x,t;x_0)}{\partial t} = \frac{\alpha}{2}\frac{\partial^2 f(x,t;x_0)}{\partial x^2} - \beta\frac{\partial f(x,t;x_0)}{\partial x} +$$
$$+\lambda p_0(t)\delta(x-1) + \lambda_N p_N(t)\delta(x - N + 1)\,,$$
$$\frac{dp_0(t)}{dt} = \lim_{x\to 0}\left[\frac{\alpha}{2}\frac{\partial f(x,t;x_0)}{\partial x} - \beta f(x,t;x_0)\right] - \lambda p_0(t)\,,$$
$$\frac{dp_N(t)}{dt} = -\lim_{x\to N}\left[\frac{\alpha}{2}\frac{\partial f(x,t;x_0)}{\partial x} - \beta f(x,t;x_0)\right] - \lambda_N p_N(t)\,, \tag{2}$$

where $p_0(t) = P[X(t) = 0]$, $p_N(t) = P[X(t) = N]$. In case of $G/G/1$ queue there is no barrier in $x = N$. The value of $f(n,t;n_0)$ approximates $p(n,t;n_0)$, the distribution of the number of customers in the system at time t with initial condition n_0. In case of steady state analysis the Eq. (2) have a simple analytic solution [12] giving $f(x)$ that approximates $p(n)$.

In case of transient state analysis, we represent the density function $f(n, t; x_0)$ by the density function $\phi(x, t; x_0)$ of another diffusion process having *absorbing* barriers at $x = 0$ and $x = N$. and having the following form, cf. [4]

$$
\phi(x, t; x_0) = \begin{cases}
\delta(x - x_0) & \text{for } t = 0 \\
\dfrac{1}{\sqrt{2\Pi \alpha t}} \displaystyle\sum_{n=-\infty}^{\infty} \Big\{ \exp\left[\dfrac{\beta x_n'}{\alpha} - \dfrac{(x - x_0 - x_n' - \beta t)^2}{2\alpha t} \right] \\
\qquad - \exp\left[\dfrac{\beta x_n''}{\alpha} - \dfrac{(x - x_0 - x_n'' - \beta t)^2}{2\alpha t} \right] \Big\} & \text{for } t > 0,
\end{cases}
$$
(3)

where $x_n' = 2nN$, $x_n'' = -2x_0 - x_n'$.

If the initial condition is defined by a function $\psi(x)$, $x \in (0, N)$, $\lim_{x\to 0} \psi(x) = \lim_{x\to N} \psi(x) = 0$, then the pdf of the process has the form

$$
\phi(x, t; \psi) = \int_0^N \phi(x, t; \xi)\psi(\xi)d\xi.
$$

The probability density function $f(x, t; \psi)$ of the diffusion process with elementary returns is composed of the function $\phi(x, t; \psi)$ which represents the influence of the initial conditions and of functions $\phi(x, t - \tau; 1)$, $\phi(x, t - \tau; N - 1)$ which are pd functions of diffusion processes with absorbing barriers at $x = 0$ and $x = N$, started at time $\tau < t$ at points $x = 1$ and $x = N - 1$ with densities $g_1(\tau)$ and $g_{N-1}(\tau)$ of jumps:

$$
f(x, t; \psi) = \phi(x, t; \psi) + \int_0^t g_1(\tau)\phi(x, t - \tau; 1)d\tau + \int_0^t g_{N-1}(\tau)\phi(x, t - \tau; N - 1)d\tau .
$$
(4)

Densities $\gamma_0(t)$, $\gamma_N(t)$ of probability that at time t the process enters to $x = 0$ or $x = N$ are

$$
\gamma_0(t) = p_0(0)\delta(t) + [1 - p_0(0) - p_N(0)]\gamma_{\psi,0}(t) + \int_0^t g_1(\tau)\gamma_{1,0}(t - \tau)d\tau
$$

$$
+ \int_0^t g_{N-1}(\tau)\gamma_{N-1,0}(t - \tau)d\tau ,
$$

$$
\gamma_N(t) = p_N(0)\delta(t) + [1 - p_0(0) - p_N(0)]\gamma_{\psi,N}(t) + \int_0^t g_1(\tau)\gamma_{1,N}(t - \tau)d\tau
$$

$$
+ \int_0^t g_{N-1}(\tau)\gamma_{N-1,N}(t - \tau)d\tau ,
$$
(5)

where $\gamma_{1,0}(t)$, $\gamma_{1,N}(t)$, $\gamma_{N-1,0}(t)$, $\gamma_{N-1,N}(t)$ are densities of the first passage time between corresponding points, e.g.

$$
\gamma_{1,0}(t) = \lim_{x\to 0}\left[\frac{\alpha}{2}\frac{\partial\phi(x, t; 1)}{\partial x} - \beta\phi(x, t; 1)\right] .
$$
(6)

For absorbing barriers

$$\lim_{x \to 0} \phi(x,t;x_0) = \lim_{x \to N} \phi(x,t;x_0) = 0 ,$$

hence $\gamma_{1,0}(t) = \lim_{x \to 0} \frac{\alpha}{2} \frac{\partial \phi(x,t;1)}{\partial x}$. The functions $\gamma_{\psi,0}(t)$, $\gamma_{\psi,N}(t)$ denote densities of probabilities that the initial process, started at $t = 0$ at the point ξ with density $\psi(\xi)$ will end at time t by entering respectively $x = 0$ or $x = N$.

Finally, we may express $g_1(t)$ and $g_N(t)$ with the use of functions $\gamma_0(t)$ and $\gamma_N(t)$:

$$g_1(\tau) = \int_0^\tau \gamma_0(t) l_0(\tau - t) dt , \qquad g_{N-1}(\tau) = \int_0^\tau \gamma_N(t) l_N(\tau - t) dt , \qquad (7)$$

where $l_0(x)$, $l_N(x)$ are the densities of sojourn times in $x = 0$ and $x = N$; the distributions of these times are not restricted to exponential ones as it is in Eq. (2). In practice, we solve the system of above equations in Laplace domain and then look for the solution in Eq. (4) inverting numerically the Laplace transform of $f(x,t;x_0)$.

2.2   Multiple Channel Models

The same approach may by applied in case of multiple channel models. The output process depends here on the number of active channels, therefore the diffusion parameters depend on the value of the diffusion process. We assume piecewise constant parameters, dividing diffusion interval into sub-intervals having different but constant diffusion parameters.

Let us see it in the case of $G/G/c/c + K$ model, i.e. having c channels and K places in queue.

The diffusion process is performed between barriers placed at $x = 0$ and $x = c + K$ and is divided into c sub-intervals

$$(0,1], [1,2] \ldots [c-2, c-1], [c-1, c+K-1], [c+K-1, c+K).$$

corresponding to different number of working channels and having different diffusion parameters; except the last two intervals that have the same diffusion parameters but are distinguished because of jumps from $c + K$ to $c + K - 1$. The parameters are defined as

$$\alpha_i = \lambda C_A^2 + i\mu C_B^2, \quad \beta_i = \lambda - i\mu \qquad (8)$$

for $i - 1 < x < i, i = 1, 2, \ldots, c - 1$, and

$$\alpha_c = \lambda C_A^2 + c\mu C_B^2, \quad \beta_c = \lambda - c\mu \qquad (9)$$

for $c - 1 < x < c + K$.

The jumps are performed from $x = 0$ to $x = 1$ with intensity λ and from $x = c + K$ to $x = c + K - 1$ with intensity $c\mu$.

Transient Solution. We should solve diffusion equations at each sub-interval separately but having in mind that there are probability flows between neighboring sub-intervals. Between sub-intervals we put fictive absorbing barriers enabling to write balance equations of flow entering a barrier and immediately reappearing on the other side in a small distance ϵ from the barrier. As in case of $G/G/1/N$ station, the system of diffusion equations and balance equations is solved in Laplace domain and then the solutions are inverted numerically.

The density functions for the intervals are as follows:

$$f_1(x,t;\psi_1) = \phi_1(x,t;\psi_1) + \int_0^t g_1(\tau)\phi_1(x,t-\tau;1)d\tau +$$

$$+ \int_0^t g_{1-\varepsilon}(\tau)\phi_1(x,t-\tau;1-\varepsilon)d\tau,$$

$$f_n(x,t;\psi_n) = \phi_n(x,t;\psi_n) + \int_0^t g_{n-1+\varepsilon}(\tau)\phi_n(x,t-\tau;n-1+\varepsilon)d\tau +$$

$$+ \int_0^t g_{n-\varepsilon}(\tau)\phi_n(x,t-\tau;n-\varepsilon)d\tau, \quad n = 2,\ldots c+K-1,$$

$$f_{c+K}(x,t;\psi_{c+K}) = \phi_{c+K}(x,t;\psi_{c+K}) +$$

$$\int_0^t g_{c+K-1+\varepsilon}(\tau)\phi_{c+K}(x,t-\tau;c+K-1+\varepsilon)d\tau +$$

$$+ \int_0^t g_{c+K-1}(\tau)\phi_{c+K}(x,t-\tau;c+K-1)d\tau \tag{10}$$

and the densities $g_i(t)$ come from balance equations similar to Eq. 6 for $G/G/1/N$ station.

This is transient solution but valid for constant model parameters. If e.g input rate λ is time dependent, also diffusion parameters become functions of time. Our engineering approach is to divide the axis of time into small intervals where parameters may be considered constant and where the above solution applies; the solution at the end of a time sub-interval defines initial conditions for the next sub-interval. It looks complex but we have mastered the numerical side of the model as it will be seen in example below.

If the queue size is unlimited, in $G/G/c$ model, of course the right barrier is removed and the last interval is extended to infinity. If there is no queue at all in $G/G/c/K+c$ model, the right barrier is placed at $x = c$. If the customers population is limited to H customers in $G/G/c/K+c/H$, $G/G/c//H$, $G/G/c/c/H$ models, it should be reflected in the diffusion parameters, e.g. In case of $G/G/c/c/H$ system with finite population, $H > c$

$$\beta_n = (H-n+1)\nu - n\mu,$$
$$\alpha_n = (H-n+1)\nu C_A^2 + n\mu C_B^2, \quad 1 \le n \le c,$$
$$\beta_n = (H-n+1)\nu - c\mu,$$
$$\alpha_n = (H-n+1)\nu C_A^2 + c\mu C_B^2, \quad n \ge c.$$

where ν and C_A^2 refer to the sojourn time of a single customer in the customers pool.

Numerical Example of $M/D/20/20$ Queue. The service time T is equal 1 time-unit. The input flow is presented in Fig. 1 and resembles a TCP flow intensity. We divide the time interval $[0,60]$ into 60 sub-intervals having time-independent parameters, each interval at the end gives initial conditions for the next interval. We also divide diffusion interval into 20 intervals having their specific constant parameters and solve diffusion equation at each of them withe together with probability balance equations for 19 barriers between intervals and for 2 barriers with jumps in $x = 0$ and $x = 20$. Figure 1 displays also the output flow computed as $\lambda_{out}(t) = \lambda(t - T)[(1 - p(N, t - T))]$ and compares it with simulation. Figures 2 and 3 present mean number of customers as a function of time and two exemplary density functions for $t = 50$ and $t = 70$ given by the model and by simulation.

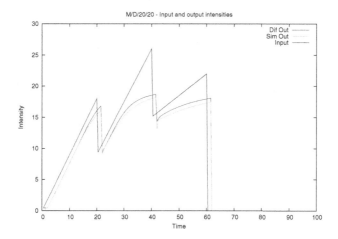

Fig. 1. Input stream intensity λ and output streams intensities

Steady-State Solution. The steady state solution of Eq. 10 has the form

$$f_i(x) = C_{1,i} + C_{2,i}e^{z_i x}, \quad \text{where} \quad z_i = \frac{2\beta_i}{\alpha_i},$$

for any interval i. To determine constants $C_{1,i}$, $C_{2,i}$ we have the conditions of continuity of the solution:

$$f_{i-1}(x_i) = f_{i+1}(x_i), \quad i = 1, \ldots c + K - 1$$

and the conservation of probability at each sub-interval

$$\frac{\alpha_i}{2} \frac{\partial f_i(x, t; \psi_i)}{\partial x} - \beta f_i(x, t; \psi_i) = 0;$$

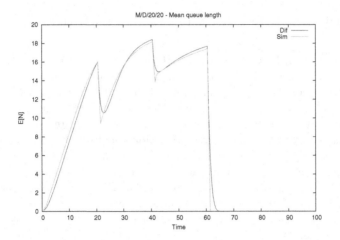

Fig. 2. Mean number of customers as a function of time

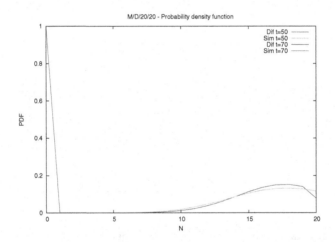

Fig. 3. Densities $f(x, t = 50; 0)$, $f(x, t = 70; 0)$ and corresponding simulation histograms

in the first and last interval we should include in these balance equations the transport of probability due to jumps.

The additional condition is the normalisation, as the integral od $f_i(x)$ over the interval $[x_{i-1}, x_i]$ is

$$\int_{x_{i-1}}^{x_i} f_i(x)dx = C_{1,i}(x_i - x_{i-1}) + C_{2,i}(1/z_i)[e^{z_i x_i} - e^{z_i x_{i-1}}],$$

then twe have

$$1 = p_0 + \sum_{i=1}^{c+K} \{ C_{1,i}(x_i - x_{i-1}) +$$

$$+ C_{2,i}(1/z_i)[e^{z_i x_i} - e^{z_i x_{i-1}}] \} + p_{c+K}.$$

Due to this, the steady state solution of the $G/G/c/c + K$ diffusion model with jumps becomes

$$f(x) = \begin{cases} \dfrac{\lambda p_0}{-\beta_1}(1 - e^{z_1 x}), & 0 < x \le 1, \\ \quad\vdots \\ \dfrac{\lambda p_0}{-\beta_1}(1 - e^{z_1 x})e^{z_2(x-1)+\cdots+z_n(x-(n-1))}, & n-1 \le x \le n, \\ \quad\vdots \\ \dfrac{\lambda p_0}{-\beta_1}(1 - e^{z_1 x})e^{z_2(x-1)+\cdots+z_c(x-(c-1))}, & c-1 \le x \le c+K-1, \\ \dfrac{c\mu p_{K+c}}{-\beta_m}(e^{z_c(x-(c+K))} - 1), & c+K-1 \le x < c+K, \end{cases}$$

$$\tag{11}$$

where p_{K+c} is the probability that the diffusion process is at the upper barrier at $x = K + c$ (the queue is saturated), and p_0 is the probability that the process is at the lower barrier at $x = 0$, i.e. the station is empty,

$$p_{K+c} = \frac{\lambda p_0 \beta_m}{c\mu \beta_1} \left[\frac{1 - e^{z_1(c+K-1)}}{e^{-z_c} - 1} \right] e^{z_2(c+K-2)+\cdots+z_c K}$$

where $\rho = \lambda/c\mu$ and p_0 is determined from the normalization condition. The probability density function (PDF) of the delay or response time is

$$f_T(t) = \frac{x_0}{\sqrt{2\pi\alpha t^3}} e^{-\left[\frac{2x_0\beta}{\alpha} + \frac{(x_0-\beta)^2}{2\alpha t} \right]}$$

$$\tag{12}$$

where x_0 is the mean queue size seen by an arriving packet, $\beta = \lambda - c\mu$ and $\alpha = \lambda C_A^2 + c\mu C_B^2$ are the parameters of the diffusion movement of the packet towards the server or transmitter.

3   Numerical Examples

Below we present an example of $G/G/c/K + c$ station allowing us to evaluate the influence of parameters on the performance of the station represented by the distribution of the number of customers and the distribution of the delay introduced by the station. Figures 4, 5, 6, 7, 8, 9, 10, 11 present the impact the of input traffic rate λ, of the squared coefficient of variation of the arrival and service times C_A^2, C_B^2, of the number of parallel channels, of the speed μ of service, and of the capacity K of the queue. It illustrates the use of the model in practical quantitative evaluations.

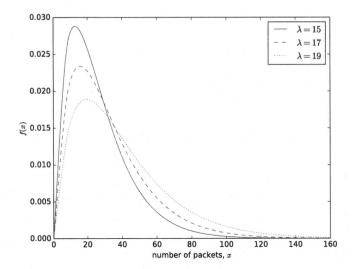

Fig. 4. The influence of input traffic rate λ on the distribution of the number of packets for $K = 150$, $c = 10$ and $\mu = 10$

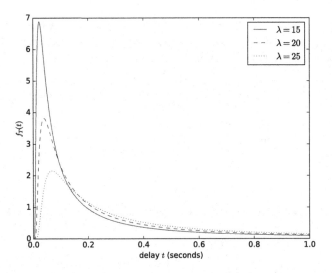

Fig. 5. The influence of input traffic rate λ on the distribution of the delay (response time) for $K = 150$, $c = 10$ and $\mu = 10$

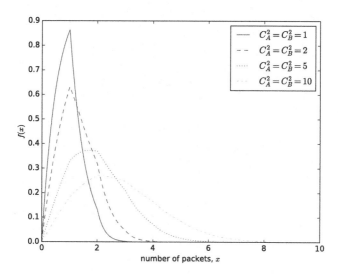

Fig. 6. The influence of the squared coefficient of variation of the arrival and service times C_A^2, C_B^2, on the distribution of the number of customers for $K = 50$, $c = 10$ and $\mu = 30$, $\lambda = 20$

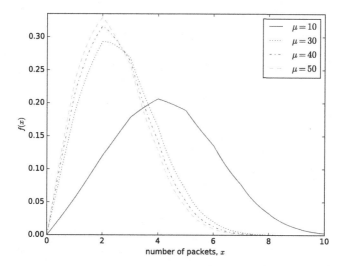

Fig. 7. The influence of the service rates μ on the distribution of the number of packets for $K = 50$, $\lambda = 20$ and $c = 10$, $C_A^2 = C_B^2 = 5$

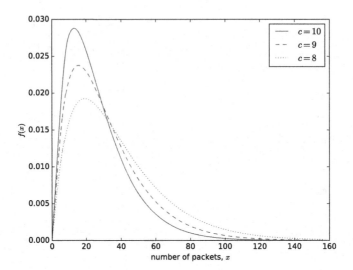

Fig. 8. The influence of the number of channels c on the distribution of the number of customer for $K = 150$, $\lambda = 15$ and $\mu = 10$

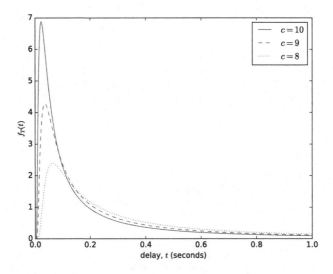

Fig. 9. The influence of the number of service channels c on the distribution of the delay (response time) for $K = 150$, $\lambda = 15$ and $\mu = 10$

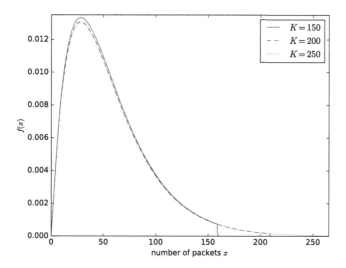

Fig. 10. The influence of the buffer size K on the distribution of the number of packets for $\lambda = 15$ and $c = 10$, $\mu = 20$

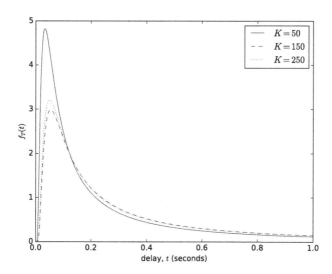

Fig. 11. The influence of the buffer size K on the distribution of the delay (response time) for $\lambda = 15$ and $c = 10$, $\mu = 20$

4   Conclusions

Multichannel queuing models are often used in evaluation of communication systems, recently for cloud/fog/edge computations. The article presents diffusion approximation models of multichannel systems having general interarrival and service time distributions and making possible transient analysis, useful in investigation of the systems dynamics. Numerical examples demonstrate that the models may be used in quantitative analysis.

References

1. Bonald, T., Roberts, J.: Internet and the Erlang formula. ACM SIGCOMM Comput. Commun. Rev. **42**(1), 25–30 (2012)
2. Chang, X., Wang, B., Muppala, J.K., Liu, J.: Modeling active virtual machines on IaaS clouds using an M/G/m/m+K Queue. IEEE Trans. Serv. Comput. **9**(3), 408–420 (2016)
3. Chromy, E., Suran, J., Kovacik, M., Kavacky, M.: Usage of Erlang formula in IP networks. Commun. Netw. **3**, 161–167 (2011)
4. Cox, R.P., Miller, H.D.: The Theory of Stochastic Processes. Chapman and Hall, London (1965)
5. Czahórski, T., Grochla, K.: A diffusion approximation moels for cloud computation with task migration. In: Proceedings of the IEEE International Conference on Fog Computing (ICFC 2019) (2019)
6. Czachórski, T., Fourneau, J., Nycz, T., Pekergin, F.: Diffusion approximation model of multiserver stations with losses. Electron. Notes Theor. Comput. Sci. **232**, 125–143 (2009)
7. Engset, T.O.: Die Wahrscheinlichkeitsrechnung zur Bestimmung der Wahlerzahl in automatischen Fernsprechamtern, Elektrotechnische Zeitschrift, Heft 31 (1918)
8. Erlang, A.K.: The theory of probabilities and telephone conversations. Nyt Tidsskrift Matematik No B **20**, 33–39 (1909)
9. Erlang, A.K.: Solutions of some problems in the theory of probabilities of significance in automatic telephone exchnges. Electroteknikeren **13**, 5–13 (1917)
10. Fiuk, M., Czachórski, T.: A queueing model and performance analysis of UPnP/HTTP client server interactions in networked control systems. In: Gaj, P., Sawicki, M., Kwiecień, A. (eds.) CN 2019. CCIS, vol. 1039, pp. 366–386. Springer, Cham (2019). https://doi.org/10.1007/978-3-030-21952-9_27
11. Fowler, S., Häll, C.H.,Yuan, D., Baravdish, G., Mellouk, A.: Analysis of vehicular wireless channel communication via queueing theory model. In: Communications (ICC) 2014 IEEE International Conference on Communications, pp. 1736–1741. IEEE (2014)
12. Gelenbe, E.: On approximate computer systems models. J. ACM **22**(2), 261–269 (1975)
13. Gelenbe, E., Pujolle, G.: The behaviour of a single queue in a general queueing network. Acta Informatica **7**(2), 123–136 (1976)
14. Glabowski, M., Kaliszan, A., Stasiak, M.: Modeling product form state dependent systems with BPP traffic. J. Perform. Eval. **67**(3), 174–190 (2010)
15. Kaufman, J.: Blocking in a shared resource environment. IEEE Trans. Commun. **COM–29**(10), 1474–1481 (1981)
16. Li, H., Thng, I.LJ.: Edge node buffer usage in optical burst switching networks. Photon. Netw. Commun. **13**, 31 (2006)
17. Ait El Mahjoub, Y., Fourneau, J.-M., Castel-Taleb, H.: Analysis of energy consumption in cloud center with tasks migrations. In: Gaj, P., Sawicki, M., Kwiecień, A. (eds.) CN 2019. CCIS, vol. 1039, pp. 301–315. Springer, Cham (2019). https://doi.org/10.1007/978-3-030-21952-9_23

18. Narman, H.S., Atiquzzaman, M.: Analysis of static partial carrier components assignment in LTE systems. In: Telecommunication & Network Research Lab School of Computer Science. The University of Oklahoma, 110 W. Boyd, Room 150, Norman, Oklahoma, pp. 73019–6151 (2014)
19. Roberts, J.W.: A service system with heterogenous user requirements - application to multi-service telecommunications systems. In: Pujolle, G. (Ed.) Proceedings of Performance of Data Communicatons Systems and their Applications, Amsterdam, North Holland, pp. 423–431 (1981)
20. Roberts, J.W.: Traffic theory and the internet. IEEE Commun. Mag. **39**(1), 94–99 (2001)
21. Stasiak, M., Glabowski, M., Wisniewski, A., Zwierzykowski, P.: Modelling and Dimensioning of Mobile Networks, from GSM to LTE. Wiley, Hoboken (2011)
22. Whitt, W.: A Diffusion Approximation for the G/GI/n/m Queue. Oper. Res. **52**(6), 922–941 (2004). INFORMS

Model and Algorithm of Next Generation Optical Switching Systems Based on 8 × 8 Elements

E. Barabanova[1]([✉])[iD], K. Vytovtov[1][iD], and V. Podlazov[2][iD]

[1] Astrakhan State Technical University, Tatischeva 16 str., 414004 Astrakhan, Russia
elizavetaalexb@yandex.ru
[2] V.A.Trapeznikov Institute of Control Sciences of RAS, Profsoyuznaya 65 str., 117997 Moscow, Russia

Abstract. One of the key elements of modern fiber-optic systems is an optical switch. So, the next-generation all optical switching systems have been presented by authors of the paper [1,2]. Those systems have been based on 4 × 4 and 8 × 8 switching elements and ones have high performance and low complexity. And here we offer the models and algorithms of the all-optical 8 × 8 and 64 × 64 switching systems with decentralized control for the first time. The presented models is based on the graph theory and they allow us to get so-call non-blocking conditions for the first time also. The analytical model for the complexity and diameter of the proposed schemes is presented for the first time. The characteristics of the offered systems are compared with well-known schemes too. And the calculation results showed a significant advantage of the presented systems in comparison with the well-known existing ones.

Keywords: Graph model · Diameter · Self-turning · Complexity · Algorithm

1 Introduction

Switching system is an important element of any information processing system designed to solve social, technical, and scientific problems [3–5]. For today there are many types of switches, which can be used for development high performance computing and telecommunication systems. Obviously that the most perspective of them are optical switching systems [6,7]. Such systems have high performance in comparison with their electron analogues and based on modern materials [6,8]. It is known that performance of optical switching systems is strong depended on a it's control scheme [6]. Optical schemes can have a central control one or they can be self-turning [1–7]. In [1,2] the authors have been offered the next-generation optical switching systems with decentralized control based on 4 × 4 and 8 × 8 switching elements. However not all problems in this direction have

The reported study was founded by RFBR according to the research project 18-37-00059/18.

been solved in that papers. In particular, the functioning algorithms and models of 8×8 and 64×64 switching systems have not been described. Here we present the solution of these important problems for the first time.

In this paper we use the graph theory for modelling our optical switching systems [1,2]. Indeed this theory is one of most applicable to describe operation and properties of switching systems [6,9–11] from our point of view. For example, new type of non-blocking generalized Close schemes with a large number of independent paths between nodes have been described by a quasi-complete graph in [10]. Probability graph method has also been used for calculating circuit blocking probability in [11]. Here we use graph models to describe non-blocking conditions of the new 8×8 and 64×64 switching systems. In this paper, we also present the accurate analytical model and the results of numerical calculation for the complexity the 8×8 and 64×64 optical switching system and diameter of them. Analysis of the results showed a significant advantage of our schemes in comparison with the existing ones.

2 The Graphs of a Next-Generation Optical Switching Systems

2.1 The 8 × 8 Switching System Graph

First of all let us consider the 8×8 optical switching system described in [2]. Obviously, the graph $G = (I, S, O, E)$ (Fig. 1) of that system is a bipartite [7,8]. The edges (E) of G correspond to the system connecting lines, the vertices correspond to the inputs (I) and outputs (O), and two switching elements (S) of the system. The switching system model consists of four sets. The first one is the set of the inputs $I = \{1, 2, 3, 4, 5, 6, 7, 8\}$, the second one is the set of the switching elements $S = \{1'', 2''\}$, the third one is the set of the outputs $O = \{1', 2', 3', 4', 5', 6', 7', 8'\}$, and the fourth one is the set of the edges $E = \{\{c_{x1}d_{1y}, \}\{c_{x2}d_{2y}\}, x = 1..8, y = 1..8\}$.

It is followed from the graph structure (Fig. 1) that the degrees of the sets $I(G)$ and $O(G)$ are equal to 2 and the degree of the set $S(G)$ is equal to 16. It also is seen (Fig. 1) that the diameter of the connected graph G is equal to 2 and it is denoted as $diam(G)$. This graph G is called as connected because every pair of the distinct vertices is joined by the paths between sets $I(G)$ and $O(G)$. Thus we can say than the switching system presented by connecting graph is full-accessible [6]. Now let us consider the paths between two pairs of the vertices: $(1, 8')$ and $(8, 1')$ (the dash lines in the Fig. 1). The first path includes the edges c_{11} and d_{18} and the second path includes the edges c_{82} and d_{21}. Obviously the switching system is non-blocking in this moment as there are no intersections at the graph vertices in this case. Intersections will be in the vertices $1''$ or $2''$ if there are additional connections between other vertices of the graph. Taking into account the fact that optical signals can be coming at all eight inputs simultaneously, the number of the intermediate vertices of the switch element set $S = \{1'', 2'', 3'', 4'', 5'', 6'', 7'', 8''\}$ must be equal to 8 for satisfaction of system non-blocking property at any time (Fig. 1).

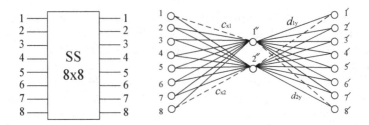

Fig. 1. The 8×8 switching system graph

2.2 The 64×64 Switching System Graph

In this subsection the graph of the 64×64-switching system [2] is considered (Fig. 2). The one $G' = (I', O', E', I'', O'')$ is a bipartite graph too. The edges (E') of G' correspond to the connecting lines. The vertices (I'), (O') correspond to the inputs and outputs of the first cascade, and the vertices (I''), (O'') correspond to the inputs and outputs of the second cascade. The diameter of the connected graph G' is equal to 4 and it is denoted as $diam(G')$.

Now let us consider the paths between the two vertex pairs (1, 57) and (6, 64) (the solid lines in Fig. 2). The first path includes the 1 and 8 vertices of the first cascade, and the 8 and 57 vertices of the second cascade. The second path includes the 6 and 8 vertices of the first cascade, and the 8 and 64 vertices of the second cascade. We can see in Fig. 2 that there are intersections at the 8 vertex of the first cascade output $set\{O'\}$ and at the 8 vertex of the second cascade input $set\{I''\}$. Therefore the switching system is blocking in this moment. Analogously if we consider the paths between the two pairs of graph vertices (58, 1) and (64, 5) (the solid lines in Fig. 2) there will be intersections at the 57 vertices. As a result collisions will be occurred. To solve the blocking problem in the system, we offer to use controlled delay lines in inputs of the 8×8 switching systems. In this way we obtain the 64×64 non-blocking optical system for the first time.

3   The Optical Switching System Algorithms

It is well-known [6,7] that a switching algorithm is an action sequence performed by unit control element or several control elements of a switching system to find a connection path between its input terminals and its output terminals. It has been shown in [1,2] that studied next generation optical switching systems are self-turning. Thus the control devices are located inside the 8×8 and 64×64 system switching elements, and light paths are determined by control information of a packet header, and these ones are established by such local devices. Here the functioning algorithms for such a type of the systems are presented for the first time. This algorithms take into account the principles that allow us to create non-blocking system too.

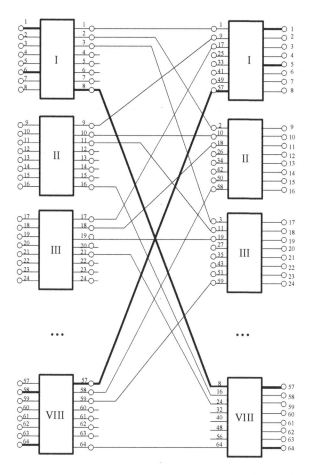

Fig. 2. The 64 × 64 switching system graph

3.1 The 8 × 8 Optical Switching System Algorithm

First of all note that there is no any data transmission protocol for all-optical systems for today. Therefore as the example we consider a data packet analogous to Ethernet one. The main fields of a packet which can be used for switching in our system are presented in the timing diagram of the transmission of the control and information signals (Fig. 3). The packet starts with a preamble (PR). The header includes output terminal address. Each output terminal is associated with its own combination of three control wavelengths. The possible combinations of the wavelengths are presented in the Table 1.

Now let us consider the algorithm of the next-generation 8 × 8-optical switching system functioning (Fig. 4). At the moment of time t_1 (Fig. 3) a communication channel is established through one of the system switching elements. If the first switching element is free then the communication channel is established

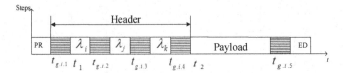

Fig. 3. The timing diagram of the transmission of control and information signals in 8×8 switching system

Table 1. The combinations of control signals for 8×8 switching system

Number of outputs	Combinations of control signals
1	$\lambda_1 \lambda_1 \lambda_1$
2	$\lambda_1 \lambda_1 \lambda_2$
3	$\lambda_1 \lambda_2 \lambda_1$
4	$\lambda_1 \lambda_2 \lambda_2$
5	$\lambda_2 \lambda_1 \lambda_1$
6	$\lambda_2 \lambda_1 \lambda_2$
7	$\lambda_2 \lambda_2 \lambda_1$
8	$\lambda_2 \lambda_2 \lambda_2$

through this one, otherwise it can be established through the second switching element but only if it is free in this moment. If both of switching elements are occupied the optical signal goes in a delay line. The output number y_n of the 8×8 switching system is determined by the control signal set λ_i, λ_j, λ_k. Thus the first output corresponds to the combination $\lambda_1 \lambda_1 \lambda_1$, the second one corresponds to $\lambda_1 \lambda_1 \lambda_2$, and etc. (Table 1).

The payload is transmitted through the 8×8-system when the communication channel between the given input and the given output is established (the moment t_3 in Fig. 4). The signals must be separated by guard intervals $t_{g.i.1}$, $t_{g.i.2}$, $t_{g.i.3}$, $t_{g.i.4}$, $t_{g.i.5}$ in order to avoid imposition of the control signals as well as the information signals and control signals. The end of the optical packet is indicated by the end-of-data-stream symbol ED.

3.2 The 64×64 Optical Switching System Algorithm

Here the algorithm of the next-generation 64×64-optical switching system is presented for the first time. The considered 64×64-switching system is based on the 8×8-switching cells [2]. Therefore the 64×64-optical switching system algorithm is generalization of the 8×8 one. Each output of 64×64-system is associated with one of own combination including the six control wavelengths from $\lambda_1 \lambda_1 \lambda_1 \lambda_3 \lambda_3 \lambda_3$ to $\lambda_2 \lambda_2 \lambda_2 \lambda_4 \lambda_4 \lambda_4$. The certain combination of three control wavelengths from $\lambda_1 \lambda_1 \lambda_1$ to $\lambda_2 \lambda_2 \lambda_2$ determines the number of the input stage 8×8 cell of the 64×64-switching system. The other combination from $\lambda_3 \lambda_3 \lambda_3$

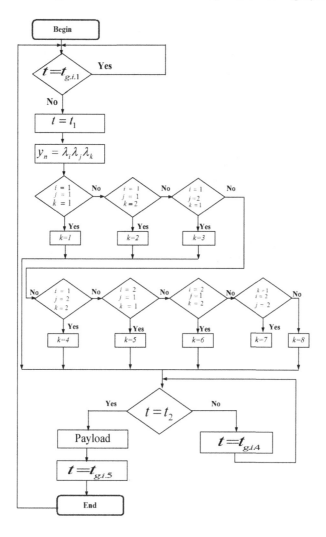

Fig. 4. The 8×8 switching system algorithm

to $\lambda_4 \lambda_4 \lambda_4$ determines the output stage 8×8 cell. Thus we must have the 64 combinations of the control signal wavelength. The possible combinations of wavelengths are presented in Table 2.

The functioning algorithm of the 64×64-switching system is double-step. At the first step (the moment t_1 (Fig. 5)) a communication channel is established through the 8×8-switching system of the first stage (IC). The output number y_n of the 8×8 cell of the first stage of the 64×64-switching system is determined by the set of the three control signals λ_i, λ_j , λ_k.

Table 2. The combinations of control signals for 64×64 switching system

Number	Combinations	Number	Combinations
1	$\lambda_1\lambda_1\lambda_1 \; \lambda_3\lambda_3\lambda_3$	33	$\lambda_2\lambda_1\lambda_1 \; \lambda_3\lambda_3\lambda_3$
2	$\lambda_1\lambda_1\lambda_1 \; \lambda_3\lambda_3\lambda_4$	34	$\lambda_2\lambda_1\lambda_1 \; \lambda_3\lambda_3\lambda_4$
...	...	...	...
31	$\lambda_1\lambda_2\lambda_2 \; \lambda_4\lambda_4\lambda_3$	63	$\lambda_2\lambda_2\lambda_2 \; \lambda_4\lambda_4\lambda_3$
32	$\lambda_1\lambda_2\lambda_2 \; \lambda_4\lambda_4\lambda_3$	64	$\lambda_2\lambda_2\lambda_2 \; \lambda_4\lambda_4\lambda_4$

During the second step (the moment t_2 (see Fig. 5)) a communication channel is established through the 8×8 switching system of the second stage (OC). The output number z_n of the 8×8- switching system of the second stage is determined by the control signals λ_i, λ_j, λ_k values. The number y_n of the second stage 8×8- switching system corresponds to the number of the output of the first stage of the 8×8- switching system. Thus it is beginning the process of payload transmitting only after these steps (the time t_3 in Fig. 5).

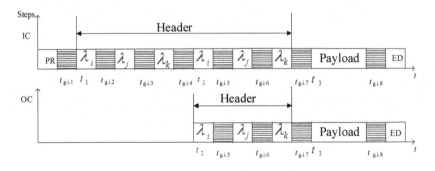

Fig. 5. The timing diagram of signal transmission in the 64×64 system

4　Calculation of Complexity and Diameter

The analysis of complexity and diameter of switching system is very important point in the problem of communication systems design [7,8]. For the considered case there are no even mathematical relations of these values for today. In this section we consider two kinds of the switching schemes complexity. The first one is circuit complexity and the other one is switching complexity. At first let us considered the circuit complexity. For today there is no any definition of the circuit complexity for such a type of the schemes, so we introduce the concept of this complexity as a total number of basic elements. Therefore in this case the 8×8 switching system is taken as a basic element. It is obtained in this work for the first time that the circuit complexity s_k of a $N_k \times N_k$ next generation switching systems based on 8×8 switching cell can be calculated by using following expressions:

$$s_k = \frac{N_k \times k}{8} \tag{1}$$

where N_k is the number of inputs (outputs) and k is the number of stages.

Indeed considering the scheme in Fig. 3 [2] we can see the regularity presented in Table 3. This regularity allows us to conclude that the circuit complexity is determined by (1) for any n.

Now let us express n from the equality $k = 2^n$ (see Table 3):

$$n = \log_2 k \tag{2}$$

Analysing the scheme and using mathematical induction method (see Table 3) we can write the expressions for number of switching elements of one stage m_k and number of inputs N_k. Substituting (2) into the $N_k = 2^{3 \cdot 2^n}$ and carrying out algebraic transformations we can finally express the circuit complexity through the number of inputs as:

$$s_k = \frac{N_k}{24} \log_2 N_k \tag{3}$$

It is well-known that switching complexity S_k is the number of switching points. For calculating number of switching points we have transformed 8×8 switching system [2] into the simple scheme (Fig. 6). The scheme consists of two 8×1 multiplexers (MUX), two 1×8 demultiplexers (DMUX), eight 2×2 multiplexers and delay lines (DL) at the inputs and outputs of the system. So the switching complexity of the 8×8 switching system is forty-eight. The switching complexity of the $N_k \times N_k$ switching system can be calculated using following expressions:

$$S_k = 6 \cdot N_k \cdot k, \tag{4}$$

The regularity of (4) is clean seen from Table 3. Analogously to (3) it can be obtained the expression for the switching complexity through the number of inputs:

$$S_k = 2N_k \log_2 N_k, \tag{5}$$

Transmission time and reliability of a switching system depend on a diameter of this system. In the switching system theory they have written that a diameter is the number of stages [6]. Now we calculated the diameter of our next generation optical system as following:

$$k = \frac{1}{3} \log_2 N_k, \tag{6}$$

The structural characteristics of Benes, Shpanke, Shpanke-Benes [7] switching systems, and next generation switching systems based on the 8×8 basic elements [2] are presented in Table 4.

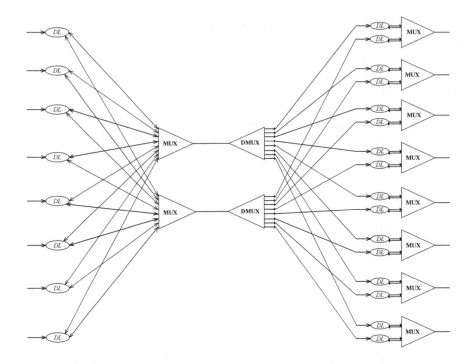

Fig. 6. The 8×8 switching system simple scheme

Table 3. The dependences of the complexity and diameter on the number of the inputs

n	0	1	2		n	$n+1$
k	1	2	4	...	2^n	2^{n+1}
m_k	1	8	512	...	$2^{3(2^n-1)}$	$2^{3(2^{n+1}-1)}$
N_k	8	64	4096	...	$2^{3 \cdot 2^n}$	$2^{3 \cdot 2^{n+1}}$
s_k	1	16	2048	...	$2^{3 \cdot 2^n + n + 3}$	$2^{3 \cdot 2^{n+1} + n + 4}$
S_k	48	768	98304	...	$3^{2 \cdot 2^n + n + 1}$	$3^{2 \cdot 2^{n+1} + n + 2}$

The dependences of the circuit complexity s_k on the number of inputs (outputs) for Shpanke, Benes, Shpanke-Benes, and the next generation optical switching systems based on the 8×8 basic elements are shown in Fig. 7.

Analysing the data of Table 4 and Fig. 7, we can conclude that the circuit complexity of the proposed optical switching circuits based on the 8×8-elements is approximately 20 times less than the circuit complexity of the well-known relatively simple multi-stage switching systems. As the number of inputs increases, the ratio of the circuit complexity of the known schemes and the proposed one increases. In particular, when the number of inputs (outputs) is 4096, the complexity s_k of the proposed scheme is 23 times less than the Benes scheme,

Table 4. The dependences of the circuit and switching complexities and diameter on the number of inputs. The scheme types: 1-Benes, 2-Shpanke, 3-Shpanke-Benes, 4-next generation optical switching systems based on the 8×8 basic elements

	s_k	S_k	k
1	$\frac{N_k}{2} \cdot (2 \log_2 N_k - 1)$	$N_k \cdot (2 \log_2 N_k - 1)$	$2 \log_2 N_k - 1$
2	$2 N_k (N_k - 1)$	$4 N_k (N_k - 1)$	$2 \log_2 N_k$
3	$\frac{N_k}{2} (N_k - 1)$	$N_k (N_k - 1)$	N_k
4	$\frac{N_k}{24} \log_2 N_k$	$2 N_k \log_2 N_k$	$\frac{1}{3} \log_2 N_k$

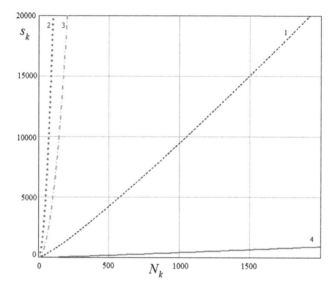

Fig. 7. The dependences of the circuit complexity on the number of inputs

16380 times less than the Shpanke scheme, and 4096 times less than the Shpanke-Benes scheme.

Analogously analysing the data presented in Table 4 and Fig. 8, we see that the switching complexity of the proposed optical switching circuits based on the 8×8 elements is approximately equal to the switching complexity of the simplest well-known multi-stage Benes scheme. But the switching complexity S_k of the proposed scheme is 683 times less than Shpanke scheme when the number of inputs (outputs) is 4096 and it is 171 times less than the Shpanke-Benes scheme.

Studying the data presented in Table 4 and in Fig. 9, we can conclude that the diameter of the proposed optical switching circuits is approximately 5 times less than the diameter of the known simple multi-stage switching systems. In particular, when the number of inputs and outputs is 4096, the diameter k of the proposed switching system based on the 8×8 cells is 5,75 times less than

Benes scheme, 6 times less than Spanke scheme and 1024 times less than Spanke-Benes scheme.

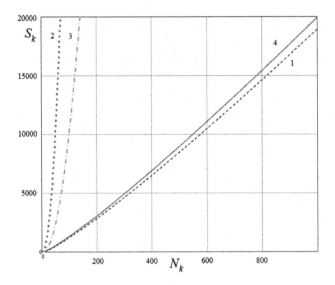

Fig. 8. The dependences of the switching complexity on the number of inputs

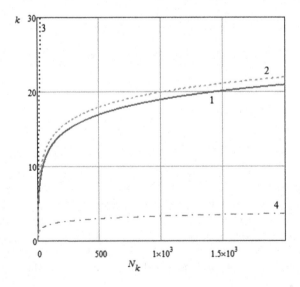

Fig. 9. The dependences of the switching systems diameter on the number of inputs

5   Conclussions

In this paper, the graph models and algorithms of the next generation all optical self-tuning 8×8 and 64×64 switching systems (Sects. 2 and 3) that have been proposed by authors in [2] are presented here for the first time. These results allow us to carry out system analysis, estimate possible cases and calculate probability of system blocking. The analytical expressions of the circuit and switching complexity, and the diameter of the next generation all optical self-tuning 8×8 and 64×64 switching systems are presented in this paper for the first time too (Sect. 4). It is shown that the developed optical switching systems have the low complexity and the small diameter in comparison with well-known systems. The numerical calculations in accordance to the development method are carried out too. Since the numerical research we can conclude that the circuit complexity of the proposed optical switching circuits based on 8×8 elements is 20 times less than the circuit complexity of relatively simple switching systems based on 2×2 elements [6]. The switching complexity of the next generation switching systems based on the 8×8 elements is approximately equal to Benes scheme but it is hundreds times less than this parameter for well-known Shpanke and Shpanke-Benes systems [7]. Additionally it is proved that the diameter of the proposed systems based on the 8×8 elements is 5 time less than the diameter of well-known switching systems based on 2×2 switches.

References

1. Vytovtov, K.A., Barabanova, E.A., Podlazov, V.S.: Model of next-generation optical switching system. Commun. Comput. Inf. Sci. **918**, 377–386 (2018)
2. Vytovtov, K.A., Barabanova, E.A., Barabanov, I.O.: Next-generation switching system based on 8×8 self-turning optical cell. In: Proceedings of International Conference on Actual Problems of Electron Devices Engineering, Saratov, pp. 306–310 (2018)
3. Golubev, A., Chechetkin, I., Parygin, D., Sokolov, A., Shcherbakov, M.: Geospatial data generation and preprocessing tools for urban computing system development. Proc. Comput. Sci. **101**, 217–226 (2016). Proceedings of the 5th International Young Scientist Conference on Computational Science, Krakow, Poland. Elsevier
4. Vishnevskiy, V., Semenova, O.: Queueing system with alternating service rates for free space optics-radio hybrid channel. In: Vinel, A., Bellalta, B., Sacchi, C., Lyakhov, A., Telek, M., Oliver, M. (eds.) MACOM 2010. LNCS, vol. 6235, pp. 79–90. Springer, Heidelberg (2010). https://doi.org/10.1007/978-3-642-15428-7_9
5. Kutuzov, D., Osovsky, A., Stukach, O., Starov, D.: CPN-based model of parallel matrix switchboard. In: Moscow Workshop on Electronic and Networking Technologies (MWENT): Proceedings of National Research University "Higher School of Economics". Russia, Moscow, 14–16 March 2018. https://ieeexplore.ieee.org/document/8337180
6. Bawab, E.I.: Optical Switching. Springer, Heidelberg (2006)
7. Kabacinski, W.: Nonblocking Electronic and Photonic Switching Fabrics. Springer, Heidelberg (2005)
8. Chai, Z., Hu, X., Wang, F., Niu, X., Xie, J., Gong, Q.: Ultrafast all-optical switching. In: Advanced Optical Materials, p. 21 (2017)

9. Podlazov, V.S.: A Comparison of system area networks: generalized extended multiring vs flattened butterfly. Autom. Remote Control **79**(3), 571–580 (2018)
10. Karavay, M.F., Podlazov, V.S.: An extended generalized hypercube as a fault-tolerant system area network for multiprocessor systems. Autom. Remote Control **76**(2), 336–352 (2015)
11. Lin, X., Sun, W., Veeraraghavan, M., Hu, W.: Time-shifted multilayer graph: a routing framework for bulk data transfer in optical circuit-switched networks with assistive storage. Opt. Soc. Am. **3**(3), 162–174 (2016)

Characterizing the Degree of LTE Involvement in Supporting Session Continuity in Street Deployment of NR Systems

Vyacheslav Begishev[1](✉) ⬤, Andrey Samuylov[1,2] ⬤, Dmitri Moltchanov[2] ⬤, and Konstantin Samouylov[1,3] ⬤

[1] Peoples' Friendship University of Russia (RUDN University),
6 Miklukho-Maklaya St., Moscow 117198, Russian Federation
{begishev-vo,samuylov-ak,samuylov-ke}@rudn.ru
[2] Tampere University, 33720 Tampere, Finland
{andrey.samuylov,dmitri.moltchanov}@tuni.fi
[3] Institute of Informatics Problems, Federal Research Center, "Computer Science and Control" of the Russian Academy of Sciences,
44-2 Vavilova St., Moscow 119333, Russian Federation

Abstract. The prospective roll out of recently standardized New Radio (NR) systems operating in millimeter wave frequency band pose unique challenges to network engineers. In this context, the support of NR-based vehicle-to-infrastructure communications is of special interest due to potentially high speeds of user equipment and semi-stochastic dynamic blockage conditions of propagation paths between UE and BR base station (BS). In this conditions even the use of advanced NR functionalities such as multiconnectivity supporting active connections to multiple BSs located nearby may not fully eliminate outages. Thus, to preserve session continuity for UEs located on vehicles a degree of LTE support might be required. In this paper, we quantify the amount of LTE support required to maintain session continuity in street deployment of NR systems supporting multiconnectivity capabilities. Particularly, we demonstrate that it is heavily affected by the traffic conditions, inter-site distance between NR BSs and the degree of multiconnectivity.

Keywords: New Radio · Millimeter wave · Street deployment · Blockage · Outage · LTE support · Multiconnectivity

1 Introduction

The New Radio (NR) technology recently standardized by 3GPP promise to bring extraordinary rates at the air interface enabling modern and future rate-

The publication has been supported by "RUDN University Program 5-100" (V.O. Begishev, simulation model development). The reported study was funded by RFBR, project numbers 18-37-00380 (A. Samuylov, numerical analysis) and 18-00-01555 (18-00-01685) (K.E. Samouylov, problem formulation and project administration).

© Springer Nature Switzerland AG 2019
V. M. Vishnevskiy et al. (Eds.): DCCN 2019, LNCS 11965, pp. 71–83, 2019.
https://doi.org/10.1007/978-3-030-36614-8_6

greedy applications. However, specifics of propagation properties at millimeter wave frequencies as well as relatively small coverage of such system induces new challenges to system designers.

The problem of dynamic blockage of propagation paths in NR systems has been deeply addressed in recent literature. The models characterizing stochastic properties of blockage and non-blockage intervals under different mobility of UEs and blockers have been reported in [1–3]. The authors in [4] employed an analytical representation 3GPP 3D cluster-based propagation model [5] provided in [6] to characterize spatial dependency of channel blockage states.

Identifying outages caused by blockage as one of the main reasons for performance degradation, the authors recently addressed performance of NR systems in dynamic blockage conditions. The upper bound of capacity of NR systems in presence of multiconnectivity is provided in [7]. The practical gains of multiconnectivity with finite number of simultaneously supported links have been reported in [8,9]. The concept of bandwidth reservation to improve session continuity in NR systems has been proposed and analyzed in [10,11]. The joint use of bandwidth reservation and multiconnectivity operation has been assessed in [10,12]. Performance of NR systems in three-dimensional deployments has been assessed in [13,14]. Recently, the studies addressing practical deployments started to appear. Particularly, the study in [15] reports on the NR deployment in square environment. The authors in [16,17] considered complex street deployment of NR systems.

One of the most challenging use-cases for NR system is the support of user equipment (UE) deployed at moving vehicles in street environments. Potentially high speeds of vehicles require frequent BS changes and fast beam alignment procedures. Furthermore, the blockage process is characterized by much complex dynamics compared to human bidy blockage that has been thoroughly studies in literature [1,2]. Finally, semi-regular environment, where mobility is limited to straight lanes but the inter-vehicles distance (IVD) and vehicle dimensions are random leads to complex environment for system analysts. As a result, compared to purely stochastic UE deployments such as those observed at squares, parks, leisure resorts, etc., service performance of vehicles in street deployments of NR systems has been loosely characterized so far.

In presence of dynamic vehicle body blockage in street deployment of NR systems, the propagation paths between vehicle-mounted UE and NR base station (BS) can be blocked by other vehicles. Depending on the distance between BS and UE blockage events may lead to outages. To alleviate these consequences of blockage 3GPP has proposed multi-connectivity option [18]. According to it, UE is allowed to support multiple connections to NR BSs located nearly. In case of outage event with the current NR BS, UE may switch one of backup connections. However, the authors in [19] have recently demonstrated that outages may still happen even when three or more connections are simultaneously supported.

One of the critical performance metrics for rate-greedy multimedia applications such as high-definition streaming or augmented/virtual reality, is session continuity. Outage events may produce a drastic negative effect on this metrics causing frequent session interruptions. Thus, to ensure uninterrupted connectivity additional backup radio access technologies such as LTE can be used during outage intervals. In this study, we characterize the required degree of LTE involvement into service process of vehicle-mounted UE in street deployment of NR systems. Introducing *street outage capacity* of NR deployment as the capacity per unit length of a street required by vehicles in outage conditions to support session continuity, we characterize it as a function of (i) street traffic conditions, (ii) degree of multiconnectivity and (iii) inter-site distance (ISD) between BSs. The obtained results can be used to determine the ISD such that LTE BS covering a certain distance of streets in a district may ensure session continuity in case of NR outage conditions.

The rest of the paper is organized as follows. In Sect. 2 we introduce the system model and metrics of interest. Further, in Sect. 3 we formalize our performance evaluation framework. We report numerical results in Sect. 4. The conclusions are drawn in the last section. Headings should be numbered. Lower level headings remain unnumbered; they are formatted as run-in headings.

2 System Model

We consider a typical street NR deployment illustrated in Fig. 1. NR BSs are located on both sides of the street, for example, at lampposts that have a constant height h_A. The distance between the access points is such that they form isosceles triangles. The number of lanes is assumed to be $N = 4$. The width of each line is constant and equal to w. We are interested in the UE associated with target vehicle which moves at a speed of v_U. We consider height UE to be constant h_U.

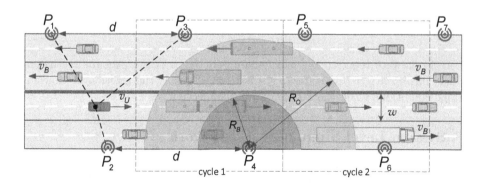

Fig. 1. The considered street NR BS deployment with multi-connectivity operation.

The line-of-sight (LoS) propagation path between NR BS and vehicle-mounted UE may be blocked by moving vehicles, so-called blockers. At each lane blockers form the Poisson process with an intensity λ_B. The vehicle speed is constant and equal to the value v_B. The length of blockers is a random variable characterized by a certain cumulative distribution function (CDF).

In our study, we assume that blockage of LoS by cars leads to a 20 dB decrease in the received signal power. When simulating the NR propagation, we apply the 3GPP urban micro (UMi) street model [5]. According to it, the path loss measured in dB is defined as

$$L(q) = \begin{cases} 52.4 + 21.0 \log q + 20 \log f_c, \text{ blocked,} \\ 32.4 + 21.0 \log q + 20 \log f_c, \text{ non-bl.,} \end{cases} \tag{1}$$

where q is the three-dimensional (3D) distance between the UE and the NR BS, f_c is the carrier frequency in GHz.

The employed propagation model divides the area around NR BS into three zones. In the first zone, limited by distance R_B, no outage may happen regardless of whether LoS is blocked or not. In the second zone, from R_B to R_O, the NR is available if LoS is not blocked. Finally, starting from the distance R_O, no communications is feasible. The values of R_B and R_O are calculated according to the propagation model in (1).

The target UE is allowed to use the 3GPP multi-connectivity operation [18]. According to it, UE maintains an active link with M neighboring NR BSs, where M is known as the "degree of multi-connectivity". We also assume that UE can instantly switch to the best NR BSs out of M available.

Irrespective of the degree of multiconnectivity there might be situations when UE experiences outage with all M available NR BSs. To ensure session continuity we assume that LTE BS located in the area provides additional support. To characterize the degree of LTE involvement into ensuring session continuity, in this study, we concentrate on the so-called *street outage capacity* defined as the bitrate per meter of a street required to ensure session continuity

$$R_O = 4Rp\theta q_i, \tag{2}$$

where is the rate required by vehicle-mounted UE, p is the probability that a vehicle has an active application, $1/\theta$ is the IVD, q_i is the fraction of time in outage conditions when the degree of multiconnectivity is i.

3   System Level Simulation Framework

To quantify the identified metric of interest we have developed a system level simulation framework (SLS) that features analytical pre-processing at the initialization face to speed up the execution. Below, we outline the basic principles of the underlining methodology including the pre-processing phase and actual structure of the simulation engines.

Analyzing the deployment illustrated in Fig. 1, one may notice that for a given degree of multiconnectivity M the connectivity patterns itself. Thus, to assess the amount of time UE spends in outage conditions, it is sufficient to concentrate on those points in time when the state of UE changes. All these points

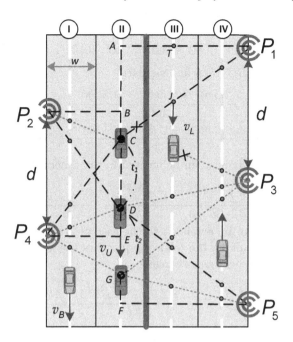

Fig. 2. Detailed illustration of deployment parameters for $M = 3$.

that define the beginning of different zones can be accurately determined leaving the blocking vehicles and their dimensions as the only stochastic variables. Particularly, for each degree of multiconnectivity, M, we calculate the points of LoS/nLoS state changes. Therefore, before initializing the simulations we first determine the geometry of the zones, where outage may occur, and then proceed with the UE dynamics.

3.1   Analytical Representation

We illustrate the proposed mixed analytical-simulation methodology using $M = 3$ as an example, see Fig. 2. To define the pattern we need to calculate the coordinates of the transition points, C, D, and G, where the set of NR BS associations changes. We denote them as $C(BC, 0, h_U)$, $D(BD, 0, h_U)$, and $G(BG, 0, h_U)$, where the origin is at $B(0, 0, h_U)$. The required distances BC, BD, and BG can be sought from associated triangles.

As an example, consider BC. Using the right triangles $\Delta P_1 AC$ and $\Delta P_4 EC$ and applying the Pythagorean theorem we define the following system:

$$\begin{cases} (P_1 A)^2 + (AC)^2 = (P_1 C)^2, \\ (P_4 E)^2 + (EC)^2 = (P_4 C)^2, \\ P_1 C = P_4 C, \\ AC + EC = 1.5d. \end{cases} \tag{3}$$

Solving the system we arrive at

$$BC = |AC - AB| = |AC - 0,5d|.$$ (4)

Distances BD and BG are found similarly

$$BD = \frac{(2.5w)^2 + (1.5d)^2 - (1.5w)^2}{3d},$$ (5)

$$BG = \frac{-(2.5w)^2 + (1.5d)^2 + (1.5w)^2}{3d} + 0.5d.$$ (6)

Now, the transition points can be written in the general form as

$$x_{n+1} = \frac{(-1)^{n+1}\left[\left(\frac{5w}{2}\right)^2 - \left(\frac{3w}{2}\right)^2\right] + \left(\frac{3d}{2}\right)^2}{3d} + \frac{d(n-1)}{2}, \quad n = 0, 1, \ldots. $$ (7)

The latter result can be extended to the case of arbitrary number of lanes $N = 2k$, $k \in \mathbb{Z}$. Particularly, the following provides the locations of the points where the set of UE associations changes

$$x_{n+1} = \frac{(-1)^{n+1}\left[\left(\frac{N+1}{2}w\right)^2 - \left(\left(\lfloor\frac{N-1}{2}\rfloor + \frac{1}{2}\right)w\right)^2\right]}{3d} + $$
$$+ \frac{(1.5d)^2}{3d} + \frac{d(n-1)}{2}, \quad n = 0, 1, \ldots. $$ (8)

Let now t be the time interval a vehicle occludes LoS path. Denote NR BS and UE coordinates by (x_{AP}, y_{AP}, h_A) and (x_{UE}, y_{UE}, h_U), respectively, and coordinates of other vehicles by (x_B, y_B, h_H). To compute t consider triangle $\Delta P_1 AC$. Denote by $P_1 A$ a perpendicular from P_1 to a point lying in the middle of the strip of motion. Let this point T have the coordinates (x_T, y_T, z_T). Now, TJ is

$$TJ = \frac{(P_1 T)(AC)}{P_1 A},$$ (9)

where AC is x-coordinate of the UE, and variables $P_1 A$ and $P_1 T$ are distances to access point P_1. Observe that the movement in our system is uniformly rectilinear. Thus, TJ and AC for any t are described by the following system

$$\begin{cases} AC = x_A + v_U t, \\ TJ = x_A + v_L t. \end{cases}$$ (10)

Using (10) we obtain v_L as

$$v_L = v_U (TJ - x_A)/(AC - x_A).$$ (11)

Observe that the time instant t, when two objects moving towards each other meet, is provided by

$$t = \frac{\sqrt{(x_J - x_B)^2 + (y_J - y_B)^2 + (z_J - h_B)^2}}{v_L + v_B},$$ (12)

where (x_J, y_J, z_J) are coordinates of J in Fig. 2.

For the cases $M = 1, M = 2$ the analysis is similar.

Table 1. Default system parameters.

Scenario	Jam	Normal	Highway
Tagged vehicle speed	20	40	120
Other vehicles speed	20	40	120
Inter-vehicle distance	2	5	20

3.2 Implementation

The simulation engine is built based on time-driven discrete simulation (TD-DES) framework. The software was developed using general-purpose programming language (Java) with multi-threaded optimizations [20]. The modeling procedure consists of two stages, simulation execution and data analysis. The main part of the simulation is to track the initial and final points in time when the marked vehicle is blocked by vehicles moving in other lanes. Each blockage event is processed and stored in an external file. Post-processing of blockage data is then used to determine outage events and their durations.

The following data collection and analysis procedure has been used to obtain the metrics of interest. For each set of input parameters, the simulations were performed for 10^3 seconds of system time. The start of the steady-state period was detected using exponentially-weighted moving average (EWMA) statistics with a weight parameter set to 0.05 [20]. Statistical data were collected in the steady-state only. We used the batch means strategy to remove residual correlations in the statistical data. According to it, the entire data set was divided into 10^5 data blocks. The means of these data blocks served as independent identically distributed observations. Classic statistical methods were further applied to obtain point estimate of the sought metrics of interest.

4 Numerical Results

In this section, we numerically study street outage capacity for a range of system parameters. Particularly, we consider three typical traffic conditions outlined in Table 1. The default system parameters are summarized in Table 2. Also, recalling the structure of expression for street outage capacity in (2), we assume that (i) the unit rate required by an application, $R = 1$ bits/s and (ii) all vehicles are associated with active sessions, i.e., $p = 1$. For non-unit rate R and p, the street outage capacity can be obtained by appropriately scaling the data presented in this section.

Figure 3 illustrates the street outage capacity as a function of the ISD between NR BSs for $M = 1$, i.e., no multiconnectivity is supported by UEs. As one may observe, the street outage capacity remains at the constant level for all considered ISDs for highway traffic conditions. The reason is that the IVD is rather high (20 m.) resulting in rather low density of vehicles per meter. However, as we consider normal traffic conditions the IVD, where IVD is four times smaller

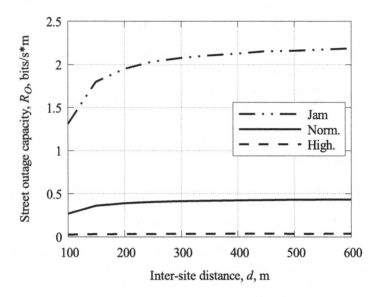

Fig. 3. Street outage capacity, M = 1

while the speed is three times lower, we start to notice differences between street outage capacity for different ISDs. These effects become even more profound when addressing traffic jam conditions. Numerically, for ISD of 100 m $R_O \approx 1.3$ and increases to $R_O \approx 2.2$ for ISD of 600 m.

Table 2. Default system parameters.

Parameter	Value
Emitted power, P	0.2 W
Carrier frequency, f_C	28 GHz
BS transmit antenna gain, G_T	14.58 dB
UE receive antenna array, G_R	8.57 dB
SNR threshold, S_T	0 dB
Blockage radius in blocked state, R_B	45 m
Outage radius in non-blocked state, R_O	172 m
Height of BS, h_A	5 m
Width of lanes, w	4 m
Number of lanes, N	4
Height of UE, h_U	1.5 m
Mean length of vehicles, L	3.5 m
Mean height of vehicles, h_H	2 m

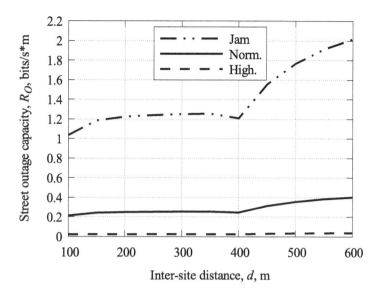

Fig. 4. Street outage capacity, $M = 2$

Consider now how the street outage capacity changes when the degree of multiconnectivity increases as indicated in Figs. 4 and 5. First comparing the results in Figs. 3 and 4 one may notice quantitative improvement in street outage capacity when switching from $M = 1$ to $M = 2$. The most dramatic effect is however, observed for traffic jam road conditions, when the street outage capacity remains almost flat up until approximately 400 m of ISD. Particularly, at ISD of 400 m. Operating using just one back up link allows to reduce the amount of traffic that needs to be supported at LTE BS twice. Increasing the degree of multiconnectivity to $M = 3$ allows to decrease LTE requirements even further, especially in the ISD range of 100–300 m. Much milder effect are observed for normal and highway road traffic conditions.

We also would like to specifically highlight the quantitative differences between the amount of traffic that needs to be temporarily supported by LTE technology in different road traffic conditions to ensure session continuity. Particularly, the street outage capacity jam road traffic conditions could be up to four times higher compared to normal conditions inducing large deviations into the amount of traffic that needs to be supported by LTE depending on the time of a day. The mean street outage capacity is highway traffic conditions is usually much smaller even for high ISDs between NR BS allowing to use a single LTE BS to cover large highway segments. These quantitative differences need to be taken into account at the NR deployment phase.

Having characterized the street outage capacity we now proceed analyzing the application related performance metrics. Particularly, we concentrate on the mean outage and non-outage times as on the critical metrics for applications. To this end, Figs. 6, 7 and 8 characterize the outage and non-outage durations as a

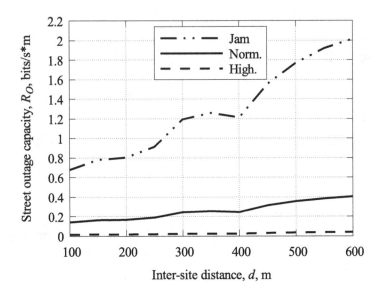

Fig. 5. Street outage capacity, M = 3

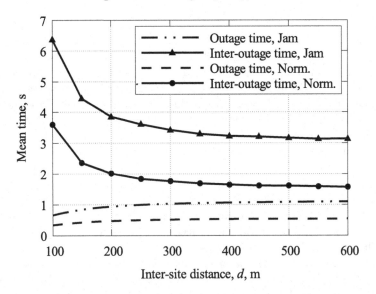

Fig. 6. Mean outage and non-outage time, M = 1

function of the distance between NR BSs for traffic jam and normal traffic conditions and different degrees of multiconnectivity. Analyzing the results, one may observe that qualitatively the metrics are characterized by similar behavior for all considered values of M. However, the use of higher degree of multiconnectivity drastically decreases the mean outage duration therefore requiring less support from LTE BS.

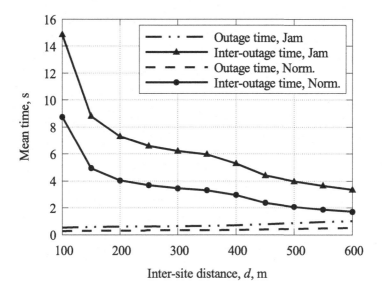

Fig. 7. Mean outage and non-outage time, M = 2

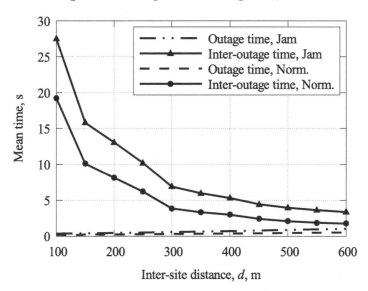

Fig. 8. Mean outage and non-outage time, M = 3

One of the important observations that can be deduced from Figs. 6, 7 and 8 is the intensity of service interruptions caused by outage events. As one may observe, not only the mean durations of outage state increases but the intensity of outage decreases when the the value of M grows.

5 Conclusions

Motivated by ensuring session continuity, in this study we formalized a system model for vehicle-mounted UEs in street deployment of 3GPP NR systems with multiconnectivity capabilities. Having introduced the street outage capacity metric defined as the rate required to ensure session continuity per unit length of a street, we characterized the degree of LTE system involvement in service process of vehicle-mounted UEs.

Our numerical results indicate that the street outage capacity is drastically affected by the street traffic conditions and the degree of multiconnectivity supported by UE. Particularly, it is mainly affected by inter-vehicle distance and speed of vehicles. Furthermore, the support of multiconnectivity option is vital for traffic jam road conditions as even then use of one back-up link allows to decrease the amount of traffic needed at LTE BSs almost twice. The developed model and presented results can be used to determine the required density of NR BSs such that LTE BS may ensure session continuity in case of outages.

References

1. Gapeyenko, M., et al.: On the temporal effects of mobile blockers in urban millimeter-wave cellular scenarios. IEEE Trans. Veh. Technol. **66**, 10124–10138 (2017)
2. Samuylov, A., et al.: Characterizing spatial correlation of blockage statistics in urban mmWave systems. In: Globecom Workshops (GC Wkshps), pp. 1–7. IEEE (2016)
3. Moltchanov, D., Ometov, A., Koucheryavy, Y.: Analytical characterization of the blockage process in 3GPP new radio systems with trilateral mobility and multi-connectivity. Comput. Commun. **146**, 110–120 (2019)
4. Gapeyenko, M., et al.: Spatially-consistent human body blockage modeling: a state generation procedure. IEEE Trans. Mob. Comput. **1**(1), 1–15 (2019)
5. 3GPP, Study on channel model for frequencies from 0.5 to 100 GHz (Release 15), 3GPP TR 38.901 V15.0.0, June 2018
6. Gapeyenko, M., et al.: An analytical representation of the 3GPP 3D channel model parameters for mmWave bands. In: Proceedings of the 2nd ACM Workshop on Millimeter Wave Networks and Sensing Systems, pp. 33–38. ACM (2018)
7. Moltchanov, D., Ometov, A., Andreev, S., Koucheryavy, Y.: Upper bound on capacity of 5G mmWave cellular with multi-connectivity capabilities. Electron. Lett. **54**(11), 724–726 (2018)
8. Gapeyenko, M., et al.: On the degree of multi-connectivity in 5G millimeter-wave cellular urban deployments. IEEE Trans. Veh. Technol. **68**(2), 1973–1978 (2019)
9. Gerasimenko, M., Moltchanov, D., Gapeyenko, M., Andreev, S., Koucheryavy, Y.: Capacity of multiconnectivity mmWave systems with dynamic blockage and directional antennas. IEEE Trans. Veh. Technol. **68**(4), 3534–3549 (2019)
10. Begishev, V., et al.: Quantifying the impact of guard capacity on session continuity in 3GPP new radio systems. IEEE Trans. Veh. Technol. **1**(1), 1–15 (2019)
11. Moltchanov, D., et al.: Improving session continuity with bandwidth reservation in mmWave communications. IEEE Wirel. Commun. Lett. **8**(1), 105–108 (2018)

12. Kovalchukov, R., et al.: Improved session continuity in 5G NR with joint use of multi-connectivity and guard bandwidth. In: 2018 IEEE Global Communications Conference (GLOBECOM), pp. 1–7. IEEE (2018)

13. Kovalchukov, R., et al.: Evaluating sir in 3D millimeter-wave deployments: direct modeling and feasible approximations. IEEE Trans. Wirel. Commun. **18**(2), 879–896 (2018)

14. Kovalchukov, R., et al.: Analyzing effects of directionality and random heights in drone-based mmWave communication. IEEE Trans. Veh. Technol. **67**(10), 10064–10069 (2018)

15. Petrov, V., et al.: Dynamic multi-connectivity performance in ultra-dense urban mmWave deployments. IEEE J. Sel. Areas Commun. **35**(9), 2038–2055 (2017)

16. Petrov, V., et al.: Achieving end-to-end reliability of mission-critical traffic in softwarized 5G networks. IEEE J. Sel. Areas Commun. **36**(3), 485–501 (2018)

17. Petrov, V., Moltchanov, D., Andreev, S., Heath Jr, R.W.: Analysis of intelligent vehicular relaying in urban 5G+ millimeter-wave cellular deployments. arXiv preprint arXiv:1908.05946 (2019)

18. 3GPP, NR; Multi-connectivity; Overall description (Release 15), 3GPP TS 37.340 V15.0.0, Jan 2018

19. Begishev, V., Samuylov, A., Moltchanov, D., Machnev, E., Koucheryavy, Y., Samouylov, K.: Connectivity properties of vehicles in street deployment of 3GPP NR systems. In: 2018 IEEE Globecom Workshops (GC Wkshps), pp. 1–7. IEEE (2018)

20. Perros, H.G.: Computer simulation techniques: the definitive introduction (2009)

Dolph-Chebyshev and Barcilon-Temes Window Functions Modification

V. P. Dvorkovich and A. V. Dvorkovich[✉]

Moscow Institute of Physics and Technology,
9, Institutkiy per., Dolgoprudny, Russian Federation
v.dvorkovich@mail.ru, dvork-alex@yandex.ru

Abstract. Well known Dolph-Chebyshev and Barcilon-Temes window functions have been considered. Possible modifications allowing for substantial improvement of window function parameters have been investigated. These modifications utilizes the characteristics of neighboring odd and even components processing.

Keywords: Digital signal processing · Fourier transform · Harmonic analysis · Window function

1 Introduction

A comprehensive review of various standard and constructed by many authors window functions for harmonic signal analysis have been done in classic article by Harris [1]. In particular, Dolph-Chebyshev and Barcilon-Temes windows have been analyzed there.

Specific feature of Dolph-Chebyshev windows is the equality of all window spectrum side lobe amplitudes. This feature causes substantial suppression of side lobes for given main lobe width, or minimal width of main lobe for given level of side lobes. Dolph-Chebyshev window functions have been developed for analysis of discrete signals with periodic spectrum [1–3].

2 Dolph-Chebyshev Window Functions

Spectral components of Dolph-Chebyshev windows are defined using the values of equidistant samples of Fourier transform:

$$W(k) = (-1)^k \frac{cos(N \arccos(\beta \cos(\pi k/N)))}{\text{ch}(N \operatorname{arcch}(\beta))}, \tag{1}$$

where $0 \le k < N$, $\beta = \text{ch}(\operatorname{arcch}(1/h)/N)$, $h = 10^{-\alpha}$,

$$\arccos(z) = \begin{cases} \dfrac{\pi}{2} - \arctan \dfrac{z}{\sqrt{1-z^2}}, & |z| \le 1, \\ \ln(z + \sqrt{1-z^2}), & |z| > 1. \end{cases}$$

© Springer Nature Switzerland AG 2019
V. M. Vishnevskiy et al. (Eds.): DCCN 2019, LNCS 11965, pp. 84–93, 2019.
https://doi.org/10.1007/978-3-030-36614-8_7

Weighting coefficients of Dolph-Chebyshev window are usually defined by inverse Fourier transform of spectral function, but they could be find straightly using formula:

$$w(k) = \sum_{n=0}^{k} \frac{(-1)^k (N-n-2)! \beta^{-2n}}{(N-k-n-1)! n! (k-n)!}. \tag{2}$$

The formula (2) defines the coefficients $w(k)$ for k from 0 to L, where $L = N/2 - 1$ for even N and $L = (N-1)/2$ for odd N. The other coefficients are defined from even symmetry of weighting function. The sum of resulting coefficients should be normalized to 1.

Normalized spectral functions of Dolph-Chebyshev windows depends on two parameters (n and h) and are defined by the formula:

$$F_{DC}(y) = F_{DC}(y, n, h) = \frac{T_n[\xi \cos(\pi y)]}{T_n[\xi]}, \tag{3}$$

where $y = f \Delta T$, ΔT – sampling interval, f – frequency $\xi = \mathrm{ch}(R/n)$, $R = \mathrm{arcch}(1/h)$, $h = 10^{-\alpha}$, $T_n(z) = \begin{cases} \cos(n \arccos(z)), & z \le 1, \\ \mathrm{ch}(n \, \mathrm{arcch}(z)), & z > 1, \end{cases}$ – Chebyshev polynomial of the first kind of n-th order.

The amplitudes of spectrum side lobes of $F_{DC}(y)$ are equal to each other. These amplitudes are defined by the value -20α (Fig. 1). So, side lobe decay rate is equal $\Delta W = 0 \, \mathrm{dB}$ (illustrated by green lines on the Fig. 1).

3   Barcilon-Temes Window Functions

Constructuion of Barcilon-Temes window functions is based on the algorithm of spectral components energy minimization outside the main lobe. This criterion gives the compromise between the criteria use for construction of Dolph-Chebyshev and Kaiser-Bessel windows [3,4].

Spectral components of Barcilon-Temes windows are defined by equidistant samples of Fourier transform:

$$W(k) = (-1)^k h \frac{\sqrt{1-h^2} \cos(y_k) + (y_k/R) \sin(y_k)}{(Rh^2 + \sqrt{1-h^2})((y_k/R)^2 + 1)}, \tag{4}$$

where $y_k = N \arccos(\beta \cos(\pi k/N))$, $R = \mathrm{arcch}(1/h)$, $\beta = \mathrm{ch}(R/N)$, $h = 10^{-\alpha}$.

Normalized spectral functions of Barcilon-Temes windows constructed with the use of Chebyshev polynomials also depends on two parameters (n and h) and are defined by formula:

$$F_{BT}(y) = F_{BT}(y, n, h) = \frac{T_n[\xi \cos(\pi y)] + (\xi/R) \cos(\pi y) U_n[\xi \cos(\pi x)]}{T_n[\xi] + (\xi/R) U_n[\xi]}, \tag{5}$$

where $U_n(z) = \dfrac{\sin(n \arccos(z))}{\sqrt{1-z^2}}$, $z \le 1$ – Chebyshev polynomial of the second kind of n-th order.

The amplitudes of spectrum side lobes of $F_{BT}(y)$ are approximately equal -20α and slightly decreases towards the borders of interval $y = -0,5$ and $y = 0,5$ (Fig. 2). The rate of side lobe decay is equal $\Delta W = 2\,\mathrm{dB}$ per octave (illustrated by green lines on the Fig. 2).

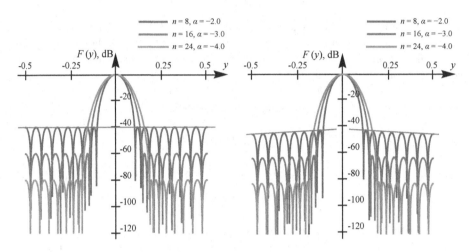

Fig. 1. Spectra of Dolph-Chebyshev windows (Color figure online)

Fig. 2. Spectra of Barcilon-Temes windows (Color figure online)

4 Properties of Dolph-Chebyshev and Barcilon-Temes Windows

The functions $F_{DC}(y)$ and $F_{BT}(y)$ (further generally specified as $F(y)$) are periodic.

For even values of the coefficients $n = 2m$:

$$F(y + k) = F(y),\ |y| \leq 1/2,\ k = ..., -2, -1, 0, 1, 2, ...\ \text{(Fig. 3)}.$$

For odd values of the coefficients $n = 2m - 1$:

$$F(y + k) = (-1)^k F(y),\ |y| \leq 1/2,\ k = ..., -2, -1, 0, 1, 2, ...\ \text{(Fig. 4)}.$$

The parameter n defines the relative frequency of cosine oscillations $ny/2$ at the interval $[-1/2, 1/2]$, and the parameter $h = 10^{-\alpha}$ defines the amplitude of the oscillations.

Figure 5 illustrates the structure of the functions $F(y)$ near zero for even value of $n = 2m$; the sequence of cosine oscillations $h\cos(\pi ny)$ is shown below.

In this case if $n = 2m$ and $m = 2k$ are even, then the intervals $[-1/2, 0]/[0, 1/2]$ contain an integer number of oscillation periods, the values

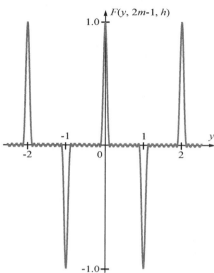

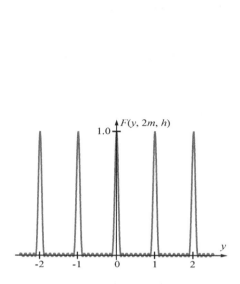

Fig. 3. Structure of window $F(y)$ with even value n

Fig. 4. Structure of window $F(y)$ with odd value n

of the function at the boarders of the interval $[-1/2, 1/2]$ is positive and equal to $F(1/2) = F(-1/2) = h$.

If $m = 2k - 1$ is an odd value, an integer number of half-periods of oscillations fits in the intervals $[-1/2, 0]/[0, 1/2]$, the values of the function at the boarders are negative and equal to $F(1/2) = F(-1/2) = -h$.

The Fig. 6 illustrates the structure of the functions $F(y)$ near zero for odd value $n = 2m - 1$. In this case an integer number of oscillation half-periods fits into interval $[-1/2, 1/2]$ and $F(1/2) = F(-1/2) = 0$.

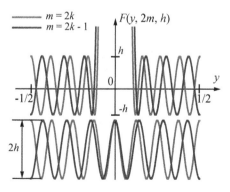

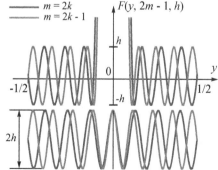

Fig. 5. Structure of window $F(y)$ near zero with even value n

Fig. 6. Structure of window $F(y)$ near zero with odd value n

Derivatives of these functions at interval $[-1/2, 1/2]$ boarders for even values $m = 2k$ are negative – $F'(1/2) = F'(-1/2) = -\pi nh$, and for odd values $m = 2k - 1$ are positive – $F'(1/2) = F'(-1/2) = \pi nh$.

Normalized spectra of Dolph-Chebyshev and Barcilon-Temes windows are identically defined by finite number of cosine components.

For $n = 2m$

$$F(y) \equiv b_0 + 2 \sum_{k=1}^{m} b_k \cos(2\pi ky), \qquad (6)$$

where $b_0 = 2 \int_0^{1/2} F(y)dy$, $b_k = 2 \int_0^{1/2} F(y)\cos(2\pi ky)dy$, $b_0 + 2 \sum_{k=1}^{m} b_k \equiv 1$.

For $n = 2m - 1$

$$F(y) \equiv 2 \sum_{k=1}^{m} d_k \cos(\pi(2k - 1)y), \qquad (7)$$

where $d_k = 2 \int_0^{1/2} F(y)\cos(\pi(2k-1)y)dy$, $2 \sum_{k=1}^{m} d_k \equiv 1$.

Relative waveforms of Dolph-Chebyshev and Barcilon-Temes windows are simply defined using Eqs. (8) and (9).

For $n = 2m$

$$u(x) = \text{Sinc}(\pi x) + \frac{1}{b_0} \sum_{k=1}^{m} b_k \left[\text{Sinc}(\pi(x - k)) + \text{Sinc}(\pi(x + k))\right]. \qquad (8)$$

For $n = 2m - 1$

$$u(x) = \frac{1}{u(0)} \sum_{k=1}^{m} d_k \left[\text{Sinc}(\pi(x - k + 1/2)) + \text{Sinc}(\pi(x + k - 1/2))\right], \qquad (9)$$

where $u(0) = \frac{4}{\pi} \sum_{k=1}^{m} \frac{(-1)^{k-1}}{2k - 1}$.

A set of main parameters [1] of Dolph-Chebyshev and Barcilon-Temes windows and its relative spectra for even values $n = 2m$ ($u[2(m + 1)x]$ and $F[y/(2(m+1))]$) and odd values $n = 2m - 1$ ($u[(2m+1)x]$ and $F[y/((2m+1))]$) is shown in the Table 1 for some values of n and for seven values of α (2.0, 2.5, 3.0, 3.5, 4.0, 4.5, 5.0).

The following designations are used in the Table: ΔW – spectrum side lobe decay rate; W_{max} – maximal side lobe level; PG – processing gain; ΔF_n – equivalent noise bandwidth; $\Delta F_{-3\,dB}$ and $\Delta F_{-6\,dB}$ – width of the window function at the levels $-3\,dB$ and $-6\,dB$; δ, % – the relative difference between the noise bandwidth and $-3\,dB$ level of window function; SL – spectrum scalloping loss; MPL – maximum processing loss; $C(r)$ – correlation of overlapping parts of the window function for overlapping values $r = 0.75$ and $r = 0.5$.

Side lobe level relative to main lobe level ratios are defined by the coefficient α and are equal -20α dB. For the values of α shown above the ratios are equal $-40, -50, -60, -70, -80, -90$ and -100 dB correspondingly.

The Eqs. (6) and (7) defines the shape of frequency response for low pass filters after the substitution of the argument y to relative frequency fT, where $f_0 = 1/(2T)$ – filter frequency cutoff, $0 \leq f \leq f_0$. Pulse response characteristic of these filters are defined by equations:

$$u_{2m}(t) = \sum_{k=-m}^{m} b_k \delta(|k|T) \text{ for } n = 2m \text{ and} \tag{10}$$

$$u_{2m-1}(t) = \sum_{k=-m}^{m} b_k \delta((|k| - 1/2)T) \text{ for } n = 2m - 1. \tag{11}$$

The Eqs. (8) and (9) defines the shape of the pulse response of these filters after the substitution of the argument x to the relative time interval t/T.

Table 1. Dolph-Chebyshev and Barcilon-Temes windows main parameters

Window function	W_{max}, dB	PG	ΔF_n, bin	ΔF_{-3dB}, bin	δ, %	ΔF_{-6dB}, bin	SL, dB	MPL, dB	$C(r)$, %	
									$r = 0.75$	$r = 0.5$
Dolph-Chebyshev, $\Delta W = 0\,dB$ per octave										
$n = 8$, $\alpha = -2.0$	-40	0.686	1.458	1.397	4.359	1.944	1.521	3.157	55.02	18.62
$n = 13$, $\alpha = -2.5$	-50	0.649	1.541	1.465	5.171	2.049	1.370	3.249	53.29	15.05
$n = 16$, $\alpha = -3.0$	-60	0.608	1.645	1.562	5.302	2.177	1.209	3.371	50.93	11.36
$n = 21$, $\alpha = -3.5$	-70	0.575	1.739	1.648	5.487	2.308	1.087	3.490	48.62	8.70
$n = 24$, $\alpha = -4.0$	-80	0.545	1.835	1.738	5.610	2.443	0.979	3.617	46.10	6.50
$n = 29$, $\alpha = -4.5$	-90	0.519	1.925	1.821	5.685	2.555	0.893	3.737	43.68	4.89
$n = 32$, $\alpha = -5.0$	-100	0.497	2.014	1.905	5.732	2.670	0.818	3.858	41.22	3.65
Barcilon-Temes, $\Delta W = 2\,dB$ per octave										
$n = 8$, $\alpha = -2.0$	-42.5	0.679	1.473	1.407	4.923	1.957	1.491	3.172	54.73	17.93
$n = 13$, $\alpha = -2.5$	-51.7	0.641	1.560	1.483	5.193	2.072	1.338	3.270	52.87	14.31
$n = 16$, $\alpha = -3.0$	-61.4	0.602	1.661	1.576	5.403	2.197	1.187	3.390	50.55	10.87
$n = 21$, $\alpha = -3.5$	-71.0	0.570	1.755	1.662	5.554	2.329	1.069	3.511	48.21	8.30
$n = 24$, $\alpha = -4.0$	-80.4	0.541	1.849	1.750	5.662	2.452	0.966	3.635	45.73	6.23
$n = 29$, $\alpha = -4.5$	-90.4	0.516	1.938	1.837	5.694	2.690	0.882	3.755	43.32	4.69
$n = 32$, $\alpha = -5.0$	-100.2	0.494	2.025	1.915	5.775	2.681	0.809	3.874	40.90	3.51

5   Modification of Dolph-Chebyshev and Barcilon-Temes Windows

Taking into account the peculiarities and structures of Dolph-Chebyshev and Barcilon-Temes windows discussed above, it is possible to propose some options for modifications that substantially change the level of spectrum side lobes.

It is possible to use, for example, two options for construction of modified window functions based on combinations of normalized spectra of Dolph-Chebyshev and Barcilon-Temes windows according to the formulas:

$$F_{\mathrm{mod1}}(y, n, h) = \frac{1}{2}[F(y, n - 2, h) + F(y, n, h)], \tag{12}$$

$$F_{\mathrm{mod2}}(y, n, h) = \frac{1}{2}\left[F(y, n, h) + \frac{F(y, n + 2, h) + F(y, n - 2, h)}{2}\right], \tag{13}$$

where $F(y, n, h)$ – standard spectra of Dolph-Chebyshev and Barcilon-Temes windows.

$F_{\mathrm{mod1}}(y, n, h)$ and $F_{\mathrm{mod2}}(y, n, h)$ can be also divided into functions of even and odd arguments when $n = 2m$ and $n = 2m - 1$.

The spectra of functions (12) and (13) are defined by Eqs. (6) or (7) depending on evenness or oddness of coefficient n value.

Table 2 shows main parameters of a set of modified window functions constructed using (12) and (13) equations.

Figures 7, 8, 9 and 10 illustrates normalized spectra of functions $F_{\mathrm{mod1}}(y)$ constructed using even $n = 2m$ (Fig. 7) and odd $n = 2m - 1$ (Fig. 8) Dolph-Chebyshev windows, as well as even (Fig. 9) and odd (Fig. 10) Barcilon-Temes windows.

Taking into account the fact that waveforms of function $F(y, n, h)$ and half-sum of functions $F(y, n + 2, h) + F(y, n - 2, h)$ in the Eq. (13) are very close in the area of main lobes, and the structures of cosine oscillations of these functions near zero are shifted by π, the equation defines new window functions practically coincides with the main lobe of function $F(y, n, h)$, but substantially decreases to the boundaries of interval $[-1/2, 1/2]$.

Figures 11 and 12 shows second type of modification of Dolph-Chebyshev and Barcilon-Temes windows using transforms (13) respectively.

Table 2. Main parameters of modified Dolph-Chebyshev and Barcilon-Temes windows

Window function	W_{max}, dB	PG	ΔF_n, bin	ΔF_{-3dB}, bin	δ, %	ΔF_{-6dB}, bin	SL, dB	MPL, dB	$C(r)$, % $r = 0.75$	$r = 0.5$
Dolph-Chebyshev - mod1, $\Delta W = 9\,\text{dB}$ per octave										
$n = 16$, $\alpha = -3.0$	-64.8	0.572	1.748	1.660	5.312	2.321	1.074	3.500	48.43	8.42
$n = 21$, $\alpha = -3.5$	-73.8	0.549	1.821	1.733	5.088	2.411	0.993	3.597	46.51	6.75
$n = 24$, $\alpha = -4.0$	-83.5	0.523	1.911	1.807	5.802	2.537	0.905	3.718	44.08	5.09
$n = 29$, $\alpha = -4.5$	-93.2	0.502	1.991	1.958	5.804	2.629	0.836	3.826	41.88	3.92
$n = 32$, $\alpha = -5.0$	-103.2	0.482	2.076	1.961	5.829	2.761	0.771	3.943	39.51	2.93
Barcilon-Temes - mod1, $\Delta W = 11\,\text{dB}$ per octave										
$n = 16$, $\alpha = -3.0$	-67.2	0.567	1.763	1.571	5.350	2.340	1.058	3.519	48.05	8.08
$n = 21$, $\alpha = -3.5$	-75.6	0.544	1.837	1.737	5.667	2.431	0.977	3.618	46.09	6.44
$n = 24$, $\alpha = -4.0$	-85.2	0.520	1.925	1.821	5.672	2.553	0.893	3.736	43.71	4.88
$n = 29$, $\alpha = -4.5$	-94.7	0.499	2.004	1.894	5.771	2.647	0.826	3.844	41.52	3.76
$n = 32$, $\alpha = -5.0$	-104.5	0.479	2.087	1.973	5.781	2.770	0.763	3.958	39.19	2.82
Dolph-Chebyshev - mod2, $\Delta W = 15\,\text{dB}$ per octave										
$n = 16$, $\alpha = -3.0$	-68.5	0.546	1.830	1.734	5.558	2.429	0.982	3.608	46.30	6.56
$n = 21$, $\alpha = -3.5$	-76.3	0.529	1.892	1.794	5.462	2.502	0.922	3.690	44.65	5.41
$n = 24$, $\alpha = -4.0$	-86.2	0.506	1.978	1.806	5.802	2.537	0.846	3.808	42.27	4.08
$n = 29$, $\alpha = -4.5$	-95.0	0.488	2.050	1.938	5.802	2.714	0.789	3.907	40.25	3.19
$n = 32$, $\alpha = -5.0$	-105.0	0.469	2.133	2.014	5.914	2.830	0.731	4.021	37.93	2.39
Barcilon-Temes - mod2, $\Delta W = 17\,\text{dB}$ per octave										
$n = 16$, $\alpha = -3.0$	-71.2	0.541	1.848	1.752	5.468	2.450	0.965	3.631	45.83	6.22
$n = 21$, $\alpha = -3.5$	-78.6	0.524	1.909	1.804	5.790	2.530	0.906	3.714	44.18	5.13
$n = 24$, $\alpha = -4.0$	-88.2	0.502	1.992	1.883	5.810	2.638	0.834	3.828	41.86	3.89
$n = 29$, $\alpha = -4.5$	-97.2	0.485	2.054	1.951	5.815	2.734	0.779	3.926	39.86	3.05
$n = 32$, $\alpha = -5.0$	-107.0	0.466	2.145	2.026	5.890	2.845	0.723	4.037	37.59	2.29

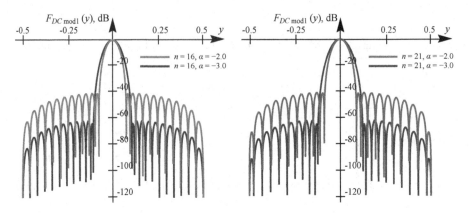

Fig. 7. Spectra of first modification of Dolph-Chebyshev windows with even value n

Fig. 8. Spectra of first modification of Dolph-Chebyshev windows with odd value n

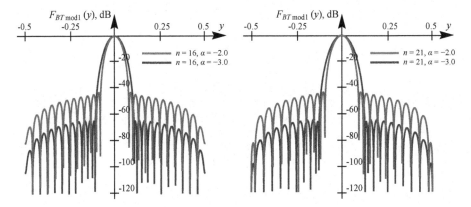

Fig. 9. Spectra of first modification of Barcilon-Temes windows with even value n

Fig. 10. Spectra of first modification of Barcilon-Temes windows with odd value n

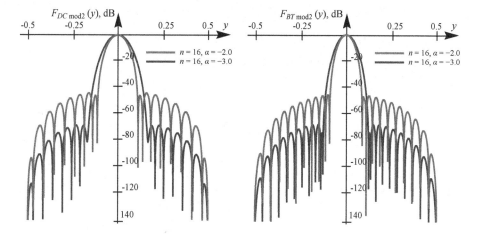

Fig. 11. Spectra of second modification of Dolph-Chebyshev windows

Fig. 12. Spectra of second modification of Barcilon-Temes windows

6 Conclusion

It was shown that normalized spectrum functions of Dolph-Chebyshev and Barcilon-Temes windows are identically defined by finite number of cosine functions with even and odd arguments.

It is easy to calculate corresponding digital low-pass filters and its pulse response characteristics.

Some variants of modification for Dolph-Chebyshev and Barcilon-Temes windows have been proposed taking into account its peculiarities depending on parameter n. These variants allows to decrease substantially side lobe levels of windows spectra.

Two options of linear transform of modified windows have been considered. Its main parameters have been calculated.

References

1. Harris, F.J.: On the use of windows for harmonic analysis with the discrete Fourier transform. Proc. IEEE **66**(1), 51–83 (1978)
2. Lynch, P.: The Dolph-Chebyshev window: a simple optimal filter. Mon. Weather. Rev. **125**, 655–660 (1997)
3. Dvorkovich, V.P., Dvorkovich, A.V.: Window Functions for Harmonic Signal Analysis, 2nd edn., Revised and Expanded, 208 p. Technosphere, Moscow (2016)
4. Barcilon, V., Temes, G.C.: Optimum impulse response and the Van Der Maas function. IEEE Trans. Circuit Theory **19**, 336–342 (1972)

Principles of Building a Power Transmission System for Tethered Unmanned Telecommunication Platforms

V. M. Vishnevsky$^{(\boxtimes)}$, B. N. Tereschenko, D. A. Tumchenok,
A. M. Shirvanyan, and Alexander Sokolov

V. A. Trapeznikov Institute of Control Sciences of Russian Academy of Sciences,
65 Profsoyuznaya Street, 117997 Moscow, Russia
vishn@inbox.ru, borist@mail.ru, dtumchenok@gmail.com, artshirvanyan@mail.ru,
aleksandr.sokolov@phystech.edu

Abstract. A description of the state of research on the creation of energy transfer systems for unmanned tethered aircraft is given. A comparative analysis of the options for constructing systems of this class is carried out. The advantages and disadvantages of implementing systems based on direct and alternating currents are shown. Formulas are given for the dependencies of the transmitted power on the structural characteristics of the cable-cable of the energy transmission system. The advantage of using alternating current for the implementation of high power ground-to-board energy transmission systems is substantiated.

Keywords: Tethered high-altitude platform · Tethered unmanned telecommunication platforms · UAV · Cable · Direct current power transmission system · Alternate current power transmission system · Comparative analysis · Wired power transmission system

1 Introduction

Currently, there is a clear tendency towards an increase in the number of unmanned aerial vehicles (UAVs), due to the need to use them in various civilian sectors and defense. The widespread introduction of UAVs in an area where unmanned aerial vehicles were not previously used was made possible thanks to significant advances in aircraft manufacturing, electronics, computer technology, global positioning systems, radio engineering, etc. [1]. A significant drawback of autonomous UAVs is the limited flight time. The tethered UAVs are free from this drawback, in which the engines and on-board telecommunication equipment are supplied with power via a thin copper cable from the ground to the board.

Due to the high demand for tethered unmanned telecommunication platforms in research centers of advanced countries of the world, intensive work on the design and implementation of such platforms has begun in recent years. Theoretical studies close to the subject of this article are conducted in England

(University of York) [2,3], France (University of Toulouse) [4], China (Institute of Genetics, Chinese Academy of Sciences) [5], USA (Hampton Institute) [6], Japan (Tohuku University) [7] and Russia (Institute for Control Problems of the Russian Academy of Sciences), etc. In these papers, various aspects of the design of using tethered drones are considered. We also note the works [7–9] and patents [10–12], devoted directly to the description of data transmission systems for tethered high-altitude unmanned platforms. However, it should be noted that the main attention in the described works is devoted to low power ground-to-board energy transmission systems. There are also no theoretical studies on a comparative analysis of how to implement an energy transfer system using direct or alternating current, depending on the required value of the power transmitted from the ground to the board. This article is devoted to the study of these critical issues in the design of such systems.

2   Cable Requirements for Tethered Unmanned High-Altitude Platform

The architecture of a system of tethered unmanned aerial vehicles includes the following: a ground power source, which can be made as a stationary or mobile device, a strength-power-communications (SPC) cable for power transmission, a winch to provide tension, to wind and release the SPC cable and an unmanned aerial vehicle with telecommunication equipment on board. For tethered aerial vehicles, the lifting height of which exceeds 20 m, ground and onboard current and voltage transducers are used.

The purpose of the SPC cable is the effective transfer of power from the ground electrical source towards the UAV of a tethered high-altitude unit, as well as providing high-speed data transfer via an airborne-to-Earth communication channel. The purpose dictates that a mobile UAV is connected by a SPC cable with a fixed ground power supply, while the impact of the wind flow of the SPC cable is transmitted to the UAV through the cable attachment point, which results in additional disturbing static and dynamic effects and power cost required to compensate for the disturbing effects of the wind flow. Moreover, the UAV must hold the weight of the lifted SPC cable, which leads to a decrease in the payload mass of the tethered high-altitude unit. Guided by these considerations, we shall, in order of importance, list the specific requirements for the SPC cable of the tethered high-altitude unit:

1. The maximum possible transmitted power;
2. The minimum possible linear mass;
3. The minimum aerodynamic resistance (the minimum outer diameter);
4. A sufficient mechanical strength to provide for a normal operation of wire and optical conductors in case of eventual SPC cable tensions;
5. A flexibility and resistance in case of multiple windings and unwindings on the winch drum;
6. The highest temperature index of the SPC cable.

The first three items on the list of requirements for the SPC cable are almost entirely determined by the SPC cable design. The last three items on the list are determined by the properties of the selected materials used in the manufacture of the SPC cable.

Let us consider the design of a 2-wire SPC cable with an optical channel and a Kevlar reinforcing core protecting conductor wires and optical fibers from stretching. If we build the SPC cable from standard, commercially available wires, then we will get Design version 1 similar to that shown in Fig. 1, consisting of 2 round insulated wires, a Kevlar flexible reinforcing rod and an optical module with 4 optical cable strands. Figure 2 shows Design version 2 of the SPC cable with sector-shaped wires, allowing for a larger cross-section of conducting wires with the same external diameter of the SPC cable.

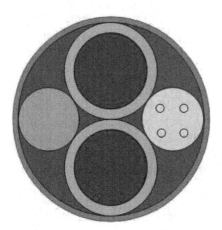
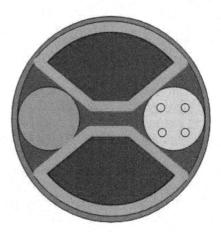

Fig. 1. Industrial conducting wire design

Fig. 2. SPC cable design with sector-shaped conductors

As shown in the versions of SPC cable designs presented in Figs. 1 and 2, there are cavities inside the cable external sheath, which should be filled with some material to make the structure resistant to creasing. The cavities are a significant part of the total cable volume, and filling them with, for example, foamed fluoroplastic will increase the linear mass of the SPC cable. Let us assume as a measure of design optimization, the filling level of the cable cross-section with the material of the conducting wire (*FL* is the parameter of the designed strength-power communications cable which is equal to the ratio of the sum of the areas of conducting wires to the total cross-sectional area of the SPC cable).

$$FL = \frac{\sum S_c}{S},\qquad(1)$$

where S_c is the cross-section of the conductive core, S is the total cross-sectional area.

Thus, for example, for the design shown in Fig. 1, $FL = 0.276$, and for the design shown in Fig. 2, $FL = 0.371$. The physical processes of power transmission over a long line (via a SPC cable) are similar and close enough both for Design version 1 and Design version 2, therefore, we will consider the results of the analysis to be valid for the SPC cables of both design versions, taking at the same time Design version 1 as basic. The requirements for the SPC cable in terms of the maximum level of transmitted power and resistance to repeated windings give origin to the need to use sheathed wires with the insulation high electrical and mechanical strength. Among the commercial conductors, the most suitable for this purpose are high-voltage wires of the HV HT MI type (high voltage high temperature with monolithic insulation) with fluoroplastic insulation and a temperature index of 155 °C, manufactured by the Special Design Office of Cable Industry (Mytishchi). The electrical strength of the PTFE insulation of these conducting wires with a thickness of 0.8 mm, equals to 10 kV, which enables to transmit high power over such wires. The list of requirements for the SPC cable for a high-altitude platform gives rise to the expediency of using such auxiliary materials as Kevlar to ensure tensile strength, PTFE for cable sheathing, and an optical module in PBTP sheath having a high temperature index. In view of the aforesaid, we can conclude that the main materials used in the manufacture of the cable are such as copper, PTFE, Kevlar, polyethylene, PBTP, glass. The density of copper is $8.9\,g/cm^3$, the density of fluoroplastic is $2.2\,g/cm^3$, the density of Kevlar is $1.4\,g/cm^3$, the density of polyethylene is $0.9\,g/cm^3$. Assuming the average density of all materials except copper equaling to $1.9\,g/cm^3$ and neglecting the mass of fiberglass strands, we can write as follows:

$$W_{\text{lin}} = 8,9 \cdot S \cdot FL + 1,9 \cdot S \cdot (1 - FL) = S(1.9 + 7 \cdot FL) \qquad (2)$$

As follows from definition (1) and from formula (2), the FL parameters and W_{lin} are functions which depend on the geometric characteristics of the cable design. These characteristics must be related to any geometric parameter(s) of the cable design. This relation seems to be the simplest in the Design version shown in Fig. 1. Indeed, the external diameter of the cable of Design version 1 is equal to the sum of the diameters of 2 wires with insulation and 2 layers of cable sheath. Assuming that the insulation thickness of high voltage high temperature conductors with monolithic insulation is equal 0.8 mm and the wall thickness of the sheath of the SPC cable is equal to 0.5 mm:

$$D = 2(d + 2 \cdot 0,8 + 0,5) = 2d + 4,2; S_c = \pi \left(\frac{D - 4,2}{4} \right)^2 \qquad (3)$$

where: D is the outer diameter of the SPC cable;

d is the diameter of the conductive core.

S_c is the cross-section area of the conductive core.

Therefore:

$$FL = \frac{2\pi \left(\frac{D-4.2}{4}\right)^2}{\pi \frac{D^2}{4}} = \frac{1}{2} \left(\frac{D-4.2}{D}\right)^2, W_{\text{lin}} = \pi \frac{D^2}{4} \left(1,9 + 3,5 \left(\frac{D-4.2}{D}\right)^2\right)$$
(4)

It is seen in expression (4) that for a SPC cable, manufactured in accordance with Design version 1, the FL value rises asymptotically as D is being increased, and approaches the value of $1/2$. Moreover, the linear mass of the SPC cable is a design characteristic of the SPC cable, is proportional to the square of the external diameter of the SPC cable, and does not depend on the transmitted power level.

Now, by calculating the cross-section of the SPC cable using (3) and FL, pertinent to the cable design version, given the allowable current density J, the value of wire-to-wire voltage U (for direct current) or the value of wave impedance $\widetilde{R}$ (for alternating current), one can estimate the maximum power transmitted via a SPC cable that will be different depending on the type of current used - DC or AC.

3   Assessment of Power in Case of Direct Current Power Transmission System

With the allowable current density J and the wire-to-wire voltage of U, a SPC cable with cable conductors with a diameter of d transmits the following power (direct current) P_-:

$$P_- = \pi \frac{d^2}{4} J \cdot U \text{ or considering } (3) P_- = \frac{\pi}{4} \left(\frac{D-4,2}{2}\right)^2 \cdot J \cdot U$$

$$\begin{cases} W_{\text{lin}} = \pi \frac{D^2}{4} \left(1,9 + \frac{7}{2} \left(\frac{D-4,2}{D}\right)^2\right) \\ P_- = \pi \left(\frac{D-4,2}{4}\right)^2 \cdot J \cdot U \end{cases}$$
(5)

where: P_- is the power transmitted via a SPC cable $[W]$
    U is the wire-to-wire voltage $[V]$
    J is the current density in the conducting wires $[A/mm^2]$
    D is the outer diameter of the SPC cable $[mm]$
    W_{lin} is the linear mass of the SPC cable $[g/m]$

In Fig. 3 below, the graphs of parametric functions (5) are presented for the values of linear mass and the level of the power transmitted via a SPC cable by direct current in relation to the external diameter of the SPC cable.

The next important characteristic of a SPC cable, describing the efficiency of power transmission is the relative total of power losses per unit of length of the SPC cable. To calculate this parameter, let us consider the scheme of power

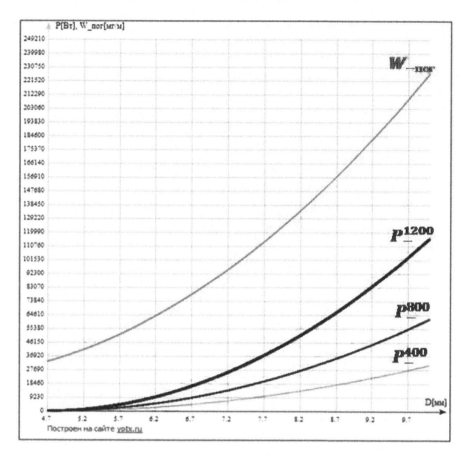

Fig. 3. Dependence of the transmitted power on the diameter

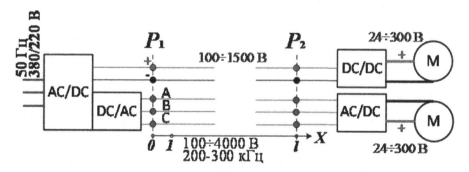

Fig. 4. Power transmission schemes using direct and alternating current types

transmission over a SPC cable from a ground source towards a converter onboard a UAV, as presented in Fig. 4.

For lapidary purpose, Fig. 4 shows at once 2 power transmission schemes: by direct current and by alternating 3-phase current.

In case of power transmission by direct current when calculating the linear power loss in the conducting wires per unit of length of the SPC cable:

$$\Delta_- = \frac{\Delta P}{l} = \frac{P_1 - P_2}{l} = \frac{I^2 \bar{r}}{l} = \frac{2plI^2}{S_c l} = \frac{2pI^2}{S_c} = 2pS_c J^2 \tag{6}$$

where: Δ_- is the linear power loss in a SPC cable
  p is the conductivity of the wire core material
  $\bar{r}$ is the impedance of 2 wires of the line
  l is the length of a 2-strand SPC cable line
  S_c is the cross-section of conductor
  I is the current intensity in the SPC cable
  J is the current density in the wire

Thus, in accordance with (6), in case of direct current, the linear power loss per unit of length of the SPC cable is proportional to the electrical impedance of conductor material, to the square of the density of the current in conducting wires, and to the cross-section of the wires of the SPC cable.

For example, let us suppose that you want to transmit power of 5 kW at a voltage of 400 V and a current density of 14 A/mm^2, using DC/DC VICOR BCM4414VD1E5135C02 converters with an output voltage of 50 V. Calculating the external diameter of the SPC cable and its cross-section area according to formulas (3) and (4), it is to note that the FL of the SPC cable amounts to 0.0487 for the data taken in account in the calculation. Given the selected initial data, the linear mass of the cable of Design version 1 will be equal, in accordance with (4), to $W_{lin} = 72.3$ g/m, and the external diameter of the SPC cable amounts to 6.33 mm. The power loss in the SPC cable, during the transmission of 5 kW, will be, in accordance with (3) and (6): $2 * 175 * 10^{-4} * 14^2 * 0.89 = 6.1$ W/m. It is to note that, due to the relatively low wire-to-wire voltage used in the example -400 V, it is possible to choose wires with thinner insulation, which will make it possible to slightly increase the FL, as well as reduce the linear mass and the diameter of the SPC cable.

4 Assessment of Power in Case of Alternate Current Power Transmission System

Let us consider the similar characteristics of the SPC cable for alternate current.

If in DC circuits the values of voltage and current intensity are described by Ohm's law which links current voltage, intensity and resistance of an electrical circuit, then in case of time-varying values of current intensity and voltage in a long conductor line, the coordinates of the observation point, of current intensity, voltage, resistance, inductance and capacity of the conductive line are interlinked by a system of telegraph differential equations:

$$\frac{\partial}{\partial x}U(x,t) = -L\frac{\partial}{\partial t}I(x,t) - rI(x,t)$$

$$\frac{\partial}{\partial x}I(x,t) = -C\frac{\partial}{\partial t}U(x,t) - gU(x,t)$$

where: L is the inductance per unit of line length;
 C is the capacity per unit of line length;
 r is the impedance per unit of line length;
 g is the insulation conductance per unit of line length.

The above described system reduces to two wave equations for current and voltage:

$$\frac{\partial^2 U}{\partial x^2} - LC\frac{\partial^2 U}{\partial t^2} = 0$$

$$\frac{\partial^2 I}{\partial x^2} - LC\frac{\partial^2 I}{\partial t^2} = 0$$

$$(7)$$

The solution (7) both for current and for voltage is a superposition of two identical twice differentiated arbitrary functions of the following arguments: $t - x\sqrt{LC}$ and $t + x\sqrt{LC}$

$$U(x,t) = C_1 f_1(t - x\sqrt{LC}) + C_2 f_1(t + x\sqrt{LC})$$

$$I(x,t) = D_1 f_2(t - x\sqrt{LC}) + D_2 f_2(t + x\sqrt{LC})$$

Here, $\frac{1}{\sqrt{LC}} = V_f$ is the phase velocity of wave propagation, and the solution itself is a combination of two waves propagating in opposite directions with a speed of V_f. In the case of alternate current, the voltage and intensity of current at the same time point are different for different points of the conductor line. The specific form of the functions f_1 and f_2 is determined basing on the initial or boundary conditions. Let us assume, for example, that a sinusoidal current and voltage source is connected to the line start. In this case, the boundary conditions will be written as follows:

$$U(0,t) = U_1 e^{i\omega t}$$

$$I(0,t) = I_1 e^{i\omega t}$$

The solution (7), under these conditions and given the absence of the reflected wave, will be written as follows:

$$U(x,t) = U_1 e^{i\omega t - \alpha x}$$

$$I(x,t) = I_1 e^{i\omega t - \alpha x}$$

$$(8)$$

Solution (8) takes place in the only case when the load impedance is equal to the wave resistance of the long line $\widetilde{R} = \sqrt{\frac{L}{C}}$, in all other cases reflected current and voltage waves occur:

$$U(x,t) = U_{ref}e^{i\omega t + \alpha x}$$
$$I(x,t) = I_{ref}e^{i\omega t + \alpha x}$$

Reflected waves back-transfer the power from the point of line-load connection to the power source. Thus, the obvious condition for the maximum efficiency of power transmission by alternating current will be the equality of the wave resistance of the line to the line's load resistance. The value of the transmitted power in this case will be as follows:

$$P_{max} = I_{lin}^2 \cdot \tilde{R} \tag{9}$$

The magnitude of wave impedance for the line including 2 round conductor wires can be calculated by the following formula:

$$\tilde{R} = \frac{120}{\sqrt{\varepsilon}} ln \frac{\overset{\frown}{D}}{r} \tag{10}$$

where: $\overset{\frown}{D}$ is the distance between conductor centers
r is the radius of conductors
ε is the dielectric permittivity between conductors

When transmitting power at frequencies above 50 kHz, it is also necessary to take into account the influence of the skin effect, which means that the current density decreases as augmenting the distance from the surface inwards the conductor. A compensation of this effect is achieved by the fact that the conductive cores of the wires are made of multiple thin wires insulated with lacquer or enamel with a diameter of 0.03 to 0.1 mm. Quantitatively, the action of this effect can be taken into account using the ratio of the use of cross-section K_s, equal to the ratio of the density of alternating current to the density of direct current in a conducting wire with the same cross-section. For example, a conductive core consisting of wires with a diameter of 0.1 mm at a frequency of 200 kHz will have $K_s = 0,9$, and at a frequency of 300 kHz - $K_s = 0,88$.

The derivation of the formula for determining the maximum power transmitted by alternate current, $P_\approx$ will be carried out with the same initial parameters of the SPC cable of Design version 1, which were used in (3). For fluoroplastic ($\varepsilon = 2$) of conducting wires of Design version 1 of the SPC cable, using (3) and (10), we will get the following:

$$\tilde{R} = 83 ln \left(2 \frac{D-1}{D-4,2} \right)$$

In accordance with (8), the maximum power transmitted via a SPC cable by alternate current is written:

$$P_\approx = 83 \left(ln \left(2 \frac{D-1}{D-4,2} \right) \right)^* \left(\pi \left(\frac{D-4,2}{4} \right)^2 \cdot K_s \cdot J \right)^2 \tag{11}$$

Figure 5 shows the dependences of the linear mass and the maximum power transmitted $P_\approx$ on the diameter of the SPC cable at a frequency of voltage and current of 300 kHz.

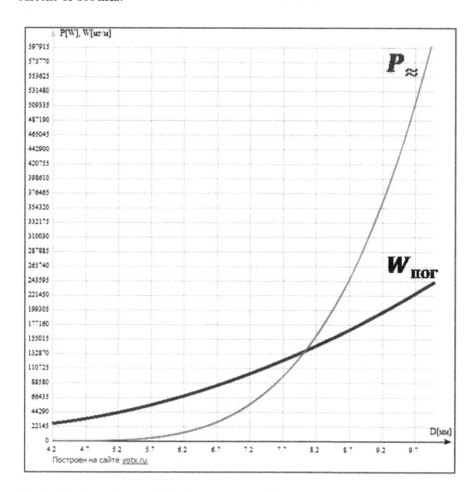

Fig. 5. The dependences of the linear mass and the maximum power transmitted on the diameter of the cable

Let us calculate the linear power loss per unit length of the SPC cable during power transmission by alternating current. In accordance with the solution of the telegraph equations for a long line (7), the amplitudes of voltage and current are attenuated in a SPC cable according to the following law:

$U = U_1 e^{-\alpha x}$, $I = I_1 e^{-\alpha x}$, therefore the power is attenuated according to the following law:

$$P = P_1 e^{-2\alpha x}$$

where: $\alpha = \dfrac{\tilde{r}}{\tilde{R}}$ is the coefficient of attenuation of the power transmission line

$\widetilde{R} = \sqrt{\frac{L}{C}}$ – is the wave impedance of the power transmission line

$\widetilde{r}$ – is the linear impedance of the power transmission line The linear power loss per length unit of the SPC cable (x = 1) shall be equal to the following:

$\Delta_{\approx} = P_1 - P_1 e^{-2\alpha x} = P_1 \left(1 - e^{-2\alpha * 1}\right)$ because $\propto$ significantly less 1, then

$$\Delta_{\approx} = 2P_1\alpha = 2P_1\frac{\widetilde{r}}{\widetilde{R}}, \; \Delta_{\approx} = 2I_1^2\widetilde{R}\frac{\widetilde{r}}{\widetilde{R}} = 2\left(K_s S_c J\right)^2 \frac{p}{S_c} = 2pS_c\left(K_s J\right)^2$$

$$\Delta_{\approx} = 2pS_c(K_s J)^2 \tag{12}$$

5   Comparative Analysis of the Optional Builds of a Wired Power Transmission System

Comparing (6) and (12), it is to note that the linear power loss during power transmission by alternating current is less than the loss when using direct current due to the skin effect.

We will compare the results of power transmission over a SPC cable by direct and alternating currents. To do this, let us impose the graphs (Fig. 6) of the dependences of the maximum transmitted power on the external diameter of the SPC cable for both types of currents.

From the summary graph shown in Fig. 6, one can see that, starting from a certain diameter of the conductor core and, accordingly, from a certain the external diameter of the SPC cable, as well as from a certain level of transmitted power, the power transmission through the SPC cable by alternating current gets a decisive advantage:- much more power can be transmitted over a thinner and lighter SPC cable. For example, the curves shown in Fig. 6 evidence that the SPC cable with a diameter of 8 mm can transmit about 48 kW of power by direct current at a voltage of 1200 V, while alternating current can transmit about 135 kW through the same cable, or transfer 48 kW through a thinner and lighter cable with an outer diameter of 7 mm. When designing tethered UAVs, the developer must select the power value consumed by the unmanned aerial vehicle, taking into account the specified payload mass, the UAV operational height, cable weight, cable tension level, and also take into consideration the impact of wind flow on the UAV and on the cable. For such calculations, it is more convenient to consider the dependence of the power transmitted via the SPC cable by direct or alternating types of current on the linear mass of the SPC cable. Since the outer diameter of the SPC cable, made in accordance with Design version 1, uniquely determines the linear mass of the SPC cable, it is possible to easily recalculate the family of dependences of the power levels given in Fig. 6 for the parametric functions of the argument of the SPC cable weight, which are shown in Fig. 7.

Figure 7 shows the zoomed in starting part of the curves shown in Fig. 6, recalculated with respect of the argument W_{lin}. The intersection points of the curves $P_{\approx}(W_{\text{lin}})$ and $P_-(W_{\text{lin}})$ determine W_{lin}, which correspond to the equisignificant use of both types of currents, although, in conformity with (6) and (12),

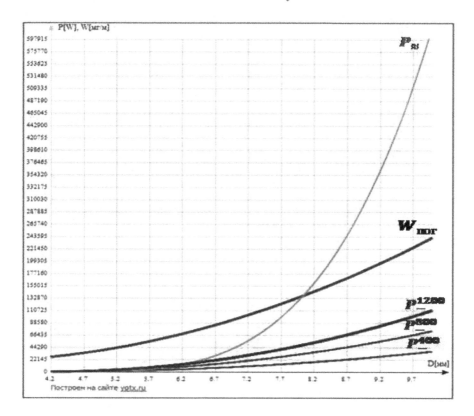

Fig. 6. The dependences of the maximum transmitted power on the external diameter of the cable for both types of currents

even if the transmitted powers are equal, the use of an alternating current will be preferable due to lower energy losses in the conductors. Let us calculate W_{lin} at which the power transmitted via a SPC cable by both kinds of currents will be equal. Using (5) and (11) we obtain the following equation:

$$P_- = \frac{\pi}{4}\left(\frac{D-4,2}{2}\right)^2 JU = P_\approx = 83\left(\ln\left(2\frac{D-1}{D-4,2}\right)\right)\left(\pi\left(\frac{D-4,2}{4}\right)^2 JK_s\right)^2$$

This equation is completely equivalent to a simpler equation with respect to the radius R of the conductive core of the SPC cable:

$$\pi R^2 JU = 83\left(\pi R^2 JK_s\right)^2 \ln\left(2 + \frac{1,6}{R}\right) \tag{13}$$

Equation (13) is not solvable with respect to R, so its solution is easier to be sought by overlaying, for example, the curves $P_-(D)$ and $P_\approx(D)$. You can also search for a solution by the method of successive approximations, or by

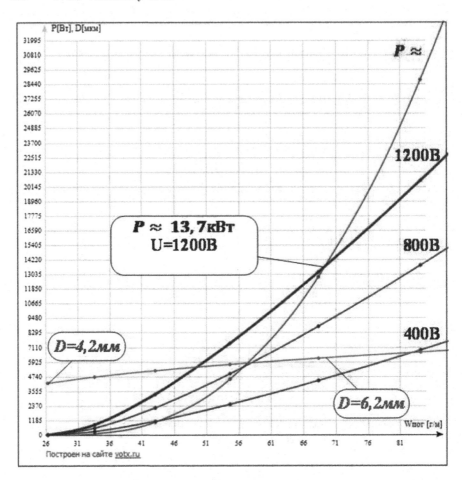

Fig. 7. The family of dependences of the power levels for the parametric functions of the argument of the SPC cable weight

approximating the expression $ln\left(2 + \dfrac{1,6}{R}\right)$ with simpler functions and then solving the resulting approximated equation. Let us perform a differentiation of R using the functions P_- and $P_\approx$.

$$\frac{\partial P_-(R)}{\partial R} = 2\pi RJU \tag{14}$$

$$\frac{\partial P_\approx(R)}{\partial R} = 83R^3(\pi K_s J)^2\left(4ln\left(2\frac{R+0,8}{R}\right) - \frac{0,8}{(R+0,8)}\right) \tag{15}$$

Considering the obtained expressions (14) and (15) for derivatives, we see that they are monotonically increasing functions of R, and this is true even starting from the values $R = 0,1$ $\dfrac{\partial P_-(R)}{\partial R} \gg 1$ and $\dfrac{\partial P_\approx(R)}{\partial R} \gg 1$ This means that, as the conductor radius increases, the increase in the power transmitted over the SPC cable is much greater than the increment in the conductor core radius. Thus, a slight increase in the conductor radius gives way to a large increase in the transmitted power, thereby reducing the proportion of consumed power required by the aerial vehicle to lift and hold the SPC cable weight. With a conductor radius of $0.5\,\text{mm}$ $\dfrac{\partial P_{\approx}(R)}{\partial R} > \dfrac{\partial P_-(R)}{\partial R}$, and with $R \geq 0,7\,\text{mm}$ $\dfrac{\partial P_{\approx}(R)}{\partial R} \gg \dfrac{\partial P_-(R)}{\partial R}$, which makes it possible to make the statement that, in the case of using stageless switching elements for direct current, the transmission of large, of the order of $14\,\text{kW}$ and more, power levels by alternating current becomes preferable on a zero-option basis. Using (13) or the graphs shown in Fig. 7, we will build a route chart to select the optimal type of current, the geometric and mass parameters of the SPC cable, depending on the level of consumed power required for the projected UAV of the tethered high-altitude platform. This will be a table of boundary values of power, of geometrical and mass characteristics of the SPC cable, in which the use of both types of currents is equivalent, and in addition, at lower power values the use of direct current will be more profitable, and at higher power values it is more expedient to use alternating current (Table 1).

Table 1. Power dependence on cable characteristics

$P[W]$	$U_-[V]$	$R[mm]$	$D[mm]$	$W_{lin}[g/m]$
1300	400	0.25	5.2	43
5400	800	0.4	5.8	57
13700	1200	0.5	6.2	69
23000	1500	0.6	6.6	83

where: P is the power transmitted through the SPC cable

U_- is the wire-to-wire voltage in the SPC cable

R is the radius of the conducting wire

D is the outer diameter of the SPC cable

W_{lin} is the linear mass of the SPC cable

Below is Table 2 showing the SPC cable characteristics during the transmission of power of various level by direct and alternating ($300\,\text{kHz}$) current types.

Table 2. Cable characteristics during the transmission of power for both types of currents

P[kW]	Direct current			Alternate current			
	400 V						
	D[mm]	W[g/m]	P[W/m]	D [mm]	W [g/m]	[W/m]	U [V]
5	6.333	72.4	6.1	5.7	55.8	2.5	869
8	6.9	91	9.8	6	61.5	3.2	1075
11	7.36	108.3	13.4	6.1	66	3.8	1239
14	7.76	124.7	17.0	6.2	70	4.4	1376
17	8.132	141.1	20.8	6.4	73	4.9	1501
20	8.472	157.3	24.6	6.5	76	5.3	1620
23	8.78	172.7	28.2	6.55	79	5.8	1724
26	9.08	188.5	32.1	6.6	82	6.2	1820
29	9.34	202.8	35.6	6.7	84	6.5	1909
	800 V						
5	5.72	55.2	3.1	5.7	55.8	2.5	869
8	6.12	66.0	4.9	6	6	3.2	1075
11	6.44	75.7	6.8	6.1	66	3.8	1239
14	6.72	84.8	8.5	6.2	70	4.4	1376
17	6.98	93.9	10.4	6.4	73	4.9	1501
20	7.22	102.8	12.3	6.5	76	5.3	1620
23	7.44	111.4	14.1	6.55	79	5.8	1724
26	7.64	119.6	15.9	6.6	82	6.2	1820
29	7.84	128.1	17.8	6.7	84	6.5	1909
	1200 V						
5	5.44	48.4	2.0	5.7	55.8	2.5	869
8	5.76	56.2	3.3	6	61.5	3.2	1075
11	6.028	63.4	4.5	6.1	66	3.8	1239
14	6.264	70.3	5.7	6.2	70	4.4	1376
17	6.472	76.7	6.9	6.4	73	4.9	1501
20	6.664	82.9	8.2	6.5	76	5.3	1620
23	6.844	89.1	9.4	6.55	79	5.8	1724
26	7.008	94.9	10.6	6.6	82	6.2	1820
29	7.168	100.9	11.9	6.7	84	6.5	1909
	1500 V						
5	5.304	45.3	1.6	5.7	55.8	2.5	869
8	5.596	52.1	2.6	6	61.5	3.2	1075
11	5.84	58.3	3.6	6.1	66	3.8	1239
14	6.044	63.8	4.6	6.2	70	4.4	1376
17	6.232	69.3	5.6	6.4	73	4.9	1501
20	6.404	74.5	6.5	6.5	76	5.3	1620
23	6.564	79.6	7.5	6.55	79	5.8	1724
26	6.712	84.6	8.5	6.6	82	6.2	1820
29	6.856	89.5	9.5	6.7	84	6.5	1909

6   Conclusion

The study has been completed of the electrical aspects of power transmission via a SPC cable of a tethered high-altitude telecommunication platform based on an unmanned aerial vehicle. The analysis and comparison have been performed for a number of the most important electrical characteristics affecting power transmission by direct and alternating currents via a SPC cable manufactured according to Design version 1. A SPC cable with sector wires, manufactured according to Design version 2 was also considered, in which a SPC cable transmits noticeably higher power while having the same external dimensions. Due to the greater technological complexity of Design version 2 compared to Design version 1, the manufacturing of such cable is much more expensive, therefore the use of such cable is justified only on a commercial scale.

A table has been compiled and justified containing the optimal geometric and mass parameters of the SPC cable for various levels of power transmitted via the SPC cable, as well as the boundaries of the areas where power transmission by direct current takes precedence over the transmission by alternating current.

It is shown and substantiated the statement that in the case of using stageless DC-to-DC converters of up to 1500 V, the transmission of power via a SPC cable, at the levels of transmitted power of 23 kW and higher, is only advisable by alternating current.

A comparison was made between the linear energy losses in the SPC cable for direct and alternating current types. It is proven that the linear loss is less during the transfer of power by alternating current. Since the study considered only the electrical aspects of power transmission, it did not cover the power consumption required to compensate for disturbing static and dynamic losses arising from the exposure of the SPC cable of the high-altitude platform to atmospheric wind flows.

Acknowledgements. The reported study was funded by RFBR, project number 19-29-06043.

References

1. Fetisov, V.S., Tagirov, M.I., Muhametzyanova, A.I.: Podzaryadka elektricheskih bespilotnyh letatel'nyh apparatov: obzor sushchestvuyushchih razrabotok i perspektivnyh reshenij. Aviakosmicheskoe priborostroenie (11), 7–26 (2013)
2. Zakaria, M.D., Grace, D., Mitchell, P.D.: Antenna array beamforming strategies for high altitude platform and terrestrial coexistence using K-means clustering. In: 2018 IEEE 13th International Conference on Communications, pp. 259–264 (2018)
3. Chandrasekharan, S., et al.: Designing and implementing future aerial communication networks. IEEE Commun. Mag. **54**(5), 26–34 (2016)
4. Tognon, M., Franchi, A.: Position tracking control for an aerial robot passively tethered to an independently moving platform. IFACPapersOnLine **50**(1), 1069–1074 (2017)
5. Wang, G., Samarathunga, W., Wang, S.: Uninterruptible power supply design for payload tethered hexaroters. Int. J. Emerg. Eng. Res. Technol. **4**(2), 16–21 (2016)

6. Morales-Perryman, Q., Lee, D.D.: Tethering system for unmanned aerial vehicles. Hampton University, Electrical Engineering, pp. 1–7 (2018)
7. Kiribayashi, S., Yakushigawa, K., Nagatani, K.: Design and development of tether-powered multirotor micro unmanned aerial vehicle system for remote-controlled construction machine. In: Hutter, M., Siegwart, R. (eds.) Field and Service Robotics. SPAR, vol. 5, pp. 637–648. Springer, Cham (2018). https://doi.org/10.1007/978-3-319-67361-5_41
8. Kiribayashi, S., Ashizawa, J., Nagatani, K.: Modeling and design of tether powered multicopter. In: 2015 IEEE International Symposium on Safety, Security, and Rescue Robotics (SSRR), pp. 1–7. IEEE (2015)
9. Wasantha, G.W., Wang, S.: Heavy payload tethered hexaroters for agricultural applications: power supply design (2015)
10. United States patent US20120112008A1. System for high altitude tethered powered flight platform (2012)
11. United States patent US6325330B1. Power generation, transmission, and distribution system for an aerostat using a lightweight tether (2001)
12. United States patent US20140183300A1. Tethered payload system and method (2014)

On the Stability of D2D Connection with the Use of Kinetic Equation for SIR Empirical Distribution

Yurii Orlov[1,2], Anastasia Ivchenko[2], Natalia Podzharaya[1],
Anastasiia Sochenkova[1], Vsevolod Shorgin[3], Aliaksandr Birukou[1],
Yuliya Gaidamaka[1,3(✉)], and Konstantin Samouylov[1,3]

[1] Applied Informatics and Probability Department,
Peoples' Friendship University of Russia (RUDN University),
Miklukho-Maklaya St. 6, Moscow 117198, Russian Federation
natalia.podzharaya@gmail.com, anasochenkova@gmail.com,
aliaksandr.birukou@springer.com,
{gaydamaka-yuv,samuylov-ke}@rudn.ru
[2] Kinetic Equations Department,
Keldysh Institute of Applied Mathematics of RAS,
Miusskaya Sq. 4, Moscow 125047, Russian Federation
yuno@kiam.ru, orlmath@keldysh.ru
[3] Federal Research Center "Computer Science and Control"
of the Russian Academy of Sciences,
Vavilov St. 44-2, Moscow 119333, Russian Federation
vshorgin@ipiran.ru

Abstract. The numerical simulation of the signal to interference ratio distribution in the wireless D2D communication is analyzed using means of kinetic equation. We consider the model of transmitters and receivers motion, which is mechanically determined, but the distribution function, mentioned above, is non-stationary. The signal to interference ratio for devices, performing a random walk, is described with the use of a non-stationary Fokker-Planck equation with respect to a sample distribution of this parameter. The empirical distribution of the periods with continuous connection is numerically constructed for the real scheme of deterministic motion of Moscow metro trains.

The publication has been prepared with the support of the "RUDN University Program 5-100" (Yurii Orlov, mathematical model development; Aliaksandr Birukou, writing, review and editing; Natalia Podzharaya, technical analysis; Anastasiia Sochenkova, technical review). The reported study was funded by RFBR, project number 17-07-00845 (Yuliya Gaidamaka, simulation model development) and 19-07-00933 (Konstantin Samouylov, problem formulation and analysis). This work has been developed within the framework of the COST Action CA15104, Inclusive Radio Communication Networks for 5G and beyond (IRACON).

© Springer Nature Switzerland AG 2019
V. M. Vishnevskiy et al. (Eds.): DCCN 2019, LNCS 11965, pp. 111–124, 2019.
https://doi.org/10.1007/978-3-030-36614-8_9

Keywords: Wireless communication · Kinetic equation ·
Non-stationary random walk · Mathematical modeling · SIR
distribution function · D2D communication · Moscow metro

1   Introduction

The analysis of the radio channel quality in wireless networks helps to understand network properties better and remains an urgent task. Early research
on this problem for the first wireless networks considered fixed transceivers.
A traditional radio channel quality metric is signal-to-interference and noise
ratio (SINR) [7,16]. For stationary transceivers, SINR is studied depending on
the density of devices for various distributions of their spatial location [2,6]. In
4G/5G networks [1], the same problem has already arisen for the case of moving transceivers. In this case, the approaches developed for a device-to-device
(D2D) network with fixed devices [14,15] are applicable, but they should be
supplemented with a mathematical apparatus for simulating the device's movement [3,4,11]. Note that in the case of moving transceivers, the SINR is no
longer a constant value; it changes in time. The task of analyzing the connection
between moving devices is the subject of a lot of research in the field of vehicle-
to-vehicle (V2V) moving transceivers, where the quality was studied depending
on the speed of devices, for regular movement on roads and railways [8,12,13,17],
as well as for underground communication (in tunnels, in underground parking
lots, in subway trains). Though there are solutions for the interaction of the
WiFi train network with the base stations located in the tunnel using propri-
etary protocols, mainly for fourth- and fifth-generation mobile networks, it is
also interesting to model the scenario when D2D communication between the
terminals located in the metro train can be established without the participation
of the WiFi metro network. In all these cases it is necessary to take into account
the different signal propagation environment [10].

In the traditional approach, SINR is studied in case of zero noise, so we finally
have signal-to-interference ratio (SIR), which is defined as a ratio of the useful
signal power at a given spatial point to the sum of signal powers from other
sources [7,16]. Zero noise means that we neglect the thermal radiation of the
receivers, and we model the signal attenuation due to a propagation environment
using a path loss model. In [2,6] the signal power is proportional to r^{-2}, where
r is a non-zero distance between the target transmitter and receiver. So the
SIR value is defined as $r^{-2}/\sum_k r_k^{-2}$, where r_k is a distance between the target
receiver and another transmitter. One can see, that SIR indicator is nonlinear
with respect to transceivers positions distribution, so the tractable pdf of SIR
may be obtained only in some relatively simple cases, e.g. for a given spatial
transceivers distribution without any moving effects. But in practice, we need
to describe the non-stationary random walk of receivers and transmitters, so
that the appropriate method for modeling of statistical characteristics of the
corresponding ensemble of trajectories should be developed.

If the SIR value becomes less then predefined minimum threshold, corresponding to the service under consideration, the wireless connection for the associated pair ceases to meet the Quality of Service (QoS) requirements. The threshold is defined by various standards and depends on the service. The possibility of outage periods in the wireless connection and an allowed duration of the corresponding intervals during a service provision is determined by the Quality of Experience (QoE) requirements, which are defined by international standards. For example, the wireless LTE technology in real-time gaming service through the handheld gaming console is a service with a guaranteed bit rate (GBR). According to the concept of QoS Class Identifier (QCI) this service has QCI = 4 and priority 3. For network performance, it means, that the allowed delay of packet transmission does not exceed 50 ms and the packet loss does not exceed 10^{-3}. In order to investigate such parameters for various models of moving transceivers, it is necessary to develop methods for the analysis of the duration of the service availability periods and the periods when the wireless connection is absent. In particular, one needs to model the link stability between several moving wireless devices.

In this paper, we consider a network, where the transceivers move along the predefined trajectories with predefined speed dynamics. In this case, the distances between the objects change in a deterministic manner. We simulate the increment of the coordinates using the kinetic approach [5]. The stability of the D2D connection is then characterized by the normalized average SIR, which is a functional of distances. The simulation of the quality of the D2D connection is done using the real data on train movements in Moscow metro.

2   Simulation of Devices Trajectories

Our method of SIR analysis is based on simulation of devices trajectories with the use of the Fokker-Planck kinetic equation for PDF of independent coordinates increments [9]. The main argument of this approach is that the SIR is a non-linear and non-monotonic functional of devices positions, so in general, the direct expression of SIR PDF is impossible. Assuming that the spatial averaging over ensemble of device trajectories and averaging positions of an arbitrary device over sufficiently large time period are asymptotically the same procedures, we can analytically investigate the SIR PDF by means of kinetic equation for device coordinate PDF. Here we mention this approach in brief.

If an independent coordinate increment has the same PDF as a coordinate itself, SIR average value can be expressed as a non-linear functional of coordinate PDF. Let

$$X(t) = X(0) + \sum_{k=1}^{T(t)} x_k, \quad Y(t) = Y(0) + \sum_{k=1}^{T(t)} y_k, \quad Z(t) = Z(0) + \sum_{k=1}^{T(t)} z_k$$

be the trajectory coordinates for a certain device in a discrete instant of time t, and $\mathbf{r}_k = \{x_k, y_k, z_k\}$ are corresponding coordinate increments for the time step k, where $T(t)$ is the number of steps depending on t.

Let PDF $f(\mathbf{r}, t)$ be described by the Fokker-Planck equation:

$$\frac{\partial f}{\partial t} + div(\mathbf{u}(\mathbf{r}, t)f) - \frac{\lambda(t)}{2}\Delta f = 0. \tag{1}$$

We consider a drift parameter $\mathbf{u}(\mathbf{r}, t)$ and a non-stationary diffusion coefficient $\lambda(t) \geq 0$ to be known. In practice they can be determined through the empirical sample mutual PDF $F(\mathbf{r}, \mathbf{v}, t)$ of values of $\mathbf{r}$ and its first differences $\mathbf{v}$ so that (see [14])

$$f(\mathbf{r}, t) = \int F(\mathbf{r}, \mathbf{v}, t)dv, \tag{2}$$

$$f(\mathbf{r}, t)\mathbf{u}(\mathbf{r}, t) = \int \mathbf{v}F(\mathbf{r}, \mathbf{v}, t)dv, \quad \lambda(t) = \frac{d\sigma^2}{dt} - 2cov_{r,v},$$

$$\sigma^2(t) = \int (\mathbf{r} - \mathbf{m}(t))^2 f(\mathbf{r}, t)dr, \quad \mathbf{m}(t) = \int \mathbf{r}f(\mathbf{r}, t)dr.$$

Let $\varphi(|\mathbf{r}|) = 1/|\mathbf{r}|^2$. Then the field of interference at the point $\mathbf{r}$ can be determined as

$$U(\mathbf{r}, t) = \int \varphi(|\mathbf{r} - \mathbf{r}'|)f(\mathbf{r}', t)dr' = \int \frac{f(\mathbf{r}', t)}{|\mathbf{r} - \mathbf{r}'|^2}dr'. \tag{3}$$

Here $\mathbf{r}$ is a position of the first device (target transmitter) with respect to second device (target receiver), and $\mathbf{r}'$ denotes the respective positions of other devices (interfering transmitters). Let the total number of devices be N. Then the average over an ensemble SIR value for two arbitrary devices is defined as

$$s(t) = \frac{1}{N} \int \frac{\varphi(\mathbf{r})}{U(\mathbf{r}, t)}f(\mathbf{r}, t)dr. \tag{4}$$

Since $U(\mathbf{r}, t)$ is defined as (3), the average SIR (4) is a non-linear functional of coordinate PDF of devices. For this model it can be obtained the evolution equation for average SIR:

$$N\frac{ds}{dt} = \int \frac{\varphi(\mathbf{r})}{U(\mathbf{r}, t)}\frac{\partial f(\mathbf{r}, t)}{\partial t}dr - \int \frac{\varphi(\mathbf{r})}{U^2(\mathbf{r}, t)}\frac{\partial U(\mathbf{r}, t)}{\partial t}f(\mathbf{r}, t)dr, \tag{5}$$

and

$$\frac{\partial U(\mathbf{r}, t)}{\partial t} = \int \frac{1}{|\mathbf{r} - \mathbf{r}'|^2}\frac{\partial f(\mathbf{r}', t)}{\partial t}dr'.$$

The term $\dfrac{\partial f}{\partial t}$ can be expressed from the Eq. (1). So after integration by parts we replace the occurring after this action derivative of the function

$\phi(|\mathbf{r} - \mathbf{r}'|)$ by $\mathbf{r}'$ to the derivative by $\mathbf{r}$. As a result we obtain the following expression:

$$\frac{\partial U}{\partial t} = \frac{\lambda}{2}\Delta U - div\mathbf{J}, \quad \mathbf{J} = \int \varphi(|\mathbf{r} - \mathbf{r}'|)\mathbf{u}(\mathbf{r}', t)f(\mathbf{r}', t)d\mathbf{r}'.$$

This means, that the average interference field changes over time in the same way as PDF, i.e. according to the diffusion equation with the same ratio as in (1). Analogously

$$\int \frac{\varphi(\mathbf{r})}{U(\mathbf{r}, t)}\frac{\partial f(\mathbf{r}, t)}{\partial t}d\mathbf{r} = -\int \frac{\varphi(\mathbf{r})}{U(\mathbf{r}, t)}div(\mathbf{u}f)d\mathbf{r} + \frac{\lambda}{2}\int \frac{\varphi(\mathbf{r})}{U(\mathbf{r}, t)}\Delta f d\mathbf{r}.$$

After integration by parts we obtain

$$\int \frac{\varphi(\mathbf{r})}{U(\mathbf{r}, t)}\frac{\partial f(\mathbf{r}, t)}{\partial t}d\mathbf{r} = \int f(\mathbf{r}, t) \cdot \left(\mathbf{u}\nabla + \frac{\lambda}{2}\Delta\right)\left(\frac{\varphi}{U}\right)d\mathbf{r}.$$

As a result the Eq. (5) has the form:

$$N\frac{ds}{dt} = \int \left(\left(\mathbf{u}\nabla + \frac{\lambda}{2}\Delta\right)\left(\frac{\varphi}{U}\right)\right)f(\mathbf{r}, t)d\mathbf{r} \\ - \int \frac{\varphi}{U^2}\left(\frac{\lambda}{2}\Delta U - div\mathbf{J}\right)f(\mathbf{r}, t)d\mathbf{r}. \tag{6}$$

This equation is non-linear with respect to distribution function, i.e., with respect to the density of the ensemble of their sample trajectories.

From (6) it follows, that if PDF is stationary, the average SIR value is constant. For the model of chaotic transceivers motion it seems to be adequate. But there is another type of movement, which is entirely realistic and for which the SIR PDF is not connected with devices positions. Such type of motion can be presented as a deterministic transport motion along the definite trajectories. The example of great interest is the following: the PDF of coordinate increments is stationary, but the SIR PDF is non-stationary. The corresponding model is described below.

3   The Moscow Metro Trains Motion Model

Let us consider a scheme of Moscow metro trains motion (10 lines, 198 stations, 186 trains). The coordinates of stations are presented in Fig. 1. The altitude is expressed in kilometers under the sea level.

We suppose that all trains are moving following their route schedules. The train stops at the platform for 25 s, then proceeds with a constant acceleration during 20 s, after that runs with a constant velocity of about 70 km per hour and then moves with continuous negative acceleration before the next stop for 15 s. The example of a 20-minutes fragment of train coordinates dependence on time is presented in Fig. 2 for each metro line.

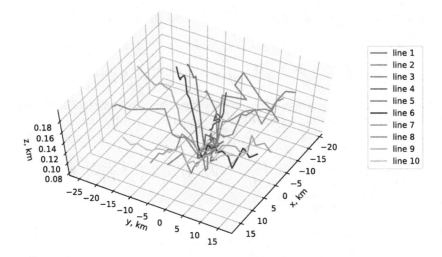

Fig. 1. Moscow metro 3D scheme

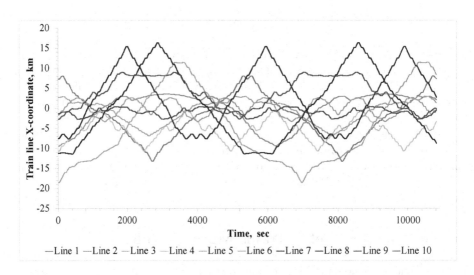

Fig. 2. The train X-coordinates, depending on time

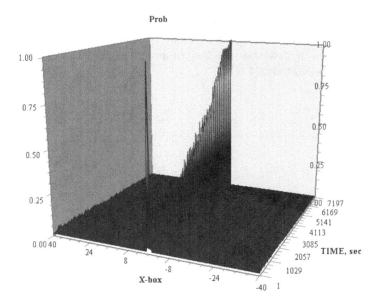

Fig. 3. PDF for X-coordinate increments of moving trains

So we have an ensemble of trains consisting of 186 individual trajectories. The corresponding PDF of coordinate increments, depending on time, is presented in Fig. 3.

Obviously, in the first 25 s, this PDF is delta-function of a zero argument. But in consequent time instants this PDF becomes non-uniform.

4   Coordinate PDF Analysis

Let us consider the distance between the PDF of coordinate increments in C-norm for various time instants:

$$\rho = ||F(x, t_1) - F(x, t_2)||_C = \sup_x |F(x, t_1) - F(x, t_2)|. \tag{7}$$

Our ensemble consists of $N = 186$ trajectories. Very useful notion is a stationary point of significance level of Kolmogorov-Smirnov statistics $\varepsilon(N)$ for sample function, which is defined as a solution of the equation

$$\varepsilon = 1 - K\left(\sqrt{\frac{N}{2}}\varepsilon\right),$$

where $K(z)$ is Kolmogorov function. Since for all stationary distributions the level $\varepsilon(N)$ is the same, the non-stationary effects can be revealed from the comparison of the empirical stationary point of the distribution function of distances

(7) between PDFs $F(x,t)$ with theoretical value of $\varepsilon(N)$, which was tabulated in [15]. For $N = 186$ this value is approximately equal to $\varepsilon(186) \approx 0{,}15$.

The time-series of distances (7) for our practical model of Moscow metro trains motion is presented in Fig. 4.

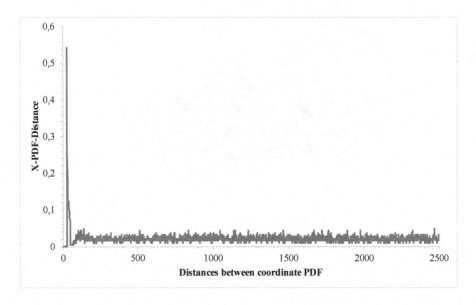

Fig. 4. Distances between coordinate PDFs

So let $G(\rho)$ be a distribution function of distances (7) between PDFs of devices coordinate increments. The numerical solution for stationary point of the significance level

$$\rho = 1 - G(\rho)$$

gives the value $\rho^* \approx 0{,}02$, which is sufficiently small with respect to $\varepsilon(186) \approx 0{,}15$. Thus the empirical PDF $F(x,t)$ should be treated as stationary. Hence the method of investigation of SIR evolution PDF, described by formulas (1)–(6), is not applicable to this situation. We need to construct the empirical SIR PDF as a result of trains' motion modeling itself.

5   The Empirical PDF for SIR

For the sake of simplicity we suppose that each train corresponds to one device – sender or receiver. We chose two arbitrary trains, which are connecting with each other, and calculate the interference field from all others trains.

Let $\mathbf{r}_i(t)$ be a position of a train number i at the time instant t in 3D scheme (see Fig. 1). Let also $i = 1; 2$ be the numbers of chosen trains, connecting to each other. The SIR value is defined as

$$s(t) = \frac{1/|\mathbf{r}_1(t) - \mathbf{r}_2(t)|^2}{\sum\limits_{k=3}^{186} 1/|\mathbf{r}_k(t) - \mathbf{r}_2(t)|^2}.$$

The result is presented in Fig. 5. We also compare the empirical SIR trajectory with technically minimal SIR level $S^* = 0{,}01$ and with SIR moving average line (SMA) over the period of 500 s. One can see, that the connection between two devices is not continuous, and SIR value rather often becomes less then S^*. One can see also, that SMA is not adequate indicator of connection discontinuity, because it lies above minimal level, while SIR has oscillating trajectory. So we need to construct an empirical PDF for SIR to investigate this function in detail.

First of all we calculate empirical SIR PDF $F_N(s,t)$ for various sample length N. After that we construct the distances (7) for neighboring samples

$$\rho(N,t) = ||F_N(x,t) - F_N(x,t+N)||_C$$

and corresponding function $G_N(\rho)$. The stationary point $\rho^*(N)$ of significance level of empirical SIR obtained numerically from the equation

$$\rho = 1 - G_N(\rho). \tag{8}$$

The example of distances time-series between SIR PDF for sample length $N = 1500$ is presented in Fig. 6, and solution of Eq. (8) is presented in Fig. 7.

We see that the empirical significance level is higher, than the theoretical one. So the SIR function is not stationary for all sample lengths. It means, that for SIR one can construct an appropriate kinetic equation, e.g. Fokker-Planck equation. Instead of (1) we must write

$$\frac{\partial g(s,t)}{\partial t} + \frac{\partial}{\partial s}(w(s,t)g(s,t)) = \frac{\mu(t)}{2}\frac{\partial^2 g(s,t)}{\partial s^2}, \tag{9}$$

where $g(s,t)$ is SIR for definite sample length, $w(s,t)$ is a corresponding velocity and $\mu(t)$ is a diffusion coefficient.

The example of estimation of kinetic equation parameters is given in Fig. 8. The range between minimal and maximal SIR values was separated onto 50 equal class intervals. The diffusion coefficient for this case according to (2) is equal to $\mu \approx 0{,}16$. So we construct the kinetic model for evolution of empirical SIR in this practical case.

Equation (9) enables to obtain evolution equation for average SIR value (see SMA curve in Fig. 5 above) and its variation. Let us designate

$$A(t) = \langle s \rangle = \int sg(s,t)ds, \sigma_A^2(t) = \langle (s - A)^2 \rangle.$$

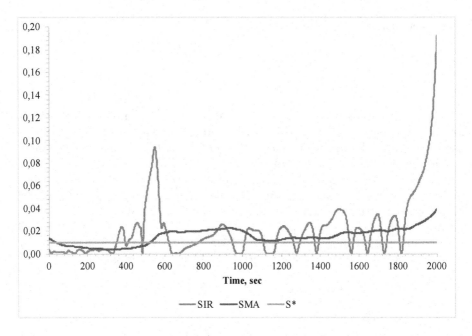

Fig. 5. SIR and SIR moving average trajectories for Moscow metro scheme of trains motion

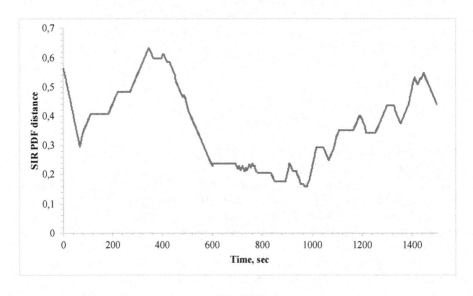

Fig. 6. Distances between SIR PDFs for sample length 1500 s

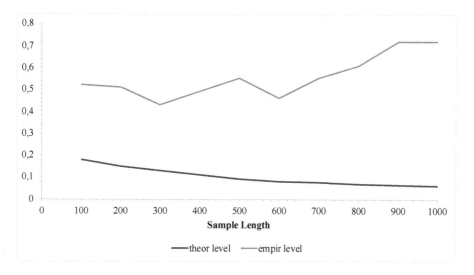

Fig. 7. Stationary point of theoretical and empirical significance levels

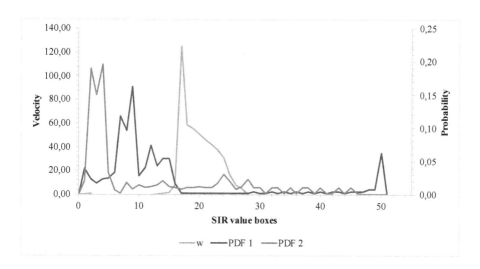

Fig. 8. Empirical PDF and corresponding velocity for sample length 500 s

We have

$$\frac{dA}{dt} = \int s \frac{\partial g(s,t)}{\partial t} ds = \int s \left(-\frac{\partial(wg)}{\partial s} + \frac{\mu}{2} \frac{\partial^2 g}{\partial s^2} \right) ds = \int w(s,t)g(s,t)ds \equiv W(t),$$

where $W(t)$ is an average velocity of variation of SIR probability density function. Analogously we obtain

$$\frac{d\sigma_A^2}{dt} = \mu(t) + 2\int (s - A(t))\,(w(s,t) - W(t))\,g(s,t)ds = \mu(t) + 2cov_{s,w}.$$

So the kinetic equation for empirical SIR PDF enables to obtain the evolution equations for moments of this distribution.

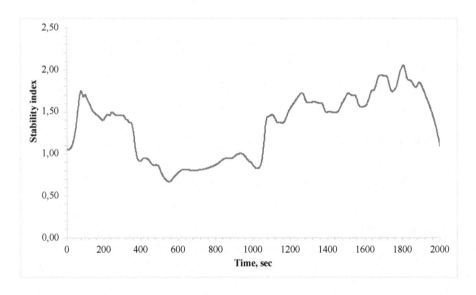

Fig. 9. Normalized average SIR value

6   Time Periods of Continuous Connection

Let us analyze the Fig. 5 again. We see, that time periods, when current SIR value is less, then S^*, have non-uniform distribution. Namely, the probability that outage period belongs to interval from 19 to 23 s, is equal to 0,84. Very short and very long time periods have rather small probabilities. This is due to hypothesis of regular trains motion, when two trains arrive to the same station simultaneously, moving in the opposite directions. It means, that, although SIR value has stochastic non-stationary behaviour, the disconnection periods may have a regular structure.

One should note, that during the last 1000 s in Fig. 5 the total time percentage of outage periods is equal approximately to 20%, while for the time period from 400 to 1000 s the analogous part equals 60%. The SIR moving average curve (SMA line) lies beyond the critical value S^* in both cases. But the normalized average value (so-called Sharp coefficient) $\xi(t) = A(t)/\sigma_A(t)$ for the last part is approximately two times higher, then for previous part (see Fig. 9). This coefficient can be treated as the index of the connection stability.

Thus the method of stochastic control may use this indicator as the main control parameter. The dispersion minimization is more effective action in comparison with the average SIR value increasing.

References

1. ITU towards "IMT for 2020 and beyond" - IMT-2020 standards for 5G. International Telecommunications Union. https://www.itu.int/en/ITU-R/study-groups/rsg5/rwp5d/imt-2020/Pages/default.aspx
2. Baccelli, F., Błaszczyszyn, B.: Stochastic geometry and wireless networks: volume I theory. Found. Trends® Netw. **3**(3–4), 249–449 (2010). https://doi.org/10.1561/1300000006
3. Bai, F., Helmy, A.: A survey of mobility models in wireless adhoc networks (2004)
4. Camp, T., Boleng, J., Davies, V.: A survey of mobility models for ad hoc network research. Wirel. Commun. Mob. Comput. **2**(5), 483–502 (2002). https://doi.org/10.1002/wcm.72
5. Fedorov, S., et al.: SIR distribution in D2D environment with non-stationary mobility of users. In: Proceedings - 31st European Conference on Modelling and Simulation, ECMS 2017, pp. 720–725 (2017). https://doi.org/10.7148/2017-0720
6. Haenggi, M.: Stochastic Geometry for Wireless Networks. Cambridge University Press, Cambridge (2012). https://doi.org/10.1017/CBO9781139043816
7. Haenggi, M., Ganti, R.K.: Interference in large wireless networks. Found. Trends® Netw. **3**(2), 127–248 (2009). https://doi.org/10.1561/1300000015
8. He, R., Molisch, A.F., Tufvesson, F., Zhong, Z., Ai, B., Zhang, T.: Vehicle-to-vehicle propagation models with large vehicle obstructions. IEEE Trans. Intell. Transp. Syst. **15**(5), 2237–2248 (2014). https://doi.org/10.1109/TITS.2014.2311514
9. Kleinhans, D., Friedrich, R.: Continuous-time random walks: simulation of continuous trajectories. Phys. Rev. E - Stat. Nonlin. Soft Matter Phys. **76**(6) (2007). https://doi.org/10.1103/PhysRevE.76.061102
10. Lauridsen, M., Gimenez, L.C., Rodriguez, I., Sorensen, T.B., Mogensen, P.: From LTE to 5G for connected mobility. IEEE Commun. Mag. **55**(3), 156–162 (2017). https://doi.org/10.1109/MCOM.2017.1600778CM
11. Musolesi, M., Mascolo, C.: Mobility models for systems evaluation. In: Garbinato, B., Miranda, H., Rodrigues, L. (eds.) Middleware for Network Eccentric and Mobile Applications, pp. 43–62. Springer, Heidelberg (2009). https://doi.org/10.1007/978-3-540-89707-1_3
12. Naumov, V.: Analysis of time and distance delays in car following models. In: 2010 International Conference on Intelligent Systems, Modelling and Simulation, pp. 296–299 (2010). https://doi.org/10.1109/ISMS.2010.62
13. Naumov, V.: New queueing approach to the vehicle platoon analysis. In: 2010 Fifth International Multi-conference on Computing in the Global Information Technology, pp. 175–179 (2010). https://doi.org/10.1109/ICCGI.2010.18
14. Orsino, A., et al.: Direct connection on the move: characterization of user mobility in cellular-assisted D2D systems. IEEE Veh. Technol. Mag. **11**(3), 38–48 (2016). https://doi.org/10.1109/MVT.2016.2550002

15. Petrov, V., Moltchanov, D., Kustarev, P., Jornet, J.M., Koucheryavy, Y.: On the use of integral geometry for interference modeling and analysis in wireless networks. IEEE Commun. Lett. **20**(12), 2530–2533 (2016). https://doi.org/10.1109/LCOMM.2016.2610435
16. Rong, Z., Rappaport, T.: Wireless Communications: Principles and Practice, Solutions Manual, 1st edn. Prentice Hall, Upper Saddle River (1996)
17. Unterhuber, P., Sand, S., Fiebig, U., Siebler, B.: Path loss models for train-to-train communications in typical high speed railway environments. IET Microwaves Antennas Propag. **12**(4), 492–500 (2018). https://doi.org/10.1049/iet-map.2017.0600

The Use of Asymmetric Numeral Systems Entropy Encoding in Video Compression

Fedor Konstantinov[(⊠)], Gennady Gryzov, and Kirill Bystrov

Moscow Institute of Physics and Technology, Institutsky lane 9, Dolgoprudny, Russia
{fedor.konstantinov,kirill.bystrov}@phystech.edu, gryzov@gmail.com

Abstract. Modern image and video compression technologies includes many compression methods both lossless and lossy. Entropy encoding has a special place among lossless compression methods - it performs final data compression before its conversion to output bitstream.

Huffman coding was used for these purposes, but arithmetic coding is widely used nowadays. This algorithm allows to achieve higher compression rate than Huffman coding, however, it is inferior in speed.

New method called ANS (Asymmetric Numeral Systems) was proposed a few years ago. This statistical coding algorithm is comparable in terms of compression ratio with arithmetic coder and in speed with Huffman coder. ANS is currently used in such data compressors as Zstandard (Google), LZFSE (Apple), DivANS (Dropbox), etc.

In this paper, it is proposed to use ANS in wavelet video codecs. The research and development of an effective compression method of wavelet transform coefficients with usage of ANS entropy encoding was performed.

Keywords: Video coding · Data compression · Asymmetric Numeral Systems · Wavelet image coding · Video codecs performance · Context-based coding

1 Introduction

High-quality video accounts for an increasing proportion of the total transmitted traffic in modern telecommunications networks. Compression methods' efficiency issues become more significant in this connection.

Entropy coder, which performs the final data compression before its conversion to output bitstream, is an important part of any codec.

Huffman coding [1] is computationally simple, but it does not allows to get close to Shannon limit in terms of compression ration because it encodes every symbol by natural number of bits $(1, 2, 3, ...)$.

Arithmetic coding [2] was introduced and began to be applied much later. Computational complexity is one of the reasons for this. Nevertheless this method allows to get close to Shannon limit. It also should be considered that multi-symbol arithmetic coder has not fast implementation – neither in software

© Springer Nature Switzerland AG 2019
V. M. Vishnevskiy et al. (Eds.): DCCN 2019, LNCS 11965, pp. 125–139, 2019.
https://doi.org/10.1007/978-3-030-36614-8_10

nor in hardware. Therefore binary variant is used more often in practice, but it requires preliminary binarization of the input symbols.

New statistical coding method was developed in the last decade – ANS (Asymmetric Numeral System) [3,4]. ANS allows to get close to Shannon limit, like arithmetic coder, being a multi-symbol coder, and it is comparable in terms of encoding speed with Huffman coder because it has tabled variant. Thus ANS has advances of arithmetic coder and Huffman coder at the same time. Also ANS is free from patent protection which increases its attractiveness for the use in applications.

ANS has already been used in such data compressors as Zstandard (Google), LZFSE (Apple), DivANS (Dropbox) and others. However, this method has not been used in video coding yet. The usage of ANS as entropy coder in video coding systems based on wavelet decomposition of images looks promising both in terms of data compression and in terms of video codec performance.

Thereby the aim of this work is the research and development of a method for compressing the wavelet coefficients of image decomposition using ANS coding.

2   The Use of ANS in Wavelet Decomposition of Images

2.1   Asymmetric Numeral System

ANS coding is based on the application of system states and transitions between them. Probabilities of occurrence in data stream are assigned to symbols of the input alphabet (the common approach is to count symbols frequencies in the input message). These probabilities are used by encoder for building state transition table which determines for each state what will be the new state when specific alphabet symbol is entered. The state can be only raised. The least significant bit is discarded (bitwise right shift) from the binary representation of the state in order to reduce the state because the table has a finite size. The encoder output stream is filled with these least significant bits.

Consider the example where input alphabet consists of three symbols (A, B and C) and the following frequencies are used for table construction: A – 9, B – 14, C – 5.

ANS state transition table is constructed with this data (Fig. 1). The new state, which determines transition direction, is placed at the intersection of row and column in this example. Numbers of states placed in rows are distributed approximately evenly over a predetermined interval (between 2 and 31 in Fig. 1) and number of cells in a row is proportional to frequency of the symbol that corresponds to this row. The range interval is determined by "precision" parameter since its upper bound is $2^{precision} - 1$ (in this example $precision = 5$).

The encoding process for input symbols sequence BABBCAB is presented in Fig. 2. Output bitstream: 1010111011.

Decoding process is performed on the ANS table in the same way but in reverse order (from right to left) by translating a sequence of bits in states and corresponding symbols.

Incoming symbol	The initial state														
	1	2	3	4	5	6	7	8	9	10	11	12	13	14	15
A	3	6	9	12	17	19	22	25	28						
B	2	4	7	8	11	13	14	18	20	23	24	27	29	30	31
C	5	10	16	21	26										

Fig. 1. Example of ANS state transition table

st0	Sym	Out	st1	st2
1 (1)	B	-	1	2
2 (10)	A	-	2	6
6 (110)	B	-	6	13
13 (1101)	B	-	13	29
29 (11101)	C	(101)	3 (11)	16
16 (10000)	A	(0)	8 (1000)	25
25 (11001)	B	(1)	12 (1100)	27
27 (11011)	end	(11011)	-	-

Fig. 2. Example of ANS encoding process

2.2   Wavelet-Based Image Decomposition

Wavelet video coding is based on decomposition of an image into frequency sub-bands, where each of them includes its own space-frequency part of the original signal (Figs. 3 and 4). This procedure is analogous of the input data convolution with a filter bank. The features of human perception of visual information [5] make possible a more flexible quantization of the original signal by using the discrete wavelet-transform (DWT) of images, thus achieving higher data compression ratios.

Multichannel DWT scheme application allows to increase the compression ratio while maintaining the quality of the reconstructed image in comparison with standard two-channel schemes in consequence of more compact packing of the signal energy over frequency subbands, as it was shown by a number of studies [6,7].

It is necessary to take into account that separate DWT frequency subbands contain a lot of wavelet coefficients, what increases the efficiency of their coding by tabled variant of ANS (tANS).

Thereby the usage of ANS in conjunction with multichannel DWT may be considered as promising approach in video coding, which should be examined in practice.

Fig. 3. Original image from tested video sequence *pedestrian area*

Fig. 4. Example of three-channel wavelet-decomposed image from tested video sequence *pedestrian area*

2.3   The Use of Contexts in the Coding of Wavelet Coefficients

Context-based approach is used in a number of algorithms, such as PPM [8], BWT [9], Associative Coder of Buyanovsky [10], for computing conditional probabilities of symbols depending on preceding symbol sequences what allows to improve encoding efficiency.

Since frequency subbands of wavelet-decomposed image have characteristic spatial features along various directions then these statistical properties can be applied for forming contexts of wavelet coefficients. In the framework of this study, it is suggested to use the set of spatially neighboring wavelet coefficients and ranges of their values for the contexts formation (Fig. 5a, b).

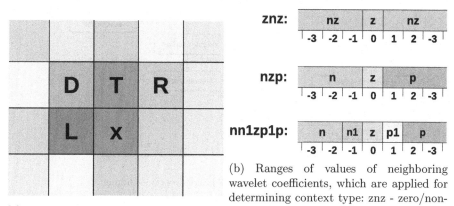

(a) Example of using spatially neighboring wavelet coefficients L, T, D, R as contexts for currently coding coefficient x

(b) Ranges of values of neighboring wavelet coefficients, which are applied for determining context type: znz - zero/non-zero; nzp - negative/zero/positive value; nn1zp1p - less than -1/-1/0/+1/more than +1

Fig. 5. Formation scheme of contexts used in the paper

The application of multi-symbol entropy coder, such as ANS, allows to significantly improve compression using contexts.

The table of frequencies of symbols occurrences in the wavelet coefficients array is constructed for each context and then separately written to output stream. On the one hand, an increase in number of contexts improves the compression ratio of the encoded data. On the other hand, as the number of contexts grows, the number of tables needed to be transferred for decoding increases, what enlarges the amount of transferred data. Hereby, finding the trade-off that minimizes the whole volume of the output data is one of this work's task.

3   Practical Implementation

In the framework of the research, the ANS method for coding the results of the wavelet decomposition of images using contexts was developed and the results of its application on test video sequences were investigated.

Multichannel wavelet video codec implementation (Fig. 6) and the three-channel filter bank 23/23/23 - 13/13/13 from [11] were chosen to perform wavelet video coding. All videos were tested in Intra mode. Tabled variant of ANS (tANS), based on Andrew Polar's ANS implementation [12], was exploited. This approach was modified for coding wavelet coefficients with contexts usage.

Various combinations of neighboring wavelet coefficients (Fig. 5a) and ranges of their values (Fig. 5b) are used for contexts formation.

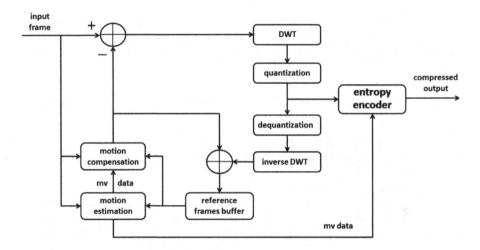

Fig. 6. Multichannel wavelet video codec scheme

Notwithstanding that ANS is multi-symbol coder, a big number of symbols decreases its efficiency. Wavelet coefficients are commonly have values between -100 and 100 in that implementation. Coefficients are decomposed into signed exponents (bit length of the coefficient plus its sign) and mantissas (all bits except the most significant 1) in order to reduce this range.

Exponents are fed to the ANS-coder after that and then the mantissas are just written to the bitstream one after another. The most significant bits of mantissa can be included into exponent, thereby increasing the number of symbols and reducing the mantissa length. Practice showed that the most optimal way is to add one high order bit of the mantissa to the exponent.

The data of each frequency subband is formed by context headers which sequentially follow each other. Context header is a frequency table of symbols with specified context, followed by coder output bitstream and all mantissas after that. The structure of output data stream is shown in the Fig. 7.

Coefficient called "precision" is responsible for the approximation accuracy of symbols probabilities and for the size of state transition table in proposed ANS method. Its increase leads to both an gain in compression ratio and an growth in the time required for building ANS tables and coding the data.

The results for the proposed ANS technique are compared with the results of binary arithmetic coder, used in video codec Schrodinger [13], applied for the same wavelet coefficients. Metric "entropy-PSNR" (PSNR - Peak Signal-to-Noise Ratio), which characterizes the ratio of the output data volume and the quality of the reconstructed signal, was applied to estimate the effectiveness of the proposed methods. Proposed technique's performance, which represents the average time spent on the entropy encoding of one frame of a video sequence, is also evaluated in the paper.

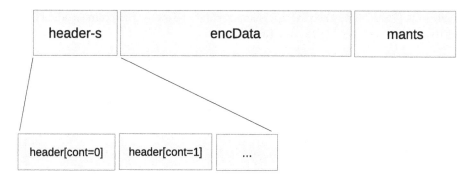

Fig. 7. Data decomposition scheme for frequency subband: *cont* - specific context; *header-s* - context headers; *encData* - encoded exponents; *mants* - mantissas

The following video sequences were selected for testing:

- 4cif (704 × 576) - city, crew, harbour, ice, soccer; [14]
- 720p (1280 × 720) - mobcal, parkrun, shields, stockholm; [15]
- 1080p (1920×1080) - blue sky, pedestrian area, rush hour, station2, sunflower, tractor; [15]

4   Results

In the framework of this research the ANS-coder and arithmetic coder were quantitatively compared using such metrics as compression ratio ("bitrate − PSNR" ratio) and speed of operation.

The results of entropy coding of wavelet coefficients were obtained for the ANS method with different values of the precision parameter and for different sets of contexts.

A contexts set LT_nzp appeared to be the best on average by compression ratio. This context set uses left and top neighbors of wavelet coefficient (Fig. 5a) and considers their values in three ranges: n − less than zero, z − equal to zero, p − greater than zero (Fig. 5b). Therefore, all the results presented in this section are obtained for this contexts set.

The most characteristic coding results for some of the video sequences are presented in Figs. 8, 9, 10, 11, 12, 13, 14, 15, 16 and 17. They represent the dependence of PNSR on bitrate for ANS and arithmetic coders, as well as the ratio of the bitrates of these entropy coders for different PSNRs. For most 1080p video sequences the proposed method gave a gain in compression compared to the arithmetic coder (for example, Figs. 8, 9, 14, 15, 16 and 17), and for some of them the gain was significant (Figs. 8 and 9). For video sequences with other resolutions (720p, 4cif) their compression results are on average almost identical (for example, Figs. 10, 11, 12 and 13).

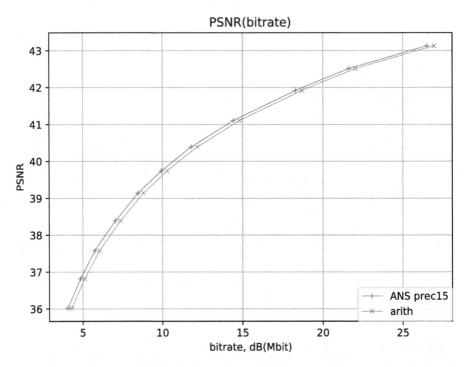

Fig. 8. Video sequence *pedestrian area* (ANS with precision = 15): bitrate – PSNR ratio

Time measurements of tables build, data encoding and other operations (splitting into mantissas and exponents, collecting statistics, selecting contexts, etc.) were also obtained (Figs. 18 and 19). These measurements reveal that the time of the coding process itself (encProc) is comparable to the time of the binary arithmetic coder operation, and the main time costs are due to the construction of ANS-tables and are highly dependent on the parameter *precision*. Thus, the *precision* increases the compression ratio, but also increases the total time spent by the ANS coder. However, these dependencies are nonlinear: with each next *precision* value, the compression ratio increases weaker (Fig. 20), but the tables construction time, as well as their size, equal to $2^{precision}$, grows exponentially.

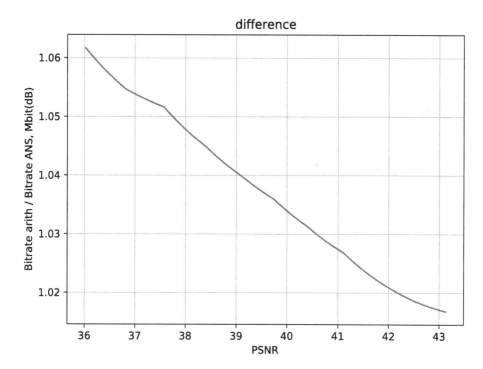

Fig. 9. *pedestrian area* (ANS with precision = 15): bitrate arith/bitrate ANS ratio

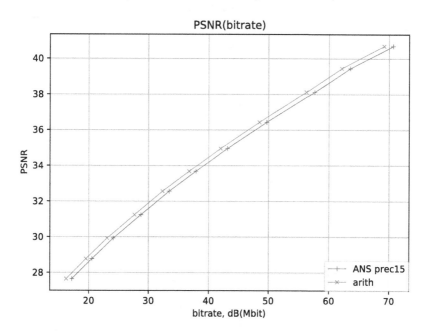

Fig. 10. Video sequence *parkrun* (ANS with precision =15): bitrate – PSNR ratio

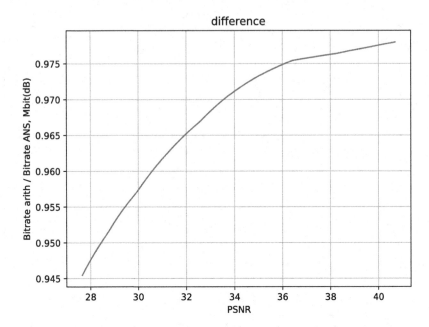

Fig. 11. *parkrun* (ANS with precision = 15): bitrate arith/bitrate ANS ratio

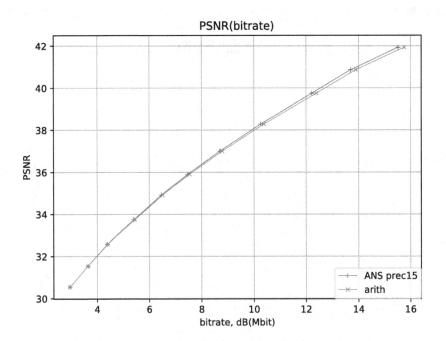

Fig. 12. Video sequence *city* (ANS with precision = 15): bitrate – PSNR ratio

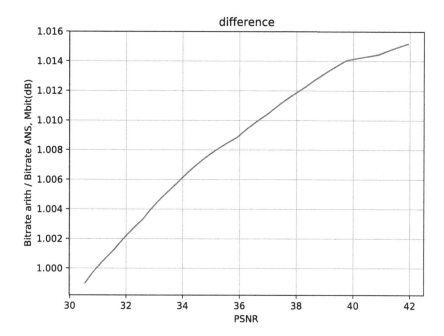

Fig. 13. *city* (ANS with precision =15): bitrate arith/bitrate ANS ratio

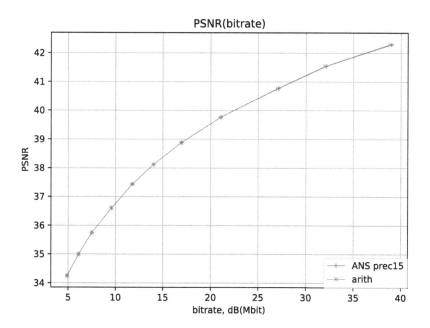

Fig. 14. Video sequence *station2* (ANS with precision =15): bitrate – PSNR ratio

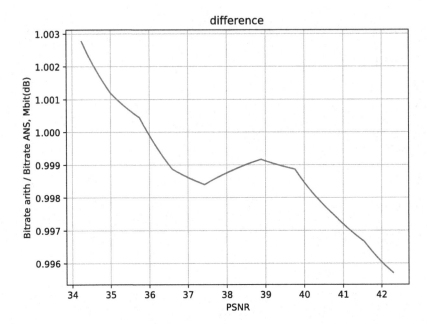

Fig. 15. *station2* (ANS with precision =15): bitrate arith/bitrate ANS ratio

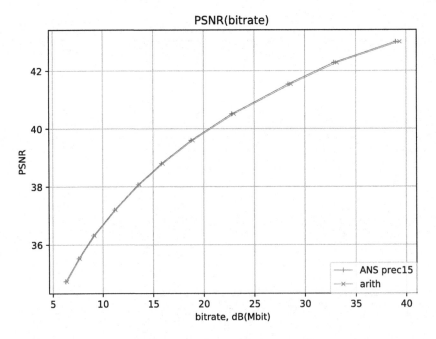

Fig. 16. Video sequence *tractor* (ANS with precision = 15): bitrate – PSNR ratio

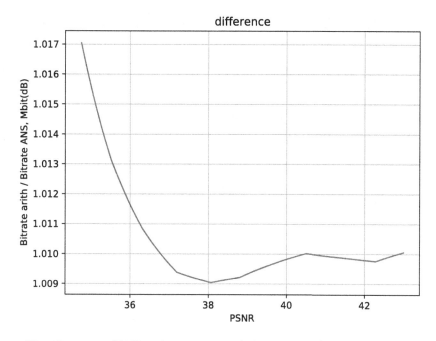

Fig. 17. *tractor* (ANS with precision =15): bitrate arith/bitrate ANS ratio

		pedestrian area		parkrun		city		station2		tractor	
arith		0.048		0.050		0.015		0.053		0.051	
ans_prec10	table	0.087		0.111		0.070		0.084		0.094	
	encProc	0.025	0.179	0.023	0.182	0.007	0.094	0.028	0.182	0.026	0.183
	other	0.067		0.048		0.016		0.069		0.064	
ans_prec15	table	2.672		3.525		2.188		2.549		2.984	
	encProc	0.039	2.777	0.052	3.627	0.016	2.221	0.043	2.660	0.048	3.100
	other	0.066		0.050		0.016		0.067		0.067	

Fig. 18. Average encoding time of one frame in seconds

		pedestrian area		parkrun		city		station2		tractor	
arith		1.00		1.00		1.00		1.00		1.00	
ans_prec10	table	1.80		2.24		4.68		1.58		1.84	
	encProc	0.52	3.70	0.46	3.66	0.48	6.22	0.53	3.42	0.50	3.59
	other	1.38		0.96		1.05		1.31		1.25	
ans_prec15	table	55.27		71.03		145.44		47.96		58.58	
	encProc	0.81	57.44	1.05	73.09	1.07	147.60	0.82	50.04	0.94	60.85
	other	1.36		1.01		1.10		1.27		1.32	

Fig. 19. The average encoding time of one frame in a fraction of the average encoding time of one frame by an arithmetic coder

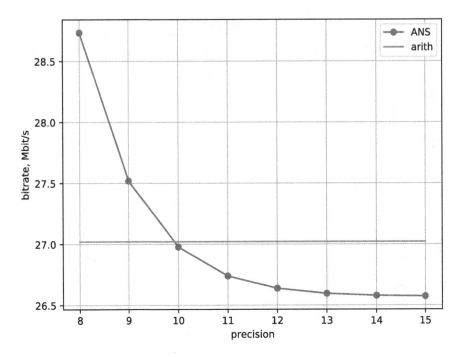

Fig. 20. Bitrate versus *precision* for video sequence *pedestrian area*

5   Conclusion

In this paper the possibility of using ANS-compression method in wavelet video coding was investigated. The results showed the perspectivity of this approach: for most 1080p video sequences it gave a gain in compression compared to the arithmetic coder, while for other resolutions (720p, 4cif) their compression ratios are on average identical. Although the current implementation is worse in speed of work, the ANS encoding process itself works faster, and the main time costs are related to support operations, which have known shortcomings and wide opportunities for optimization and improvement.

Software implementation of ANS method, the researched coder is based on, has algorithmic nonoptimalities in ANS-table construction. Eliminating them can significantly increase the speed of work and slightly improve the compression ratio.

One of the areas of improvement of proposed technique is to reduce the amount of data that carries ANS-table information, required for decoding. Now this role is performed by headers with the symbol frequencies. The reduction of their volume can be achieved by reducing the accuracy of frequency representation, transmitting headers not for all frames, or by other methods, including adaptive ones. Reducing the amount of this data will not only improve the compression ratio, but will also increase the number of contexts that could be used, that leads to reduction of volume of encoded data itself.

The adaptive selection of the precision parameter for different frames and contexts may be another approach for improving the proposed ANS method. Adaptive precision will allow to adjust compression/speed ratio, that depends on this parameter (Sect. 4). This approach will increase the speed of work and possibly reduce the volume of headers.

Since each frequency subband of wavelet-transform has its own correlation specificity between its wavelet coefficients, it is assumed that for each subband there is an optimal contexts set that minimizes output data volume. Finding such sets, and possibly their adaptive selection, is another direction of future research.

References

1. Huffman, D.A.: A method for the construction of minimum-redundancy codes. Proc. Inst. Radio Eng. **40**(9), 1098–1101 (1952)
2. Rissanen, J.J., Langdon, G.G.: Arithmetic coding. IBM J. Res. Dev. **23**, 146–162 (1979)
3. Duda, J.: Asymmetric numeral systems, arXiv:0902.0271, pp. 1–47 (2009)
4. Duda, J.: Asymmetric numeral systems: entropy coding combining speed of Huffman coding with compression rate of arithmetic coding, arXiv:1311.2540, pp. 1–24 (2014)
5. Sonal, D.K.: A study of various image compression techniques. COIT, RIMT-IET, Hisar (2007)
6. Gryzov, G.Y., Dvorkovich, A.V.: Three-channel wavelet transform for video compression applications. In: Proceedings of 2017 6th Mediterranean Conference on Embedded Computing (MECO), 11–15 June 2017, pp. 1–4 (2017)
7. Bystrov, K.S., Gryzov, G.Y., Dvorkovich, A.V., Dvorkovich, V.P.: Wavelet-based video coding: modem implementations and prospects of coding efficiency increase. In: 2017 IVth International Conference on Engineering and Telecommunication (EnT), Moscow, pp. 38–41 (2017). https://doi.org/10.1109/ICEnT.2017.15
8. Moffat, A.: Implementing the PPM data compression scheme. IEEE Trans. Commun. **38**(11), 1917–1921 (1990)
9. Burrows, M., Wheeler, D.J.: A block sorting lossless data compression algorithm. Technical report 124, Digital Equipment Corporation (1994)
10. Buyanovsky, G.: Asociative coding. Monitor **8**, 10–19 (1994). (in Russian)
11. Bystrov, K., Dvorkovich, A., Dvorkovich, V., Gryzov, G.: Usage of video codec based on multichannel wavelet decomposition in video streaming telecommunication systems. In: Vishnevskiy, V.M., Samouylov, K.E., Kozyrev, D.V. (eds.) DCCN 2017. CCIS, vol. 700, pp. 108–119. Springer, Cham (2017). https://doi.org/10.1007/978-3-319-66836-9_10
12. http://ezcodesample.com/abs/abs_article.html
13. Schrodinger. http://schrodinger.sourceforge.net/schrodinger_faq.php
14. ftp://ftp.tnt.uni-hannover.de/pub/svc/testsequences/
15. http://media.xiph.org/video/derf/

Statistical Model of Computing Experiment on Digital Color Correction

E. V. Borevich, S. V. Mescheryakov$^{(\boxtimes)}$, and V. E. Yanchus

St. Petersburg Polytechnic University,
Polytechnicheskaya, 29, 195251 St. Petersburg, Russia
serg-phd@mail.ru
https://www.spbstu.ru/

Abstract. In this paper, we describe new program application for the experimental investigation of a color scheme influence on a film frame human perception and statistical model of data results treatment. Computing experiment is carried out using new software module hosted on a network resource. New approach allows receiving reliable results given large number of observers.

Keywords: Color scheme · Film frame · Human visual system · Statistical model · Digital color correction

1 Introduction

Digital post-processing in cinematograph integrates the latest computer technologies. Film frames are initially created with artistic choice and the technology of 3D virtual scenes and then are processed using digital color correction [1].

There are two reasons why methods for working with colors in cinema differ from painting. The first reason is that cinema uses an additive color model while painting uses a subtractive color model. The second reason is a limited duration of the film demonstration [2]. The goal is to deliver the film content to a viewer without data loss. The color is important in terms of film frame readability and emotional mood in a scene [3,4].

The scientific novelty of the study is that a correlation between color scheme factors and the human visual perception of the film frame is demonstrated via computing experiments and further statistical analysis, which was never studied before.

Eye-tracking technology [5,7] is modern technique for measuring and recording the human eye movement activity. The advantage of this technology is in objective technical indicators of human perception of video frames like the number and duration of eye fixations, etc., in comparison with subjective experts' evaluation methods [2,6]. For this purpose, a series of computing experiments were prepared and carried out.

© Springer Nature Switzerland AG 2019
V. M. Vishnevskiy et al. (Eds.): DCCN 2019, LNCS 11965, pp. 140–150, 2019.
https://doi.org/10.1007/978-3-030-36614-8_11

2   Computing Experiment

Florian I. Yu is a professor, an artist, a colorist who has Ph.D. in History of Arts. Figure 1 shows the model of the human visual system for image recognition based on the Yuryev's color perception model [8]. The human saccadic system generates almost equal number of saccades per time unit. However, the parameters of saccade automaticity are modulated by a specific situation depending on external and internal conditions [9].

As a result, three stages to recognize an image are identified. The first stage is mechanical scanning of the image with eyes. It occurs unconsciously and consists of a quick look on the image. The second stage is a recognition of the image. The viewer uses his (hers) brain and experience of what he (she) had seen earlier. The third stage is an emotional response and personal impression of the viewer.

The experiments demonstrated that during the first stage of image recognition, there is no significant effect of the color scheme factor on the parameters of the viewing pattern. This effect appears during the second stage. Analysis of the viewing pattern reveals its dependence on the color scheme of the picture frame.

During the final stage of the image perception, the viewer assesses attractiveness of the image and viewer subjective opinion about the picture frame. In addition to emotional impression the person's life experience, education, erudition, his mood and health of the viewer, etc., should be taken into account. In the presence of the human factor, the eye-tracker becomes ineffective to objective assess of the influence of the color scheme. This is due to two reasons:

– Interview a large number of observers.
– Introduce a conditional index of frame attractiveness.

To increase the amount of observers for statistical processing, special software application hosted on a network resource is developed and corresponding methodology for carrying out the experiment is worked out.

2.1   Preparing the Stimulus Material

The experiment was conducted on 30 photo-realistic stimuli and 30 animation stimuli (see Figs. 2 and 3). All the stimuli for computing experiment were prepared using the computing methodology described earlier by the authors of this paper in [2] and [6].

2.2   Software Application

The developed software application is hosted on network resources and its architecture is shown in Fig. 4.

The software application was hosted on a server and used three databases: the observers database, the stimuli database and the statistical results database. The developed software supports two modes of operation – for administrator

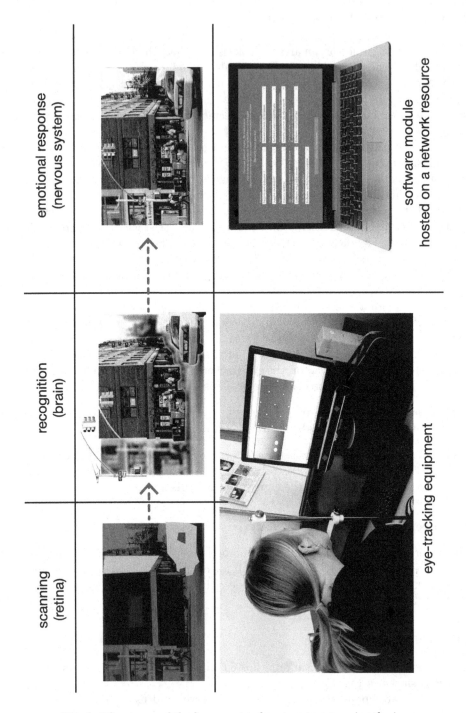

Fig. 1. The stages of the human visual system to recognize the image.

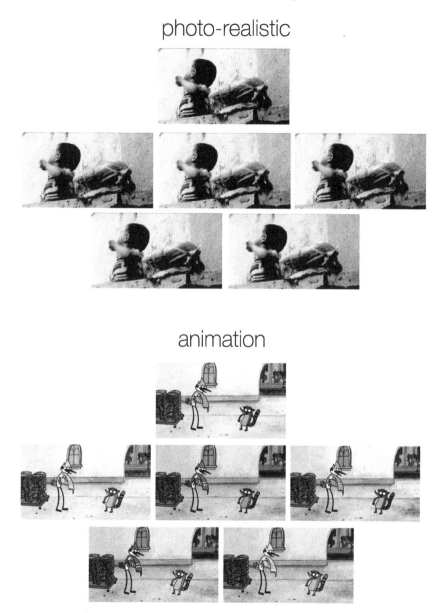

Fig. 2. Example stimulus material for experiments.

and a user. Admin functions include the load of stimulus material, view and edit the database of observers, export of the experimental data results. The users' functionality allows input personal data and pass the computing experiment using the predefined stimulus video frames. In the process of passing the online test, the observers need to fill out a user registration form.

photo-realistic

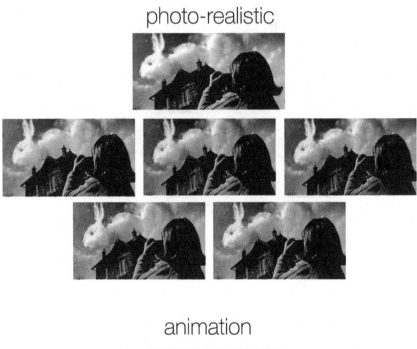

animation

Fig. 3. Example stimulus material for experiments.

Figure 5 shows sample AJAX code, which is executed when the user completed the registration form and started the test. At the same time, the user personal data, such as age, sex, education (art or not), selected film (frame or movie), is saved into the database for future statistical analysis.

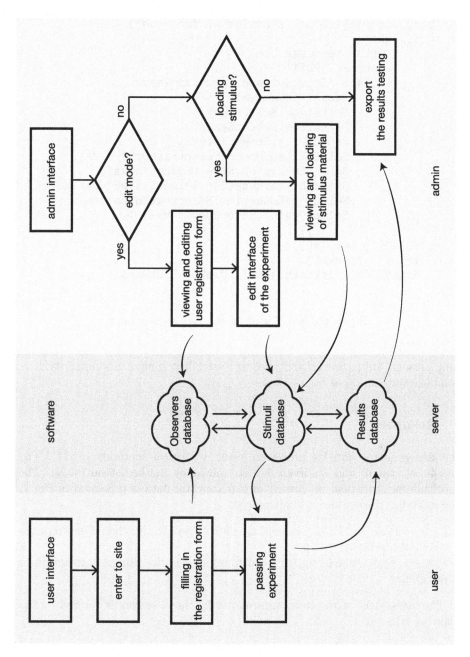

Fig. 4. Architecture of the experimental software application.

The task for observer is to view each of the colored frames and select the most attractive one (see Fig. 6). Observer evaluates the attractiveness of a frame

```
$('body').on('click', '.startTest', function() {
    $.ajax({
        url: 'sendUser.php',
        method: 'POST',
        data : {InputSurname:InputSurnameVal.,
            IdUser:IdUserVal,
            SelectAge:SelectAgeVal,
            SelectSex:SelectSexVal,
            SelectFilm:SelectFilmVal,
            SelectFilmOrMovie:SelectFilmOrMovieVal,
            SelectEducation:SelectEducationVal,
            SelectStyleEducation:SelectStyleEducationVal,
            SelectArtEducation:SelectArtEducationVal,
            SelectDevice:SelectDeviceVal}
            });
    $('div').remove();
    $('hi').remove();
    $.getJSON("Pictures.json", displayPictures);
});
```

Fig. 5. Sample code of the software application.

using personal subjective criteria such as readability, composition integrity, harmonious combination of colors, etc.

3   Results

The experimental data is processed using statistical methods [10,11]. The amount of stimuli and observers is determined by mathematical model. The algorithm for statistical processing of experimental data is presented in Fig. 7. The following conventions are introduced:

- F1 is variable and its parameters are entered to identify the influence of the color scheme factor.
- F2 is a color scheme factor, which influences the human perception is researched.

The assessment of the task performance by the observers is verified by the following criteria:

- The correctness and completeness of filling out the registration form.
- Completeness of the task of the experiment - testing all stimuli.

The important point of statistical analysis is to check data samples for normal distribution for more accuracy of the results, otherwise the amount of tests is not enough and statistical methods are not applicable.

Distribute the pictures from the most to the least liked

Reset   Next

Fig. 6. Interface of the program module created for online testing. (Color figure online)

Table 1. Example of average conventional coefficients.

Color scheme	Conditional coefficient of photo-realistic stimuli	Conditional coefficient of animated stimuli
Black and white	**2.69**	**2.24**
Complementary	**2.18**	**3.23**
Cold monochrome	**3.63**	**3.38**
Warm monochrome	**3.49**	**3.10**
Triadic	**3.01**	**3.05**

Data samples are compared using traditional methods of mathematical statistics. For this purpose, the Student's T-test is applied in case of normally distributed data, otherwise the Wilcoxon test is used. The sample of 300 users is representative enough but further will be extended using web application worldwide. The number of viewers in computing experiment accumulated 300 people, who observed 60 stimuli in each of 5 digital color correction scheme.

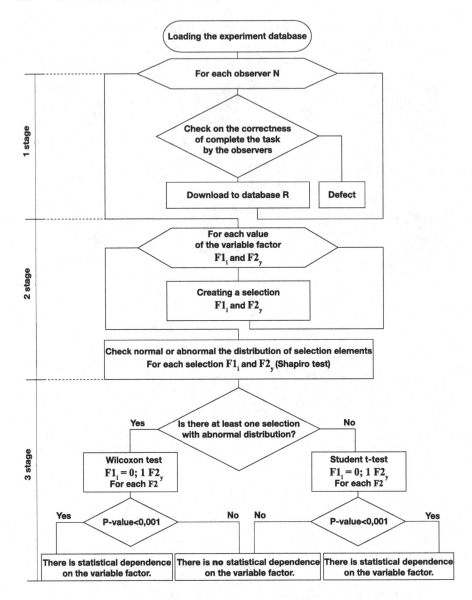

Fig. 7. Algorithm of statistical processing of experimental data.

As an output result, the experimental data of the average conditional coefficient for both photo and animated stimulus material is given for five color schemes (Table 1) and is reflected statistically in Fig. 8.

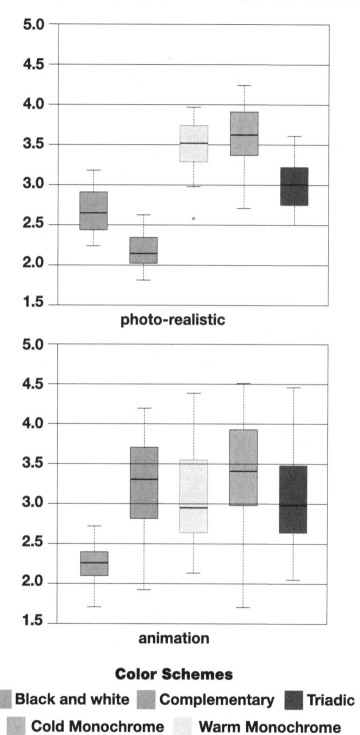

Fig. 8. The visualization of the experiment results. (Color figure online)

4 Conclusion

This article describes new approach for experimental research of the influence of digital color correction methods on the observer's subjective visual perception of video frames. Special software application for online testing is developed and is hosted on a network resource. The proposed approach has advantages against well-known objective eye-tracking technique due to attracting much more people for computing experiment, which makes statistical data more reliable. In addition, the proposed statistical model of computing experiment allows evaluating the dependence of the human visual system on the following factors:

- Gender.
- Age.
- Artistic experience.
- Special education.
- Etc.

References

1. Borevich, E.V., Mescheryakov, S.V., Yanchus, V.E.: Effective methods and models for digital processing of film material. In: International Conference on Computer Graphics and Machine Vision (Graphicon-2017), Perm, vol. 27, pp. 51–54 (2017). http://2017.graphicon.ru/
2. Mescheryakov, S.V., Yanchus, V.E., Borevich, E.V.: Experimental research of digital color correction models and their impact on visual fixation of video frames. Humanit. Sci. Univ. J. **27**, 15–24 (2017). http://uni-journal.ru/technics/archive/
3. Tarkovsky, A.A.: Talk about Color. Film Stud. Notes **1**, 147–160 (1998)
4. Chirimuta, M., Kingdom, F.A.A.: The uses of color vision: ornamental, practical, and theoretical. Minds Machi. **25**, 213–229 (2015)
5. Barabanshchikov, V.A., Milad, M.M.: Oculography Methods in the Study of Cognitive Processes and Activities, Moscow (1994)
6. Yanchus, V.E., Borevich, E.V.: Investigation of the meaning of color solution in the process of film frame harmonization. St. Petersburg State Polytech. Univ. J. **4**, 53–68 (2016). https://infocom.spbstu.ru/en/archive/
7. Orlov, P.A., Laptev, V.V., Ivanov, V.M.: Concerning the use of eye-tracking systems. St. Petersburg State Polyt. Univ. J. **5**, 84–94 (2014). https://infocom.spbstu.ru/en/archive/
8. Yuriev, F.I.: Harmony of Spheres. V. 2. Color Imagery of Information, Kiev (2007)
9. Filin, V.A.: Automation of Saccades, Moscow (2002)
10. Glanz, S.: Biomedical Statistics, Moscow (1999)
11. Kabakov, R.I.: R in Action. Analysis and Visualization of Data in the Program R. DMK Press, Moscow (2014)

Synthesis of High-Performance Window Functions Using Minimization of Difference Between Its Waveform and Spectrum

V. P. Dvorkovich and A. V. Dvorkovich[(✉)]

Moscow Institute of Physics and Technology,
9, Institutkiy per., Dolgoprudny, Russian Federation
v.dvorkovich@mail.ru, dvork-alex@yandex.ru

Abstract. A method for synthesis of high-performance window functions is under consideration. The method is based on the minimization of difference between the temporal and spectral shapes of window function. Such window may be represented by a set of cosine components. This method includes the cases with both odd and even number of cosine half-periods in window duration. A number of window functions have been calculated using the method under consideration. Depending on the number of cosine components the maximum relative difference between the waveform and spectrum of window varies from 0.17 to 1.3×10^{-12}. Main parameters of synthesized windows have been calculated showing high performance of such windows. Maximum sidelobe level decreases from -23 dB to -275 dB while overlap correlation for 0.5 overlapping decreases from 32% to 0.01%. The waveforms, spectra and their differences are illustrated for the windows with the number of cosine components varying from 2 to 21.

Keywords: Digital signal processing · Fourier transform · Harmonic analysis · Window function

1 Introduction

A comprehensive review of utilization of window functions for harmonic signal analysis has been given in classic article by Harris [1]. Since that time a set of new window functions have been constructed by many authors. Detailed analysis of window functions and its parameters along with new methods for window function synthesis was done in [2].

One of the most effective ways to synthesize window functions is based on the calculation of special finite functions, which waveforms are close to their own spectrum shapes with a given accuracy [2]. The method for synthesis of such functions have been proposed for the signals used for the evaluation of the linear characteristics of the communication channel [3]. Normalized cosine basis functions with odd and even number of half-periods may be used for window function representation [2].

© Springer Nature Switzerland AG 2019
V. M. Vishnevskiy et al. (Eds.): DCCN 2019, LNCS 11965, pp. 151–161, 2019.
https://doi.org/10.1007/978-3-030-36614-8_12

2   Synthesis of Window Functions

The window waveforms may be represented by a set of cosine components with odd number of half-periods:

$$
u_o(x) = \begin{cases} 2 \sum\limits_{k=1}^{K} d_k \cos(\pi(2k-1)x), & |x| \le 1/2, \\ 0, & |x| > 1/2, \end{cases} \tag{1}
$$

where $2 \sum\limits_{k=1}^{K} d_k = 1$, $x = t/T$ – normalized time, T – time duration of the window function.

The normalized spectra of these functions are:

$$
F_o(y) = \frac{1}{S_0} \sum_{k=1}^{K} d_k \left[\mathrm{Sinc}\left(\pi\left(y + \frac{2k-1}{2}\right)\right) + \mathrm{Sinc}\left(\pi\left(y - \frac{2k-1}{2}\right)\right) \right], \tag{2}
$$

where $S_0 = \dfrac{4}{\pi} \sum\limits_{k=1}^{K} (-1)^{k-1} \dfrac{d_k}{2k-1}$, $\mathrm{Sinc}(z) = \dfrac{\sin(z)}{z}$, $y = fT$ – normalized frequency.

The window spectrum $F_o(y)$ contains $2K$ components, has a highest maximum equal to 1 while $y = 0$, and the remaining ones are shifted from the central by relative frequency intervals: $\dots, -3/2, -1/2, 1/2, 3/2, \dots$ $F_o(y)$ is always equal to zero at points offset from $y = 0$ by integer value smaller then $-(K + 1/2)$ and large then $(K + 1/2)$.

Another variant of the window functions representation uses the normalized orthogonal cosine basic functions with even number of half-periods:

$$
u_e(x) = \begin{cases} b_0 + 2 \sum\limits_{m=1}^{M} b_m \cos(2\pi m x), & |x| \le 1/2, \\ 0, & |x| > 1/2, \end{cases} \tag{3}
$$

where $b_0 + 2 \sum\limits_{m=1}^{M} b_m = 1$.

The normalized spectra of such functions are presented in the form:

$$
F_e(y) = \mathrm{Sinc}(\pi y) + \frac{1}{b_0} \sum_{m=1}^{M} b_m \left[\mathrm{Sinc}\left(\pi\left(y + m\right)\right) + \mathrm{Sinc}\left(\pi\left(y - m\right)\right) \right]. \tag{4}
$$

The spectrum $F_e(y)$ consists of $(2M+1)$ components, the central maximum is equal to 1 and the other maxima are shifted from the central by relative frequency intervals $-M, -(M-1), \dots, -2, -1, 1, 2, \dots, (M-1), M$. $F_e(y)$ is always equal to zero at integer points offset from zero by intervals smaller then $-(M + 1)$ and large then $(M + 1)$. After conversion of the arguments of functions (1) and

(3) respectively: $x_o \rightarrow \dfrac{y}{2K+1}$ and $x_e \rightarrow \dfrac{y}{2M+1}$, the window functions can be represented as follows:

$$
u_o(y) = \begin{cases} 2 \sum\limits_{k=1}^{K} d_k \cos(\pi y \dfrac{2k-1}{2K+1}), & |y| \le (K+1/2), \\ 0, & |y| > (K+1/2), \end{cases} \tag{5}
$$

$$
u_e(y) = \begin{cases} b_0 + 2 \sum\limits_{m=1}^{M} b_m \cos(2\pi y \dfrac{m}{M+1}), & |y| \le (M+1), \\ 0, & |y| > (M+1). \end{cases} \tag{6}
$$

In such cases, the functions defined by relations (5) and (6) and the corresponding spectra (2) and (4) coincide at least at three points of the argument corresponding to the boundaries of the functions $u(y)$, and also for $y = 0$, when $u(0) = F(0) = 1$. The functions $F(y)$, which closely coincide with the windows $u(y)$ and possess the minimum level of side lobes (in the mean-square sense), may be calculated using the condition:

$$
\int\limits_{-\infty}^{\infty} [F(y) - u(y)]^2 \Rightarrow \min. \tag{7}
$$

Another, the simplest calculation option is associated with minimization of the maximum level of the modulus of the difference $\Delta(y) = |F(y) - u(y)|$:

$$
\Delta = \min_{\Lambda(y)} \left[\max_y |F(y) - u(y)| \right]. \tag{8}
$$

Algorithm (8) is quite simply implemented by selecting the positions of the points y_i at the intervals given by relations (5) and (6), in which the levels of the signals $u(y_i)$ and their spectra $F(y_i)$ are equal.

3   Calculation of Window Functions

For calculation of the coefficients of window functions, the system of equations can be supplemented by the equality of $u(y_i)$ and $F(y_i)$ either for non-integer values of y_i or equality to zero of higher derivatives of $u(y_i)$ and/or $F(y_i)$ for given values of y_i, which can lead to an increase in the rate of decay of side lobes of the Fourier spectrum of the window function.

Tables 1 and 2 shows the calculated coefficients for window functions constructed using odd and even number of half-periods of cosine components using the minimization of the differences in their waveform and spectrum for $K = 1, \ldots, 10$ and $M = 1, \ldots, 10$ respectively.

The data in Table 3 illustrates the maximum relative difference $\Delta_{max} = \max |F(y) - u(y)|$ between the window waveform and its spectrum for various values K and M.

Table 1. Calculated window function coefficients for odd number of half-periods of cosine components

K	Coefficients
1	$d_1 = 0.5$
2	$d_1 = 0.3892$; $d_2 = 0.1108$
3	$d_1 = 0.34049$; $d_2 = 0.14158$; $d_3 = 0.01793$
4	$d_1 = 0.3068902$; $d_2 = 0.1554273$; $d_3 = 0.0352948$; $d_4 = 0.0023877$
5	$d_1 = 0.2815810$; $d_2 = 0.1612808$; $d_3 = 0.0498$; $d_4 = 0.0070468$; $d_5 = 0.0002914$
6	$d_1 = 0.261678605$; $d_2 = 0.163269802$; $d_3 = 0.061249031$; $d_4 = 0.012611507$; $d_5 = 0.001163423$; $d_6 = 0.000027631$
7	$d_1 = 0.245432755692$; $d_2 = 0.162924261385$; $d_3 = 0.070157826608$; $d_4 = 0.018578663919$; $d_5 = 0.002725088038$; $d_6 = 0.0001784341$; $d_7 = 0.000002970256$
8	$d_1 = 0.2318807295$; $d_2 = 0.16142499755$; $d_3 = 0.07702917197$; $d_4 = 0.02433857247$; $d_5 = 0.00478052134$; $d_6 = 0.0005206685$; $d_7 = 0.00002504139$; $d_8 = 0.00000029725$
9	$d_1 = 0.22035033559259$; $d_2 = 0.15928444397114$; $d_3 = 0.08232220671813$; $d_4 = 0.02969799212722$; $d_5 = 0.00717149516321$; $d_6 = 0.00108045209631$; $d_7 = 0.0000898091786$; $d_8 = 0.00000323756839$; $d_9 = 0.0000000275844$
10	$d_1 = 0.210380589355285$; $d_2 = 0.156781655236038$; $d_3 = 0.086377787943023$; $d_4 = 0.034584640630848$; $d_5 = 0.009775712134041$; $d_6 = 0.001862568553082$; $d_7 = 0.00022204471831$; $d_8 = 0.000014585668318$; $d_9 = 0.000000413031197$; $d_{10} = 0.00000000272986$

As follows from the Table 3 the difference between window waveforms and its spectra is less than one percent for $K > 1$ and $M > 1$.

Table 4 gives the main parameters [1,2] of calculated window functions. The following notation is used in this table:

W_{max} – maximum side lobe level of the window function spectrum, dB;

PG – processing gain;

ΔF_n – normalized equivalent noise bandwidth per bin;

ΔF_{-3dB} and ΔF_{-6dB} – the width of the window function at the levels -3 dB and -6 dB, bin;

δ, % – the relative difference between the noise bandwidth and -3 dB level of window function;

SL – scalloping loss, dB;

MPL – maximum processing loss, dB;

$C(r)$ – correlation of overlapping parts of the window function for overlapping values $r = 0.75$ and $r = 0.5$, %.

It should be noted that for $K = 1$ and $M = 1$ the waveforms of calculated window functions are very close to Hanning windows [1] for $\alpha = 1$ and $\alpha = 2$ respectively.

Table 2. Calculated window function coefficients for even number of half-periods of cosine components

M	Coefficients
1	$b_0 = 0.5; \; b_1 = 0.25$
2	$b_0 = 0.40825; \; b_1 = 0.25; \; b_2 = 0.04588$
3	$b_0 = 0.3535534; \; b_1 = 0.2433409; \; b_2 = 0.0732234; \; b_3 = 0.0066592$
4	$b_0 = 0.3162278; \; b_1 = 0.2340885; \; b_2 = 0.091129; \; b_3 = 0.0159115;$ $b_4 = 0.0007571$
5	$b_0 = 0.28867513; \; b_1 = 0.22427983; \; b_2 = 0.1027973; \; b_3 = 0.02563747;$ $b_4 = 0.00286514; \; b_5 = 0.00008270$
6	$b_0 = 0.267261241912; \; b_1 = 0.215025885011; \; b_2 = 0.110419487754;$ $b_3 = 0.034520473055; \; b_4 = 0.005941357185; \; b_5 = 0.000453642016;$ $b_6 = 0.000008533919$
7	$b_0 = 0.25; \; b_1 = 0.2064935569; \; b_2 = 0.1153196771; \; b_3 = 0.0422896777;$ $b_4 = 0.0096113666; \; b_5 = 0.0012157562; \; b_6 = 0.0000689563; \; b_7 = 0.0000010093$
8	$b_0 = 0.2357022604; \; b_1 = 0.19876281594; \; b_2 = 0.11845357806;$ $b_3 = 0.04890775218; \; b_4 = 0.01347256112; \; b_5 = 0.00231992505;$ $b_6 = 0.00022262843; \; b_7 = 0.00000950683; \; b_8 = 0.00000010219$
9	$b_0 = 0.22360679775; \; b_1 = 0.19176998614; \; b_2 = 0.12039677757;$ $b_3 = 0.05451263749; \; b_4 = 0.017312798; \; b_5 = 0.00368256504;$ $b_6 = 0.00048598955; \; b_7 = 0.00003480499; \; b_8 = 0.00000103601;$ $b_9 = 0.00000000634$
10	$b_0 = 0.213200716355; \; b_1 = 0.185367372455; \; b_2 = 0.121402105607;$ $b_3 = 0.059229846305; \; b_4 = 0.021095283392; \; b_5 = 0.005308960272;$ $b_6 = 0.000897004824; \; b_7 = 0.000093702304; \; b_8 = 0.000005247490;$ $b_9 = 0.000000118663; \; b_{10} = 0.000000000509$

Table 3. Maximum relative difference between window waveform and spectrum

	Number of cosine component half-periods		
Odd		Even	
K	$\Delta_{max}, \%$	M	$\Delta_{max}, \%$
1	17	1	2.35
2	4.8×10^{-1}	2	7.5×10^{-2}
3	2.0×10^{-2}	3	2.8×10^{-3}
4	6.2×10^{-4}	4	3.0×10^{-4}
5	4.2×10^{-5}	5	2.0×10^{-5}
6	2.2×10^{-6}	6	4.0×10^{-7}
7	4.6×10^{-8}	7	2.0×10^{-8}
8	3.8×10^{-9}	8	1.1×10^{-9}
9	2.0×10^{-10}	9	2.8×10^{-11}
10	1.7×10^{-11}	10	1.3×10^{-12}

Table 4. Main parameters of calculated window functions

K	M	W_{max}, dB	PG	ΔF_n, bin	ΔF_{-3dB}, bin	δ, %	ΔF_{-6dB}, bin	SL, dB	MPL, dB	$C(r)$, % $r=0.75$	$C(r)$, % $r=0.5$
1	1	−23.0	0.6306	1.2337	1.1884	3.8118	1.6394	2.0981	3.0103	75.54	31.83
	1	−31.5	0.5	1.5	1.4406	4.1245	2.0	1.4236	3.1845	65.92	16.67
2	2	−46.5	0.4485	1.6282	1.5517	4.9295	2.1687	1.2319	3.349	60.53	11.88
	2	−63.4	0.4082	1.7753	1.6886	5.1342	2.3632	1.0417	3.5343	54.47	7.75
3	3	−77.2	0.378	1.9078	1.8103	5.3896	2.5391	0.9078	3.7132	49.63	5.15
	3	−91.1	0.3536	2.0339	1.9275	5.5228	2.7058	0.8016	3.835	44.8	3.38
4	4	−104.0	0.3335	2.1526	2.0378	5.6357	2.8633	0.7179	4.048	40.68	2.23
	4	−110.8	0.3162	2.2671	2.1452	5.6833	3.0153	0.648	4.2028	36.78	1.44
5	5	−128.8	0.3015	2.3722	2.2416	5.8279	3.1546	0.594	4.3456	33.34	0.97
	5	−140.5	0.2887	2.4768	2.3404	5.8299	3.2935	0.5452	4.4841	30.14	0.62
6	6	−154.6	0.2774	2.5752	2.4327	5.858	3.424	0.5052	4.6132	27.29	0.41
	6	−169.9	0.2673	2.6696	2.5198	5.9434	3.5492	0.4709	4.7354	24.72	0.27
7	7	−186.1	0.2582	2.7617	2.6059	5.9782	3.6719	0.4404	4.8522	22.27	0.18
	7	−194.1	0.25	2.8503	2.6886	6.0142	3.789	0.414	4.9627	20.26	0.12
8	8	−213.4	0.2425	2.9368	2.7687	6.0697	3.9013	0.3902	5.0689	18.33	0.08
	8	−219.9	0.2357	3.0202	2.8474	6.0682	4.0146	0.3693	5.1697	16.6	0.05
9	9	−238.6	0.2294	3.1024	2.9235	6.1204	4.1243	0.3501	5.2671	15.01	0.03
	9	−252.8	0.2236	3.1823	2.9995	6.0914	4.2299	0.3202	5.3602	13.58	0.02
10	10	−255.6	0.2182	3.2591	3.0706	6.1378	4.3322	0.3177	5.4486	12.31	0.01
	10	−275.2	0.2132	3.3356	3.132	6.5011	4.4369	0.3033	5.535	11.12	0.01

Figure 1 illustrates the waveform and spectra of calculated window functions for $K = 1, ..., 5$ and $M = 1, ..., 5$.

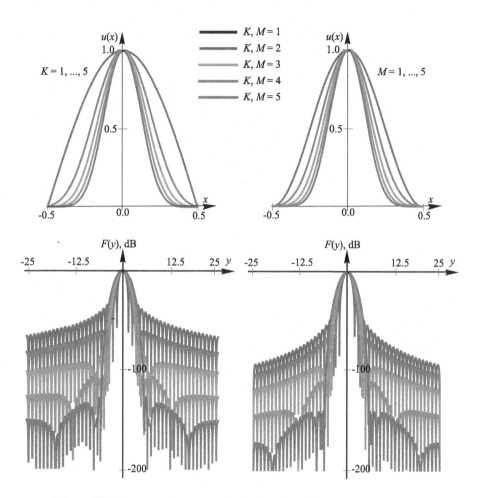

Fig. 1. Waveforms and spectra of calculated windows; $K, M = 1, ..., 5$

158   V. P. Dvorkovich and A. V. Dvorkovich

Figure 2 shows the relative differences between waveform and spectrum of calculated window functions for $K = 1, ..., 5$ and $M = 1, ..., 5$.

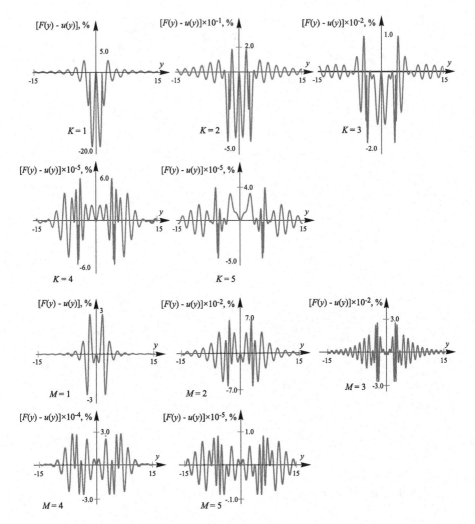

Fig. 2. Relative differences between calculated window waveform and spectrum; $K, M = 1, ..., 5$

Figures 3 and 4 illustrate the same for $K = 6, ..., 10$ and $M = 6, ..., 10$.

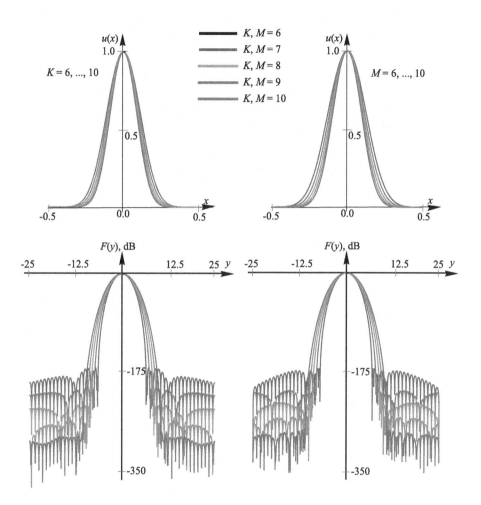

Fig. 3. Waveforms and spectra of calculated windows; K, $M = 6$, ..., 10

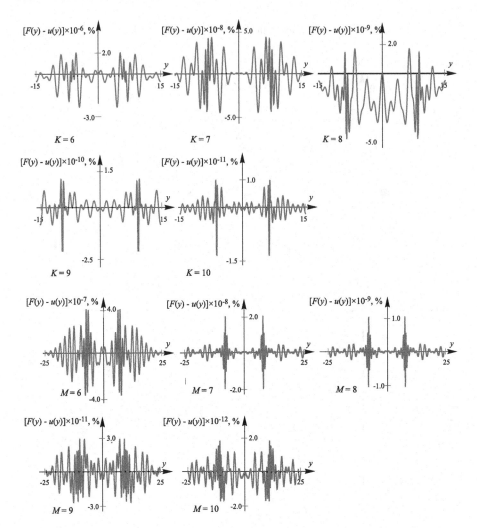

Fig. 4. Relative differences between calculated window waveform and spectrum; $K, M = 6, ..., 10$

4   Conclusion

The method for window function synthesis using minimization of difference between the window waveform and spectrum have been investigated. Two options of synthesis have been shown: utilization of cosine components with even and odd number of half-periods on window duration. The results of such function calculation for the number of cosine components varying from 2 to 23 were presented, including their spectra and main parameters. It was shown that the method can be used for construction of window functions with any given maximum side lobe level.

References

1. Harris, F.J.: On the use of windows for harmonic analysis with the discrete Fourier transform. Proc. IEEE **66**(1), 51–83 (1978)
2. Dvorkovich, V.P., Dvorkovich, A.V.: Window functions for harmonic signal analysis, 2nd edn., revised and expanded, p. 208. Technosphere, Moscow (2016). (in Russian)
3. Dvorkovich, V.P.: A method for the measurement of the linear characteristics of the television communication channel, 4 January 1988, Russian Federation Patent No. 1518924 (1988)

Analytical Modeling of Distributed Systems

Optimal Control by the Queue with Rate and Quality of Service Depending on the Amount of Harvested Energy as a Model of the Node of Wireless Sensor Network

Alexander Dudin[1,2(✉)], Chesoong Kim[3], and Sergey Dudin[1,2]

[1] Department of Applied Mathematics and Computer Science,
Belarusian State University, 220030 Minsk, Belarus
dudin-alexander@mail.ru, dudin85@mail.ru
[2] Peoples' Friendship University of Russia (RUDN University),
6 Miklukho-Maklaya St, Moscow 117198, Russia
[3] Sangji University, Wonju, Kangwon 26339, Republic of Korea
dowoo@sangji.ac.kr

Abstract. A queueing model with energy harvesting and multi-threshold control by service regimes is analysed. The available service regimes are characterized by the different service rate, requirements to the number of energy units for a request service and the probability of error occurrence during service. Error accounting is vital for adequate modelling wireless networks due to existence of an interference in a transmission thread. The increase of the number of energy units for service of a request implies an opportunity to send a stronger signal what implies the higher transmission rate and a lower probability of error occurrence during transmission. Error occurrence causes the repeated transmission and, therefore, consumption of more energy. Under the fixed parameters of the control strategy, the system dynamics is described by a continuous-time six-dimensional Markov chain. This allows to compute the steady-state distribution of the system states and, then, formulate and solve optimization problems. Numerical results are presented.

Keywords: Queueing system · Energy harvesting · Service rate control · Reliability · Optimization

1 Introduction

Wireless sensor networks are widely used for many purposes (in environment protection, health-care, security, military systems, etc.) and represent a popular subject for research. An important direction in this research is the account of possibility of the sensors operation without the centralized energy provisioning. The operation of a sensor, which collects information about a certain object and sends it via gateway sensor node to the network manager, is not possible without the use of energy. Sensors have to extract (harvest) energy from the surrounding

© Springer Nature Switzerland AG 2019
V. M. Vishnevskiy et al. (Eds.): DCCN 2019, LNCS 11965, pp. 165–178, 2019.
https://doi.org/10.1007/978-3-030-36614-8_13

environment (wind, solar, radio-frequency, etc.) themselves and, therefore, this energy has to consumed in the optimal way. In this paper, we present the results of analysis of a queueing model with energy harvesting and multi-threshold control by the use of available service regimes. The service regimes are distinguished by the *service rate*, the number of *energy units* required for a single request service and the *error occurrence* probability. Error accounting is vital for adequate modelling wireless networks due to existence of an interference in a transmission thread. The increase of the number of energy units for service of a request implies an opportunity to send a stronger signal what implies the higher transmission rate and a lower probability of error occurrence during transmission. Error occurrence may cause the repeated transmission and, therefore, consumption of more energy. Thus, if one would like to save the energy via the choice of regimes with low consumption of energy, indeed he/she will spend energy due to its use for retransmission of information after the failed transmission. We assume that the available service regimes are enumerated in the increasing order of provided quality of service and the better regimes require more energy to be used. The optimal strategy for choosing the operation regime depending on the amount of accumulated energy is assumed to be of multi-threshold type.

A similar to the analysed in this paper model was recently considered in the paper [1]. That paper can be recommended for review of the state of the art in application of queueing theory to investigation of systems with energy harvesting. More recent papers in this topic [2–11] can be also recommended. In [1], the model is investigated under quite general assumptions about the request arrival process and service process. Essential differences of the model considered in our paper in comparison to the model [1] are as follows. We assume that the capacity of the buffer for requests is infinite, while the finite buffer capacity was assumed in [1]. We additionally take into account the existing in real-world possibility of energy leakage from the buffer. The effect of leakage is taken into account in [8] for the model with simpler assumptions about distributions, which characterize the system, and without control by the system. The necessity of account of energy leakage is mentioned in survey [12]. Mention also that, since the processes of requests arrival and energy harvesting in the real-world system can be independent, in this paper we consider the separate Markovian arrival processes of requests and energy. In [1], these processes are assumed dependent. They have a common state space and underlying process.

The reminder of the paper consists of the following. Section 2 is devoted to detailed description of queueing model under study and the strategy of control by the system operation. This operation, under the fixed parameters of the control strategy, is described by a continuous-time multi-dimensional Markov chain which is formally defined in Sect. 3. In this section, the generator of this chain, which is a level-dependent Quasi-Birth-and-Death process, is presented and recommendations relating to computation of the steady-state distribution of the chain are given. It is worth to note that, due to the infinite state space of the Markov chain, the problem of computation of the steady-state distribution of this chain is more difficult than such a problem for the model considered in the paper [1] where the corresponding Markov chain has the finite state space. Expressions for computation of the values of the main performance indices of the system via the known stationary distribution of the states of the Markov chain are given in Sect. 4. Section 5 is devoted to presentation of some numerical illustrations of the obtained results. Finally, Sect. 6 contains some concluding remarks.

2   Mathematical Model

We consider a single-server queueing system with the buffer of an infinite capacity for requests and a finite buffer for energy of capacity N.

The structure of the analysed system is depicted in Fig. 1.

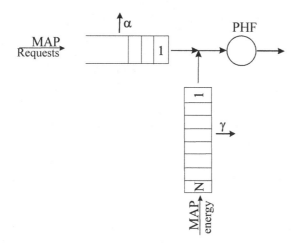

Fig. 1. Structure of the system

The arrival flows of requests and energy are described by two independent Markov Arrival Processes (MAP), see [13]. The MAP arrival flow of requests is defined via an irreducible continuous-time Markov chain v_t, $t \geq 0$, having a finite state space $\{0, 1, ..., V\}$ and the matrices D_0 and D_1. The average intensity of request arrival is denoted as λ. The MAP arrival flow of energy units is defined via an irreducible continuous-time Markov chain w_t, $t \geq 0$, having a finite state space $\{0, 1, ..., W\}$ and the matrices H_0 and H_1. The average intensity of energy unit arrival (rate of harvesting) is λ_h. The MAP is a quite general arrival process popular for modelling real traffic in modern communication networks.

We assume the following:

- The system has R possible operation regimes;
- The regimes are distinguished by the number of energy units required for service of one request, by the service rate and the probability of error-free service;

- Exactly n_r energy units are required to provide service of one request in the rth regime, $r = \overline{1,R}$;
- Required energy units immediately disappear from the buffer for energy at the moment of service beginning;
- The existing operation regimes are ordered in such a way as the following inequalities hold true: $n_1 \geq 1$, $n_r > n_{r-1}$, $r = \overline{2,R}$, $n_R \leq N$. This means that operation in the first regime requires the minimal consumption of the energy, ..., service in the Rth regime requires the maximal consumption of the energy;
- The service time of a request in the rth regime has the phase-type with failures (PHF) distribution with parameters $(\boldsymbol{\beta}^{(r)}, S^{(r)}, \mathbf{S}_1^{(r)}, q^{(r)})$.

The phase-type with failures distribution is introduced in [14]. The random time having such a distribution is briefly interpreted as follows. Let $m_t^{(r)}$, $t \geq 0$, be an irreducible continuous-time Markov chain having a finite set of transient states $\{1, 2, \ldots, M^{(r)}\}$ and two absorbing states $\{M^{(r)} + 1, M^{(r)} + 2\}$. This Markov chain is called as underlying process of PHF distribution. The dynamics of this process is defined by the set of parameters $(\boldsymbol{\beta}^{(r)}, S^{(r)}, \mathbf{S}_1^{(r)}, \mathbf{S}_2^{(r)}, q^{(r)})$ having the following probabilistic meaning. The row vector $\boldsymbol{\beta}^{(r)} = (\beta_1^{(r)}, \ldots, \beta_{M^{(r)}}^{(r)})$ defines the choice of the state of the process $m_t^{(r)}$ from the set $\{1, 2, \ldots, M^{(r)}\}$ at the service beginning instant. The sub-generator $S^{(r)}$ defines the transition rates of the chain within this set. The column vector $\mathbf{S}_1^{(r)}$ defines the rates of transition to the absorbing state $M^{(r)} + 1$ that corresponds to the successful service completion. The column vector $\mathbf{S}_2^{(r)}$, $\mathbf{S}_2^{(r)} = -S^{(r)}\mathbf{e} - \mathbf{S}_1^{(r)}$, defines the rates of transition to the absorbing state $M^{(r)} + 2$ that corresponds to a failure occurrence. Here, $\mathbf{e}$ is a column vector of appropriate size consisting of 1's. After the failure occurrence, service is interrupted and the request leaves the system forever with probability $q^{(r)}$ or tries to restart service with probability $1 - q^{(r)}$.

In the rth operation regime, the mean service rate $\mu^{(r)}$ is defined by the formula $\mu^{(r)} = (-\boldsymbol{\beta}^{(r)}(S^{(r)})^{-1}\mathbf{e})^{-1}$. The probability $P_{error-free}^{(r)}$ that an arbitrary request will receive service without error occurrence in the rth regime is computed by the formula $P_{error-free}^{(r)} = -\boldsymbol{\beta}^{(r)}(S^{(r)})^{-1}\mathbf{S}_1^{(r)}$, $r = \overline{1,R}$. The mean rate of successful completions of service in the rth regime is equal to $\hat{\mu}^{(r)} = \mu^{(r)} P_{error-free}^{(r)}$. Since we enumerated the operation regimes in the increasing order of energy consumption, it is natural to assume that the values of rates $\hat{\mu}^{(r)}$ are ordered in the increasing order of the mean rate of successful service completions, i.e., these values satisfy the inequalities

$$\hat{\mu}^{(1)} < \hat{\mu}^{(2)} < \cdots < \hat{\mu}^{(R)}.$$

Our model with several operation regimes is close to various controlled queueing models with the MAP arrival process previously considered in the literature, see, e.g., [15,16]. In those models, a multi-threshold strategy of control by choosing the appropriate service regime was used. When the queue length increases and exceeds some threshold value, the system switches to the regime with higher service rate (and higher cost of the use). Oppositely, when the queue length decreases and becomes less than some threshold value, the system switches to the cheaper regime with lower service rate. Thus, the thresholds were chosen in the state space of queue length. In the model under study, we stress the critical role of the available (harvested) energy and apply the multi-threshold strategy of control by choosing the operation regime depending of the amount of energy.

The moments of control are the service beginning epochs. The strategy of control is defined by the thresholds N_r, $r = \overline{1, R}$, satisfying the inequalities $N_r \geq n_r$, $r = \overline{1, R}$, $N_r > N_{r-1}$, $r = \overline{2, R}$, $N_R \leq N$. If the request arrives when the server is idle and the number of energy units in the energy buffer is not less than N_1, the request immediately starts service. If the number of energy units in the buffer at the service beginning moment belongs to the interval $[N_r, N_{r+1})$, $r = \overline{1, R - 1}$, service is provided in the rth regime. If the number of energy units is greater than or equal to N_R, the service is provided in the Rth regime. If the request arrives when the server is busy or the server is idle but the number of available units of energy is less than N_1, the request joins the buffer.

Let during service a failure occur and the request, which received service, is not immediately lost but should restart service. In such a situation, if the number of available units of energy is less than N_1, the request is lost. If the number of available units of energy is greater or equal than N_1, the request restarts service in the regime determined according to the threshold strategy defined above.

The requests waiting in the buffer are *impatient*. Each request departs from the system, independently of other waiting requests, after an exponentially distributed with the parameter α, $0 < \alpha < \infty$, time. The leakage of the energy units staying in the buffer takes place. The non-used energy unit leaves the buffer after an exponentially distributed sojourn time with the rate γ, $0 \leq \gamma < \infty$. It is worth to note that the leakage of energy may occur due to the natural discharge of the battery as well as due to the use of the energy units not only for the signals transmission purpose, but also for certain background service, e.g., sensing the object of monitoring.

3   Process of System States and Stationary Distribution

Let us consider the six-dimensional stochastic process that describes the dynamics of the system

$$\xi_t = \{k_t, r_t, n_t, v_t, w_t, m_t^{(r_t)}\}, \ t \geq 0.$$

Here,

- k_t is the current number of requests in the buffer, $k_t \geq 0$;
- r_t, $r_t = \overline{0, R}$, defines the state of the server: $r_t = 0$ means that the server is idle, and $r_t = r$ when the server operates in the rth regime, $r = \overline{1, R}$;
- n_t is the number of available energy units in the buffer, $n_t = \overline{0, N}$;
- v_t is the state of the underlying process of the MAP arrival of requests, $v_t = \overline{0, V}$;

- w_t is the state of the underlying process of the MAP arrival of energy, $w_t = \overline{0, W}$;
- $m_t^{(r_t)}$ is the state of the underlying process of the PHF service process, $m_t^{(r_t)} = \overline{1, M^{(r_t)}}$

at the moment t, $t \geq 0$.

It is easy to verify that under any fixed set of the thresholds of control strategy this process is an irreducible continuous-time Markov chain.

Let us enumerate the states of the chain $\xi_t, t \geq 0$, in the direct lexicographic order of its components.

Theorem 1. The generator Q of the Markov chain ξ_t has the following form:

$$Q = \begin{pmatrix} Q_{0,0} & Q_{0,1} & O & O & \cdots \\ Q_{1,0} & Q_{1,1} & Q_{1,2} & O & \cdots \\ O & Q_{2,1} & Q_{2,2} & Q_{2,3} & \cdots \\ \vdots & \vdots & \vdots & \vdots & \ddots \end{pmatrix}.$$

The non-zero blocks $(Q_{k,k}^{r,r'})_{r,r'=\overline{0,R}}$ of the matrix $Q_{k,k}$, $k \geq 0$, are the following matrices:

$$Q_{0,0}^{0,0} = I_{N+1} \otimes (D_0 \oplus H_0) + E^+ \otimes I_{\bar{V}} \otimes H_1 - (\gamma C_N - \gamma C_N E^-) \otimes I_{\bar{V}\bar{W}},$$

$$Q_{0,0}^{0,r'} = E_{r'}^- \otimes D_1 \otimes I_{\bar{W}} \otimes \beta^{(r')}, \ r' = \overline{1,R},$$

$$Q_{k,k}^{0,0} = I_{N_1} \otimes (D_0 \oplus H_0) + E_1^+ \otimes I_{\bar{V}} \otimes H_1 - (k\alpha I_{N_1} + \gamma C_{N_1-1} - \gamma C_{N_1-1}\tilde{E}) \otimes I_{\bar{V}\bar{W}},$$

$$Q_{k,k}^{r,r} = I_{N+1} \otimes (D_0 \oplus H_0 \oplus S^{(r)}) + E^+ \otimes I_{\bar{V}} \otimes H_1 \otimes I_{M_r} + E_r^- \otimes I_{\bar{V}\bar{W}} \otimes (1 - q^{(r)})\mathbf{S}_2^{(r)}\beta^{(r)}$$

$$- (k\alpha I_{(N+1)} + \gamma C_N - \gamma C_N E^-) \otimes I_{\bar{V}\bar{W}M_r}, r = \overline{1,R}, \ k \geq 0,$$

$$Q_{k,k}^{r,r'} = E_{r'}^- \otimes I_{\bar{V}\bar{W}} \otimes (1 - q^{(r)})\mathbf{S}_2^{(r)}\beta^{(r')}, \ r, r' = \overline{1,R}, r \neq r',$$

$$Q_{0,0}^{r,0} = I_{(N+1)\bar{V}\bar{W}} \otimes (\mathbf{S}_1^{(r)} + q^{(r)}\mathbf{S}_2^{(r)}) + \hat{I} \otimes I_{\bar{V}\bar{W}} \otimes (1 - q^{(r)})\mathbf{S}_2^{(r)}, \ r = \overline{1,R},$$

$$Q_{k,k}^{r,0} = \bar{I} \otimes I_{\bar{V}\bar{W}} \otimes (\mathbf{S}_1^{(r)} + \mathbf{S}_2^{(r)}), \ r = \overline{1,R}.$$

The non-zero blocks $(Q_{k,k-1}^{r,r'})_{r,r'=\overline{0,R}}$ of the matrix $Q_{k,k-1}$, $k \geq 1$, are the following matrices:

$$Q_{1,0}^{0,0} = \alpha \tilde{I} \otimes I_{\bar{V}\bar{W}},$$

$$Q_{k,k-1}^{0,0} = k\alpha I_{N_1\bar{V}\bar{W}}, \ k \geq 2,$$

$$Q_{k,k-1}^{r,r} = k\alpha I_{(N+1)\bar{V}\bar{W}M_r} + E_r^- \otimes I_{\bar{V}\bar{W}} \otimes (\mathbf{S}_1^{(r)} + q^{(r)}\mathbf{S}_2^{(r)})\beta^{(r)}, \ r = \overline{1,R}, \ k \geq 1,$$

$$Q_{k,k-1}^{0,1} = E_0^- \otimes I_{\bar{V}} \otimes H_1 \otimes \beta^{(1)}, \ k \geq 1,$$

$$Q_{k,k-1}^{r,r'} = E_{r'}^- \otimes I_{\bar{V}\bar{W}} \otimes (\mathbf{S}_1^{(r)} + q^{(r)}\mathbf{S}_2^{(r)})\beta^{(r')}, \ r, r' = \overline{1,R}, r \neq r', \ k \geq 1.$$

The matrix $Q_{k,k+1}$, $k \geq 0$, has the following form

$$Q_{k,k+1} = \text{diag}\{Q_{k,k+1}^{r,r}, \ r = \overline{0,R}\},$$

where

$$Q_{0,r}^{0,0} = \bar{I} \otimes D_1 \otimes I_{\bar{W}},$$
$$Q_{k,k+1}^{0,0} = I_{N_1} \otimes D_1 \otimes I_{\bar{W}},$$
$$Q_{k,k+1}^{r,r} = I_{N+1} \otimes D_1 \otimes I_{\bar{W}M_r}, \ r = \overline{1,R}, \ k \geq 0.$$

Here, I (O) is the identity (zero) matrix of an appropriate size; $\bar{V} = V + 1$, $\bar{W} = W + 1$; the symbols $\otimes$ and $\oplus$ denote the Kronecker product and the sum of matrices, correspondingly; C_r is the diagonal matrix with the diagonal entries $\{0, 1, \ldots, r\}$, $r = \overline{1,R}$.

Some other matrices are used above to describe transition probabilities of the number of available energy units at various transition moments of the Markov chain ξ_t. The majority of the entries of these matrices are equal to zero with the rest of entries equal to 1. In Table 1, we present the denotations for these matrices, their dimensions and indication which of their elements are equal to 1.

Table 1. Auxiliary matrices

Denotation	Size	Entries equal to 1
E^+	$(N+1) \times (N+1)$	$(E^+)_{n,n+1}$, $n = \overline{0, N-1}$, and $(E^+)_{N,N}$
E^-	$N_1 \times N_1$	$(E^-)_{n,n-1}$, $n = \overline{1, N}$
$\tilde{E}$	$(N+1) \times (N+1)$	$\tilde{E}_{n,n-1}$, $n = \overline{1, N_1 - 1}$
E_1^+	$N_1 \times N_1$	$((E^+)_1)n, n+1$, $n = \overline{0, N_1 - 2}$
$E_r^-, r = \overline{1, R-1}$	$(N+1) \times (N+1)$	$(E_r^-)_{m,m-n_r}$, $m = \overline{N_r, N_{r+1} - 1}$
E_R^-	$(N+1) \times (N+1)$	$(E_R^-)_{m,m-n_R}$, $m = \overline{N_R, N}$
E_0^-	$N_1 \times (N+1)$	$(E_0^-)_{N_1-1, N_1-n_1}$
$\bar{I}$	$(N+1) \times N_1$	$(\bar{I})_{n,n}$, $n = \overline{0, N_1 - 1}$
$\tilde{I}$	$N_1 \times (N+1)$	$(\tilde{I})_{n,n}$, $n = \overline{0, N_1 - 1}$
$\hat{I}$	$(N+1) \times (N+1)$	$(\hat{I})_{n,n}$, $n = \overline{0, N_1 - 1}$

The proof of Theorem 1 essentially follows the lines of the detailed proof presented for a simpler model in [1]. Therefore, it is omitted there. The differences are related with the separate account of underlying processes of arrival of request and energy (this adds one more component to the Markov chain under study), presence of the infinite buffer (this makes the generator having an infinite size) and opportunity of energy leakage (this makes the behavior of the number of available energy units between service beginning moments similar not to the Birth process, but to the Birth-and-Death process).

From the structure of the generator, it is clear that the Markov chain ξ_t is a level-dependent Quasi-Birth-and-Death process. Also, it can be shown that the Markov chain ξ_t belongs to the class of continuous-time asymptotically quasi-Toeplitz Markov chains ($AQTMC$), see [17]. The results from [17] can be used to check the ergodicity of the system. It can be shown that because $\alpha > 0$ the system is ergodic for any set of system parameters.

Let us form the stationary probability vectors of the system states $\boldsymbol{\pi}_k$, $k \geq 0$, according to introduced lexicographical enumeration of the states of the chain. It is well known that the probability vectors $\boldsymbol{\pi}_k$, $k \geq 0$, satisfy the following infinite system of linear algebraic equations:

$$(\boldsymbol{\pi}_0, \boldsymbol{\pi}_1, \dots)Q = \mathbf{0}, \quad (\boldsymbol{\pi}_0, \boldsymbol{\pi}_1, \dots)\mathbf{e} = 1.$$

The problem of solving such a type of systems is quite difficult. The majority of the papers where such systems appear offer to solve the system via some kind of truncation. This is quite rough method and it does not give any guarantee that the obtained solution is close to the real one. Moreover, even the truncated system can have big size and it is difficult to obtain the approximated solution. In [17], the method is elaborated for finding the solution of this system via the replacement of this system by the another infinite system derived via sequential implementation of censoring of the original Markov chain with varying levels of censoring. This method exploits the use of the level dependent extension of the famous matrix G by M. Neuts. This numerically stable method is recently refined in [18]. Here, we used algorithms from [18] to compute the row vectors $\boldsymbol{\pi}_k$, $k \geq 0$. Each vector $\boldsymbol{\pi}_k$ has the structure $\boldsymbol{\pi}_k = (\boldsymbol{\pi}(k, 0), \dots, \boldsymbol{\pi}(k, R))$ where, in turn, the vector $\boldsymbol{\pi}(k, r)$ has the structure $(\boldsymbol{\pi}(k, r, 0), \dots, \boldsymbol{\pi}(k, r, N))$, $r = \overline{0, R}$.

4   Main Performance Measures of the System

Using the known results concerning the limits of the functionals defined on the trajectories of a Markov chain, it is possible to derive formulas for computation of various performance characteristics of the system under study. We present only a few of them.

The mean number K of requests presenting in the buffer at an arbitrary moment is calculated using the formula

$$K = \sum_{k=1}^{\infty} k \boldsymbol{\pi}_k \mathbf{e}.$$

The variance $\hat{K}$ of the number of requests presenting in the buffer at an arbitrary moment is calculated using the formula

$$\hat{K} = \sum_{k=1}^{\infty} k^2 \boldsymbol{\pi}_k \mathbf{e} - K^2.$$

The probability that at an arbitrary moment the server does not work is computed by

$$P_{idle} = \sum_{k=1}^{\infty} \sum_{n=0}^{N_1-1} \boldsymbol{\pi}(k, 0, n)\mathbf{e} + \sum_{n=0}^{N} \boldsymbol{\pi}(0, 0, n)\mathbf{e}.$$

The probability that at an arbitrary moment the server works in the rth regime is computed by

$$P_r = \sum_{k=0}^{\infty} \boldsymbol{\pi}(k,r)\mathbf{e}, \quad r = \overline{1,R}.$$

The probability that an arbitrary request is lost in the system due to impatience is calculated using the formula

$$Q^{(loss-imp)} = \lambda^{-1}\alpha K.$$

The probability that an arbitrary request is lost due to the failure occurrence is calculated using the formula

$$Q^{(loss-fail)} = \lambda^{-1}\sum_{k=0}^{\infty}\sum_{r=1}^{R} q^{(r)}\boldsymbol{\pi}(k,r)(I_{(N+1)\bar{V}\bar{W}} \otimes \mathbf{S}_2^{(r)})\mathbf{e}.$$

The probability that an arbitrary request is lost due to the lack of energy at the moment when service has to be restarted is calculated using the formula

$$Q^{(loss-lack)} = \lambda^{-1}\sum_{k=0}^{\infty}\sum_{r=1}^{R}\sum_{n=0}^{N_1-1}(1-q^{(r)})\boldsymbol{\pi}(k,r,n)(I_{\bar{V}\bar{W}} \otimes \mathbf{S}_2^{(r)})\mathbf{e}.$$

The probability that an arbitrary request is lost is calculated using the formula

$$Q^{(loss)} = Q^{(loss-imp)} + Q^{(loss-fail)} + Q^{(loss-lack)}.$$

The probability that an arbitrary unit of energy is lost due to the buffer overflow is calculated using the formula

$$Q_{h-overflow}^{(loss)} = \lambda_h^{-1}\left[\boldsymbol{\pi}(0,0,N)(I_{\bar{V}} \otimes H_1)\mathbf{e} + \sum_{k=0}^{\infty}\sum_{r=1}^{R}\boldsymbol{\pi}(k,r,N)(I_{\bar{V}} \otimes H_1 \otimes I_{M_r})\mathbf{e}\right].$$

The probability that an arbitrary unit of energy is lost due to the leakage is calculated using the formula

$$Q_{h-leakage}^{(loss)} = \frac{\gamma}{\lambda_h}\left(\sum_{k=0}^{\infty}\sum_{r=1}^{R}\sum_{n=1}^{N} n\boldsymbol{\pi}(k,r,n)\mathbf{e} + \sum_{k=1}^{\infty}\sum_{n=1}^{N_1-1} n\boldsymbol{\pi}(k,0,n)\mathbf{e} + \sum_{n=1}^{N} n\boldsymbol{\pi}(0,0,n)\mathbf{e}\right).$$

The probability that an arbitrary unit of energy is lost is calculated using the formula

$$Q_h^{(loss)} = Q_{h-overflow}^{(loss)} + Q_{h-leakage}^{(loss)}.$$

As soon as the main performance measures have been computed, one can formulate and solve various performance optimization problems.

5   Numerical Results

We assume that the MAP arrival flow of requests is defined by the following matrices:

$$D_0 = \begin{pmatrix} -0.67582 & 0 \\ 0 & -0.021935 \end{pmatrix}, \ D_1 = \begin{pmatrix} 0.671327 & 0.004493 \\ 0.012217 & 0.009718 \end{pmatrix}.$$

This flow has the average arrival rate λ equal to 0.5.

The MAP arrival flow of energy units is defined by the following matrices:

$$H_0 = \begin{pmatrix} -4.056541 & 0 \\ 0 & -0.131687 \end{pmatrix}, \ H_1 = \begin{pmatrix} 4.029533 & 0.027008 \\ 0.073321 & 0.058366 \end{pmatrix}.$$

This flow has the average arrival rate λ_h equal to 3.

We assume that service of an arbitrary request in the first regime requires $n_1 = 1$ energy unit. The PHF service time distribution in the first regime is defined by the parameters

$$\beta^{(1)} = (0.6, 0.4), \ S^{(1)} = \begin{pmatrix} -1 & 0.3 \\ 1 & -3 \end{pmatrix}, \ \mathbf{S}_1^{(1)} = (0.6, 1.9)^T, \ q^{(1)} = 0.5.$$

The mean service time in the first regime is equal to 1.0296, and the probability of successful service is equal to 0.897.

Service of an arbitrary request in the second regime requires $n_2 = 5$ energy units. The PHF service time distribution in the second regime is defined by the parameters

$$\beta^{(2)} = (0.8, 0.2), \ S^{(2)} = \begin{pmatrix} -4 & 1 \\ 1 & -3 \end{pmatrix}, \ \mathbf{S}_1^{(2)} = (2.9, 1.9)^T, \ q^{(2)} = 0.6.$$

The mean service time in the second regime is equal to 0.3818, and the probability of successful service is equal to 0.9618.

The capacity of the buffer for energy is equal to $N = 20$. The intensity of requests impatience is equal to $\alpha = 0.02$. The leakage intensity $\gamma = 0.01$.

Let us vary the parameter N_2 in the interval $[n_2, N]$ and the parameter N_1 in the interval $[n_1, N_2 - 1]$.

Figures 2, 3 and 4 illustrate the dependence of the mean number K of requests presenting in the buffer, the probability $Q^{(loss)}$ that an arbitrary request is lost, and the probability $Q_{h-leakage}^{(loss)}$ that an arbitrary unit of energy is lost due to the leakage on the values N_1 and N_2.

Let us consider the problem of the choice of the thresholds (N_1, N_2) providing the minimal value of the probability $Q^{(loss)}$ that an arbitrary request is lost. Using the results of implemented computations for various sets of (N_1, N_2) to draw Fig. 2, it can be verified that the minimum of $Q^{(loss)}$ is equal to 0.07527375 and is achieved when $N_1 = 1$ and $N_2 = 16$. Therefore, to minimize the probability $Q^{(loss)}$, it is necessary to start new service when the buffer of energy is not empty and use the second regime if the number of energy units in the buffer is greater or equal to 16.

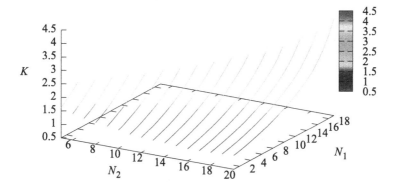

Fig. 2. Dependence of the mean number K of requests presenting in the buffer on the values N_1 and N_2

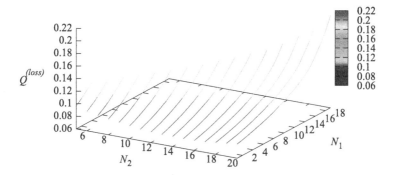

Fig. 3. Dependence of the probability $Q^{(loss)}$ that an arbitrary request is lost on the values N_1 and N_2

Since we assumed that the probability of error occurrence is quite small, there exists almost linear dependence between K and $Q^{(loss)}$. Both these characteristics essentially increase when the threshold N_1 grows (the server stays idle until a quite large number of requests accumulates in the buffer). The probability $Q^{(loss)}_{h-leakage}$ essentially depends on the threshold N_1 and only slightly depends on N_2.

Figures 5, 6 and 7 illustrate the dependence of the probability $Q^{(loss)}_h$ that an arbitrary unit of energy is lost, the probability P_1 that at an arbitrary moment the server works in the first regime, and the probability P_2 that at an arbitrary moment the server works in the second regime on the values N_1 and N_2. Oppositely to the probability $Q^{(loss)}_{h-leakage}$, the probability $Q^{(loss)}_h$ more essentially depends on the value of N_2 namely because this value determines whether or not the number of energy units in the buffer will reach the buffer capacity N. As it is intuitively clear, the probability P_1 increases while the probability P_2 decreases when the threshold N_2 grows.

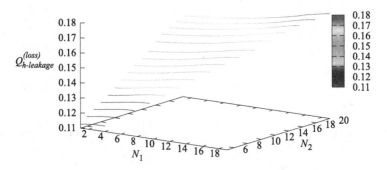

Fig. 4. Dependence of the probability $Q^{(loss)}_{h-leakage}$ that an arbitrary unit of energy is lost due to the leakage on the values N_1 and N_2

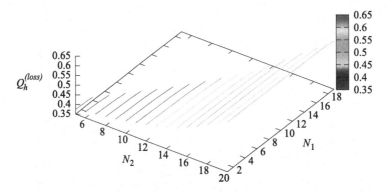

Fig. 5. Dependence of the probability $Q^{(loss)}_h$ that an arbitrary unit of energy is lost on the values N_1 and N_2

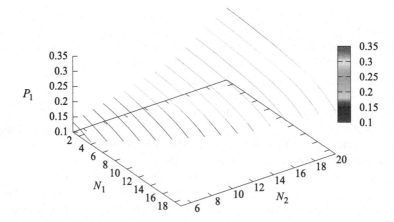

Fig. 6. Dependence of the probability P_1 that at an arbitrary moment the server works in the first regime on the values N_1 and N_2

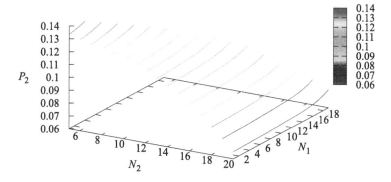

Fig. 7. Dependence of the probability P_2 that at an arbitrary moment the server works in the second regime on the values N_1 and N_2

6 Conclusion

In this paper, we analysed the extension of the model considered in [1] to the cases of account of independent flows of requests and energy units, an infinite buffer for arriving requests and the leakage of energy. The system has a finite number of available regimes of operation distinguished by the service rate, the probability of error occurrence and energy consumption. The choice of the operation regime is implemented according to the threshold strategy depending on the currently available number of harvested energy units. The behavior of the system under the fixed set of the thresholds is defined by a six-dimensional Markov chain. This chain is analysed. Expressions for computation of some performance indices of the system are derived. Illustrations of their dynamics depending on the values of the thresholds are presented. The results can be extended to the cases of the hysteresis strategy of control (see, e.g., [19]) which allows to reduce the frequency of changing operation regimes of the system.

Acknowledgments. The work by Chesoong Kim has been supported by the Basic Science Research Program through the National Research Foundation of Korea (NRF) funded by the Ministry of Education (Grant No. 2018K2A9A1A06072058). The work by Alexander Dudin and Sergei Dudin has been supported by "RUDN University Program 5-100".

References

1. Kim, C., Dudin, S., Dudin, A., Samouylov, K.: Multi-threshold control by a single-server queuing model with a service rate depending on the amount of harvested energy. Perform. Eval. **127–128**, 1–20 (2018)
2. Ashraf, N., Hasan, A., Qureshi, H., Lestas, M.: Combined data rate and energy management in harvesting enabled tactile IoT sensing devices. IEEE Trans. Ind. Inform. **15**(5), 3006–3015 (2019)

3. Bae, Y., Baek, J.: Optimal design of RF energy-harvesting network: throughput and delay perspective. Sensors **19**(1), 1–20 (2019)
4. Chunlin, L., Chen, W., Tang, J., Luo, Y.: Radio and computing resource allocation with energy harvesting devices in mobile edge computing environment. Comput. Commun. **145**, 193–202 (2019)
5. Di Lorenzo, P., Battiloro, C., Banelli, P., Barbarossa, S.: Dynamic resource optimization for decentralized signal estimation in energy harvesting wireless sensor networks. In ICASSP 2019 IEEE International Conference on Acoustics, Speech and Signal Processing (ICASSP), pp. 4454–4458 (2019)
6. El Shafie, A., Al-Dhahir, N., Ding, Z., Duong, T., Hamila, R.: Wiretap TDMA networks with energy-harvesting rechargeable-battery buffered sources. IEEE Access **7**, 17215–17229 (2019)
7. Guo, L., Chen, Z., Zhang, D., Liu, J., Pan, J.: Sustainability in body sensor networks with transmission scheduling and energy harvesting. IEEE Internet Things J. (2019). https://doi.org/10.1109/JIOT.2019.2930076
8. Gelenbe, E., Zhang, Y.: Performance optimization with energy packets. IEEE Syst. J. (2019). https://doi.org/10.1109/JSYST.2019.2912013
9. Knorn, S., Dey, S., Ahlen, A., Quevedo, D.: Optimal energy allocation in multi sensor estimation over wireless channels using energy harvesting and sharing. IEEE Trans. Autom. Control (2019). https://doi.org/10.1109/TAC.2019.2896048
10. Patil, K., De Turck, K., Fiems, D.: Optimal data collection in wireless sensor networks with correlated energy harvesting. Ann. Telecommun. **74**(5–6), 299–310 (2019)
11. Zhang, Y.: Optimal energy distribution with energy packet networks. Probab. Eng. Inform. Sci. (2019). https://doi.org/10.1017/S0269964818000566
12. Ray, P.: Energy packet networks: an annotated bibliography. SN Comput. Sci. **1**, 6 (2020). https://doi.org/10.1007/s42979-019-0008-x
13. Chakravarthy, S.: The batch Markovian arrival process: a review and future work. In: Krishnamoorthy, A., Raju, N., Ramaswami (eds.) Advances in Probability Theory and Stochastic Processes, pp. 21–29. Notable Publications Inc., New Jersey (2001)
14. Dudin, A., Dudin, S.: Analysis of a priority queue with phase-type service and failures. Int. J. Stochast. Anal. Article ID 9152701, 1–11 (2016)
15. Dudin, A.: Optimal multithreshold control for the $BMAP/G/1$ queue with N service modes. Queueing Syst. **30**, 41–55 (1998)
16. Kim, C., Klimenok, V., Birukov, A., Dudin, A.: Optimal multi-threshold control by the $BMAP/SM/1$ retrial system. Ann. Oper. Res. **141**(1), 193–210 (2006)
17. Klimenok, I., Dudin, A.: Multi-dimensional asymptotically quasi-Toeplitz Markov chains and their application in queueing theory. Queueing Syst. **54**, 245–259 (2006)
18. Dudin, S., Dudina, O.: Retrial multi-server queueing system with PHF service time distribution as a model of a channel with unreliable transmission of information. Appl. Math. Modell. **65**, 676–695 (2019)
19. Kim, C.S., Dudin, A., Dudin, S., Dudina, O.: Hysteresis control by the number of active servers in queueing system with priority service. Perform. Eval. **101**, 20–33 (2016)

On Optimal Control Policy of $MAP(t)/M/2$ Queueing System with Heterogeneous Servers and Periodic Arrival Process

Dmitry Efrosinin[1,2]($\boxtimes$) and Natalia Stepanova[3]

[1] Johannes Kepler University Linz, Altenbergerstrasse 69, 4040 Linz, Austria
dmitry.efrosinin@jku.at
[2] Peoples' Friendship University of Russia (RUDN University),
Miklukho-Maklaya St. 6, 117198 Moscow, Russia
[3] V.A. Trapeznikov Institute of Control Sciences of RAS,
Profsoyuznaya St., 65, 117997 Moscow, Russia
natalia0410@rambler.ru
https://www.jku.at, https://www.eng.rudn.ru, https://www.ipu.ru/en

Abstract. In this paper we consider an optimal control problem for the $MAP(t)/M/2$ queueing system with heterogeneous servers is introduced. The Markov arrival process (MAP) has time-dependent and periodic rates for phase transitions. We built a continuous time finite-horizon Markov decision process (MDP) with the aim to minimize a cost function. We solve a Bellman equation as a system of ordinary differential equations with time-dependent coefficients. We show that the optimal policy is of threshold type with threshold levels depending on the phases of arrival process. Moreover, the periodic variation of arrival attributes makes a threshold control policy piecewise constant time-dependent and periodic. We study numerically the speed of convergence of the policy to a periodic pattern. For the fixed control policy we calculate a transient solution. and provide a sensitivity analysis to determine how sensitive the performance measures are to changes in parameter values and in inter-arrival time correlation.

Keywords: Periodic Markov arrival process · Heterogeneous servers · Markov decision process · Time-dependent threshold policy

1 Introduction

For many queueing system that functions as stochastic models for commerce call centers, modern banking, traffic, computer and telecommunication systems, the

The publication has been prepared with the support of the "RUDN University Program 5-100" (recipients D. Efrosinin, mathematical model development). The reported study was funded by RFBR, project number 17-07-00142 (recipient D. Efrosinin, numerical implementation).

© Springer Nature Switzerland AG 2019
V. M. Vishnevskiy et al. (Eds.): DCCN 2019, LNCS 11965, pp. 179–194, 2019.
https://doi.org/10.1007/978-3-030-36614-8_14

assumptions about stationary arrival or service attributes can not be accepted as adequate any more. Time dependence is important due to possible seasonality or periodic behaviour of stochastic process which can be observed in many real world service systems. Time-dependent call volumes, time varying air traffic, non-stationary track arrival rates, periodic packet and massage volumes are only some examples where time dependence can have a sufficient contribution to a performance characteristics of the corresponding queueing system. Therefore, the time-dependent and periodic arrivals must be definitive taken into account by designing, developing and controlling the service systems. Another problem point by modelling the modern service system is a correlation of the inter-arrival times. Markov Arrival Processes (MAPs) have been used broadly in recent years to model the possible correlation of lengths of successive inter-arrival times. For details about description and implementation of such stochastic processes the reader can refer to number of very famous papers, e.g. [1,12,14].

The assumption of homogeneity of servers is one more typical restriction. The queues of the type $MAP/M/c$ with c heterogeneous servers are scarce in the scientific literature dedicated to queueing theory. The $MAP/M/2/B$ queueing system with finite buffer, two heterogeneous servers and Random Server Selection allocation policy was analysed in [3]. Performance analysis for queueing systems with several heterogeneous servers and Markov modulated arrivals has been provided in [2], where the dependency of threshold policy on the arrival phases was demonstrated. The dynamic optimization problems for the queues of the type $MAP/PH/K$ were studied in [7], where multi-threshold and monotonicity properties of the optimal allocation policy were numerically confirmed. A number of papers study queueing systems in a random environment. Some special cases and literature overview for the multi-server queues with MAP can be found in [5]. [8] studies the influence of a correlation in a stationary queue with correlated arrivals, heterogeneous servers and random environment.

The queues with time-dependent parameters were analysed in different variations. [20] provides a general overview of the main results obtained for time-dependent queueing systems where mostly the non-stationary Poisson arrival processes were implemented. We have found the only one paper [4], which examines a single-server queue with time-varying MAP. In contrast to the previous results, the combination of time-dependent $MAP(t)$ and heterogeneous servers in the context of correlation effect on the optimal control and performance measures in a finite planning horizon is relatively new problematic. In this paper we model and solve the problem of an optimal control in a queueing system $MAP(t)/M/2$ with a periodic phase transition rates. We examine the effect of periodic correlation on optimal threshold levels with piecewise constant periodic structure and will see how fast they converge to a periodic pattern.

The rest of the paper is organized as follows. In Sect. 2, we describe the mathematical model. In Sect. 3, we describe the control problem in details and provide an MDP model. In Sect. 4 we obtain a transient solution for the model with fixed threshold levels which can be treated as a quasi-birth-and-death process. Numerical examples and conclusions are in Sects. 5.

For use in sequel, let $\mathbf{e}(r)$, $\mathbf{e}_j(r)$, and $I(r)$ denote, respectively, the (column) vector of dimension r consisting of 1's, column vector of dimension r with 1 in the jth position and 0 elsewhere, and an identity matrix of dimension r. When there is no need to emphasize the dimension of these vectors we will suppress the suffix. Thus, $\mathbf{e}$ and $\mathbf{e}_j$ will denote a column vectors of appropriate dimension. The notation $\mathbf{e}'(r)$ appearing in a vector will stand for the vector transpose. The notation $\otimes$ stand for the Kronecker product. For more details on Kronecker product, we refer the reader to [13]. At last, the notation $1_{\{A\}}$ specifies the indicator function, which takes the value 1 if the event A holds, and 0 otherwise.

2   The $MAP(t)/M/2/B$ Queueing Model

In this paper we will focus on the continuous-time finite puffer $MAP(t)/M/2/B$ queueing system with heterogeneous servers, where the components of the Markov Arrival Process could be time-varying, or in particular, periodic. The n-phase $MAP(t)$ process will be denoted by two matrices $D_0(t)$ and $D_1(t)$, where $D(t) = D_0(t) + D_1(t)$ is the generator of the continuous-time Markov chain $\{N(t)\}_{t \geq 0}$ associated with an arrival process on the state space $E_N = \{1, \ldots, n\}$, i.e. $D(t)\mathbf{e}(n) = \mathbf{0}$. Here $D_0(t) = [\xi_{ij}(t)]_{1 \leq i,j \leq n}$ contains transitions without arrival and $D_1(t) = [\nu_{ij}(t)]_{1 \leq i,j \leq n}$ includes transition rates accompanying with a new arrival. Note that $\xi_{ii}(t) = -\xi_i(t) = -(\sum_j \nu_{ij}(t) + \sum_{j \neq i} \xi_{ij}(t))$. The average arrival rate

$$\lambda(t) = \boldsymbol{\alpha}(t)D_1(t)\mathbf{e}(n), \tag{1}$$

where $\boldsymbol{\alpha}(t)$ is the solution of the linear system $\boldsymbol{\alpha}(t)D(t) = \mathbf{0}$, $\boldsymbol{\alpha}(t)\mathbf{e}(n) = 1$. The lag-1 correlation coefficient r of the inter-arrival times is computed by

$$r(t) = \frac{\lambda(t)\boldsymbol{\alpha}(t)(-D_0(t))^{-1}D_1(t)(-D_0(t))^{-1}\mathbf{e}(n) - 1}{c_v^2(t)}, \tag{2}$$

where $c_v(t)$ is a coefficient of variation given by

$$c_v^2(t) = 2\lambda(t)\boldsymbol{\alpha}(t)(-D_0(t))^{-1}\mathbf{e}(n) - 1. \tag{3}$$

The customers are served by two heterogeneous servers with rates μ_1 and μ_2, where $\mu_1 > \mu_2$. The cost structure consists of the waiting cost c_0 for each unit time a customer spends in a queue and the usage costs c_1 and c_2 for each unit of time a customer stays at server 1 and 2 respectively.

As it was shown in [6], the optimal stationary allocation policy between $K \geq 2$ heterogeneous servers in $MAP/M/K$ queueing system which minimizes the long-run average cost per unit of time is of threshold type. According to this policy the fastest server must be used whenever it is free and there are waiting customers in the queue. The kth service speed server must be activated in a state x when $k - 1$ faster server are busy and the number of customers in a queue $q(x)$ exceeds some specified threshold level $q_{n(x)}$ for each modulating arrival phase $n(x)$. In this paper we discuss that the periodic variation of the attributes of the arrival process makes the optimal threshold-based control policy for the assigning of customers between servers time-dependent and periodic. Moreover, as it will be shown, the optimal policy is piecewise constant.

As it was discussed above, according to an optimal policy it is always optimal to assign a customer to the fast server. In our case it is a server 1. The second server is used according to a threshold policy, where threshold depends on a state of the $MAP(t)$. In this context, we denote by $Q(t) \in E_Q = \mathbb{N}_0 \cap [0, B]$ the number of customers in the queue plus one customer at server 1, $S(t) \in E_S = \{0, 1\}$ denotes the number of customers at server 2. The state of the system is described by a continuous-time Markov chain

$$\{X(t)\}_{t \geq 0} = \{Q(t), S(t), N(t)\}_{t \geq 0} \tag{4}$$

on the state space $E = E_Q \times E_S \times E_N$. Hence, a state $x \in E$ is a triplet of the form, $x = (q(x), s(x), n(x))$, where $q(x)$ denotes the number of customers in the queue and server 1, $s(x)$ is the status of server 2 in state $x \in E$, and $n(x)$ is a state of the arrival process.

We formulate the optimal allocation problem as a Markov decision problem with a finite planning horizon and use dynamic programming approach to calculate optimal threshold levels as well as performance measures under optimal control policy for correlated inter-arrival times.

3   Markov Decision Process

The model considered is a continuous-time Markov decision process (MDP). The elements of MDP will be explained in this section in detail as well as algorithm for solving optimization problem. We would like to obtain the optimal allocation policy for an interval $(0, T_H]$, where $T_H < \infty$ is a finite planning horizon. The controllable finite-horizon model associated with a Markov chain (4) is a five-tuple

$$\{E, A, \{A(x), x \in E\}, \lambda_{xy}(t; a(t)), c(x)\}, \tag{5}$$

where

- E is a *state space*,
- $A = \{1, 2\}$ is an *action space* with elements $a(t) \in A$, where $a(t) = 1$ means "to send a customer arrived to the queue" and $a(t) = 2$ - "to send a customer to server 2" at time t.
- $A(x) \subseteq A$ denotes *subset of actions* defined only for state x where the action is occurred at a decision epochs including the moments just after an arrival and just after a service completion at server 2, i.e. $A(q, 0, n) = A, q < B, A(B, 0, n) = \{2\}$, $A(q, 1, n) = \{1\}, q < B$.
- The system's dynamics is described by a probability low depending on an infinitesimal generator $\Lambda^f(t) = [\lambda_{xy}(t; a(t))]_{x,y \in E}$. $\lambda_{xy}(t; a(t))$ is a time-dependent *transition rate* to go from x to state y under a control action $a(t)$ at time t associated with a state occurred just after an arrival in state x or just after a service completion in state x. The model is conservative, i.e.

$$\lambda_{xy}(t; a(t)) \geq 0, \ y \neq x, \ \lambda_{xx}(t; a(t)) = -\lambda_x(t; a(t)) = -\sum_{y \neq x} \lambda_{xy}(t; a(t)),$$

$$\lambda_x(t; a(t)) < \infty,$$

where for $y \neq x$ and $\mathbf{e}_j := \mathbf{e}_j(3)$ we have

$$\lambda_{xy}(t; a(t)) = \begin{cases} \nu_{n(x)n(y)}(t), & y = (n(y) - n(x))\mathbf{e}_3 + \mathbf{e}_1, \ q(x) = 0, \\ \nu_{n(x)n(y)}(t), & y = (n(y) - n(x))\mathbf{e}_3 + \mathbf{e}_{a(t)}, \ q(x) > 0 \\ & a(t) \in A((n(y) - n(x))\mathbf{e}_3), \\ \xi_{n(x)n(y)}(t), & y = x + (n(y) - n(x))\mathbf{e}_3, \ n(y) \in E_N \setminus \{n(x)\}, \\ \mu_1, & y = x - \mathbf{e}_1, \ q(x) > 0, \\ \mu_2, & y = x - \mathbf{e}_2, \ s(x) = 1, \ q(x) = 0, \\ \mu_2, & y = x - \mathbf{e}_2 - \mathbf{e}_1 + \mathbf{e}_a, \ s(x) = 1, \ q(x) > 0, \\ & a(t) \in A(x - \mathbf{e}_2 - \mathbf{e}_1). \end{cases}$$

– In each state x we receive pro unit of time an immediate cost

$$c(x) = [c_0(q(x) - 1) + c_1]1_{\{q(x)>0)\}} + c_2 1_{\{s(x)=1\}}$$

which depends only on the current state but not on the action taken. In the paper we fix $c_i = 1, i = 0, 1, 2$, i.e. the function $c(x)$ represents the number of customers in the system in state x.

A controller or decision maker chooses an action according to the following decision rule which will refer to as time-dependent control policy.

Definition 1. *A time dependent control policy is a function $f : E \times (0, T_H] \to A(x)$ which prescribes a selection at time t of a control action $f(x, t) = a(t) \in A(x)$ whenever the process $\{X(t)\}_{t \geq 0}$ is in state $x \in E$ at time t just after an arrival and just after a service completion at server 2 if $q(x) > 0$ and $s(x) = 1$.*

In other words, the usage of a policy f means that if the system is in state x at time t in case of a new arrival the action chosen is $f(x, t)$, in case of service completion at server 2 in state x with $q(x) > 1$ the action is $f(x - \mathbf{e}_2 - \mathbf{e}_1, t)$.

Theorem 1. *The total average cost $V(x, t)$ up to time t given initial state is $x \in E$ is a solution of the Bellman equation*

$$\frac{\partial V(x, t)}{\partial t} = c(x) - (\xi_{n(x)}(t) + \mu_1 + \mu_2)V(x, t) \tag{6}$$

$$+ \sum_{i \in E_N} \nu_{n(x)i}(t)[V((i - n(x))\mathbf{e}_3 + \mathbf{e}_1, t)1_{\{q(x)=0\}}$$

$$+ TV(x + (i - n(x))\mathbf{e}_3, t)1_{\{q(x)>0\}}]$$

$$+ \sum_{i \in E_N \setminus \{n(x)\}} \xi_{n(x)i}(t)V(x + (i - n(x))\mathbf{e}_3, t)$$

$$+ \mu_1 V(x - \mathbf{e}_1, t)1_{\{q(x)>0\}} + \mu_2 V(x - \mathbf{e}_2, t)1_{\{s(x)=1,q(x)=0\}}$$

$$+ \mu_2 TV(x - \mathbf{e}_2 - \mathbf{e}_1, t)1_{\{s(x)=1,q(x)>0\}},$$

with $V(x,0) = 0, x \in E$, where operator T stands for decision making at a decision epoch and is of the form

$$TV(x,t) = \min\{V(x + \mathbf{e}_1, t), V(x + \mathbf{e}_2, t)\}, \ s(x) = 0.$$

Proof. The minimal total average cost up to time time t given initial state is $x \in E$ is defined as

$$V(x,t) = \inf_{f \in \Pi} \mathbb{E}^f \left[\int_0^t \Big([c_0(Q(t) - 1) + c_1] 1_{\{Q(t) > 0\}} + S_1(t) \Big) dt | X(0) = x \right],$$

where Π is the set of all admissible policies $f \in \Pi = A(0,0,1) \times A(0,0,2) \times \cdots \times A(B,1,1) \times A(B,1,2)$. According to a Markov decision theory [18], necessary and sufficient condition for a measurable policy f to be optimal is that for some small interval of length Δt the following equation holds,

$$V(x, t + \Delta t) = c(x)\Delta t + (1 - (\xi_{n(x)}(t) + \mu_1 + \mu_2)\Delta t)V(x,t) \qquad (7)$$
$$+ \sum_{i \in E_N} \nu_{n(x)i}(t)\Delta t [V((i - n(x))\mathbf{e}_3 + \mathbf{e}_1, t)1_{\{q(x)=0\}}$$
$$+ \min\{V(x + (i - n(x))\mathbf{e}_3 + \mathbf{e}_1, t), V(x + (i - n(x))\mathbf{e}_3 + \mathbf{e}_2, t)\}1_{\{q(x)>0\}}]$$
$$+ \sum_{i \in E_N \setminus \{n(x)\}} \xi_{n(x)i}(t)\Delta t V(x + (i - n(x))\mathbf{e}_3, t)$$
$$+ \mu_1 \Delta t V(x - \mathbf{e}_1, t)1_{\{q(x)>0\}} + \mu_2 \Delta t V(x - \mathbf{e}_2, t)1_{\{s(x)=1, q(x)=0\}}$$
$$+ \mu_2 \Delta t \min\{V(x - \mathbf{e}_2, t), V(x - \mathbf{e}_1, t)\}1_{\{s(x)=1, q(x)>0\}}.$$

In this equation the first term in the right hand side represents the costs during a time interval of duration Δt, the second term represents the total processing cost of all customers being in the system during the subsequent time interval of duration t in case that there are no arrivals, no departures and no arrival phase changing, and the remaining terms represent respectively the total processing cost in case of an arrival, in case of arrival phase changing without arrival, in case of service completion at server 1 and 2. After some elementary manipulation, and passing to the limit when $\Delta t \to 0$, we get a Bellman equation in form (6) with initial condition $V(x, 0) = 0, x \in E$ for all $t \in (0; T_H]$.

Remark 1. The system of differential equations of the type

$$\frac{d}{dt}\mathbf{V}(t) = \min_{f \in \Pi}(\mathbf{c} + \Lambda^f(t)\mathbf{V}(t)), \ \mathbf{V}(0) = 0, \qquad (8)$$

where $\mathbf{V}(t)$ and $\mathbf{c}$ the corresponding column vectors with components $V(x,t)$ and $c(x)$ respectively, can be solve by different methods, e.g. using a Runge-Kutta algorithm [19]. In this paper we use the LSODA approach which is a version of LSODE (the Livermore Solver for Ordinary Differential Equations) and is a switching between a nonstiff Adama (multistep Adams-Moulton) and a stiff Gear BDF (Backward Difference Formula) method [17].

Remark 2. We note the the optimal control policy f is completely defined by a time-dependent threshold levels $(q_1(t), q_2(t) \dots, q_n(t))$. This fact can be proved rigorously by proving the properties of superconvexity and supermodularity [6, 16] for the increments $\frac{\partial}{\partial t} V(x, t)$. The proof is omitted to conserve space.

Moreover, it is possible to prove that the optimal control policy f and, as a consequence, the corresponding threshold levels $q_i(t), i = 1, 2, \dots, n$, defining the policy are piecewise constant in the following sense.

Definition 2. *A function $q_i(t), 0 \leq t \leq T_H$, is a piecewise constant if for any finite $t^*, t^* \leq T_H$, the interval $[0, t^*]$ can be divided into a finite number of intervals $(0, t_1), (t_1, t_2), \dots, (t_{m-1})$ such that $q_j(t)$ is constant on $(t_j, t_{j+1}), 0 \leq j \leq m - 1$.*

The ambiguity at the endpoints, which we refer to as switching points, is solved by assuming that $q_i(t)$ is continuous on the left.

Theorem 2. *There are piecewise constant (from the left) threshold levels $q_i(t)$, $i = 1, 2, \dots, n$, defined on $[0, T_H]$ which minimizes Eq. (6) everywhere. These thresholds are optimal.*

Proof. As it was proved in [11], for the continuous-time Markov decision process there is a piecewise constant policy f which minimizes Eq. (8) everywhere and this policy exists. Due to Remark 2 the optimal policy f is defined through a an optimal threshold level specified for each arrival phase and hence these thresholds must be also piecewise constant.

4   Transient Solution and Performance Metrics

Consider a special case of a system $MAP(t)/M/2/B$ with two arrival phases, i.e. $n = 2$. As it was discussed in a previous section, the optimal policy f belongs to a class of a piecewise constant threshold polices defined by two threshold levels $(q_1(t), q_2(t))$ defined for each state of an arrival process. Under the assumption that average arrival rate from phase 1 is larger as from phase 2, i.e. $\lambda_1(t) \geq \lambda_2(t)$, where $(\lambda_1(t), \lambda_2(t)) = \boldsymbol{\alpha}(t) D_1(t)$, we may expect that the incentive to make an assignment of a customer to server 2 will be larger in state $(q, 0, 1)$ as in state $(q, 0, 2)$ and hence we can set $q_1(t) \leq q_2(t)$. Under this assumption we partition the state space E according to level $\mathbf{q}$ as follows, $E = \bigcup_{q=0}^{B} \mathbf{q}$, where

$$\mathbf{q} = \{(q, 0, 1), (q, 0, 2), (q, 1, 1), (q, 1, 2)\}, \ 0 \leq q \leq q_2(t), \tag{9}$$
$$\mathbf{q} = \{(q, 1, 1), (q, 1, 2)\}, \ q_2(t) + 1 \leq q \leq B.$$

Theorem 3. *The Markov chain $\{X(t)\}_{t\geq 0}$ belongs to a class of quasi-birth-and-death (QBD) processes with three-diagonal threshold dependent block infinitesimal generator $\Lambda(t; q_1(t), q_2(t))$ defined as*

$$\Lambda(t; q_1(t), q_2(t)) = [\lambda_{xy}(t; q_1(t), q_2(t))]_{x,y \in E} = \tag{10}$$

$$\left(\begin{array}{cccccccccccc}
B_{00}(t) & B_{01}(t) & 0 & 0 & 0 & 0 & 0 & 0 & 0 & \cdots & 0 \\
B_{10} & B_{11}(t) & B_{01}(t) & 0 & 0 & 0 & 0 & 0 & 0 & \cdots & 0 \\
0 & B_{10} & B_{11}(t) & B_{01}(t) & 0 & 0 & 0 & 0 & 0 & \cdots & 0 \\
\ddots & \ddots & \ddots & \ddots & & & & & & & \\
0 & \cdots & B_{10} & B_{22}(t) & B_{02}(t) & 0 & 0 & 0 & 0 & \cdots & 0 \\
0 & \cdots & 0 & B_{20} & B_{33}(t) & B_{02}(t) & 0 & 0 & 0 & \cdots & 0 \\
\ddots & \ddots & \ddots & \ddots & & & & & & \ddots & \\
0 & \cdots & 0 & 0 & B_{20} & B_{44}(t) & B_{03}(t) & 0 & 0 & \cdots & 0 \\
0 & \cdots & 0 & 0 & 0 & B_{30} & B_{55}(t) & B_{04}(t) & 0 & \cdots & 0 \\
0 & \cdots & 0 & 0 & 0 & 0 & B_{40} & B_{55}(t) & B_{04}(t) & \cdots & 0 \\
\ddots & \ddots & \ddots & \ddots & & & & & & \ddots & \\
0 & \cdots & 0 & 0 & 0 & 0 & 0 & B_{40} & B_{55}(t) & B_{04}(t) \\
0 & \cdots & 0 & 0 & 0 & 0 & 0 & 0 & B_{40} & B_{66}(t)
\end{array}\right)
\left.\begin{array}{l} \\ \\ \\ \end{array}\right\} q_1(t)+1
\left.\begin{array}{l} \\ \\ \\ \\ \end{array}\right\} q_2(t)-q_1(t)+1,
\left.\begin{array}{l} \\ \\ \\ \end{array}\right\} B-q_2-1$$

where the coefficient matrices appearing in (10) on the main diagonal and including transitions within a certain level $\mathbf{q}$ are given by

$$B_{00}(t) = \begin{pmatrix} D_0(t) & 0 \\ \mu_2 I_2 & D_0(t)-\mu_2 I_2 \end{pmatrix}, \; B_{11}(t) = \begin{pmatrix} D_0(t)-\mu_1 I_2 & 0 \\ \mu_2 I_2 & D_0(t)-(\mu_1+\mu_2)I_2 \end{pmatrix},$$

$$B_{22}(t) = \begin{pmatrix} D_0(t)-\mu_1 I_2 & D_1(t)(\mathbf{e}_1(2)\otimes \mathbf{e}_1'(2)) \\ \mu_2 I_2 & D_0(t)-(\mu_1+\mu_2)I_2 \end{pmatrix}, \; B_{33}(t) = \begin{pmatrix} D_0(t)-\mu_1 I_2 & D_1(t)(\mathbf{e}_1(2)\otimes \mathbf{e}_1'(2)) \\ \mu_2(\mathbf{e}_2(2)\otimes \mathbf{e}_2'(2)) & D_0(t)-(\mu_1+\mu_2)I_2 \end{pmatrix},$$

$$B_{44}(t) = \begin{pmatrix} D_0(t)-\mu_1 I_2 & D_1(t) \\ \mu_2(\mathbf{e}_2(2)\otimes \mathbf{e}_2'(2)) & D_0(t)-(\mu_1+\mu_2)I_2 \end{pmatrix}, \; B_{55}(t) = D_0(t)-(\mu_1+\mu_2)I_2,$$

$$B_{55}(t) = \begin{pmatrix} -\xi_{12}(t) & \xi_{12}(t) \\ \xi_{21}(t) & -\xi_{21}(t) \end{pmatrix} - (\mu_1+\mu_2)I_2.$$

The lower diagonal has coefficient matrices with transitions from upper level $\mathbf{q}+1$ to $\mathbf{q}$,

$$B_{10} = \mu_1 I_4, \; B_{20} = \mu_1 I_4 + \mu_2(\mathbf{e}_3(4)\otimes \mathbf{e}_3'(4)),$$
$$B_{30} = (\mu_1+\mu_2)(\mathbf{e}_1(2)\otimes \mathbf{e}_3'(4)+\mathbf{e}_2(2)\otimes \mathbf{e}_4'(4)), \; B_{40} = (\mu_1+\mu_2)I_2.$$

The upper diagonal has coefficient matrices with transition from lower level $\mathbf{q}-1$ to $\mathbf{q}$,

$$B_{01}(t) = I_2 \otimes D_1(t), \; B_{02}(t) = \begin{pmatrix} D_1(t)(\mathbf{e}_2(2)\otimes \mathbf{e}_2'(2)) & 0 \\ 0 & D_1(t) \end{pmatrix},$$
$$B_{03}(t) = \mathbf{e}_2(2) \otimes D_1(t), \; B_{04}(t) = D_1(t).$$

Proof. Let $\boldsymbol{\pi}(t)$, partitioned as $\boldsymbol{\pi}(t) = (\boldsymbol{\pi}_0(t), \boldsymbol{\pi}_1(t), \ldots, \boldsymbol{\pi}_B(t))$, denote the stationary probability vector of the generator $\Lambda(t; q_1(t), q_2(t))$. Hence, $\boldsymbol{\pi}(t)$ satisfies

$$\frac{d}{dt}\boldsymbol{\pi}(t) = \boldsymbol{\pi}(t)\Lambda(t; q_1(t), q_2(t)), \ \boldsymbol{\pi}(t)\mathbf{e}(|E|) = 1, \ \boldsymbol{\pi}(0) = \mathbf{e}'_1(|E|), \ t \in (0, T_H]. \tag{11}$$

For a fixed threshold levels $(q_1(t), q_2(t))$ the system of Kolmogorov differential equations is of the form:

$$\frac{d}{dt}\pi_{(0,0,n)}(t) = \xi_{nn}(t)\pi_{(0,0,n)}(t) + \xi_{3-nn}(t)\pi_{(0,0,3-n)}(t) + \mu_1\pi_{(1,0,n)}(t) + \mu_2\pi_{(0,1,n)}(t),$$

$$\frac{d}{dt}\pi_{(0,1,n)}(t) = (\xi_{nn}(t) - \mu_2)\pi_{(0,1,n)}(t) + \xi_{3-nn}(t)\pi_{(0,1,3-n)}(t) + \mu_1\pi_{(1,1,n)}(t),$$

$$\frac{d}{dt}\pi_{(k,0,n)}(t) = (\xi_{nn}(t) - \mu_1)\pi_{(k,0,n)}(t) + \xi_{3-nn}(t)\pi_{(k,0,3-n)}(t) + \nu_{nn}(t)\pi_{(k-1,0,n)}(t)$$
$$+ \nu_{3-nn}(t)\pi_{(k-1,0,3-n)}(t) + \mu_1\pi_{(k+1,0,n)}(t) + \mu_2\pi_{(k,1,n)}(t), \ 1 \le k \le q_1(t),$$

$$\frac{d}{dt}\pi_{(k,1,n)}(t) = (\xi_{nn}(t) - (\mu_1 + \mu_2))\pi_{(k,1,n)}(t) + \xi_{3-nn}(t)\pi_{(k,1,3-n)}(t) + \nu_{nn}(t)\pi_{(k-1,1,n)}(t)$$
$$+ \nu_{3-nn}(t)\pi_{(k-1,1,3-n)}(t) + \mu_1\pi_{(k+1,1,n)}(t) + \mu_2\pi_{(k+1,1,n)}(t)1_{\{k=q_1 \wedge n=2\}}, \ 1 \le k \le q_1(t),$$

$$\frac{d}{dt}\pi_{(k,0,n)}(t) = (\xi_{nn}(t) - \mu_1)\pi_{(k,0,n)}(t) + \xi_{3-nn}(t)\pi_{(k,0,3-n)}(t) + \mu_1\pi_{(k+1,0,n)}(t)1_{\{k<q_2(t)\}}$$
$$+ (\nu_{nn}(t)\pi_{(k-1,0,n)}(t) + \nu_{3-nn}(t)\pi_{(k-1,0,3-n)}(t) + \mu_2\pi_{(k,1,n)}(t))1_{\{n=2\}}, \ q_1(t) + 1 \le k \le q_2(t),$$

$$\frac{d}{dt}\pi_{(k,1,n)}(t) = (\xi_{nn}(t) - (\mu_1 + \mu_2))\pi_{(k,1,n)}(t) + \xi_{3-nn}(t)\pi_{(k,1,3-n)}(t) + \nu_{nn}(t)\pi_{(k-1,1,n)}(t)$$
$$+ \nu_{3-nn}(t)\pi_{(k-1,1,3-n)}(t) + \mu_1\pi_{(k+1,1,n)}(t) + (\nu_{nn}(t)\pi_{(k,0,n)}(t) + \nu_{3-nn}(t)\pi_{(k,0,3-n)}(t)$$
$$+ \mu_2\pi_{(k+1,1,n)}(t))1_{\{n=1 \wedge k=q_2(t)\}}, \ q_1(t) + 1 \le k \le q_2(t),$$

$$\frac{d}{dt}\pi_{(k,1,n)}(t) = (\xi_{nn}(t) - (\mu_1 + \mu_2))\pi_{(k,1,n)}(t) + \xi_{3-nn}(t)\pi_{(k,1,3-n)}(t) + \nu_{nn}(t)\pi_{(k-1,1,n)}(t)$$
$$+ \nu_{3-nn}(t)\pi_{(k-1,1,3-n)}(t) + (\mu_1 + \mu_2)\pi_{(k+1,1,n)}(t), \ q_2(t) + 1 \le k \le B - 1,$$

$$\frac{d}{dt}\pi_{(B,1,n)}(t) = -(\xi_{12}(t) + \mu_1 + \mu_2)\pi_{(B,1,n)}(t) + \xi_{3-nn}(t)\pi_{(B,1,3-n)}(t) + \nu_{nn}(t)\pi_{(B-1,1,n)}(t)$$
$$+ \nu_{3-nn}(t)\pi_{(B-1,1,3-n)}(t). \tag{12}$$

By expressing (12) in matrix form according to partition (9) we obtain infinitesimal generator $\Lambda(t; q_1(t), q_2(t))$ in form (10).

For solving (12) we can use different methods, e.g. the fourth order Runge-Kutta method. Once values for $\boldsymbol{\pi}(t)$ are calculated, we can compute a variety of performance measures such as mean number of customers in the system $\bar{L}(t)$ and in the queue $\bar{Q}(t)$, blocking P_B and empty P_0 probabilities.

Remark 3. Due to correlated inter-arrival times the PASTA (Poisson arrivals see time averages) property does not hold, since the queue length and the modulating phase (and hence also the arrival rate) are correlated. This correlation implies that the distribution seen by an arriving customer is different from $\boldsymbol{\pi}(t)$. Therefore instead of $\boldsymbol{\pi}$ we use the distribution just prior to an arrival which can be expressed as

$$\hat{\boldsymbol{\pi}}_q(t) = \begin{cases} \eta^{-1}\boldsymbol{\pi}_q(t)(I_2 \otimes D_1(t)) & 0 \le q \le q_2(t), \\ \eta^{-1}\boldsymbol{\pi}_q(t)D_1(t) & q_2(t) + 1 \le q \le B \end{cases}, \tag{13}$$

where $\eta^{-1} = \sum_{q=0}^{q_2(t)} \boldsymbol{\pi}_q(t)(I_2 \otimes D_1(t))\mathbf{e}(4) + \sum_{q=q_2(t)+1}^{B} \boldsymbol{\pi}_q(t)D_1(t)\mathbf{e}(2)$.

Corollary 1. *The performance measures can be obtained as follows*

$$\bar{L}(t) = \sum_{q=0}^{q_2(t)} \hat{\boldsymbol{\pi}}_q(t)[q(\mathbf{e}_1(4) + \mathbf{e}_2(4)) + (q+1)(\mathbf{e}_3(4) + \mathbf{e}_4(4))]$$

$$+ \sum_{q=q_2(t)+1}^{B} (q+1)\hat{\boldsymbol{\pi}}_q(t)\mathbf{e}(2) \tag{14}$$

$$\bar{Q}(t) = \sum_{q=1}^{q_2(t)} (q-1)\hat{\boldsymbol{\pi}}_q(t)\mathbf{e}(4) + \sum_{q=q_2(t)+1}^{B} (q-1)\hat{\boldsymbol{\pi}}_q(t)\mathbf{e}(2), \tag{15}$$

$$P_B = \hat{\boldsymbol{\pi}}_B \mathbf{e}(2), \ P_0 = \hat{\boldsymbol{\pi}}_0(\mathbf{e}_1(4) + \mathbf{e}_2(4)). \tag{16}$$

The probabilities $\boldsymbol{\pi}(t)$ can be used also to derive the waiting and sojourn time distributions of customers that will be our next research aim.

Remark 4. The infinite buffer queueing system can bes approximated by a finite buffer equivalent system. For bounded puffer size B the size of the set space

$$|E| = 4n(q_n + 1) + 2n(B - q_n).$$

The buffer size B will be chosen in such a way that it satisfies the condition

$$\boldsymbol{\pi}_{q_n+1}(t^*)(I - R^{B-q_2})(I - R)\mathbf{e}(|E|) < \varepsilon,$$

obtained from stationary state distribution for $t^* = \text{argmax}_t\{\lambda(t), r(t)\}$ and for given threshold-based policy with threshold levels $(q_1, \ldots, q_n)$. The computation of the stationary state distribution can be performed according to algorithm for the quasi-birth-and-death (QBD) processes [15]. R is a minimal non-negative solution to the matrix equation $R^2 B_{40} + RB_{55}(t^*) + B_{04}(t^*) = 0$ for the homogeneous part of the QBD-process.

5   Numerical Results

In this section we provide a series of test cases to assess the influence of the sinusoidal $MAP(t)$ arrival process with $E_N = \{1, 2\}$ on optimal control policy and the system performance. The cases include problems with constant average arrival rate, variation coefficient and periodic correlation function which exhibits different amplitudes, periodic average arrival rate with almost constant and periodic correlation functions. In all cases we keep some parameters constant, so we fix $c_i = 1, i = 0, 1, 2, \mu_1 = 20, \mu_2 = 1, \nu_{12}(t) = 0.9, \xi_{12}(t) = 0.8, \xi_{21}(t) = 0.1$.

Case 1: The constant average arrival rate, the variation coefficient and periodic correlation function,

$$\lambda(t) = 5, \ r(t) = 0.18(0.95 + 0.9\sin(0.0173t - 2)), \ c_v^2(t) = 4.$$

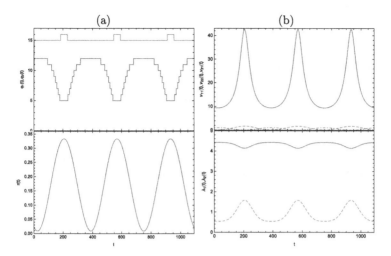

Fig. 1. $(q_1(t), q_2(t))$ versus $r(t)$ (a) $\nu_{11}(t)$, $\nu_{22}(t)$ (*dashed*), $\nu_{21}(t)$ (*dot dashed*) versus $\lambda_1(t)$, $\lambda_2(t)$ (*dashed*) (b)

The parameters $\nu_{11}(t), \nu_{22}(t)$ and $\nu_{21}(t)$ for each $t \in [0, T_H]$ were calculated from the system (1)–(3). Figure 1(a) illustrates the influence of the sinusoidal inter-arrival correlation $r(t)$ on the optimal threshold policy $(q_1(t), q_2(t))$ and Fig. 1(b) shows the effect of the periodic rates , $\nu_{22}(t)$ and $\nu_{21}(t)$ on average arrival rate $\lambda_i(t)$ from phase i calculated by $\lambda_i(t) = \boldsymbol{\alpha}(t)D_1(t)\mathbf{e}_i(2), i \in E_N$. In this example $\lambda_1(t) > \lambda_2(t)$, i.e. the incentive to make an assignment to server 2 must be higher in states x with $n(x) = 1$ which is confirmed by the fact that $q_1(t) < q_2(t)$. First of all we observe that the policy match the periodic piecewise constant pattern after a very short time moving away from starting point of time which shows how fast the reaching of equilibrium happens. We also observe that the higher correlation is resulted in lower values of threshold level $q_1(t)$ and higher values of $q_2(t)$ which coincides with behaviour of corresponding average rates $\lambda_1(t)$ and $\lambda_2(t)$. This result seems to be very interesting, since for queues with uncorrelated inter-arrival times, with increasing average arrival rate threshold levels normally decrease. In Fig. 2, we display performance functions $\bar{L}(t)$, $\bar{Q}(t)$, $P_0(t)$ and $P_B(t)$. We see again a periodic structure of the functions. We notice also that the high correlation makes sufficient contribution into a system load, so the average number of customers in the system at the moments of time with maximal values of correlation can be ten times higher as for moments with uncorrelated arrivals. The empty and blocking probability takes respectively its minimum and maximum when the correlation function peaks.

Case 2: The same as in Case 1 but the periodic correlation function has a smaller amplitude,

$$\lambda(t) = 5, \ r(t) = 0.15(0.95 + 0.9\sin(0.0173t - 2)), \ c_v^2(t) = 4.$$

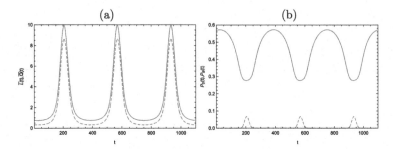

Fig. 2. $\bar{L}(t), \bar{Q}(t)$ (*dashed*) (a) and $P_0(t), P_B(t)$ (*dashed*) for optimal control policy $(q_1(t), q_2(t))$

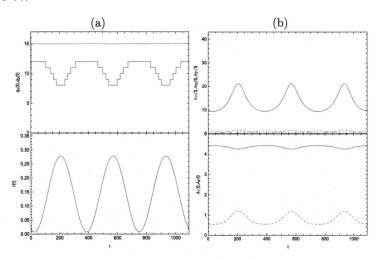

Fig. 3. $(q_1(t), q_2(t))$ versus $r(t)$ (a) $\nu_{11}(t), \nu_{22}(t)$ (*dashed*), $\nu_{21}(t)$ (*dot dashed*) versus $\lambda_1(t), \lambda_2(t)$ (*dashed*) (b)

Similarly to Case 1 in Fig. 3(a) we observe the periodic threshold level $q_1(t)$ while $q_2(t)$ is a constant. Here we can report that the amplitude of threshold levels decrease with decreasing amplitude of the correlation function, whereas we notice that as shown in 3(b) the average rates $\lambda_1(t)$ and $\lambda_2(t)$ take almost the same values as in previous case. Figure 4(a,b) illustrate the effect of a decreased correlation on the performance measures. We observe that even a small reduction of the inter-arrival time correlation is resulted in sufficient decreasing of the mean number of customers in the system and in the queue as well as in very small blocking probability.

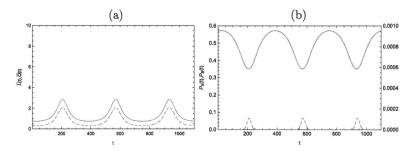

Fig. 4. $\bar{L}(t), \bar{Q}(t)$ (*dashed*) (a) and $P_0(t), P_B(t)$ (*dashed*) for optimal control policy $(q_1(t), q_2(t))$

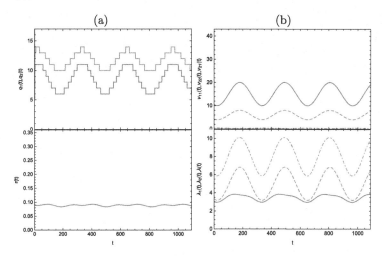

Fig. 5. $(q_1(t), q_2(t))$ versus $r(t)$ (a) $\nu_{11}(t), \nu_{22}(t)$ (*dashed*), $\nu_{21}(t)$ (*dot dashed*) versus $\lambda_1(t), \lambda_2(t)$ (*dashed*), $\lambda(t)$ (*dot dashed*) (b)

Case 3: Periodic average arrival rate and almost constant correlation function. Here we choose the following functions for the arrival phase transition rates,

$$\nu_{11}(t) = 10(1.5 + 0.5\sin(0.02t - 2)),$$
$$\nu_{22}(t) = 4(1.5 + 0.5\sin(0.02t - 2)),$$
$$\nu_{21}(t) = 0.2(1.5 + 0.5\sin(0.02t + 1)).$$

The functions $r(t)$ and $\lambda(t)$ calculated respectively by (1) and (2) are illustrated in Fig. 5. In this case the correlation function takes a quite small (<0.1) constant value and the main impact to the periodical pattern of the control policy is done by average arrival rate. As shown in Fig. 5, both $q_1(t)$ and $q_2(t)$ varies periodically with the same period as the average arrival rate. The maximum values of $\lambda(t)$ corresponds to minimum of $q_1(t)$ and $q_2(t)$ that is expected for the system with a high load. In Fig. 6, we see that the variation of the mean perfor-

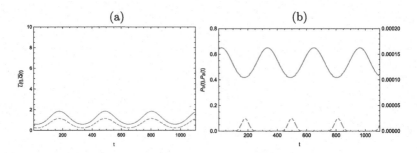

Fig. 6. $\bar{L}(t), \bar{Q}(t)$ (*dashed*) (a) and $P_0(t), P_B(t)$ (*dashed*) for optimal control policy $(q_1(t), q_2(t))$

mance measures is quite modest, smaller as 2, although the value of the average arrival rate fluctuates in interval $\lambda(t) \in (5.9, 10)$ which is sufficient higher as in Case 1, where $\lambda(t) = 5$. This example confirm again the statement how important to take into account the correlation by designing the queueing system and estimating its performance characteristics.

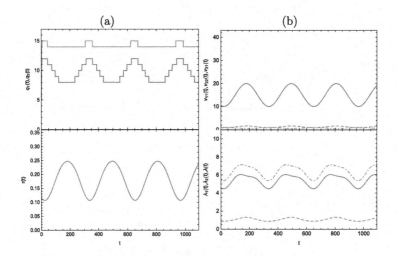

Fig. 7. $(q_1(t), q_2(t))$ versus $r(t)$ (a) $\nu_{11}(t)$, $\nu_{22}(t)$ (*dashed*), $\nu_{21}(t)$ (*dot dashed*) versus $\lambda_1(t)$, $\lambda_2(t)$ (*dashed*), $\lambda(t)$ (*dot dashed*) (b)

Case 4: Periodic average arrival rate and correlation function. The arrival rates take now the following values:

$$\nu_{11}(t) = 10(1.5 + 0.5 \sin(0.02t - 2)),$$
$$\nu_{22}(t) = 0.8(1.5 + 0.5 \sin[0.02t - 2)),$$
$$\nu_{21}(t) = 0.5(1.5 + 0.5 \sin(0.02t + 1)).$$

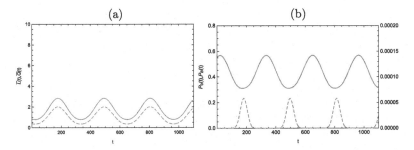

Fig. 8. $\bar{L}(t), \bar{Q}(t)$ (*dashed*) (a) and $P_0(t), P_B(t)$ (*dashed*) for optimal control policy $(q_1(t), q_2(t))$

The functions $r(t)$ and $\lambda(t)$ are plotted in Fig. 7. The effect of the periodic variations for correlation function and average arrival rate on the optimal threshold policy is illustrated in Fig. 7. We observe, that for time periods with a high correlation and average arrival rate, threshold levels decrease and for low correlation and average arrival rate – vice versa. In last Fig. 8, we display the mean performance measures and probabilities of an empty and blocked system. The higher variation of the correlation is resulted in increasing of the mean number of customers in the system and in the queue as well as the growth of the blocking probability although the values remain negligible.

Summarizing the aforementioned results we make the following comments. As it is well known, see for example [9,18], in time homogeneous irreducible and positive-recurrent MDPs with infinite planning horizon the time-dependent policy converges to an optimal stationary policy which exists. Therefore, for the long-run average cost criterion the controller chooses an action $f(x) = a$ whenever the controllable Markov process $\{X(t)\}_{t \geq 0}$ is in state x independently on time moment t. But in Markov processes with time-dependent transition rates the situation is different, since the optimal control action $f(x, t) = a(t)$ in some state x varies with a time. In periodic Markov chains there are periodic states with a property that the chain can return to these states only in number of transitions which is multiples of some integer p larger than 1. In our case it is obvious that the MDP is periodic one with repetitive control actions in the long-run. According to a discussion made in [21], which in turn is based on the convergence properties of the periodic Markov chain [10] and on the fact that there exists a periodic control for periodic continuous-time Markov chain with an optimal total average cost criterion, we may conclude that the control policy discussed here also exhibits periodicity property with the same period as a Markov chain.

References

1. Andersen, A.T., Nielsen, B.F.: A markovian approach for modeling packet traffic with long-range dependence. IEEE J. Sel. Areas Commun. **16**(5), 719–732 (1998)

2. Boel, R.K., Talat, N.K.: Performance analysis and optimal threshold policies for queueing systems with several heterogeneous servers and Markov modulated arrivals. Matrix-Anal. Methods Stochast. Models **183**, 117–136 (1997)
3. Chakravarthy, S.R., Chukova, S., Dimitrov, B.: Analysis of $MAP/M/2/K$ queueing model with infinite resequencing buffer. Perform. Eval. **31**, 211–228 (1998)
4. Dormuth, D.W., Alfa, A.S.: Two finite-difference methods for solving $MAP(t)/PH(t)/1/K$ queueing models. Queueing Syst. **27**, 55–78 (1997)
5. Dudina, O., Dudin, S.: Queueing system $MAP/M/N/N + K$ operating in random environment as a model of call center. In: Dudin, A., Klimenok, V., Tsarenkov, G., Dudin, S. (eds.) BWWQT 2013. CCIS, vol. 356, pp. 83–92. Springer, Heidelberg (2013). https://doi.org/10.1007/978-3-642-35980-4_10
6. Efrosinin, D.: Controlled queueing systems with heterogeneous servers. VDM Verlag, Saarbrücken (2004)
7. Efrosinin, D., Breuer, L.: Threshold policies for controlled retrial queues with heterogeneous servers. Ann. Oper. Res. **141**, 139–162 (2006)
8. Efrosinin, D., Krishnamoorthy, A., Vishnevskiy, V., Kozyrev, D.: The effect of a MAP flow on performance measures of controllable queueing system with heterogeneous servers in a random environment. In: 21st International Conference on Distributed Computer and Communication Networks (DCCN2018), pp. 415–433. RUDN University (2018)
9. Howard, R.A.: Dynamic Programming and Markov Processes. Wiley, Hoboken (1960)
10. Levin, D.A., Peres, Y., Wilmer, E.L.: Markov Chains and Mixing Times. American Mathematical Society, Providence (2006)
11. Linssen, J.: Continuous-time Markov decision processes. Bachelor thesis. University of Utrecht (2016)
12. Lucantoni, D.: New results in the single server queue with batch Markovian process. Commun. Stat. Stochast. Models **7**, 1–46 (1991)
13. Marcus, M., Minc, H.: A Survey of Matrix Theory and Matrix Inequalities. Allyn and Bacon, Boston (1964)
14. Neuts, M.: A versitile Markovian point process. J. Appl. Probab. **16**, 764–779 (1979)
15. Neuts, M.F.: Matrix-Geometric Solutions in Stochastic Models: An Algorithmic Approach. The John Hopkins University Press, Baltimore (1981)
16. Özkan, E., Kharoufeh, J.P.: Optimal control of a two-server queueing system with failure. Probab. Eng. Inform. Sci. **28**(04), 489–527 (2014)
17. Polyanin, A.D., Zaitsev, V.F.: Handbook of Ordinary Differential Equations: Exact Solutions, Methods, and Problems. CRC Press/Chapman and Hall, Boca Raton/London (2017)
18. Puterman, M.L.: Markov Decision Processes: Discrete Stochastic Dynamic Programming. Wiley, Hoboken (2005)
19. Schaumburg, K., Wasniewski, J., Zlatev, Z.: Solution of ordinary differential equations with time dependent coefficients. Development of a semiexplicit Runge-Kutta algorithm and application to a spectroscopic problem. Comput. Chem. **3**(2–4), 57–63 (1979)
20. Schwarz, J.A., Selinka, G., Stolletz, R.: Performance analysis of time-dependent queueing systems: survey and classification. Omega **63**, 170–189 (2016)
21. Tirdad, A., Grassmann, W.K., Tavakoli, J.: Optimal policies of $M(t)/M/c/c$ queues with two different levels of servers. Eur. J. Oper. Res. **249**, 1124–1130 (2016)

Estimation of the Parameters of Continuous-Time Finite Markov Chain

Alexander Andronov[1,2(✉)], Irina Jackiva (Yatskiv)[2], and Diana Santalova[3]

[1] Saint-Petersburg State University of Civil Aviation,
Pilotov 38, 196210 Saint-Petersburg, Russia
aleksander.andronov1@gmail.com
[2] Transport and Telecommunication Institute,
Lomonosova 1, Riga 1019, Latvia
jackiva.i@tsi.lv
[3] Møreforsking Molde AS, Britvegen 4, 6410 Molde, Norway
diana.santalova@himolde.no

Abstract. A simple approach to the estimation of finite continuous-time Markov chain's parameters is suggested. It is supposed that the trajectory of a considered Markov chain is fully observed, where each observation is about one jump of the chain. The observations are disjoint on k sub-samples, with respect to initial state of the chain. This allows to simplify the parameters estimation process.

Therefore for the case of small samples distribution of the estimates is expressed by Erlang and binomial distributions; however, for the big sample case the approximation by normal distribution is used. Asymptotical distribution of the estimates is investigated as well. The conducted experimental study shows good results.

Keywords: Markov chain · Parameters estimation · Density approximation · Multivariate normal distribution

1 Introduction

Finite continuous-time Markov chains have enormous applications in theoretical and applied stochastic investigations [5–7,10,11]. Statistical aspects of corresponding probabilistic models are investigated insufficiently [1–3,8,13,15]. In this paper we suggest a simple approach to the estimation of finite continuous-time Markov chain's parameters.

Parameters of Markov chain are given by matrix $\lambda = (\lambda_{i,j})$, $i, j = \{1, \ldots, k\}$, where k is the number of states. It is supposed that we have n observations, each of which is concerned with one jump of the chain. For each observation a ternary (i, j, t), where i and j are the initial and the following state of the chain, t is the elapsed time in the state i, is fixed. This case takes place, for example, when all trajectory of a considered Markov chain is fully observed.

These observations can be disjoint on k sub-samples, with respect to initial state of the chain. Therefore the problem reduces to estimation parameters $\lambda_{i,j}$

© Springer Nature Switzerland AG 2019
V. M. Vishnevskiy et al. (Eds.): DCCN 2019, LNCS 11965, pp. 195–206, 2019.
https://doi.org/10.1007/978-3-030-36614-8_15

for fixed i basing on samples (j_r, t_r), $r = 1, 2, \ldots, n_i$. Let $N_{i,j}$ be a number of an occurrence of the state j in sub-sample (j_r, t_r). Obviously

$$n_i = N_{i,1} + N_{i,2} + \cdots + N_{i,k}. \tag{1}$$

Let $\Lambda_i = \lambda_{i,1} + \lambda_{i,2} + \cdots + \lambda_{i,k}$, $\mu_{i,j} = \frac{1}{\lambda_{i,j}}$, $M_i = \frac{1}{\Lambda_i}$. Corresponding estimates will be noted by the asterisk: $\lambda_{i,j}^\star$, $\Lambda_i^\star$, $\mu_{i,j}^\star$, $M_i^\star$.

2 Density of Distribution for $\{\lambda_{i,j}^\star\}$

We have:

$$M_i^\star = \frac{1}{n_i} \sum_{r=1}^{n_i} t_r. \tag{2}$$

It gives us estimates of interest:

$$\Lambda_i^\star = \frac{1}{M_i^\star}. \quad \lambda_{i,j}^\star = \frac{N_{i,j}}{n_i} \Lambda_i^\star, \quad \mu_{i,j}^\star = \frac{1}{\lambda_{i,j}^\star}. \tag{3}$$

It is known that sum $\sum_{r=1}^{n_i} t_r$ has Erlang distribution with parameters Λ_i and n_i, having density:

$$\widetilde{f}_i(z) = \frac{1}{(n_i - 1)!} \Lambda_i (z\Lambda_i)^{n_i - 1} \exp(-z\Lambda_i), \quad z \geq 0. \tag{4}$$

Therefore estimate $M_i^\star$ has the following density:

$$f_i(z) = n_i \widetilde{f}_i(zn_i) = \frac{n_i}{(n_i - 1)!} \Lambda_i (zn_i\Lambda_i)^{n_i - 1} \exp(-zn_i\Lambda_i), \quad z \geq 0. \tag{5}$$

Now we can represent cumulative $P\{\Lambda_i^\star \leq z\}$ and density $g_i(z)$ functions for $\Lambda_i^\star$:

$$P\{\Lambda_i^\star \leq z\} = P\left\{\frac{1}{M_i^\star} \leq z\right\} = P\left\{M_i^\star \geq \frac{1}{z}\right\} = 1 - P\left\{M_i^\star \leq \frac{1}{z}\right\} = 1 - \int_0^{\frac{1}{z}} f_i(x) dx,$$

$$g_i(z) = z^{-2} f_i(z^{-1}) = z^{-2} \frac{n_i}{(n_i - 1)!} \Lambda_i \left(n_i\Lambda_i\frac{1}{z}\right)^{n_i - 1} \exp\left(-n_i\Lambda_i\frac{1}{z}\right), z \geq 0. \tag{6}$$

Moments of the order $r \leq n_i$ exist and are calculated by formula

$$E(\Lambda_i^{\star r}) = (n_i\Lambda_i)^r \prod_{j=1}^r \frac{1}{(n_i - j)!}. \tag{7}$$

In particular

$$E(\Lambda_i^\star) = \frac{n_i}{n_i - 1} \Lambda_i, \quad E\left((\Lambda_i^\star)^2\right) = \frac{n_i^2}{(n_i - 1)(n_i - 2)} \Lambda_i^2,$$

$$Var(\Lambda_i^\star) = \frac{n_i^2}{(n_i - 1)^2(n_i - 2)} \Lambda_i^2. \tag{8}$$

Random vector $(N_{i,1}, N_{i,2}, \ldots, N_{i,k})$ has polynomial distribution with parameters n_i and $\frac{\lambda_{i,1}}{\Lambda_i}, \frac{\lambda_{i,2}}{\Lambda_i}, \ldots, \frac{\lambda_{i,k}}{\Lambda_i}$:

$$P\Big\{N_{i,1} = n_{i,1}, \ldots, N_{i,k} = n_{i,k}\Big\} = \frac{n_i!}{n_{i,1}! \ldots n_{i,k}!} \Big(\frac{\lambda_{i,1}}{\Lambda_i}\Big)^{n_{i,1}} \cdots \Big(\frac{\lambda_{i,k}}{\Lambda_i}\Big)^{n_{i,k}} =$$

$$\frac{n_i!}{n_{i,1}! \ldots n_{i,k}!} \Big(\frac{M_1}{\mu_{i,1}}\Big)^{n_{i,1}} \cdots \Big(\frac{M_k}{\mu_{i,k}}\Big)^{n_{i,k}}, n_{i,1} + \cdots + n_{i,k} = n_i, i = 1, \ldots, k. \quad (9)$$

Here $n_{i,1}, \ldots, n_{i,k}$ are natural numbers from the set $\{0, 1, \ldots, n_i\}$, whose sum equals n_i. Let us denote $\Omega(n_i, k)$ a set of all such vectors $\widetilde{n}_i = (n_{i,1}, \ldots, n_{i,k})$. Note that partial variable $N_{i,j}$ has binomial distribution with parameters n_i and $\frac{\lambda_{i,j}}{\Lambda_i}$.

Therefore estimate $\lambda^\star_{i,j}$ has the following density:

$$h_{i,j}(z) = \sum_{\eta=1}^{n_i} \binom{n_i}{\eta} \Big(\frac{\lambda_{i,j}}{\Lambda_i}\Big)^{\eta} \Big(1 - \frac{\lambda_{i,j}}{\Lambda_i}\Big)^{n_i-\eta} \frac{n_i}{\eta} g\Big(z \frac{n_i}{\eta}\Big), \quad z \geq 0. \quad (10)$$

This distribution has a singular component:

$$P\{\lambda^\star_{i,j} = 0\} = \Big(1 - \frac{\lambda_{i,j}}{\Lambda_i}\Big)^{n_i}. \quad (11)$$

Obviously if $\lambda^\star_{i,j} = 0$ then $\mu^\star_{i,j} = \infty$. Therefore the distribution of $\mu^\star_{i,j}$ is a singular one.

If a sample size is big, formulas (6) and (10) work improperly. In this case a normal approximation of the Erlang and binomial distributions can be used.

In case of approximation of (6), one observation for initial state i gives a result having expectation $\frac{1}{\Lambda_i}$ and variance $\frac{1}{\Lambda_i^2}$. Empirical mean (2) has the same expectation and variance $\frac{1}{n_i \Lambda_i^2}$. Therefore density (6) can be approximated by normal distribution

$$\widetilde{g}_i(z) = z^{-2} \Lambda_i \sqrt{\frac{n_i}{2\pi}} \exp\Big(-\frac{1}{2}\Big(\Lambda_i \sqrt{n_i}\Big(\frac{1}{z} - 1/\Lambda_i\Big)\Big)^2\Big) =$$

$$z^{-2} \Lambda_i \sqrt{\frac{n_i}{2\pi}} \exp\Big(-z^{-2} \frac{1}{2}\Big(\Lambda_i \sqrt{n_i}(1 - z/\Lambda_i)\Big)^2\Big), \quad z \in (-\infty, \infty). \quad (12)$$

The density $\widetilde{h}_{i,j}(z)$ of estimation $\lambda^\star_{i,j}$ is calcuated by (10) where $\widetilde{g}_i(z)$ must be used instead of $g_i(z)$.

In respect to the binomial distribution, that is used in formula (10). It is approximated by normal distribution with mean $n_i \frac{\lambda_{i,j}}{\Lambda_i}$ and a variance $n_i \frac{\lambda_{i,j}}{\Lambda_i}\Big(1 - \frac{\lambda_{i,j}}{\Lambda_i}\Big)$. Let $dnorm\Big(u, n_i \frac{\lambda_{i,j}}{\Lambda_i}, \sqrt{n_i \frac{\lambda_{i,j}}{\Lambda_i}\Big(1 - \frac{\lambda_{i,j}}{\Lambda_i}\Big)}\Big)$ be a corresponding density function. Now instead of densities $h_{i,j}(z)$ and $\widetilde{h}_{i,j}(z)$, we have the following density:

$$\tilde{\tilde{h}}_{i,j}(z) = \int_0^n \frac{n_i}{u} \tilde{g}\left(z\frac{n_i}{u}\right) dnorm\left(u, n_i\frac{\lambda_{i,j}}{\Lambda_i}, \sqrt{n_i\frac{\lambda_{i,j}}{\Lambda_i}\left(1 - \frac{\lambda_{i,j}}{\Lambda_i}\right)}\right) du$$

$$= \int_0^n \frac{n_i}{u}\left(z\frac{n_i}{u}\right)^{-2} \Lambda_i\sqrt{\frac{n_i}{2\pi}} \exp\left(-\left(z\frac{n_i}{u}\right)^{-2}\frac{1}{2}\left(\Lambda_i\sqrt{n_i}\left(1 - \frac{zn_i}{u\Lambda_i}\right)\right)^2\right)$$

$$\times dnorm\left(u, n_i\frac{\lambda_{i,j}}{\Lambda_i}, \sqrt{n_i\frac{\lambda_{i,j}}{\Lambda_i}\left(1 - \frac{\lambda_{i,j}}{\Lambda_i}\right)}\right) du, \quad -\infty \le z \le \infty. \quad (13)$$

3   Expectation, Variance and Covariance for $\{\lambda_{i,j}^\star\}$

Densities $h_{i,j}(z)$, $\tilde{h}_{i,j}(z)$ and $\tilde{\tilde{h}}_{i,j}$ allow to calculate the expectation and the variance of the estimate $\lambda_{i,j}^\star$. Moments of the estimate $\lambda_{i,j}^\star$ can be calculated in the following way:

$$E(\lambda_{i,j}^{\star r}) = E\left(\left(\frac{N_{i,j}}{n_i}\right)^r\right) E(\Lambda_i^{\star r}), \quad r = 1, 2, \dots. \quad (14)$$

In particular:

$$E(\lambda_{i,j}^\star) = \frac{\lambda_{i,j}}{\Lambda_i} E(\Lambda_i^\star) = \frac{\lambda_{i,j}}{\Lambda_i}\frac{n_i}{n_i - 1}\Lambda_i = \frac{n_i}{n_i - 1}\lambda_{i,j}, \quad (15)$$

$$E(\lambda_{i,j}^{\star 2}) = E\left(\left(\frac{N_{i,j}}{n_i}\right)^2\right) E(\Lambda_i^{\star 2}) =$$

$$\left(\left(\frac{\lambda_{i,j}}{\Lambda_i}\right)^2 + \frac{1}{n_i}\frac{\lambda_{i,j}}{\Lambda_i}\left(1 - \frac{\lambda_{i,j}}{\Lambda_i}\right)\right)\frac{n_i^2}{(n_i - 1)(n_i - 2)}\Lambda_i^2 =$$

$$\left(\lambda_{i,j}^2 + \frac{1}{n_i}\lambda_{i,j}(\Lambda_i - \lambda_{i,j})\right)\frac{n_i^2}{(n_i - 1)(n_i - 2)}.$$

$$Var(\lambda_{i,j}^\star) = E(\lambda_{i,j}^{\star 2}) - \left(E(\lambda_{i,j}^\star)\right)^2 =$$

$$\lambda_{i,j}^2\frac{n_i^2}{(n_i - 1)^2(n_i - 2)} + \lambda_{i,j}(\Lambda_i - \lambda_{i,j})\frac{n_i}{(n_i - 1)(n_i - 2)} =$$

$$\lambda_{i,j}^2\frac{n_i}{(n_i - 1)^2(n_i - 2)} + \lambda_{i,j}\Lambda_i\frac{n_i}{(n_i - 1)(n_i - 2)}. \quad (16)$$

Now we calculate a covariance between parameters $\lambda^\star_{i,j}$ and $\lambda^\star_{i,j^\star}$ for different j and $j^\star$. The second mixed moment between $N_{i,j}$ and $N_{i,j^\star}$ is the following:

$$E\big(N_{i,j}N_{i,j^\star}\big) = \sum_{n_{i,j}=0}^{n_i} P\{N_{i,j}=n_{i,j}\}n_{i,j}E\big(N_{i,j^\star} \mid N_{i,j}=n_{i,j}\big) =$$

$$\sum_{n_{i,j}=0}^{n_i} \frac{n_i!}{n_{i,j}!(n_i-n_{i,j})!}\left(\frac{\lambda_{i,j}}{\Lambda_i}\right)^{n_{i,j}}\left(1-\frac{\lambda_{i,j}}{\Lambda_i}\right)^{n_i-n_{i,j}} n_{i,j}\big(n_i-n_{i,j}\big)\frac{\lambda_{i,j^\star}}{\Lambda_i-\lambda_{i,j}} =$$

$$\frac{\lambda_{i,j^\star}}{\Lambda_i-\lambda_{i,j}}\left(n_i^2\frac{\lambda_{i,j}}{\Lambda_i}-\left(\left(n_i\frac{\lambda_{i,j}}{\Lambda_i}\right)^2+n_i\frac{\lambda_{i,j}}{\Lambda_i}\left(1-\frac{\lambda_{i,j}}{\Lambda_i}\right)\right)\right) =$$

$$n_i(n_i-1)\frac{\lambda_{i,j}\lambda_{i,j^\star}}{\Lambda_i^2}. \tag{17}$$

It allows us to get the second mixed moment and the covariance for $\lambda^\star_{i,j}$ and $\lambda^\star_{i,j^\star}$:

$$E\big(\lambda^\star_{i,j}\lambda^\star_{i,j^\star}\big) = E\left(\frac{N_{i,j}}{n_i}\Lambda_i^\star\frac{N_{i,j^\star}}{n_i}\Lambda_i^\star\right) = n_i^{-2}E\big(N_{i,j}N_{i,j^\star}\big)E\big(\Lambda_i^{\star 2}\big) =$$

$$= \frac{(n_i-1)}{n_i}\frac{\lambda_{i,j}\lambda_{i,j^\star}}{\Lambda_i^2}\frac{n_i^2}{(n_i-1)(n_i-2)}\Lambda_i^2 = \frac{n_i}{n_i-2}\lambda_{i,j}\lambda_{i,j^\star}, \tag{18}$$

$$Cov\big(\lambda^\star_{i,j}\lambda^\star_{i,j^\star}\big) = E\big(\lambda^\star_{i,j}\lambda^\star_{i,j^\star}\big) - E\big(\lambda^\star_{i,j}\big)E\big(\lambda^\star_{i,j^\star}\big) =$$

$$\frac{n_i}{n_i-2}\lambda_{i,j}\lambda_{i,j^\star} - \frac{n_i}{n_i-1}\lambda_{i,j}\frac{n_i}{n_i-1}\lambda_{i,j^\star} = \lambda_{i,j}\lambda_{i,j^\star}\frac{n_i}{(n_i-1)^2(n_i-2)}. \tag{19}$$

4   Joint Distribution of Estimates $\{\lambda^\star_{i,j}\}$

Now we consider the distribution function of estimates vector $\lambda_i^\star = \big(\lambda^\star_{i,1},\ldots,\lambda^\star_{i,k}\big)$:

$$F_i(x) = P\big(\lambda^\star_{i,1}\le x_1,\lambda^\star_{i,2}\le x_2,\ldots,\lambda^\star_{i,k}\le x_k\big), \quad x=(x_1,x_2,\ldots,x_k)\ge 0.$$

Let $N_i = \big(N_{i,1},\ldots,N_{i,k}\big)$ be a vector having the distribution (9), $\tilde{n}_i = \big(n_{i,1},\ldots,n_{i,k}\big)\in\Omega(n_i,k)$ be some possible value of N_i. Then

$$F_i(x) = \sum_{\tilde{n}_i\in\Omega(n_i,k)} P\{N_i=\tilde{n}_i\}\prod_{j\in\{1,2,\ldots,k\}}P\{\lambda^\star_{i,j}\le x_j \mid N_{i,j}=n_{i,j}\} =$$

$$\sum_{\tilde{n}_i\in\Omega(n_i,k)} P\{N_i=\tilde{n}_i\}\prod_{j\in\{1,2,\ldots,k\}}P\left\{\frac{n_{i,j}}{n_i}\Lambda_i^\star\le x_j\right\} =$$

$$\sum_{\tilde{n}_i\in\Omega(n_i,k)} P\{N_i=\tilde{n}_i\}\prod_{j\in\{1,2,\ldots,k\}}P\left\{\Lambda_i^\star\le x_j\frac{n_i}{n_{i,j}}\right\}.$$

If $\varphi(x, \tilde{n}_i) = \min\limits_{j \in \{1,2,\ldots,k\}} x_j \frac{n_i}{n_{i,j}}$ then the necessary formula is the following:

$$F_i(x) = \sum_{\tilde{n}_i \in \Omega(n_i, k)} P\{N_i = \tilde{n}_i\} G_i(\varphi(x, \tilde{n}_i)), \quad x = (x_1, x_2, \ldots, x_k) \geq 0, \quad (20)$$

where G_i is a cumulative distribution function for the density (6).

5   Asymptotic Distributions

If sample size is big then well-known results of theory of big sample can be used [4,9,12,14]. The results state that vector $\frac{1}{n_i}(N_{i,1}, N_{i,2}, \ldots, N_{i,k})$ and scalar $\Lambda_i^\star$ are consistent and asymptotically unbiased estimates of $\frac{1}{\Lambda_i}(\lambda_{i,1}, \lambda_{i,2}, \ldots, \lambda_{i,k})$ and Λ_i. Now it is necessary to prove (following [4,9,12]) that $\lambda_i^\star = (\lambda_{i,1}^\star, \ldots, \lambda_{i,k}^\star)$ has asymptotically multivariate normal distribution too.

Let $p_j = \dfrac{\lambda_{i,j}}{\Lambda_i}$, $j = 1, \ldots, k$, and

$$V_i^T = \left(\frac{N_{i,1} - n_i p_1}{\sqrt{n_i p_1}} \cdots \frac{N_{i,k} - n_i p_k}{\sqrt{n_i p_k}} \right) = \left(\frac{\frac{1}{n_i} N_{i,1} - p_1}{\sqrt{\frac{1}{n_i} p_k}} \cdots \frac{\frac{1}{n_i} N_{i,k} - p_k}{\sqrt{\frac{1}{n_i} p_k}} \right),$$

$$\varphi_i = \left(\sqrt{p_1} \cdots \sqrt{p_k} \right)^T. \tag{21}$$

In accordance with ([12], formula (6a.1.5)) an asymptotic distribution of the vector V_i is multivariate normal distribution with zero mean and a covariance matrix

$$Cov(V_i) = I - \varphi_i \varphi_i^T, \tag{22}$$

where I is a unit k-dimension matrix.

Therefore, for $j \neq j^\star$

$$Cov\left(\frac{1}{n_i} N_{i,j}, \frac{1}{n_i} N_{i,j^\star} \right) =$$

$$\sqrt{\frac{1}{n_i} p_j} \sqrt{\frac{1}{n_i} p_{j^\star}} E\left(\frac{\frac{1}{n_i} N_{i,j} - p_j}{\sqrt{\frac{1}{n_i} p_j}} \frac{\frac{1}{n_i} N_{i,j^\star} - p_{j^\star}}{\sqrt{\frac{1}{n_i} p_{j^\star}}} \right) + o\left(\frac{1}{n_i} \right) =$$

$$\sqrt{\frac{1}{n_i} p_j} \sqrt{\frac{1}{n_i} p_{j^\star}} (-1) \sqrt{p_j} \sqrt{p_{j^\star}} + o\left(\frac{1}{n_i} \right) = -\frac{1}{n_i} p_j p_{j^\star} + o\left(\frac{1}{n_i} \right), \tag{23}$$

$$Var\left(\frac{1}{n_i} N_{i,j} \right) = \frac{1}{n_i} p_j E\left(\frac{\frac{1}{n_i} N_{i,j} - p_j}{\sqrt{\frac{1}{n_i} p_j}} \right)^2 + o\left(\frac{1}{n_i} \right) =$$

$$\frac{1}{n_i} p_j (1 - p_j) + o\left(\frac{1}{n_i} \right) \tag{24}$$

that corresponds to (17).

With regard to asymptotical distribution of estimate $\Lambda_i^\star = \frac{1}{M_i^\star}$, we know that random variable $\sqrt{n_i}\left(M_i^\star - \frac{1}{\Lambda_i}\right)$ has asymptotically normal distribution with zero mean and variance $\sigma^2(\Lambda_i) = \frac{1}{\Lambda_i^2}$. With respect to the formula (6a.2.1) of [12], $\widetilde{\Lambda_i} = \sqrt{n_i}(\Lambda_i^\star - \Lambda_i)$ has asymptotically normal distribution with zero mean and variance

$$Var\left(\widetilde{\Lambda_i}\right) = \sigma^2(\Lambda_i)\left(\frac{\partial}{\partial \alpha}\frac{1}{\alpha}\right)^2\Big|_{\alpha=\frac{1}{\Lambda_i}} = \left(\frac{1}{\Lambda_i^2}\right)\Lambda_i^4 = \Lambda_i^2. \tag{25}$$

Consequently, asymptotical variance of $\Lambda_i^\star$ equals $\frac{1}{n_i}\Lambda_i^2 + o\left(\frac{1}{n_i}\right)$.

From (15) and (16) we get the following:

$$E\left(\lambda_{i,j}^\star\right) = \lambda_{i,j} + o\left(\frac{1}{n_i}\right), \quad Var\left(\lambda_{i,j}^\star\right) = \frac{1}{n_i}\lambda_{i,j}\Lambda_i + o\left(\frac{1}{n_i}\right).$$

Finally, asymptotical distribution of estimates $\lambda_i^\star = \left(\lambda_{i,1}^\star, \lambda_{i,2}^\star, \dots, \lambda_{i,k}^\star\right)$ is considered.

As for $(k+1)$-vector $U_i = \left(\Lambda_i^\star \frac{1}{n_i}N_{i,1} \dots \frac{1}{n_i}N_{i,k}\right)^T$, it has been established that components $\Lambda_i^\star$ and $\left(\frac{1}{n_i}N_{i,1} \dots \frac{1}{n_i}N_{i,k}\right)$ are independent, scalar $(\Lambda_i^\star - \Lambda_i)\sqrt{n_i}$ has asymptotically normal distribution with zero mean and variance Λ_i^2, k-vector $\left(\frac{1}{n_i}N_{i,1} - p_1 \dots \frac{1}{n_i}N_{i,k} - p_k\right)\sqrt{n_i}$ has asymptotically multivariate normal distribution with zero mean and covariance matrix $n_i \times Cov$, where matrix Cov is defined by (23) and (24).

Therefore vector

$$\sqrt{n_i}\left(U_i - E(U_i)\right) = \sqrt{n_i}\left(\left(\Lambda_i^\star \frac{1}{n_i}N_{i,1} \dots \frac{1}{n_i}N_{i,k}\right)^T - \left(\Lambda_i p_{i,1} \dots p_{i,k}\right)^T\right)$$

has asymptotically multivariate normal distribution with zero mean and covariance matrix

$$\Omega = \begin{pmatrix} \Lambda_i^2 & 0 \\ 0 & n_i Cov \end{pmatrix}. \tag{26}$$

Let $h_1(u), \dots, h_k(u)$ be functions of real $(k + 1)$-vector u, having continuous partial derivative in the neighborhood of point $u_0 = E(U_i)$. Further let $\nabla h_j(u_0)$ be a gradient-column of $h_j(u)$ in the point u_0, $\nabla h(u_0)$ be $k \times (k + 1)$-matrix of the gradients: $\nabla h(u_0)^T = \left(\nabla h_1(u_0) \dots \nabla h_k(u_0)\right)$.

With respect to the *Theorem on differentiation transformation* ([4], Theorem 2.6; [9], Theorem 3.1.3) the vector $\sqrt{n}\left(h_1(U_1) - h_1(\Lambda_i p_{i,1} \dots p_{i,k})^T, \dots, h_k(U_k) - h_k(\Lambda_i p_{i,1} \dots p_{i,k})^T\right)$ has asymptotically multivariate normal distribution with zero mean and covariance matrix $\Sigma = \nabla h(u_0)\Omega\nabla h(u_0)^T$.

In our case $u_0 = \left(\Lambda_i p_1 \dots p_k\right)^T$; $h_j(u_0) = \lambda_{i,j} = \Lambda_i p_j$, $j = 1, \dots, k$; $\nabla h_j(u_0) = \left(p_j 0 \dots 0 \Lambda_i 0 \dots 0\right)^T$. Therefore,

$$\Sigma = \left(\nabla h_1(u_0) \dots \nabla h_k(u_0)\right)^T \begin{pmatrix} \Lambda_i^2 & 0 \\ 0 & n_i Cov \end{pmatrix} \cdot \left(\nabla h_1(u_0) \dots \nabla h_k(u_0)\right).$$

Table 1. Expectations $E(\lambda_1^*)$, $E(\lambda_2^*)$, $E(\lambda_3^*)$ for various observation's number n.

n	5	10	15	20	30	40	50	60	65
$E(\lambda_1^*)$	0.606	0.555	0.536	0.526	0.517	0.513	0.510	0.508	0.508
$E(\lambda_2^*)$	1.159	1.107	1.071	1.053	1.034	1.026	1.020	1.017	1.016
$E(\lambda_3^*)$	1.637	1.643	1.604	1.579	1.552	1.538	1.535	1.525	1.523

According to (23) and (24)

$$n_i Cov\left(\frac{1}{n_i}N_{i,j}, \frac{1}{n_i}N_{i,j^\star}\right) = -p_j p_{j^\star} + o(1) = -\frac{\lambda_{i,j}}{\Lambda_i}\frac{\lambda_{i,j^\star}}{\Lambda_i} + o(1),$$

$$n_i Var\left(\frac{1}{n_i}N_{i,j}\right) = p_j(1 - p_j) + o(1) = \frac{\lambda_{i,j}}{\Lambda_i}\left(1 - \frac{\lambda_{i,j}}{\Lambda_i}\right) + o(1).$$

Therefore,

$$\Sigma_{j,j^\star} = p_j p_{j^\star}\Lambda_i^2 - \Lambda_i^2\frac{\lambda_{i,j}}{\Lambda_i}\frac{\lambda_{i,j^\star}}{\Lambda_i} = \frac{\lambda_{i,j}}{\Lambda_i}\frac{\lambda_{i,j^\star}}{\Lambda_i}\Lambda_i^2 - \Lambda_i^2\frac{\lambda_{i,j}}{\Lambda_i}\frac{\lambda_{i,j^\star}}{\Lambda_i} = 0,$$

$$\Sigma_{j,j} = p_j^2\Lambda_j^2 + \Lambda_i^2\frac{\lambda_{i,j}}{\Lambda_i}\left(1 - \frac{\lambda_{i,j}}{\Lambda_i}\right) = \Lambda_i\lambda_{i,j}. \qquad (27)$$

Our final conclusion is represented in the following theorem.

Theorem 1. *Asymptotical vector $\sqrt{n_i}(\lambda_i^* - \lambda_i)$ has multivariate normal distribution with zero mean and covariance matrix Σ, defined by (27).*

6   Example

Let us illustrate previous results considering the three-dimensional chain ($k = 3$), where the initial state $i = 1$ and the vector of transition intensities $\lambda_i = (\lambda_{i,1}\lambda_{i,2}\lambda_{i,3}) = (0.5\ 1\ 1.5)$. Further the index i will be omitted. A number of observations n will be various.

An intensity of a jump from considered state $\Lambda = 3$. Figure 1 contains a graphics of the distribution density (6) for empirical value Λ^*, calculated for $n = 6$ and $n = 20$ observations. Graphics of distribution densities (10) for parameter's estimates λ_1^*, λ_2^*, λ_3^* are presented in Figs. 2, 3 and 4, where $\lambda Our = (0.5\ 1\ 1.5)^T$. Table 1 contains expectations $E(\lambda_1^*)$, $E(\lambda_2^*)$, $E(\lambda_3^*)$ for various values of observation's number n. Values of mean errors of these estimates are presented in Table 2. Presented results for the densities were calculated by formulas (6) and (10). This formula doesn't work where observation' number n exceeds 65–80. In this case we use normal approximation (12) and (13). Figure 5 contains graphics of distribution densities (10) and (13) for parameter's estimate λ_2^* and $n = 10$. We see that they coincide. Tables 3 and 4 are analogous to Tables 1 and 2, but are calculated using the normal approximation (12) and (13).

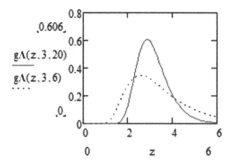

Fig. 1. Graphics of distribution densities (6).

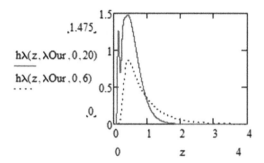

Fig. 2. Graphics of distribution densities (10) for λ_1^*.

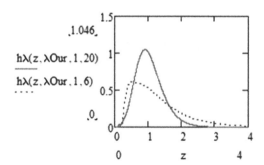

Fig. 3. Graphics of distribution densities (10) for λ_2^*.

Table 2. Mean errors of the estimates λ_1^*, λ_2^*, λ_3^* for various observation's number n.

n	5	10	15	20	30	40	50	60	70
$Err(\lambda_1^*)$	0.666	0.417	0.332	0.288	0.234	0.202	0.179	0.163	0.150
$Err(\lambda_2^*)$	0.878	0.640	0.505	0.425	0.337	0.287	0.231	0.231	0.213
$Err(\lambda_3^*)$	0.993	0.776	0.622	0.525	0.414	0.353	0.312	0.283	0.261

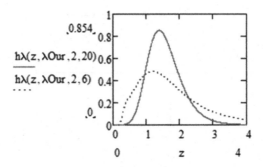

Fig. 4. Graphics of distribution densities (10) for $\lambda_3^\star$.

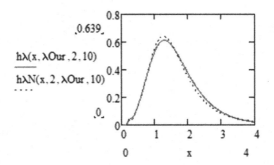

Fig. 5. Graphics of distribution densities (10) and (13) for $\lambda_3^\star$.

Table 3. Expectations $E(\lambda_1^\star)$, $E(\lambda_2^\star)$, $E(\lambda_3^\star)$ for a normal approximation.

n	60	70	80	100	120	140	160	300	400
$E(\lambda_1^\star)$	0.509	0.507	0.507	0.505	0.504	0.504	0.503	0.502	0.501
$E(\lambda_2^\star)$	1.018	1.015	1.013	1.010	1.009	1.007	1.006	1.003	1.003
$E(\lambda_3^\star)$	1.526	1.522	1.520	1.515	1.513	1.511	1.510	1.505	1.504

Table 4. Mean errors of the estimates $\lambda_1^\star$, $\lambda_2^\star$, $\lambda_3^\star$ for a normal approximation.

n	70	80	100	150	200	400	700	1000	2000
$Err(\lambda_1^\star)$	0.151	0.141	0.125	0.101	0.087	0.062	0.046	0.039	0.024
$Err(\lambda_2^\star)$	0.215	0.200	0.178	0.144	0.124	0.087	0.066	0.055	0.039
$Err(\lambda_3^\star)$	0.265	0.247	0.219	0.177	0.152	0.107	0.081	0.067	0.048

If number n of observation is big, then the considered densities are approximated by normal density very well. For example let $n = 40$. Using formulas (15) and (16) we calculate the expectation, the variance, and the standard deviation of each estimate: $E(\lambda_1^\star) = 0.513$, $\sigma(\lambda_1^\star) = 0.201$, $E(\lambda_2^\star) = 1.026$, $\sigma(\lambda_2^\star) = 0.286$, $E(\lambda_3^\star) = 1.538$, $\sigma(\lambda_3^\star) = 0.351$.

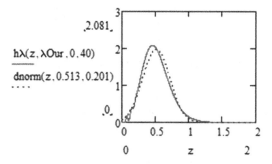

Fig. 6. Graphics of density of estimate λ_1^* and approximated normal density.

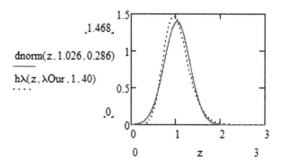

Fig. 7. Graphics of density of estimate λ_2^* and approximated normal density.

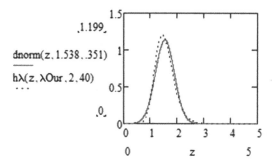

Fig. 8. Graphics of density of estimate λ_3^* and approximated normal density.

Figures 6, 7 and 8 contain two graphics: densities (10) for considered estimate λ_i^* and the density of the normal distribution $dnorm$ with calculated mean and standard deviation. We see that all the compared graphics coincide. If sample size n is bigger than 60 then these graphics coincide fully.

7 Conclusion

We considered a problem of parameter's estimation for finite continuous-time Markov chain. Distributions of the estimates were gotten for small and very big samples. It allowed us to make various statistical inferences about hypothesis and confidence interval on the parameters. What is more, it is possible to analyze how the size of the sample influences reliability of inferences about indices defined for considered Markov chain. We plan to consider such problems in our following publications.

References

1. Andronov, A.: Maximum likelihood estimates for Markov-additive processes of arrivals by aggregated data. In: Kollo, T. (ed.) Multivariate Statistics: Theory and Applications, pp. 17–33. World Scientific Publishing Company, New Jersey - London (2013)
2. Andronov, A.: Markov-modulated samples and their applications. In: Melas, V., Mignani, S., Monari, P., Salmaso, L. (eds.) Topics in Statistical Simulation, pp. 29–35. Springer, New York (2014). https://doi.org/10.1007/978-1-4939-2104-1_3
3. Andronov, A.: On reward rate estimation for finite irreducible continuous-time Markov chain. J. Stat. Theory Pract. Adv. Stat. Simul. **11**(3), 407–417 (2017)
4. Bandorf-Nielsen, O.E., Cox, D.R.: Asymptotic Techniques for Use in Statistics. Chapmen and Hall, London, New York, Tokyo, Melbourne, Madras (1989)
5. Feller, W.: An Introduction to Probability Theory and Its Applications, vol. 2. Wiley, New York (1971)
6. Gnedenko, B.W., Kovalenko, I.N.: Introduction to Queueing Theory. Nauka, Moscow (1987). (in Russian)
7. Kijima, M.: Markov Processes for Stochastic Modeling. Chapman & Hall, London (1997)
8. Kleinhof, M., Paramonov, J., Paramonova, A.: Regression model based on Markov chain theory for composite curve approximation. ACTA ET Commentationes Universitatis Tartuensis de Mathematica **8**, 143–153 (2008)
9. Kollo, T., von Rosen, D.: Advanced Multivariate Statistics with Matrices. Springer, Dordrecht (2005). https://doi.org/10.1007/1-4020-3419-9
10. Neuts, M.F.: Matrix-geometric Solutions in Stochastic Models. The Johns Hopkins University Press, Baltimore (1981)
11. Pacheco, A., Tang, L.C., Prabhu, N.U.: Markov-Modulated Processes & Semiregenerative Phenomena. World Scientific, Hoboken (2009)
12. Rao, C.R.: Linear Statistical Inference and Its Aplication. Wiley, New York-London-Sydney (1966)
13. Shorack, G.R., Welner, J.A.: Empirical Processes with Applications in Statistics. Wiley, New York (1986)
14. Srivastava, M.S.: Methods of Multivariate Statistics. Wiley-Interscience, New York (2002)
15. van de Green, S.A.: Empirical Processes in M-Estimation. Cambridge University Press, Cambridge (2000)

Asymptotic-Diffusion Analysis for Retrial Queue with Batch Poisson Input and Multiple Types of Outgoing Calls

Anatoly Nazarov[1], Tuan Phung-Duc[2], Svetlana Paul[1(✉)], and Olga Lizura[1]

[1] Institute of Applied Mathematics and Computer Science,
National Research Tomsk State University, 36 Lenina ave., Tomsk 634050, Russia
`nazarov.tsu@gmail.com, paulsv82@mail.ru, oliztsu@mail.ru`
[2] Faculty of Engineering Information and Systems, University of Tsukuba,
1-1-1 Tennodai, Tsukuba, Ibaraki 305-8573, Japan
`tuan@sk.tsukuba.ac.jp`

Abstract. In this paper, we consider a retrial queue with batch Poisson input process, arbitrarily distributed service times and arbitrarily distributed number of calls in the batch. Upon arrival, a call from the batch occupies the server if it is idle. The other calls from the batch join the orbit. If the server is busy, all calls from the batch go to the orbit. In the orbit, incoming calls stay for an exponentially distributed random delay and repeat their request for service. Besides incoming calls, the server can also make outgoing calls when idle. We assume that there are several types of outgoing calls in the system. The aim of the current research is to obtain asymptotic probability distribution of the number of incoming calls in the system using an asymptotic-diffusion analysis method.

Keywords: Batch Poisson process · Incoming calls · Outgoing calls · Retrial queue · Asymptotic-diffusion analysis method

1 Introduction

Retrial queues are mathematical models of real telecommunication systems. These systems are characterized by feature that call finding the server busy upon arrival doesn't leave the system but joins the orbit and tries to occupy the server again after some random delay. This customer behaviour arises in various communication systems with random access where multiple customers use single communication channel [1,7], such as CSMA (Carrier Sending Medium Access), ALOHA etc. Moreover, retrial behaviour occurs in real systems, especially in call-centers, where customers who failed to communicate with operator

This study (research grant No. 8.1.16.2019) was supported by The Tomsk State University competitiveness improvement programme.

© Springer Nature Switzerland AG 2019
V. M. Vishnevskiy et al. (Eds.): DCCN 2019, LNCS 11965, pp. 207–222, 2019.
https://doi.org/10.1007/978-3-030-36614-8_16

immediately repeat their calls later [4–6]. Also, retrials are common in cellular communication networks, and therefore the investigation of this phenomenon is crucial for analysis of aforementioned systems [13].

In call-centers an operator's idle time must be reduced to increase the effectiveness. This idea is reflected in blended call-centers, where an operator provides service to incoming calls and also makes outgoing calls. This server behaviour can be modelled using retrial queues with two-way communication, where the server deals with both incoming and outgoing calls [2,3,10].

Falin was the first who proposed this kind of systems [8]. He considered a single server model, where the service rates of incoming and outgoing calls are equal. Artalejo and Phung-Duc researched the models where incoming and outgoing calls have different service rates [3]. Sakurai and Phung-Duc obtained the limit formulae under low rate of retrials, high load of the system and high rate of outgoing calls conditions [11]. Sakurai and Phung-Duc also obtained the stationary joint distribution of the number of customers in the orbit and the state of the server for single server model with Poisson input and multiple types of outgoing calls [12].

Situations where incoming calls arrive in batches or groups frequently arise. Herewith, the server deals with one call at a time. On the other hand, the rest of calls from the batch are waiting in the orbit. This situation can be modelled with batch input processes.

In the current paper, we consider retrial queue with batch Poisson input, multiple types of outgoing calls and arbitrary distribution of service durations. We obtain the probability distribution of the number of incoming calls in the system using asymptotic-diffusion analysis method.

The rest of our paper is organized as follows. Section 2 presents the model and problem definition. Section 3 shows the system of equations for the underlying Markov process of the system. In Sect. 4, we present our main results where we derive the diffusion limit for the distribution of the number of customers in the system. This result is then used to obtain an approximation to the distribution of the number of customers in the system. In Sect. 5, we present some numerical results to show the accuracy of the proposed approximation method.

2   Mathematical Model and Problem Definition

We consider single server retrial queue with batch Poisson input characterized by rate λ. The number of customers in a batch is a random variable following the discrete distribution r_n $(n > 0)$ and $r_0 = 0$.

Upon arrival, a call from the batch starts its service if the server is idle. The duration of service is a random variable and follows an arbitrary distribution $B(x)$. The other calls from the batch join the orbit and stay for a random delay for an exponentially distributed time with rate σ independently of other customers. If the server is busy all calls from the batch enter the orbit. Upon retrial, a repeated call occupies the server if it is idle, otherwise the call joins the orbit again.

On the other hand, the server makes outgoing calls in its idle time. We consider the system with several types of outgoing calls. The server makes an outgoing call of type l with rate α_l, $l = \overline{2, L}$. The duration of outgoing call of type l is arbitrarily distributed with probability function $V_l(x)$, $l = \overline{2, L}$.

Let $i(t)$ denote the number of incoming calls in the system at the moment t. The main problem is to obtain the stationary probability distribution of the process $i(t)$.

3   Kolmogorov System of Equations

We introduce the following probabilities

$$P(i, t) = P\{i(t) = i\}. \tag{1}$$

Process $k(t)$ characterizes the state of the server at the moment t as follows.

$$k(t) = \begin{cases} 0, & \text{if the server is idle,} \\ 1, & \text{if an incoming call is in service,} \\ l, & \text{if an outgoing call of type } l \text{ is in service, } l = \overline{2, L}. \end{cases}$$

Process $z(t)$ is a residual time of call serving at the moment t. We denote

$$P_0(i, t) = P\{k(t) = 0, i(t) = i\},$$
$$P_k(i, z, t) = P\{k(t) = k, i(t) = i, z(t) < z\}, \quad k = 1, \dots, L \tag{2}$$

the probability distribution of the three-dimensional process $\{i(t), k(t), z(t)\}$ with variable number of components depending on the server state. As the process $\{i(t), k(t), z(t)\}$ is Markovian then the probability distribution (2) is the unique solution of Kolmogorov system of equations

$$\frac{\partial P_0(i, t)}{\partial t} = -\left(\lambda + i\sigma + \sum_{l=2}^{L} \alpha_l\right) P_0(i, t) + \frac{\partial P_1(i+1, 0, t)}{\partial z} + \sum_{l=2}^{L} \frac{\partial P_l(i, 0, t)}{\partial z},$$

$$\frac{\partial P_1(i, z, t)}{\partial t} = \frac{\partial P_1(i, z, t)}{\partial z} - \frac{\partial P_1(i, z, t)}{\partial z} - \lambda P_1(i, z, t) + i\sigma B(z) P_0(i, t)$$

$$+ \lambda \sum_{n=0}^{i-1} r_n P_1(i-n, z, t) + \lambda B(z) \sum_{n=0}^{i} r_n P_0(i-n, t), \quad i \geqslant 1,$$

$$\frac{\partial P_l(i, z, t)}{\partial t} = \frac{\partial P_l(i, z, t)}{\partial z} - \frac{\partial P_l(i, 0, t)}{\partial z} - \lambda P_l(i, z, t)$$

$$+ \lambda \sum_{n=0}^{i} r_n P_l(i-n, z, t) + \alpha_l V_l(z) P_0(i, t). \tag{3}$$

We define the partial characteristic functions

$$H_0(u,t) = \sum_{i=0}^{\infty} e^{jui} P_0(i,t), \ H_1(u,z,t) = \sum_{i=0}^{\infty} e^{jui} P_1(i,z,t),$$

$$H_l(u,z,t) = \sum_{i=0}^{\infty} e^{jui} P_l(i,z,t), \ r(u) = \sum_{n=1}^{\infty} e^{jun} r_n, \tag{4}$$

and transform the system (3) using (4) in the following form

$$\frac{\partial H_0(u,t)}{\partial t} = -\left(\lambda + \sum_{l=2}^{L} \alpha_l\right) H_0(u,t) + j\sigma\frac{\partial H_0(u,t)}{\partial u} + e^{-ju}\frac{\partial H_1(u,0,t)}{\partial z}$$

$$+ \sum_{l=2}^{L} \frac{\partial H_l(u,0,t)}{\partial z},$$

$$\frac{\partial H_1(u,z,t)}{\partial t} = \frac{\partial H_1(u,z,t)}{\partial z} - \frac{\partial H_1(u,0,t)}{\partial z} - j\sigma B(z)\frac{\partial H_0(u,t)}{\partial u}$$

$$- \lambda(1 - r(u))H_1(u,z,t) + \lambda B(z)r(u)H_0(u,t),$$

$$\frac{\partial H_l(u,z,t)}{\partial t} = \frac{\partial H_l(u,z,t)}{\partial z} - \frac{\partial H_l(u,0,t)}{\partial z} - \lambda(1 - r(u))H_l(u,z,t)$$

$$+ \alpha_l V_l(z)H_0(u,t). \tag{5}$$

We denote

$$H_k(u,t) = H_k(u,\infty,t), \ k = \overline{1,L},$$

then the characteristic function of the number of incoming calls in the system can be expressed as follows

$$H(u,t) = H_0(u,t) + H_1(u,t) + \sum_{l=2}^{L} H_l(u,t). \tag{6}$$

Taking the limit as $z \to \infty$ in the system (5) and summing up the equations we obtain

$$\frac{\partial H(u,t)}{\partial t} = \lambda(r(u) - 1)H(u,t) + (e^{-ju} - 1)\frac{\partial H_1(u,0,t)}{\partial z}.$$

Finally, we have the system of equations

$$\frac{\partial H_0(u,t)}{\partial t} = -\left(\lambda + \sum_{l=2}^{L} \alpha_l\right) H_0(u,t) + j\sigma\frac{\partial H_0(u,t)}{\partial u} + e^{-ju}\frac{\partial H_1(u,0,t)}{\partial z}$$

$$+ \sum_{l=2}^{L} \frac{\partial H_l(u,0,t)}{\partial z},$$

$$\frac{\partial H_1(u,z,t)}{\partial t} = \frac{\partial H_1(u,z,t)}{\partial z} - \frac{\partial H_1(u,0,t)}{\partial z} - j\sigma B(z)\frac{\partial H_0(u,t)}{\partial u}$$

$$- \lambda(1 - r(u))H_1(u,z,t) + \lambda B(z)r(u)H_0(u,t),$$

$$\frac{\partial H_l(u,z,t)}{\partial t} = \frac{\partial H_l(u,z,t)}{\partial z} - \frac{\partial H_l(u,0,t)}{\partial z} - \lambda(1-r(u))H_l(u,z,t)$$
$$+ \alpha_l V_l(z)H_0(u,t)$$
$$\frac{\partial H(u,t)}{\partial t} = \lambda(r(u)-1)H(u,t) + (e^{-ju}-1)\frac{\partial H_1(u,0,t)}{\partial z}. \tag{7}$$

The system of Eq. (7) is our main object for the following analysis. We will solve it using asymptotic-diffusion analysis method under low rate of retrials limit condition ($\sigma \to 0$).

4   Asymptotic-Diffusion Analysis Method

Denoting $\sigma = \varepsilon$ we introduce the following notations in the system (7)

$$\tau = t\varepsilon, \quad u = \varepsilon w, \quad H_0(u,t) = F_0(w,\tau,\varepsilon), \quad H_k(u,z,t) = F_k(w,z,\tau,\varepsilon), \quad k = \overline{1,L},$$

to obtain the system of Eq. (8)

$$\varepsilon\frac{\partial F_0(w,\tau,\varepsilon)}{\partial \tau} = -\left(\lambda + \sum_{l=2}^{L}\alpha_l\right)F_0(w,\tau,\varepsilon) + j\frac{\partial F_0(w,\tau,\varepsilon)}{\partial w}$$

$$+ e^{jw\varepsilon}\frac{\partial F_1(w,0,\tau,\varepsilon)}{\partial z} + \sum_{l=2}^{L}\frac{\partial F_l(w,0,\tau,\varepsilon)}{\partial z},$$

$$\varepsilon\frac{\partial F_1(w,z,\tau,\varepsilon)}{\partial \tau} = \frac{\partial F_1(w,z,\tau,\varepsilon)}{\partial z} - \frac{\partial F_1(w,0,\tau,\varepsilon)}{\partial z} - jB(z)\frac{\partial F_0(w,\tau,\varepsilon)}{\partial w}$$
$$- \lambda(1-r(w\varepsilon))F_1(w,z,\tau,\varepsilon) + \lambda B(z)r(w\varepsilon)F_0(w,\tau,\varepsilon),$$

$$\varepsilon\frac{\partial F_l(w,z,\tau,\varepsilon)}{\partial \tau} = \frac{\partial F_l(w,z,\tau,\varepsilon)}{\partial z} - \frac{\partial F_l(w,0,\tau,\varepsilon)}{\partial z} -$$
$$- \lambda(1-r(w\varepsilon))F_l(w,z,\tau,\varepsilon) + \alpha_l V_l(z)F_0(w,\tau,\varepsilon),$$

$$\varepsilon\frac{\partial F(w,\tau,\varepsilon)}{\partial \tau} = \lambda(r(w\varepsilon)-1)F(w,\tau,\varepsilon) + (e^{-jw\varepsilon}-1)\frac{\partial F_1(w,0,\tau,\varepsilon)}{\partial z}. \tag{8}$$

We solve the system (8) by taking the limit as $\varepsilon \to 0$ and present the result in the following theorem.

Theorem 1. *In our retrial queue, the stationary probability distribution of the system states is given as follows*

$$R_0 = \left\{1 + (\lambda + x(\tau))b + \sum_{l=2}^{L}\alpha_l v_l\right\}^{-1},$$

$$R_1 = (\lambda + x(\tau))bR_0,$$

$$R_l = \alpha_l v_l R_0, \quad l = \overline{2,L}, \tag{9}$$

where b and v_l are the mean values of the corresponding distributions $B(x)$ and $V_l(x)$, $l = \overline{2, L}$. Here the function $x(\tau)$ is a solution of a differential equation

$$x'(\tau) = \lambda r - R_1'(0), \tag{10}$$

where r is the mean number of calls in the batch. The function $R_1'(0)$ is defined by equality

$$R_1'(0) = (\lambda + x(\tau)) \left\{ 1 + (\lambda + x(\tau)) b + \sum_{l=2}^{L} \alpha_l v_l \right\}^{-1}. \tag{11}$$

Proof. Taking the limit as $\varepsilon \to 0$ in the system (8) we denote $\lim\limits_{\varepsilon \to 0} F_0(w, \tau, \varepsilon) = F_0(w, \tau)$, $\lim\limits_{\varepsilon \to 0} F_k(w, z, \tau, \varepsilon) = F(w, z, \tau)$, $k = \overline{1, L}$. Hence we obtain

$$-\left(\lambda + \sum_{l=2}^{L} \alpha_l \right) F_0(w, \tau) + j \frac{\partial F_0(w, \tau)}{\partial w} + \frac{\partial F_1(w, 0, \tau)}{\partial z} + \sum_{l=2}^{L} \frac{\partial F_l(w, 0, \tau)}{\partial z} = 0,$$

$$\frac{\partial F_1(w, z, \tau)}{\partial z} - \frac{\partial F_1(w, 0, \tau)}{\partial z} - jB(z) \frac{\partial F_0(w, \tau)}{\partial w} + \lambda B(z) F_0(w, \tau) = 0,$$

$$\frac{\partial F_l(w, z, \tau)}{\partial z} - \frac{\partial F_l(w, 0, \tau)}{\partial z} + \alpha_l V_l(z) F_0(w, \tau) = 0,$$

$$\frac{\partial F(w, \tau)}{\partial \tau} = \lambda j w r F(w, \tau) - j w \frac{\partial F_1(w, 0, \tau)}{\partial z}. \tag{12}$$

We assume that the solution of the system of Eq. (12) has the following form

$$F_0(w, \tau) = R_0 e^{j w x(\tau)}, \quad F_k(w, z, \tau) = R_k(z) e^{j w x(\tau)}.$$

Denoting

$$x(\tau) = x,$$

then we have system

$$-\left(\lambda + \sum_{l=2}^{L} \alpha_l + x \right) R_0 + R_1'(0) + \sum_{l=2}^{L} R_l'(0) = 0,$$

$$R_1'(z) - R_1'(0) + (\lambda + x) B(z) R_0 = 0,$$

$$R_l'(z) - R_l'(0) + \alpha_l V_l(z) R_0 = 0,$$

$$x'(\tau) = \lambda r - R_1'(0). \tag{13}$$

Here r is average number of calls in the batch.

We introduce Laplace-Stieltjes transforms

$$\int_0^\infty e^{-\gamma z} dR_k(z) = R_k^*(\gamma),$$

$$\int_0^\infty e^{-\gamma z} dR_k'(z) = \gamma R_k^*(\gamma) - R_k^*(0), \quad k = \overline{1, L}. \tag{14}$$

Then we rewrite second and third equations of the system (13) using Laplace-Stieltjes transforms

$$\gamma R_1^*(\gamma) - R_1'(0) + (\lambda + x)B^*(\gamma)R_0 = 0,$$
$$\gamma R_l^*(\gamma) - R_l'(0) + \alpha_l V_l^*(\gamma)R_0 = 0. \tag{15}$$

Eventually we obtain following expressions by substituting $\gamma = 0$ in equalities (15)

$$R_1'(0) = (\lambda + x)R_0,$$
$$R_l'(0) = \alpha_l R_0,$$

which we substitute in (15)

$$\gamma R_1^*(\gamma) = (\lambda + x)\left(1 - B^*(\gamma)\right)R_0,$$
$$\gamma R_l^*(\gamma) = \alpha_l\left(1 - V_l^*(\gamma)\right)R_0. \tag{16}$$

We substitute $\gamma = 0$ in equalities (16):

$$R_1 = (\lambda + x)bR_0,$$

$$R_l = \alpha_l v_l R_0,$$

where b and v_l are the mean values of the corresponding probability distributions $B(x)$ and $V_l(x)$, $l = \overline{2, L}$. Using the normalization condition from the last system we derive

$$R_0 = \left\{1 + (\lambda + x)b + \sum_{l=2}^{L} \alpha_l v_l\right\}^{-1}.$$

The Theorem is proved.

Denoting

$$x'(\tau) = a(x) = \lambda r - R_1'(0)$$

$$= \lambda r - (\lambda + x(\tau))\left\{1 + (\lambda + x(\tau))b + \sum_{l=2}^{L} \alpha_l v_l\right\}^{-1}, \tag{17}$$

and making the following replacements in the system (7)

$$H_0(u,t) = H_0^{(2)}(u,t)e^{j\frac{u}{\sigma}x(\sigma t)}, \quad H_k(u,z,t) = H_k^{(2)}(u,z,t)e^{j\frac{u}{\sigma}x(\sigma t)}, \tag{18}$$

we have

$$\frac{\partial H_0^{(2)}(u,t)}{\partial t} + jua(x)H_0^{(2)}(u,t) = -\left(\lambda + \sum_{l=2}^{L}\alpha_l + x\right)H_0^{(2)}(u,t) + j\sigma\frac{\partial H_0^{(2)}(u,t)}{\partial u}$$

$$+ e^{-ju}\frac{\partial H_1^{(2)}(u,0,t)}{\partial z} + \sum_{l=2}^{L}\frac{\partial H_l^{(2)}(u,0,t)}{\partial z},$$

$$\frac{\partial H_1^{(2)}(u,z,t)}{\partial t} + jua(x)H_1^{(2)}(u,z,t) = \frac{\partial H_1^{(2)}(u,z,t)}{\partial z} - \frac{\partial H_1^{(2)}(u,0,t)}{\partial z}$$

$$- j\sigma B(z)\frac{\partial H_0^{(2)}(u,t)}{\partial u} - \lambda(1-r(u))H_1^{(2)}(u,z,t) + B(z)(\lambda r(u) + x)H_0^{(2)}(u,t),$$

$$\frac{\partial H_l^{(2)}(u,z,t)}{\partial t} + jua(x)H_l^{(2)}(u,z,t) = \frac{\partial H_l^{(2)}(u,z,t)}{\partial z} - \frac{\partial H_l^{(2)}(u,0,t)}{\partial z}$$

$$- \lambda(1-r(u))H_l^{(2)}(u,z,t) + \alpha_l V_l(z)H_0^{(2)}(u,t)$$

$$\frac{\partial H^{(2)}(u,t)}{\partial t} + jua(x)H^{(2)}(u,t) = \lambda(r(u)-1)H^{(2)}(u,t)$$

$$+ (e^{-ju}-1)\frac{\partial H_1^{(2)}(u,0,t)}{\partial z}. \tag{19}$$

Denoting $\sigma = \varepsilon^2$ we introduce the following notations in the system (19)

$$\tau = t\varepsilon^2, \ u = w\varepsilon, \ H_0^{(2)}(u,t) = F_0^{(2)}(w,\tau,\varepsilon), \ H_k^{(2)}(u,z,t) = F_k^{(2)}(w,z,\tau,\varepsilon), \ k = \overline{1,L}$$

to obtain the following system of equations

$$\varepsilon^2\frac{\partial F_0^{(2)}(w,\tau,\varepsilon)}{\partial \tau} + jw\varepsilon a(x)F_0^{(2)}(w,\tau,\varepsilon) = -\left(\lambda + \sum_{l=2}^{L}\alpha_l + x\right)F_0^{(2)}(w,\tau,\varepsilon)$$

$$+ j\varepsilon\frac{\partial F_0^{(2)}(w,\tau,\varepsilon)}{\partial w} + e^{-jw\varepsilon}\frac{\partial F_1^{(2)}(w,z,\tau,\varepsilon)}{\partial z} + \sum_{l=2}^{L}\frac{\partial F_l^{(2)}(w,0,\tau,\varepsilon)}{\partial z},$$

$$\varepsilon^2\frac{\partial F_1^{(2)}(w,z,\tau,\varepsilon)}{\partial \tau} + jw\varepsilon a(x)F_1^{(2)}(w,z,\tau,\varepsilon) = \frac{\partial F_1^{(2)}(w,z,\tau,\varepsilon)}{\partial z} - \frac{\partial F_1^{(2)}(w,0,\tau,\varepsilon)}{\partial z}$$

$$+ \lambda jw\varepsilon r F_1^{(2)}(w,z,\tau,\varepsilon) + B(z)(\lambda + \lambda jw\varepsilon r + x)F_0^{(2)}(w,\tau,\varepsilon) - B(z)j\varepsilon\frac{\partial F_0^{(2)}(w,\tau,\varepsilon)}{\partial w},$$

$$\varepsilon^2\frac{\partial F_l^{(2)}(w,z,\tau,\varepsilon)}{\partial \tau} + jw\varepsilon a(x)F_l^{(2)}(w,z,\tau,\varepsilon) = \frac{\partial F_l^{(2)}(w,z,\tau,\varepsilon)}{\partial z} - \frac{\partial F_l^{(2)}(w,0,\tau,\varepsilon)}{\partial z}$$

$$- \lambda jw\varepsilon r F_l^{(2)}(w,z,\tau,\varepsilon) + \alpha_l V_l(z)F_0^{(2)}(w,\tau,\varepsilon),$$

$$\varepsilon^2\frac{\partial F^{(2)}(w,\tau,\varepsilon)}{\partial \tau} + jw\varepsilon a(x)F^{(2)}(w,\tau,\varepsilon)$$

$$= \lambda(r(w\varepsilon)-1)F^{(2)}(w,\tau,\varepsilon) + (e^{-jw\varepsilon}-1)\frac{\partial F_1^{(2)}(w,0,\tau,\varepsilon)}{\partial z}. \tag{20}$$

Solving the system (20) in the limit as $\varepsilon \to 0$ we obtain the following theorem.

Theorem 2. *Probability density of normalized number of incoming calls in the system given as follows*

$$\Pi(s) = \frac{C}{b(s)}\exp\{\frac{2}{\sigma}\int_0^s \frac{a(x)}{b(x)}dx\}, \tag{21}$$

where C is normalization factor, function $a(x)$ is defined by equality (17), function $b(x)$ has the following form

$$b(x) = \lambda r^{(2)} + R'_1(0) - \{(\lambda + x(\tau))g_0 + (\lambda r - a(x))R_1 + \lambda r R_0\}. \qquad (22)$$

Here

$$g_0 = \frac{R_0}{2} \frac{(a(x) - \lambda r)\{(\lambda + x(\tau))b^{(2)} + \sum_{l=2}^{L} \alpha_l v_l^{(2)}\} - 2\lambda r b}{1 + (\lambda + x(\tau))b + \sum_{l=2}^{L} \alpha_l v_l^{(2)}}, \qquad (23)$$

function $R'_1(0)$ is defined by equality (11), values $r^{(2)}$, $b^{(2)}$, $v_l^{(2)}$ are second moments of corresponding probability distributions r_n, $B(x)$, $V_l(x)$.

Proof. We rewrite the first, second and third equations of the system (20) up to $O(\varepsilon^2)$

$$jw\varepsilon a(x)F_0^{(2)}(w, \tau, \varepsilon) = -\left(\lambda + \sum_{l=2}^{L} \alpha_l + x\right) F_0^{(2)}(w, \tau, \varepsilon) + j\varepsilon \frac{\partial F_0^{(2)}(w, \tau, \varepsilon)}{\partial w}$$

$$+(1 - jw\varepsilon)\frac{\partial F_1^{(2)}(w, 0, \tau, \varepsilon)}{\partial z} + \sum_{l=2}^{L} \frac{\partial F_l^{(2)}(w, 0, \tau, \varepsilon)}{\partial z} + O(\varepsilon^2),$$

$$jw\varepsilon a(x)F_1^{(2)}(w, z, \tau, \varepsilon) = \frac{\partial F_1^{(2)}(w, z, \tau, \varepsilon)}{\partial z} - \frac{\partial F_1^{(2)}(w, 0, \tau, \varepsilon)}{\partial z} + \lambda jw\varepsilon r F_1^{(2)}(w, z, \tau, \varepsilon)$$

$$+B(z)(\lambda + \lambda jw\varepsilon r + x)F_0^{(2)}(w, \tau, \varepsilon) - B(z)j\varepsilon \frac{\partial F_0^{(2)}(w, \tau, \varepsilon)}{\partial w} + O(\varepsilon^2),$$

$$jw\varepsilon a(x)F_l^{(2)}(w, z, \tau, \varepsilon) = \frac{\partial F_l^{(2)}(w, z, \tau, \varepsilon)}{\partial z} - \frac{\partial F_l^{(2)}(w, 0, \tau, \varepsilon)}{\partial z}$$

$$+\lambda jw\varepsilon r F_l^{(2)}(w, z, \tau, \varepsilon) + \alpha_l V_l(z)F_0^{(2)}(w, \tau, \varepsilon) + O(\varepsilon^2).$$

We find the solution of this system in the following form

$$F_0^{(2)}(w, \tau, \varepsilon) = \Phi(w, \tau)\{R_0 + jw\varepsilon f_0\} + O(\varepsilon^2),$$

$$F_k^{(2)}(w, z, \tau, \varepsilon) = \Phi(w, \tau)\{R_k(z) + jw\varepsilon f_k(z)\} + O(\varepsilon^2), \quad k = \overline{1, L}, \qquad (24)$$

where $\Phi(w, \tau)$ is some scalar function, which we obtain in following derivations.
  We have

$$jw\varepsilon a(x)R_0 = -\left(\lambda + \sum_{l=2}^{L} \alpha_l + x\right) \{R_0 + jw\varepsilon f_0\} + j\varepsilon \frac{\partial \Phi(w, \tau)/\partial w}{\Phi(w, \tau)} R_0$$

$$+(1 - jw\varepsilon)R'_1(0) + jw\varepsilon f'_1(0) + \sum_{l=2}^{L} \{R'_l(0) + jw\varepsilon f'_l(0)\} + o(\varepsilon^2),$$

$$jw\varepsilon a(x)R_1(z) = R_1'(z) - R_1'(0) + jw\varepsilon\{f_1'(z) - f_1'(0)\} + \lambda jw\varepsilon r R_1(z)$$

$$+B(z)(\lambda + \lambda jw\varepsilon r + x)\{R_0 + jw\varepsilon f_0\} - B(z)j\varepsilon\frac{\partial\Phi(w,\tau)/\partial w}{\Phi(w,\tau)}R_0 + o(\varepsilon^2),$$

$$jw\varepsilon a(x)R_l(z) = R_l'(z) - R_l'(0) + \lambda jw\varepsilon r R_l(z) + jw\varepsilon\{f_l'(z) - f_l'(0)\}$$

$$+\alpha_l V_l(z)\{R_0 + jw\varepsilon f_0\} + o(\varepsilon^2).$$

Taking (13) into account we obtain

$$jw\varepsilon a(x)R_0 = -jw\varepsilon\left(\lambda + \sum_{l=2}^{L}\alpha_l + x\right) + j\varepsilon\frac{\partial\Phi(w,\tau)/\partial w}{\Phi(w,\tau)}R_0$$

$$+jw\varepsilon f_1'(0) - jw\varepsilon R_1'(0) + jw\varepsilon\sum_{l=2}^{L}f_1'(0) + o(\varepsilon^2),$$

$$jw\varepsilon a(x)R_1(z) = jw\varepsilon\{f_1'(z) - f_1'(0)\} + \lambda jw\varepsilon r R_1(z) + \lambda jw\varepsilon r B(z)R_0$$

$$+jw\varepsilon f_0 B(z)(\lambda + x) - B(z)j\varepsilon\frac{\partial\Phi(w,\tau)/\partial w}{\Phi(w,\tau)}R_0 + o(\varepsilon^2),$$

$$jw\varepsilon a(x)R_l(z) = jw\varepsilon\{f_l'(z) - f_l'(0)\} + \lambda jw\varepsilon r R_l(z) + jw\varepsilon f_0\alpha_l V_l(z) + o(\varepsilon^2).$$

Dividing the equations of the last system by $j\varepsilon$ and taking the limit as $\varepsilon \to 0$ we obtain

$$a(x)R_0 = -f_0\left(\lambda + \sum_{l=2}^{L}\alpha_l + x\right) + \frac{\partial\Phi(w,\tau)/\partial w}{w\Phi(w)}R_0 + f_1'(0) - R_1'(0)$$

$$+ \sum_{l=2}^{L}f_l'(0),$$

$$a(x)R_1(z) = f_1'(z) - f_1'(0) + \lambda r R_1(z) + \lambda r B(z)R_0 + f_0 B(z)(\lambda + x)$$

$$- B(z)\frac{\partial\Phi(w,\tau)/\partial w}{w\Phi(w,\tau)}R_0,$$

$$a(x)R_l(z) = f_l'(z) - f_l'(0) + \lambda r R_l(z) + f_0\alpha_l V_l(z). \tag{25}$$

We rewrite the systems of Eq. (25) in the form

$$- f_0\left(\lambda + \sum_{l=2}^{L}\alpha_l + x\right) + f_1'(0) + \sum_{l=2}^{L}f_l'(0) = a(x)R_0 + R_1'(0)$$

$$- \frac{\partial\Phi(w,\tau)/\partial w}{w\Phi(w,\tau)}R_0,$$

$$f_1'(z) - f_1'(0) + f_0 B(z)(\lambda + x) = a(x)R_1(z) - \lambda r R_1(z) - \lambda r B(z)R_0$$

$$+ B(z)\frac{\partial\Phi(w,\tau)/\partial w}{w\Phi(w,\tau)}R_0,$$

$$f_l'(z) - f_l'(0) + f_0\alpha_l V_l(z) = a(x)R_l(z) + \lambda r R_l(z). \tag{26}$$

For the solution f_k, $k = \overline{0,L}$ of the system (26) we present in following form

$$f_0 = CR_0 + g_0 - \varphi_0 \frac{\partial \Phi(w,\tau)/\partial w}{w\Phi(w,\tau)},$$

$$f_k(z) = CR_k(z) + g_k - \varphi_k \frac{\partial \Phi(w,\tau)/\partial w}{w\Phi(w,\tau)} \qquad (27)$$

which we substitute in the system (26), to obtain systems of equations

$$- g_0 \left(\lambda + \sum_{l=2}^{L} \alpha_l + x \right) + g'_1(0) + \sum_{l=2}^{L} g'_l(0) = a(x)R_0 + R'_1(0),$$

$$g'_1(z) - g'_1(0) + g_0 B(z)(\lambda + x) = a(x)R_1(z) - \lambda r R_1(z) - \lambda r B(z)R_0,$$

$$g'_l(z) - g'_l(0) + g_0 \alpha_l V_l(z) = a(x)R_l(z) + \lambda r R_l(z). \qquad (28)$$

$$- \varphi_0 \left(\lambda + \sum_{l=2}^{L} \alpha_l + x \right) + \varphi'_1(0) + \sum_{l=2}^{L} \varphi'_l(0) = R_0,$$

$$\varphi'_1(z) - \varphi'_1(0) + \varphi_0 B(z)(\lambda + x) = -B(z)R_0,$$

$$\varphi'_l(z) - \varphi'_l(0) + \varphi_0 \alpha_l V_l(z) = 0. \qquad (29)$$

Taking (29) into account we write important equalities

$$\varphi_0 = \frac{dR_0}{dx}, \quad \varphi_k(z) = \frac{dR_k(z)}{dx}, \quad \varphi_k(0) = \frac{dR_k(0)}{dx}. \qquad (30)$$

We apply the Laplace-Stieltjes transform to the second and third equations of the system (28) to obtain the equalities

$$\gamma g^*_1(\gamma) - g'_1(0) + g_0 B^*(\gamma)(\lambda + x) = a(x)R^*_1(\gamma) - \lambda r R^*_1(\gamma) - \lambda r B^*(\gamma)R_0,$$

$$\gamma g^*_l(\gamma) - g'_l(0) + g_0 \alpha_l V^*_l(\gamma) = a(x)R^*_l(\gamma) + \lambda r R^*_l(\gamma). \qquad (31)$$

In the second and third of the system (28) we take the limit as $z \to \infty$, denoting $R_k(\infty) = R_k$, to get the equalities

$$g'_1(0) = g_0(\lambda + x) - a(x)R_1 + \lambda r R_1 + \lambda r R_0,$$

$$g'_l(0) = g_0 \alpha_l - a(x)R_l + \lambda r R_l. \qquad (32)$$

From the last two systems (31) and (32) of equations we obtain

$$\gamma g^*_1(\gamma) = g_0[1 - B^*(\gamma)](\lambda + x) + (\lambda r - a(x))[R_1 - R^*_1(\gamma)] + \lambda r R_0[1 - B^*(\gamma)],$$

$$\gamma g^*_l(\gamma) = g_0 \alpha_l[1 - V^*_l(\gamma)] + (\lambda r - a(x))[R_l - R^*_l(\gamma)].$$

Dividing the equations by γ and taking the limit as $\gamma \to 0$ we have

$$g_1 = g_0 b(\lambda + x) + (a(x) - \lambda r)R^{*\prime}_1(\gamma)\Big|_{\gamma=0} + \lambda r R_0 b,$$

$$g_l = g_0 \alpha_l v_l + (a(x) - \lambda r) R^{*\prime}_l(\gamma) \Big|_{\gamma=0}.$$

From (27), the functions g_k are particular solutions of the inhomogeneous system (28). Therefore, they satisfy some additional condition, which we will choose in the form $\sum_{k=0}^{L} g_k = 0$. Then a solution of system (28) satisfying the condition is determined uniquely. This allows us to find the expression for g_0

$$g_0 = \frac{R_0}{2} \frac{(a(x) - \lambda r) \left\{ (\lambda + x(\tau)) b^{(2)} + \sum_{l=2}^{L} \alpha_l v_l^{(2)} \right\} - 2\lambda r b}{1 + (\lambda + x(\tau)) b + \sum_{l=2}^{L} \alpha_l v_l^{(2)}}.$$

Here the values $r^{(2)}$, $b^{(2)}$, $v_l^{(2)}$ are the second moments of the corresponding distributions r_n, $B(x)$, $V_l(x)$. Then we consider the last equation of the system (20), transforming it

$$\varepsilon^2 \frac{\partial F^{(2)}(w, \tau, \varepsilon)}{\partial \tau} + jw\varepsilon a(x) F^{(2)}(w, \tau, \varepsilon) = \lambda \left(jw\varepsilon r + \frac{(jw\varepsilon)^2}{2} r^{(2)} \right) F^{(2)}(w, \tau, \varepsilon)$$

$$+ \left(-jw\varepsilon + \frac{(jw\varepsilon)^2}{2} \right) \frac{\partial F_1^{(2)}(w, 0, \tau, \varepsilon)}{\partial z} + o(\varepsilon^3).$$

In the last equality we substitute solutions (24):

$$\varepsilon^2 \frac{\partial \Phi(w, \tau)}{\partial \tau} + jw\varepsilon a \Phi(w, \tau) \{1 + jw\varepsilon f\} = \frac{(jw\varepsilon)^2}{2} \Phi(w, \tau) \left(\lambda r^{(2)} + R_1'(0) \right)$$

$$+ jw\varepsilon \left\{ \lambda r \Phi(w, \tau)(1 + jw\varepsilon f) - \Phi(w, \tau) (R_1'(0) + jw\varepsilon f_1'(0)) \right\} + o(\varepsilon^3).$$

Here, $f = \sum_{k=0}^{L} f_k$. Then using the equality (14) we obtain

$$\varepsilon^2 \frac{\partial \Phi(w, \tau)}{\partial \tau} + (jw\varepsilon)^2 a \Phi(w, \tau) f = \frac{(jw\varepsilon)^2}{2} \Phi(w, \tau) \left(\lambda r^{(2)} + R_1'(0) \right)$$

$$+ jw\varepsilon \left\{ jw\varepsilon \lambda r \Phi(w, \tau) f - jw\varepsilon f_1'(0) \Phi(w, \tau) \right\} + o(\varepsilon^3).$$

Dividing both sides of the last equation by ε and taking the limit as $\varepsilon \to 0$ we have

$$\frac{\partial \Phi(w, \tau)}{\partial \tau} + (jw)^2 a \Phi(w, \tau) f = \frac{(jw)^2}{2} \Phi(w, \tau) \left(\lambda r^{(2)} + R_1'(0) \right)$$

$$+ (jw)^2 \left\{ \lambda r \Phi(w, \tau) f - f_1'(0) \Phi(w, \tau) \right\}.$$

Substituting solutions (27) we obtain

$$\frac{\partial \Phi(w, \tau)}{\partial \tau} = \frac{(jw)^2}{2} \Phi(w, \tau) \left(\lambda r^{(2)} + R_1'(0) \right)$$

$$-(jw)^2\Phi(w,\tau)g_1'(0) - w\varphi_1'(0)\frac{\partial\Phi(w,\tau)}{\partial w}.$$

Then we transform the last equation

$$\frac{\partial\Phi(w,\tau)}{\partial\tau} = -w\varphi_1'(0)\frac{\partial\Phi(w,\tau)}{\partial w} + \frac{(jw)^2}{2}\Phi(w,\tau)\left(\lambda r^{(2)} + R_1'(0) - 2g_1'(0)\right).$$

We denote

$$b(x) = \lambda r^{(2)} + R_1'(0) - 2g_1'(0),$$

and rewrite the last equation in the following form

$$\frac{\partial\Phi(w,\tau)}{\partial\tau} = -w\varphi_1'(0)\frac{\partial\Phi(w,\tau)}{\partial w} + \frac{(jw)^2}{2}b(x)\Phi(w,\tau). \tag{33}$$

Consider the expression $\varphi_1'(0)$ separately. Using (30), we have

$$-\frac{\partial R_1'(0)}{\partial x} = -\varphi_1'(0). \tag{34}$$

Differentiating (17) with respect to x we obtain

$$a'(x) = -\frac{\partial R_1'(0)}{\partial x}.$$

Comparing this equality with (34) we rewrite (33) in the form

$$\frac{\partial\Phi(w,\tau)}{\partial\tau} = a'(x)w\frac{\partial\Phi(w,\tau)}{\partial w} + b(x)\frac{(jw)^2}{2}\Phi(w,\tau). \tag{35}$$

This equation is a Fourier transform of the Fokker-Planck equation for the probability density $P(y,\tau)$ of the centered and normalized number of calls in orbit. Applying inverse Fourier transform to (35) we obtain

$$\frac{\partial P(y,\tau)}{\partial\tau} = -\frac{\partial}{\partial y}\{a'(x)yP(y,\tau)\} + \frac{1}{2}\frac{\partial^2}{\partial y^2}\{b(x)P(y,\tau)\}. \tag{36}$$

This is the Fokker-Planck equation for the function $P(y,\tau)$. Hence, the function $P(y,\tau)$ is a probability density of some diffusion process, which we denote as $y(\tau)$ with drift coefficient $a'(x)y$ and diffusion coefficient $b(x)$. This process is a solution of the stochastic differential equation

$$dy(\tau) = a'(x)yd\tau + \sqrt{b(x)}dw(\tau). \tag{37}$$

We consider random process of the normalized number of calls in the system

$$s(\tau) = x(\tau) + \varepsilon y(\tau), \tag{38}$$

where $\varepsilon = \sqrt{\sigma}$. We have equality (17), then using $dx(\tau) = a(x)d\tau$ we obtain

$$ds(\tau) = d(x(\tau) + \varepsilon y(\tau)) = (a(x) + \varepsilon ya'(x))d\tau + \varepsilon\sqrt{b(x)}dw(\tau). \tag{39}$$

We consider a process

$$a(s) = a(x + \varepsilon y) = a(x) + \varepsilon y a'(x) + o(\varepsilon^2),$$

$$\varepsilon\sqrt{b(s)} = \varepsilon\sqrt{b(x + \varepsilon y)} = \varepsilon\sqrt{b(x) + o(\varepsilon)} = \varepsilon\sqrt{b(x)} + o(\varepsilon^2).$$

Then the equation (39) we rewrite in the following form up to $o(\varepsilon^2)$

$$ds(\tau) = a(s)d\tau + \sqrt{\sigma b(s)}dw(\tau). \tag{40}$$

We introduce the probability density for the process $s(\tau)$

$$\pi(s,\tau) = \frac{\partial P\{s(\tau) < s\}}{\partial s}.$$

As the process $s(\tau)$ is a solution of the stochastic differential equation (40), then it is a diffusion process and for its probability density $\pi(s,\tau)$ we can write a Fokker-Planck equation

$$\frac{\partial \pi(s,\tau)}{\partial \tau} = -\frac{\partial}{\partial s}\{a(s)\pi(s,\tau)\} + \frac{1}{2}\frac{\partial^2}{\partial s^2}\{\sigma b(s)\pi(s,\tau)\}.$$

We assume that the retrial queue is functioning in stationary regime, then

$$\pi(s,\tau) = \pi(s).$$

Therefore, the Fokker-Planck equation for the stationary probability distribution $\pi(s)$ has following form

$$(-a(s)\pi(s))' + \frac{\sigma}{2}(b(s)\pi(s))'' = 0,$$

$$-a(s)\pi(s) + \frac{\sigma}{2}(b(s)\pi(s))' = 0.$$

Solving this differential equation we obtain the probability density $\pi(s)$ of the normalized number of calls in the system

$$\pi(s) = \frac{C}{b(s)}\exp\left\{\frac{2}{\sigma}\int_0^s \frac{a(x)}{b(x)}dx\right\}. \tag{41}$$

The Theorem is proved.

From the obtained probability density $\pi(s)$ we build discrete probability distribution

$$P_1(i) = \frac{\pi(\sigma i)}{\sum\limits_0^\infty \pi(\sigma i)}, \tag{42}$$

which we called a diffusion approximation of the probability distribution $P(i) = P(i(t) = i)$ of the number of calls in retrial queue, batch input and two-way communication, functioning in the stationary mode.

5 Numerical Example

We fix probability functions $B(x)$ as a gamma distribution with shape parameter s_1 and scale parameter γ_1, $s_1 = \gamma_1 = 2$.

The server makes an outgoing call with rates $\alpha_2 = 0.5$, $\alpha_3 = 0.3$, $\alpha_4 = 0.1$ and serves it for an arbitrary distributed time with probability function $V_l(x)$, $l = \overline{2,4}$ as a gamma distribution with shape parameter s_l, $l = \overline{2,4}$ and scale parameter γ_l, $l = \overline{2,4}$ and $s_2 = \gamma_2 = 0.5$, $s_3 = \gamma_3 = 1.5$, $s_4 = \gamma_4 = 3$.

The arrival rate of the batch Poisson input flow is $\lambda = 0.4$ and probability distribution of the number of calls in the batch is shifted geometric distribution with parameter $q = 0.5$.

The accuracy of an approximation we will determine using Kolmogorov range

$$\Delta = \max_{0 \leqslant i < \infty} \left| \sum_{n=0}^{i} (P(n) - P_1(n)) \right|. \tag{43}$$

Naturally, it is necessary to know the initial probability distribution $P(i)$ to find the values of Δ [9] (Table 1).

Table 1. Kolmogorov range

Δ	$\sigma = 5$	$\sigma = 1$	$\sigma = 0.5$	$\sigma = 0.2$	$\sigma = 0.1$
$\rho = 0.5$	0.080	0.056	0.033	0.022	0.017
$\rho = 0.75$	0.044	0.024	0.018	0.012	0.009
$\rho = 0.9$	0.017	0.010	0.008	0.005	0.002

Assuming that the approximation $P_1(i)$ is acceptable when its accuracy $\Delta < 0.05$. The proposed diffusion approximation $P_1(i)$ is acceptable in almost the entire spectrum of parameter values σ. The accuracy of the approximation increases (Δ decreases) with decreasing values of the parameter σ. This is quite natural due to the limiting condition $\sigma \to 0$.

An unobvious result is that the proposed approximation $P_1(i)$ is also acceptable for sufficiently large values of $0.1 \leqslant \sigma \leqslant 5$. As the load increases, the accuracy of the approximation $P_1(i)$ also increases, which indicates a very high accuracy of the proposed method.

References

1. Artalejo, J.R., Gómez-Corral, A.: Retrial queueing systems: a computational approach (2008). https://doi.org/10.1007/978-3-540-78725-9
2. Artalejo, J.R., Phung-Duc, T.: Markovian retrial queues with two way communication. J. Ind. Manag. Optim. **8**(4), 781–806 (2012)
3. Artalejo, J.R., Phung-Duc, T.: Single server retrial queues with two way communication. Appl. Math. Model. **37**(4), 1811–1822 (2013)

4. Bhulai, S., Koole, G.: A queueing model for call blending in call centers. IEEE Trans. Autom. Control **48**(8), 1434–1438 (2003)

5. Choi, B.D., Choi, K.B., Lee, Y.W.: M/G/1 retrial queueing systems with two types of calls and finite capacity. Queueing Syst. **19**(1–2), 215–229 (1995)

6. Deslauriers, A., L'Ecuyer, P., Pichitlamken, J., Ingolfsson, A., Avramidis, A.N.: Markov chain models of a telephone call center with call blending. Comput. Oper. Res. **34**(6), 1616–1645 (2007)

7. Falin, G., Templeton, J.G.: Retrial Queues, vol. 75. CRC Press, Boca Raton (1997)

8. Falin, G., Artalejo, J.R., Martin, M.: On the single server retrial queue with priority customers. Queueing Syst. **14**(3–4), 439–455 (1993)

9. Nazarov, A., Paul, S., Lizyura, O.: Retrial queue with batch input and multiple types of outgoing calls. In: ITMM 2019, pp. 245–249 (2019)

10. Nazarov, A.A., Paul, S., Gudkova, I.: Asymptotic analysis of Markovian retrial queue with two-way communication under low rate of retrials condition. In: Proceedings 31st European Conference on Modelling and Simulation, pp. 678–693 (2017)

11. Sakurai, H., Phung-Duc, T.: Scaling limits for single server retrial queues with two-way communication. Ann. Oper. Res. **247**(1), 229–256 (2015). https://doi.org/10.1007/s10479-015-1874-9

12. Sakurai, H., Phung-Duc, T.: Two-way communication retrial queues with multiple types of outgoing calls. TOP **23**(2), 466–492 (2015). https://doi.org/10.1007/s11750-014-0349-5

13. Tran-Gia, P., Mandjes, M.: Modeling of customer retrial phenomenon in cellular mobile networks. IEEE J. Sel. Areas Commun. **15**(8), 1406–1414 (1997)

A Multistage Queueing Model
with Priority for Customers Become Fit

Dhanya Babu[✉], V.C. Joshua, and A. Krishnamoorthy

Department of Mathematics, CMS College, Kottayam 686001, Kerala, India
{dhanyababu,vcjoshua,krishnamoorthy}@cmscollege.ac.in
http://www.cmscollege.ac.in

Abstract. We consider a multistage single server queueing model with
one infinite queue and two finite buffers. The primary customer on arrival
joins an infinite queue. The head of the queue finding an idle server, enter
into Stage I of service. A customer who found fit, pass on to Stage II with
probability p, $0 \leq p \leq 1$ and those who found unfit pass on to Buffer I
with complimentary probability $1 - p$. The customer from Buffer I after
an exponential duration of time move to Buffer II. A customer in Buffer II
who seems unfit again move to Buffer I. Buffer I and Buffer II individually
and collectively should not exceed capacity M. On every service completion epoch, the head of Buffer II is selected for service. We also assume
customer reneges from both infinite queue and from Buffer II within an
exponential duration of time intervals. Customers arrive according to a
Markovian arrival process (MAP). The service time follows phase type
distribution. Stability condition of the system is established. Steady-state
system size distribution is obtained. Some performance measures of the
system are evaluated.

Keywords: Multistage queues · Buffers · Reneging

1 Introduction

Queueing theory is the study of waiting in all various guises. Practically, some
important queueing models fails to answer queueing losses and so a new class
called retrial queueing systems arise. This class of queues is characterized by
the following feature: a customer on arrival, when all servers are busy leaves the
service area but after some random time repeat his demand. This field have many
applications in computer and communication networking, aircraft landing and
take-off, and in several other areas. The first mathematical result about retrial
queues were published in 1950s and applications in teletraffic theory presented
in the monograph of L. Kosten.

Queueing theory can be applied to the analysis of waiting lines in healthcare
settings. Most of healthcare systems have excess capacity to accommodate random variations, so queueing analysis can be used as short term measures, or for
facilities and resource planning. The need for application of queueing theory in

© Springer Nature Switzerland AG 2019
V. M. Vishnevskiy et al. (Eds.): DCCN 2019, LNCS 11965, pp. 223–233, 2019.
https://doi.org/10.1007/978-3-030-36614-8_17

healthcare settings is very important because the well being and life of someone is concerned. The time spent by a patient while waiting to be attended by a doctor is critical to the patient and to the image of the hospital before the public.

In recent years, many researchers came forth with the study of organ transplantation problem. Organ transplantation is an essential therapy for the treatment of many patients. [17] discussed a multi-class queueing system with reneging and the allocation policies take the form of scheduling rules for servers. The analysis of this paper is focused on two objectives one is to identify the major causes for the observed differences in the waiting time between various demographic groups and second objective is to identify policies that can eliminate such differences. A queueing model with postponed work is analyzed in [10] and a comparative study of classical and retrial queueing models is studied in [11]. Queueing analysis in healthcare systems are discussed in [5].

A multi-server tandem queues where both stations have a finite buffer and all service times are phase-type distributed discussed in [1]. Literature of tandem queueing systems can be found in [8] and [9]. A multi-server queueing system with service interruption model with two buffers are discussed in [3]. A self-promoting priority queueing model for patient waiting times which takes into account changes in health status over time is studied in [4]. The mean and distribution of the time until transplant are also obtained in that paper. [16] presents a model for restricted cross-transplantation which indicates how comparable waiting times for all blood types could be achieved.

Neuts in [13] developed the theory of PH-distributions and related point processes. In stochastic modelling, PH-distributions lend themselves naturally to algorithmic implementation and have nice closure properties with a related matrix formalism that makes them attractive for practical use. Steady-state probabilities are computed using Matrix Geometric methods [14] by Neuts. The rate matrix is computed using Ramaswami's Logarithmic reduction Algorithm by [12]. There are several methods of solving level dependent QBD. Ramaswami and Taylor presented the generalization of matrix equations to the level dependent case with possibly infinitely many phases in [15]. But in [2], equilibrium distributions and several algorithms in level dependent QBD processes are calculated. [6] defined discrete and continuous-time versions of the level dependent QBD. Klim described multi dimensional asymptotically quasi-toeplitz markov chains and their applications in queueing theory in [7].

One of the main challenges facing in organ transplantation is to find a suitable matching for an available organ. One of the main factor in matching is the fitness of a person for several reasons. Patients waiting for organ transplantation, at the time of surgery seems to be unfit for several reasons mainly of physical unfitness. This type of patients are shifted to a waiting station for the recovery. After recovered from that stage such patients again wait for surgery. But within that waiting time the patients may again unfit. This situation motivated us to model this problem. In this model we give priority to the patients who are recovered and waiting for surgery. In this paper steady- state analysis of the Level

Dependent Quasi Birth Death model is done by Neuts-Rao matrix geometric approximation [14].

2   Model Description

We consider a multistage single server queueing model (see Fig. 1) with one infinite queue and two finite buffers say Buffer I and Buffer II. Buffer I and Buffer II individually and collectively should not exceed capacity M. Customer on arrival finding a busy server joins the infinite queue. There are two stages for service station, say Stage I and Stage II. Stage I is a check- in counter and Stage II is a main counter. Each customer enters into Stage I for fitness checking and spends an exponential duration of time with parameter μ. A customer who found fit, pass on to Stage II with a probability p for service, otherwise shifted to first recovery stage say Buffer I with complimentary probability $1-p$. Each customer in Buffer I after an exponential duration of time join into second recovery stage say Buffer II with rate $j_1\mu_1$ where j_1 denotes the number of customers in Buffer I. A customer in Buffer II who found unfit again, shifted to Buffer I with an exponential rate $j_2\mu_2$ where j_2 denotes the number of customers in Buffer II. At every service completion epoch, one in Buffer II enter into service. Customers from both infinite queue and Buffer I are assumed to renege (either leads to death or in an unrecoverable stage) the system which are exponentially distributed with intensity equal to $i\theta$ where i denotes the number of customers in the queue, and $j_1\theta_1$ respectively. Service time is assumed to follow phase distribution with representation $PH(\alpha, T)$ of order s_1 where the vector T^0 is given by $T^0 = -Te$.

The MAP, a special class of tractable Markov renewal process, is a rich class of point processes that includes many well-known processes such as Poisson, PH-renewal processes, and Markov-Modulated Poisson Process. One of the most significant features of the MAP is the underlying Markov structure and fits ideally in the context of matrix-analytic solutions to stochastic models. Matrix analytic methods were first introduced and studied by Neuts. Poisson processes are the simplest and most tractable one used extensively in stochastic modelling. The idea of the MAP is to significantly generalize the Poisson processes and still keep the tractability for modelling purposes. In many practical applications, mainly in communications engineering, production and manufacturing engineering, the arrivals do not usually form a renewal process. So, MAP is a convenient tool to model both renewal and non-renewal arrivals. The customers arrive to the system with a stochastic process $\{\nu_t, t \geq 0\}$ with a state space $\{0, 1, 2,...W\}$. The sojourn time of the chain in the state i is exponentially distributed with the positive finite parameter λ_i. When the sojourn time in the state i expires, with probability $p_0(i, j)$, the process ν_t jumps to the state j without generation of customers where $i, j = \{0, 1, 2, ...W\}; i \neq j$ and with probability $p_1(i, j)$ the process ν_t jumps to the state j with generation of customers where $i, j = \{0, 1, 2, ...W\}$.

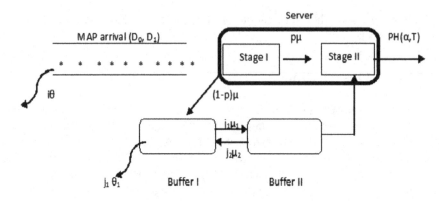

Fig. 1. Proposed queueing model

The MAP process is completely characterized by the matrices D_0 and D_1 defined by

$(D_0)_{i,i} = -\lambda_i, i = 0, 1, 2, ...W$

$(D_0)_{i,j} = \lambda_i p_0(i, j); i, j = 0, 1, 2, ...W, i \neq j$

$(D_1)_{i,j} = \lambda_i p_1(i, j); i, j = 0, 1, 2, ...W$

The point process described by the MAP is a special class of Semi-Markov processes with transition probability matrix given by

$$\int_0^x e^{(D_0 t)} dt D_1$$

By assuming D_0 to be a non-singular matrix, the inter arrival times will be finite with probability one and the arrival process does not terminate. Hence, we see that D_0 is a stable matrix. The matrix $D(1) = D_0 + D_1$ represents the generator of the process $\{\nu_t, t \geq 0\}$. The average arrival rate λ is given by

$$\lambda = \boldsymbol{\theta} D_1 \mathbf{e}$$

where $\boldsymbol{\theta}$ is the invariant vector of the stationary distribution of the Markov chain $\{\nu_t, t \geq 0\}$. The vector $\boldsymbol{\theta}$ is the unique solution to the system

$$\boldsymbol{\theta} D(1)\mathbf{e} = 0, \boldsymbol{\theta}\mathbf{e} = 1.$$

Here $\mathbf{e}$ is a column vector of appropriate size consisting of 1' s and $\mathbf{0}$ is a row vector of appropriate size consisting of zeros. The squared integral coefficient of variation of intervals between successive arrivals is given by $C_{var} = 2\lambda\boldsymbol{\theta}(-D_0)^{-1}\mathbf{e} - 1$.

Notations

- $N(t)$ be the number of customers in the infinite queue at time t
- $C(t)$ be the status of the server

$$C(t) = \begin{cases} 0, & \text{if the server is idle} \\ 1, & \text{if the server is in Stage I} \\ 2, & \text{if the server is in Stage II} \end{cases}$$

- $N_1(t)$ be the number of customers in Buffer I at time t
- $N_2(t)$ be the number of customers in Buffer II at time t
- $J_1(t)$ be the phase of the service process at time t
- $J_2(t)$ be the phase of the arrival process at time t

The above model can be represented by the Markov process

$$X^* = \{X(t)/t \geq 0\} = \{(N(t), C(t), N_1(t), N_2(t), J_1(t), J_2(t)); t \geq 0\}$$

The state space is
$\Omega = l^* \cup (\cup_{i=1}^{\infty} l(i))$ where $l^* = \{(0, 0, j_1, s_2) : j_1 \leq M, s_2 = 1, 2, ...n\}$
and
$l(i) = \{(i, 1, j_1, j_2, s_1, s_2) \cup (i, 2, j_1, j_2, s_1, s_2); i \geq 0, j_1 \leq M, j_2 \leq M, j_1 + j_2 \leq M, s_1 = 1, 2, ...m, s_2 = 1, 2, ...n\}$
This model is a level dependent quasi birth and death process (LDQBD). Quasi birth death process can be conveniently and efficiently solved by the classical matrix analytic method.

3 Steady-State Analysis

Enumerating the states of a continuous time Markov chain in lexicographic order, the infinitesimal generator of the Markov chain is of the form:

$$Q = \begin{pmatrix} A_{10} & A_{01} & & & \\ A_{21} & A_{11} & A_0 & & \\ & A_{22} & A_{12} & A_0 & \\ & & A_{23} & A_{13} & A_0 \\ & & & \ddots & \ddots & \ddots \\ & & & & \ddots & \ddots & \ddots \end{pmatrix}$$

$$A_0 = \begin{pmatrix} D_1 \otimes e_1(r) & O \\ O & D_1 \otimes e_1(r) \end{pmatrix}$$

For $i \geq 1$,

$$A_{1i} = \begin{pmatrix} F_1 & F_2 \\ O & F_4 \end{pmatrix}$$

$$F_1 = \begin{pmatrix} F_{00}^1 & F_{01}^1 & \cdots F_{0M}^1 \\ F_{10}^1 & F_{11}^1 & \cdots F_{1M}^1 \\ \vdots & & \\ F_{M0}^1 & F_{M1}^1 & \cdots F_{MM}^1 \end{pmatrix}$$

where

$$F_{00}^1 = \begin{pmatrix} (\mu - i\theta I_n) & & & \\ & (\mu - (i\theta + j_2\mu_2)I_n) & & \\ & & \ddots & (\mu - (i\theta + j_2\mu_2)I_n) \end{pmatrix}$$

$$F_{01}^1 = \begin{pmatrix} j_2\mu_2 I_n & & \\ & \ddots & \\ & & j_2\mu_2 I_n \end{pmatrix}$$

$$F_{j_1 j_1 - 1}^1 = \begin{pmatrix} j_1\theta_1 I_n & j_1\mu_1 I_n & & \\ & \ddots & \ddots & \\ & & \ddots & \ddots \\ & & & j_1\theta_1 I_n & j_1\mu_1 I_n \end{pmatrix}$$

$$F_{0M}^1 = O_{((M+1),1)}, F_{M0}^1 = O_{(1,(M+1))}$$

$$F_2 = \begin{pmatrix} F_{00}^2 & F_{01}^2 & \ldots F_{0M}^2 \\ F_{10}^2 & F_{11}^2 & \ldots F_{1M}^2 \\ \vdots & & \\ F_{M0}^2 & F_{M1}^2 & \ldots F_{MM}^2 \end{pmatrix}$$

where

$$F_{j_1 j_1}^2 = (\alpha \otimes p\mu I_n) \otimes (I_{(M+1)} - j_1)$$
$$F_{j_1 j_1 + 1}^2 = (1 - p)\mu \otimes \alpha \otimes e_{((M+1)-j_1),((M+1)-j_1-1)}(r),$$

where $r = (M, M - 1)$

$$F_4 = \begin{pmatrix} C_1 & C_{01} & O \\ C_{10} & C_{11} & O \\ \ddots & \ddots & \\ O & C_{M(M-1)} & C_{MM} \end{pmatrix}$$

$$C_1 = \begin{pmatrix} (T_0 - j_1\theta_1 I_n) \otimes I_m & & \\ T_0 \otimes \alpha & (T_0 - (j_1\theta_1 + j_2\mu_2)I_n) \otimes I_m & \\ & & T_0 \otimes \alpha \; (T_0 - (j_1\theta_1 + j_2\mu_2)I_n) \otimes I_m \end{pmatrix}$$

$$C_{01} = \begin{pmatrix} j_2\mu_2 I_n \otimes I_m & & \\ & \ddots & \\ & & \ddots \\ & & & j_2\mu_2 I_n \otimes I_m \end{pmatrix}$$

$$C_{10} = \begin{pmatrix} j_1\mu_1 I_n \otimes I_m \\ \theta_1 I_n \otimes I_m & j_1\mu_1 I_n \otimes I_m \\ & & \ddots & & \ddots \\ & & & \theta_1 I_n \otimes I_m & \mu_1 I_n \otimes I_m \end{pmatrix}$$

$$C_{11} = \begin{pmatrix} (T_0 - j_1\theta_1 I_n) \otimes I_m \\ T_0 \otimes \alpha & (T_0 - (j_1\theta_1 + j_2\mu_2)I_n) \otimes I_m \\ & & T_0 \otimes \alpha & (T_0 - (j_1(\mu_1 + \theta_1) + i\theta)I_n) \otimes I_m \end{pmatrix}$$

$$C_{M(M-1)} = \begin{pmatrix} \theta_1 I_n \otimes I_m & \mu_1 I_n \otimes I_m \end{pmatrix}$$

$$C_{MM} = \begin{pmatrix} (T - j_1(\mu_1 + \theta_1) + i\theta)I_n \otimes I_m \end{pmatrix}$$

For $i \geq 1$,

$$A_{2i} = \begin{pmatrix} E & O \\ diag(T_0 \otimes I_{M+1}, T_0 \otimes I_M, ...T_0) & i\theta_1 \otimes I_n \end{pmatrix}$$

where

$$E = \begin{pmatrix} i\theta I_n \otimes I_{M+1} & (1-p)\mu \otimes e_1(M+1, M) & & & O \\ & i\theta I_n \otimes I_{M+1} & (1-p)\mu \otimes e_1(M, M-1) \\ & & \ddots & & \ddots \\ & & & & & (1-p)\mu \otimes e_1(M, 1) \\ & & & & \ddots \\ O & & & & & i\theta I_n \otimes I_{M+1} \end{pmatrix}$$

where $e_j(r, s)$ denotes the matrix of dimension (r, s) with 1 in the jth position and zeros elsewhere. A_{10}, A_{01} and A_{21} are transition matrices in the boundary levels. This is a Level Dependent Quasi Birth Death Process (LDQBD). So we use an algorithmic solution based on Neuts Rao Truncation Method for further analysis. Let K denote the cut-off point for this truncation method. The infinitesimal generator for the process X^* is modified as

$$Q^* = \begin{pmatrix} A_{10} & A_{01} \\ A_{21} & A_{11} & A_0 \\ & A_{22} & A_{12} & A_0 \\ & & A_{23} & A_{13} & A_0 \\ & & & \cdots & \cdots & \cdots \\ & & & & \cdots & \cdots & \cdots \\ & & & & & A_{2,K-1} & A_{1,K-1} & A_0 \\ & & & & & & A_2 & A_1 & A_0 \\ & & & & & & & \cdots \cdots \cdots \end{pmatrix} \qquad (1)$$

where for $i \geq K$, $A_{1i} = A_1$ and $A_{2i} = A_2$.

3.1 Stability Condition for the Truncated System

Let π denote the steady-state probability vector of the generator $A = A_0 + A_1 + A_2$.

We see that A is an irreducible infinitesimal generator matrix and so there exists the stationary vector π of A such that $\pi A = 0$ and $\pi \mathbf{e} = 1$ where

$$\pi = (\pi_0, \pi_1)$$

The vector π, partitioned as

$$\pi = (\pi_0, \pi_1)$$

is computed by solving the equations

$$\pi_0(D_0 - K\mu I) + \pi_1(S^0 \otimes I_m) = 0$$

$$\pi_0(\beta \otimes D_1) + \pi_1(S \oplus D) = 0$$

subject to

$$\pi_0 + \pi_1 = 1$$

The system X^* is stable if and only if

$$\pi A_0 \mathbf{e} < \pi A_2 \mathbf{e}$$

ie.

$$(\pi_1 + \pi_2)[D_1 \otimes e_1(r)] < \pi_1 E + \pi_2[diag(T_0 \otimes I_{M+1}, T_0 \otimes I_M, ...T_0) + \theta_1 \otimes I_{\sum_{i \leq M} p(i)}],$$

where $p(i)$ denotes the number of partitions of i.

3.2 Computation of the Steady-State Vector

Let $\mathbf{x}$ be the steady-state probability vector of the truncated system say Q^*. Partition this vector as:
$\mathbf{x} = (\mathbf{x}_0, \mathbf{x}_1, \mathbf{x}_2, \dots)$, where $\mathbf{x}_i = (\mathbf{x}(i, 0), \mathbf{x}(i, 1), \mathbf{x}(i, 2))$ for $i \geq 0$.
Under the stability condition the steady-state probability vector of the truncated system is obtained as

$$\mathbf{x}_{(K-1)+i} = \mathbf{x}_{(K-1)} R^i, i \geq 0,$$

where K is the truncation level. where R is the minimal non negative solution to the matrix quadratic equation $R^2 A_2 + R A_1 + A_0 = 0$.

4   Performance Measures

1. Expected Number of customers in the queue

$$E[N] = \sum_{i=0}^{\infty} i\mathbf{x}(i)\mathbf{e}$$

2. Expected Number of customers in Buffer I

$$E[N_1] = \sum_{i=0}^{\infty} \sum_{j_1=0}^{M} \sum_{j_2=0}^{M-j_1} j_1 \sum_{k=0}^{2} \mathbf{x}_{ik}(j_1 j_2)\mathbf{e}$$

3. Expected Number of customers in Buffer II

$$E[N_2] = \sum_{i=0}^{\infty} \sum_{j_2=0}^{M} \sum_{j_1=0}^{M-j_2} j_2 \sum_{k=0}^{2} \mathbf{x}_{ik}(j_1 j_2)\mathbf{e}$$

4. Probability that the server is idle

$$P_0 = \sum_{0 \le j_1 \le M} \mathbf{x}_{00 j_1}(0)\mathbf{e}$$

5. Probability that the server is in Stage I

$$P_1 = \sum_{i_1=0}^{\infty} \mathbf{x}_{i1}(j_1)\mathbf{e}$$

6. Probability that the server is in Stage II

$$P_2 = \sum_{i=0}^{\infty} \mathbf{x}_{i2}(j_2)\mathbf{e}$$

7. The probability that a customer is blocked from entering into Buffer I

$$P_b = \sum_{i=0}^{\infty} (\sum_{j_1} \sum_{j_2})_{j_1+j_2=M} \mathbf{x_i}\mathbf{e}$$

8. The rate at which a customer reneging from the infinite queue

$$\eta = i\theta \sum_{i=0}^{\infty} \mathbf{x}_i\mathbf{e}$$

9. The rate at which a customer reneging from Buffer I

$$\zeta = j_1\theta_1 \sum_{i=1}^{\infty} \mathbf{x}_i\mathbf{e}$$

4.1   Conclusion

We focused on giving priority to those customers who seem to be unfit at the time of service. We target to giving immediate priority service to those customers who are successfully recovered to reduce their waiting time in the buffer before reaching their common lifetime. We intend to extend the distribution of time spent in each stage of the server by Erlang- type and to find the waiting time distribution of unfit customers.

Acknowledgement. The work of the first author is supported by Maulana Azad National Fellowship $[F1 - 17.1/2015 - 16/MANF - 2015 - 17 - KER - 65493]$ of University Grants Commission, India.

References

1. Baumann, H., Sandmann, W.: Multi-server tandem queue with Markovian arrival process, phase-type service times, and finite buffers. Eur. J. Oper. Res. **256**(1), 187–195 (2017)
2. Bright, L., Taylor, P.G.: Calculating the equilibrium distribution in level dependent quasi-birth-and-death processes. Commun. Statist.-Stochastic Models **11**(3), 497–525 (1995)
3. Dudin, A.N., Jacob, V., Krishnamoorthy, A.: A multi-server queueing system with service interruption, partial protection and repetition of service. Ann. Oper. Res. **233**(1), 101–121 (2005)
4. Drekic, S., Stanford, D.A., Woolford, D.G., McAlister, M.D.V.C.: A model for deceased-donor transplant queue waiting times. Queueing Syst. **79**(1), 87–115 (2015)
5. Green, L.: Queueing analysis in healthcare. In: Hall, R.W. (ed.) Patient Flow Reducing Delay in Healthcare Delivery. ISOR, vol. 91, pp. 281–307. Springer, Boston (2006). https://doi.org/10.1007/978-0-387-33636-7_10
6. Kharoufeh, J.P.: Level-dependent quasi-birth-and-death processes. Wiley Encycl. Oper. Res. Manag. Sci. (2011). https://doi.org/10.1002/9780470400531.eorms0460
7. Klimenok, V., Dudin, A.: Multi-dimensional asymptotically quasi-Toeplitz Markov chains and their application in queueing theory. Queueing Syst. **54**(4), 245–259 (2006)
8. Klimenok, V., Dudin, A., Vishnevsky, V.: On the stationary distribution of tandem queue consisting of a finite number of stations. In: Kwiecień, A., Gaj, P., Stera, P. (eds.) CN 2012. CCIS, vol. 291, pp. 383–392. Springer, Heidelberg (2012). https://doi.org/10.1007/978-3-642-31217-5_40
9. Klimenok, V., Dudin, A., Vishnevsky, V.: Tandem queueing system with correlated input and cross-traffic. In: Kwiecień, A., Gaj, P., Stera, P. (eds.) CN 2013. CCIS, vol. 370, pp. 416–425. Springer, Heidelberg (2013). https://doi.org/10.1007/978-3-642-38865-1_42
10. Krishnamoorthy, A., Deepak, T.G., Joshua, V.C.: Queues with postponed work. Top **12**, 375–398 (2004)
11. Krishnamoorthy, A., Joshua, V.C.: Excursion between classical and retrial queue. Inf. Syst. Technol. 315–323 (2002)
12. Latouche, G., Ramaswami, V.: Introduction to Matrix analytic Methods in Stochastic Modelling. ASA/SIAM Series on Statistics and Applied Probability (1999)

13. Neuts, M.F.: Probability distributions of phase type. In: Liber Amicorum Prof. Emeritus H. Florin, Department of Mathematics, University of Louvain, pp. 173–206 (1975)
14. Neuts, M.F.: Matrix-Geometric Solutions in Stochastic Models - An Algorithmic Approach. The Johns Hopkins University Press, Baltimore and London (1981)
15. Ramaswami, V., Taylor, P.G.: Some properties of the rate perators in level dependent quasi-birth-and-death processes with countable number of phases. Commun. Statist. Stochast. Models **12**(1), 143–164 (1996)
16. Stanford, D.A., Lee, J.M., Chandok, N., McAlister, V.: A queuing model to address waiting time inconsistency in solid-organ transplantation. Oper. Res. Health Care **3**(1), 40–45 (2014)
17. Zenios, A.S.: Modeling the transplant waiting list: a queueing model with reneging. Queueing Syst. **31**(3–4), 239–251 (1999)

Renewal Redundant Systems Under the Marshall-Olkin Failure Model. Sensitivity Analysis

Vladimir Rykov[1,2](✉) and Boyan Dimitrov[3]

[1] Gubkin Russian State University of Oil and Gas, 65 Leninsky Ave.,
Moscow 119991, Russian Federation
[2] Peoples' Friendship University of Russia (RUDN University),
6 Miklukho-Maklaya St, Moscow 117198, Russian Federation
rykov-vv@rudn.ru
[3] Kettering University, Flint, MI 48504, USA
bdimitro@kettering.edu

Abstract. Stability of various systems' characteristics with respect to changes in initial states or external factors are the key problems in all natural sciences. For stochastic systems stability often means insensitivity or low sensitivity of their output characteristics subject to changes in the shapes of some input distributions. In Kozyev et al. (2018) the reliability function for a two-components standby renewable system operating under the Marshall-Olkin failure model has been found and its asymptotic insensitivity to the shapes of its component times' distributions has been proved. In the recent paper the problem of asymptotic insensitivity of stationary and quasi-stationary probabilities for the same model are considered.

Keywords: Marshall-Olkin failure model · Sensitivity analysis · Stationary and quasi-stationary probabilities

1 Introduction and Motivation

Stability of various systems' characteristics with respect to changes in initial states or external factors are the key problems in all natural sciences. For stochastic systems stability often means insensitivity or weak sensitivity of their output characteristics subject to changes in the shapes of some input distributions.

One of the earliest results concerning insensitivity of systems' characteristics with respect to changes in the shape of service time distribution was obtained in Sevast'yanov (1957), who proved the insensitivity of Erlang's formulas to the shape of service time distribution with fixed mean value for loses in queuing systems with Poisson input flow. In Kovalenko (1976) the necessary and sufficient conditions for insensitivity of stationary reliability characteristics of a redundant renewable system with exponential life times and general repair time distributions of its components with respect to the shape of the repair time distribution has been found. These conditions consist in sufficient amount of repairing facilities, i.e. in possibility of immediate start of repair of any of the failed components. The sufficiency of this condition in the case of general life and repair time distributions was found in Rykov (2013) with the help of the multi-dimensional alternative processes theory. However, in the case of limited possibilities for restoration this result do not hold, as it was shown, for example, in Koenig a.o. (1979) with the help of additional variable method.

On the other hand in series of works Gnedenko (1964a,b) and Solov'ev (1970) it was shown that under "quick" restoration the reliability function of a cold standby double redundant renewable system tends to the exponential distribution for any life and repair time distributions of the components. This result also means the asymptotic insensitivity of reliability function of such system to the shapes of life and repair times distributions of its components. The problem of the convergence rate was not enough investigated yet. In Kalashnikov (1997) an evaluation of the convergence rate has been done in terms of moments of appropriate distributions. At that, the numerical investigations, performed in Kozyrev (2011), Rykov et al. (2017) for such systems under rare components failures demonstrated the quick enough appearance of practical insensitivity of the time dependent as well as stationary reliability characteristics with respect to the shapes of life and repair time distributions.

In series of our publications (Efrosinin and Rykov (2014), Efrosinin et al. (2014), Rykov et al. (2017), Rykov and Ngia (2014)), the review and results of which can be found in the paper Rykov (2018), the problem of systems' stationary probabilities sensitivity to the shape of life and repair time distributions of components for the same type of systems has been considered, for the case, when one of the input distributions (either of life or repair time lengths) is exponential. For these models explicit expressions for stationary probabilities have been obtained which show their evident dependence on the non-exponential distributions. Most of these investigations deal with systems where components fail independently.

In Marshall and Olkin (1967) a bivariate distribution, henceforth (MO), with dependent components defined via three independent Poisson processes, which represent three types of shocks: individual to each component and commons to both has been proposed. The model became a very popular and many publications devoted to it. However almost all of these investigations are devoted to the bivariate distributions and their properties, use the MO model only for the first failure and do not include it into the reliability process model. In the paper Kozyrev at al. (2018) a short review about MO model distribution has been done. In this paper the *stationary* and *quasi-stationary* probabilities for such a system with MO renewable failure model of its components are derived and their asymptotic insensitivity to the shapes of the repair times distributions of its components is investigated.

The paper is organized as follows. In the next section the problem set-up and some notations are introduced. In Sect. 4 the system *state stationary probabilities* (s.s.p.'s) are considered. In Sect. 5 the so called system state *quasi-stationary probabilities* (q.s.p.'s) are given. Finally, the last Sect. 6 is devoted to an investigation of the asymptotic insensitivity of stationary and the quasi-stationary probabilities with respect to the shapes of the system components' repair time distributions. The paper ends up with conclusions and description of some additional problems.

2   The Problem Setting and Notations

Consider a heterogeneous two-component redundant hot standby renewable system, wherein components fail according to the MO model. For lifetimes T_1 and T_2, the MO model is specified by the representation

$$(T_1, T_2) = (\min(A_1, A_3), \min(A_2, A_3)), \tag{1}$$

where non-negative continuous random variables A_1 and A_2 are the times to occurrence of independent "individual risk strikes" affecting the two devices. The first risk strike affects only the first component, the second one affects only the second one, while the third type of risk strike represents the time to occurrence of the "common failure" A_3 that affects both components simultaneously and leads to the failure of the entire system. It is supposed that the risk strikes are governed by independent homogeneous Poisson processes, i.e., A_i's in (1) are exponentially distributed with parameters α_i $(i = 1, 2, 3)$.

In dealing with a renewable model, we need to consider the system's renovation after its partial and/or complete failure. Here it is supposed that after a partial failure (when only one component say i, fails) the repair of type i, with random duration B_i $(i = 1, 2)$ begins.

This means that the system continues to function. After a complete system failure a repair of the whole system begins, and lasts some random time, say B_3. It is assumed that the repair times B_k $(k = 1, 2, 3)$ have absolutely continuous distributions with cumulative distribution functions (c.d.f.) $B_k(x)$ $(k = 1, 2, 3)$ and probability density functions (p.d.f.) $b_k(x)$ $(k = 1, 2, 3)$, respectively[1]. All repair times are assumed independent from the other random durations.

The system state space can be symbolically represented by $E = \{0, 1, 2, 3\}$, where 0 means that both components are working; 1 shows that the first component is being repaired, and the second one is working; 2 indicates that the second component is being repaired, and the first one is working; 3 says that both components are in down states, the system has failed and is being repaired. To describe the system's behavior we introduce a random process $\{J(t), \ t \geq 0\}$ which takes values in the phase space E, such that

$$J(t) = j, \quad \text{if at time } t \text{ the system is in state } j = 1, 2, 3.$$

---

[1] The assumption about absolute continuity of c.d.f. is used only for convenient representation of the hazard rate functions $\beta_k(x)$ and can be omitted.

Further, for the sake of shortness, we will use the following notations:

- $\alpha = \alpha_1 + \alpha_2 + \alpha_3$ is the summary risk intensity of the system failure;
- $\bar{\alpha}_1 = \alpha_1 + \alpha_3, \ \bar{\alpha}_2 = \alpha_2 + \alpha_3$;
- $b_k = \int_0^\infty x \, dB_k(x))$, $(k = 1, 2)$ is the mean repair time of a k-th component and of the whole system when $k = 3$;
- $\tilde{b}_k(s) = \int_0^\infty e^{-sx} dB_k(x) dx$ $(k = 1, 2, 3)$ is the LST of the repair time c.d.f. of a k-th component and the whole system when $k = 3$;
- $\beta_k(x) = (1 - B_k(x))^{-1} b_k(x)$ $(k = 1, 2)$ is the conditional repair intensity (hazard rate function) of a k-th component and the whole system (when $k = 3$) given that elapsed repair time is x;
- $T = \inf\{t : J(t) = 3\}$ is the system lifetime, $\quad F(t) = \mathbf{P}\{T \le t\}$ and $\tilde{f}(s)$ its LST;
- W = the system regeneration period, which represents the time interval when the system starts after a whole repair until it ends, $G(t) = \mathbf{P}\{W \le t\}$ and $\tilde{g}(s)$ its LST;
- also to shorter some formulas the following notation is used

$$\phi_i(s) = \alpha_i(1 - \tilde{b}_i(s + \bar{\alpha}_{i*})) \ (i = 1, 2), \ \ \psi(s) = \phi_1(s) + \phi_2(s), \qquad (2)$$

where $i^* = 2$ for $i = 1$ and vice versa.

3 Time Dependent Probabilities of the System During Its Lifetime Cycle

For the system behavior study the method of *additional variable* or the so-called Markovization method will be used. It consists of introducing an additional variables in order to describe the system's behavior via a Markov processes. In the case considered here, we use as such additional variable the time, spent by the state component in its J-th state subject to its last entry in it (the so-called elapsed time). We thus consider a two-dimensional Markov process $Z = \{Z(t), t \ge 0)\}$, with $Z(t) = (J(t), X(t))$ where $J(t)$ is the system state at time t, and $X(t)$ represents the elapsed time of the process in the $J(t)$-th state after its last entering in it. The process phase space is given by $\mathcal{E} = \{0, (1, x), (2, x), (3, x)\}$. Corresponding probabilities (densities with respect to additional variables) are denoted by $\pi_0(t), \pi_1(t; x), \pi_2(t; x), \pi_3(t; x)$ and we will refer to them as to the process (and the system) *micro-state* probabilities. The probabilities $\pi_i(t) = \mathbf{P}\{J(t) = j\}$ $(j = 0, 1, 2, 3)$ are called as *macro-state* process (and system) probabilities.

To calculate the time dependent system state probabilities during its life cycle the Markov process Z with absorbing state 3 should be used. Under the above assumptions, the following statement is true.

Theorem 1. *The LT $\tilde{\pi}_i(s)$ of the time dependent system state probabilities $\pi_i(t)$, $(i = 0, 1, 2\}$ and LT $\tilde{R}(s)$ of the reliability function $R(t)$ for the considered system are*

$$\tilde{\pi}_0(s) = \frac{1}{s + \alpha_3 + \psi(s)},$$

$$\tilde{\pi}_i(s) = \frac{\phi_i(s)}{(s + \bar{\alpha}_{i*})(s + \alpha_3 + \psi(s))} \quad (i = 1, 2),$$

$$\tilde{\pi}_3(s) = \frac{\bar{\alpha}_1(s + \bar{\alpha}_2)\phi_2(s) + \bar{\alpha}_2(s + \bar{\alpha}_1)\phi_1(s) + \alpha_3(s + \bar{\alpha}_1)(s + \bar{\alpha}_2)}{s(s + \bar{\alpha}_1)(s + \bar{\alpha}_2)(s + \alpha_3 + \psi(s))},$$

$$\tilde{R}(s) = \frac{(s + \bar{\alpha}_1)(s + \bar{\alpha}_2) + (s + \bar{\alpha}_2)\phi_1(s) + (s + \bar{\alpha}_1)\phi_2(s)}{(s + \bar{\alpha}_1)(s + \bar{\alpha}_2)(s + \alpha_3 + \psi(s))}, \quad (3)$$

where notations (2) is used.

Proof. By the usual method the system of Kolmogorov forward partial differential equations for the process time dependent state probabilities can be obtained:

$$\frac{d}{dt}\pi_0(t) = -\alpha\pi_0(t) + \int_0^t \pi_1(t, x)\beta_1(x)dx$$

$$+ \int_0^t \pi_2(t, x)\beta_2(x)dx;$$

$$\left(\frac{\partial}{\partial t} + \frac{\partial}{\partial x}\right)\pi_i(t; x) = -(\bar{\alpha}_{i*} + \beta_i(x))\pi_i(t; x) \quad (i = 1, 2);$$

$$\frac{d}{dt}\pi_3(t) = \alpha_3\pi_0(t) + \bar{\alpha}_1 \int_0^t \pi_2(t; x)dx$$

$$+ \bar{\alpha}_2 \int_0^t \pi_1(t; x)dx, \quad (4)$$

jointly with the initial $\pi_0(0) = 1$ and boundary conditions

$$\pi_i(t, 0) = \alpha_i\pi_0(t), \quad (i = 1, 2). \quad (5)$$

To solve this system the method of characteristics for solving the first-order partial differential equations (see Petrovskiy (1952) is used. Accordingly to this method the equations for $\pi_i(t; x)$ $(i = 1, 2)$ in (4) is determined by the system of the system of ordinary differential equations, which in symmetric form are

$$dt = dx = -\frac{d\pi_i(\cdot)}{(\alpha_{i*} + \beta_i(x))\pi_i(x)} \quad (i = 1, 2).$$

The solution of this system along characteristics $t = x$ for $x \leq t$ is

$$\pi_i(t; x) = h_i(t - x)e^{-\bar{\alpha}_{i*}x}(1 - B_i(x)) \quad (i = 1, 2), \quad (6)$$

where $h_1(\cdot)$ is some function, which is constant along characteristic and can be found from the boundary conditions (The functions $h_i(\cdot)$ in Eq. (6) are the result of application of the characteristics method to the second one of Eq. (4). However, these functions have a clear probabilistic interpretation. The states $(i, 0)$ $(i = 1, 2)$ of the process Z can be considered as partially regenerative states (the state 0 is the state of full regeneration). Times of entering into these states are consequently the times of partial and full regeneration. Thus, the functions $h_i(\cdot)$ can be considered as renewal densities of the process Z for these partial regenerative times, while the other two multipliers in formula (6) show that during time x neither failure, nor repair occurs.).

Further, from boundary conditions (5) it holds

$$\pi_i(t; 0) = h_i(t) = \alpha_i \pi_0(t) \quad (i = 1, 2). \tag{7}$$

Substitution of these solutions to the first of equation in (4) gives

$$\frac{d}{dt}\pi_0(t) = -\alpha\pi_0(t) + \int_0^t h_1(t - x)e^{-\bar{\alpha}_2 x}b_1(x)dx$$

$$+ \int_0^t h_2(t - x)e^{-\bar{\alpha}_1 x}b_2(x)dx.$$

In terms of LT with $\pi_0(0) = 1$ it holds that

$$(s + \alpha)\tilde{\pi}_0(s) - 1 = \tilde{h}_1(s)\tilde{b}_1(s + \bar{\alpha}_2) + \tilde{h}_2(s)\tilde{b}_2(s + \bar{\alpha}_1).$$

By substituting into this equation the Laplace transform $\tilde{h}_i(s) = \alpha_i \tilde{\pi}_0(s)$ of functions $h_i(t)$ $(i = 1, 2)$ from (7) after some algebra one can obtain

$$(s + \alpha)\tilde{\pi}_0(s) - \alpha_1\tilde{b}_1(s + \bar{\alpha}_2)\tilde{\pi}_0(s) - \alpha_2\tilde{b}_2(s + \bar{\alpha}_1)\tilde{\pi}_0(s) = 1.$$

From this equality taking into account that $\alpha = \alpha_1 + \alpha_2 + \alpha_3$ and the notations (2) the following representation for $\tilde{\pi}_0(s)$ follows:

$$\tilde{\pi}_0(s) = [s + \alpha_3 + \psi(s)]^{-1}. \tag{8}$$

Applying LT to functions $\pi_i(t) = \int_0^t \pi_i(t; x)dx$ in equations (6) with the help of expressions for $\tilde{h}_i(s)$ from (7) one can obtain

$$\tilde{\pi}_i(s) = \int_0^\infty e^{-st} \int_0^t \pi_i(t; x)dxdt$$

$$= \int_0^\infty e^{-st} \int_0^t h_i(t - x)e^{-\bar{\alpha}_{i*} x}(1 - B_i(x))dxdt$$

$$= \tilde{h}_i(s)\frac{1 - \tilde{b}_i(s + \bar{\alpha}_{i*})}{s + \bar{\alpha}_{i*}} = \frac{\phi_i(s)}{s + \bar{\alpha}_{i*}}\tilde{\pi}_0(s) \quad (i = 1, 2)$$

that coincides with the second expression in (3).

To find $\tilde{\pi}_3(s)$ we apply LT to the last equation of system (4). Usage of the expressions (6) for probabilities $\pi_i(t; x)$ $(i = 1, 2)$ leads as above to the following equality

$$s\tilde{\pi}_3(s) = \alpha_3\tilde{\pi}_0(s) + \bar{\alpha}_1\tilde{h}_2(s)\frac{1 - \tilde{b}_2(s + \bar{\alpha}_1)}{s + \bar{\alpha}_1}$$

$$+ \bar{\alpha}_2\tilde{h}_1(s)\frac{1 - \tilde{b}_1(s + \bar{\alpha}_2)}{s + \bar{\alpha}_2}.$$

Substitution of representations for $\tilde{h}_i(s)$ from (7) in terms of $\tilde{\pi}_0(s)$ gives

$$
\begin{aligned}
s\tilde{\pi}_3(s) &= \tilde{\pi}_0(s)\left(\frac{\bar{\alpha}_1\phi_2(s)}{s+\bar{\alpha}_1} + \frac{\bar{\alpha}_2\phi_1(s)}{s+\bar{\alpha}_2} + \alpha_3\right) \\
&= \frac{\bar{\alpha}_1(s+\bar{\alpha}_2)\phi_2(s) + \bar{\alpha}_2(s+\bar{\alpha}_1)\phi_1(s) + \alpha_3(s+\bar{\alpha}_1)(s+\bar{\alpha}_2)}{(s+\bar{\alpha}_1)(s+\bar{\alpha}_2)(s+\alpha_3+\psi_(s))},
\end{aligned}
$$

from which the expression for $\tilde{\pi}_3(s)$ in (3) follows.

Finally, taking into account that

$$
R(t) = 1 - \mathbf{P}\{T \le t\} = 1 - \pi_3(t)
$$

after some cumbersome calculations one can find

$$
\begin{aligned}
\tilde{R}(s) &= \frac{1}{s} - \tilde{\pi}_3(s) \\
&= \frac{1}{s}\left[1 - \pi_0\left(\frac{\bar{\alpha}_1\phi_2(s)}{s+\bar{\alpha}_1} + \frac{\bar{\alpha}_2\phi_1(s)}{s+\bar{\alpha}_2} + \alpha_3\right)\right] \\
&= \frac{(s+\bar{\alpha}_1)(s+\bar{\alpha}_2) + (s+\bar{\alpha}_2)\phi_1(s) + (s+\bar{\alpha}_1)\phi_2(s)}{(s+\bar{\alpha}_1)(s+\bar{\alpha}_2)(s+\alpha_3+\psi_(s))},
\end{aligned}
$$

which ends the proof.

As a corollary, by a substitution $s = 0$ one can find the mean time to the system failure.

Corollary 1. *The mean system life time with the hwlp of notations (2) can be represented as follows:*

$$
\begin{aligned}
\mathbf{E}[T] = \tilde{R}(0) &= \frac{\bar{\alpha}_1\bar{\alpha}_2 + \bar{\alpha}_2\phi_1(0) + \bar{\alpha}_1\phi_2(0)}{\bar{\alpha}_1\bar{\alpha}_2(\alpha_3+\psi_(0))} \\
&= \frac{\bar{\alpha}_1\bar{\alpha}_2 + \bar{\alpha}_2\alpha_1(1-\tilde{b}_1(\bar{\alpha}_2)) + \bar{\alpha}_1\alpha_2(1-\tilde{b}_2(\bar{\alpha}_1))}{\bar{\alpha}_1\bar{\alpha}_2(\alpha_3 + \alpha_1(1-\tilde{b}_1(\bar{\alpha}_2)) + \alpha_2(1-\tilde{b}_2(\bar{\alpha}_1)))}.
\end{aligned} \tag{9}
$$

4   Stationary Probabilities

Calculate the stationary probabilities of the considered system in terms of its initial parameters with the help of markovization method. To do that we consider introduced above two-dimensional Markov process Z, which is completely renovated after its failure, as it was mentioned in Sect. 2.

For the reason that the state 0 is a positive atom of the process Z, it is a positive recurrent (Harris) one and, therefore, has the limiting probabilities as $t \to \infty$, which coincide with the stationary ones:

$$\pi_0 = \lim_{t \to \infty} \pi_0(t), \quad \pi_i(x) = \lim_{t \to \infty} \pi_i(t; x) \ (i = 1, 2, 3).$$

Theorem 2. *The stationary micro-state probabilities of the system under consideration has the form*

$$\pi_i(x) = \alpha_i e^{-\bar{\alpha}_{i*} x}(1 - B_i(x))\pi_0 \ (i = 1, 2),$$
$$\pi_3(x) = [\alpha_1(1 - \tilde{b}_1(\bar{\alpha}_2))$$
$$+ \alpha_2(1 - \tilde{b}_2(\bar{\alpha}_1)) + \alpha_3](1 - B_3(x))\pi_0 \tag{10}$$

with appropriate macro-states probabilities

$$\pi_i = \frac{\alpha_i}{\bar{\alpha}_{i*}}(1 - \tilde{b}_i(\bar{\alpha}_{i*}))\pi_0 = \frac{\phi_i(0)}{\alpha_{i*}}\pi_0 \ (i = 1, 2),$$
$$\pi_3 = [\alpha_1(1 - \tilde{b}_1(\bar{\alpha}_2)) + \alpha_2(1 - \tilde{b}_2(\bar{\alpha}_1)) + \alpha_3]b_3\pi_0 \tag{11}$$

where π_0 is given by

$$\pi_0 = \left[1 + \alpha_1(1 - \tilde{b}_1(\bar{\alpha}_2))\left(b_3 + \frac{1}{\bar{\alpha}_2}\right)\right.$$
$$\left. + \alpha_2(1 - \tilde{b}_2(\bar{\alpha}_1))\left(b_3 + \frac{1}{\bar{\alpha}_1}\right) + \rho_3\right]^{-1}. \tag{12}$$

Proof. Analogously to the case of Markov processes with discrete states space one can write down the following system of balance equations for stationary probabilities of the process Z:

$$\alpha\pi_0 = \int_0^\infty \pi_1(x)\beta_1(x)dx + \int_0^\infty \pi_2(x)\beta_1(x)dx$$
$$+ \int_0^\infty \pi_3(x)\beta_3(x))dx,$$
$$\dot{\pi}_i(x) = -(\bar{\alpha}_{i*} + \beta_i(x))\pi_i(x) \ (i = 1, 2),$$
$$\dot{\pi}_3(x) = -\beta_3(x)\pi_3(x) \tag{13}$$

with boundary conditions

$$\pi_i(0) = \alpha_i\pi_0 \ (i = 1, 2),$$
$$\pi_3(0) = \alpha_3\pi_0 + \bar{\alpha}_1 \int_0^\infty \pi_2(x)dx + \bar{\alpha}_2 \int_0^\infty \pi_1(x)dx. \tag{14}$$

Solutions of the last two of Eq. (13) are

$$\pi_i(x) = C_i e^{-\bar{\alpha}_{i*}x}(1 - B_i(x)) \ (i = 1, 2),$$
$$\pi_3(x) = C_3(1 - B_3(x)). \tag{15}$$

Using boundary conditions (14) to find unknown constants C_i gives

$$C_i = \pi_i(0) = \alpha_i \pi_0 \ (i = 1, 2),$$
$$C_3 = \pi_3(0) = [\alpha_1(1 - \tilde{b}_1(\bar{\alpha}_2)) + \alpha_2(1 - \tilde{b}_2(\bar{\alpha}_1) + \alpha_3]\pi_0.$$

Substitution of these constants to the formulas (15) leads to (10).

Simple integration of the expressions in formulas (15) with respect to x and using the values of constants from the above equations allows to find appropriate stationary macro-state probabilities that are represented by formulas (11). The normalizing conditions gives

$$1 = \pi_0 + \pi_1 + \pi_2 + \pi_3$$
$$= \left[1 + \frac{\alpha_1(1 - \tilde{b}_1(\bar{\alpha}_2))}{\bar{\alpha}_1} + \frac{\alpha_2(1 - \tilde{b}_2(\bar{\alpha}_1))}{\bar{\alpha}_2}\right.$$
$$\left. + \alpha_1(1 - \tilde{b}_1(\bar{\alpha}_2)) + \alpha_2(1 - \tilde{b}_2(\bar{\alpha}_1)) + \alpha_3 b_3\right]\pi_0,$$

from which the formula (12) for π_0 follows that completes proof of the theorem.

5 Quasi-stationary Probabilities

For studying the system behavior at its life time more interesting than its stationary probabilities are so called quasi stationary probabilities (q.s.p.'s) which is defined as limits of conditional probabilities to be in any state given the system is not fail yet,

$$\bar{\pi}_i = \lim_{t\to\infty} \mathbf{P}\{J(t) = i | t \le T\}$$
$$= \lim_{t\to\infty} \frac{\mathbf{P}\{J(t) = i, \ t \le T\}}{\mathbf{P}\{t \le T\}} = \lim_{t\to\infty} \frac{\pi_i(t)}{R(t)}. \tag{16}$$

Because the life time coincide with the time, when the system occurs in the third state, the q.s.p's. make sense only for states $0, 1, 2$. In order to calculate the q.s.p. for these states it is possible to use LT of appropriate functions, which have been represented in the Theorem 1. Based on the above theorem the following result can be proved.

Theorem 3. *The q.s.p.'s of the model under consideration have the form*

$$\bar{\pi}_i = \lim_{t\to\infty} \frac{\pi_i(t)}{R(t)} = \frac{A_i}{A_R} \ (i = 0, 1, 2), \tag{17}$$

where values A_i, A_R are residuals of the functions $\tilde{\pi}_i(s)$ and $\tilde{R}(s)$ in the point $-\gamma$, which is the maximal root of the equation

$$\psi(s) = -s - \alpha_3. \tag{18}$$

Proof. Instead of calculation of the q.s.p.'s with the help of formula (16) we use their LT's directly. Note that the behavior of the functions $\pi_i(t)$ $(i = 0, 1, 2)$ and $R(t)$ when $t \to \infty$ depends on the roots of their LT denominators. Note also that the denominators of these functions LT are almost the same, and the behavior of the functions $\pi_i(t)$ $(i = 0, 1, 2)$ and $R(t)$ for $t \to \infty$ depends mostly on the maximal (with minimal absolute value) root of their LT denominators.

The denominator of the function $\tilde{R}(s)$ have only negative roots: $s_1 = -\bar{\alpha}_1$, $s_2 = -\bar{\alpha}_2$ and the root of the equation (18)[2]. Show that the last root is negative, which will be denoted by $-\gamma$, and minimal in absolute value between the aboves, $\gamma < \min\{\bar{\alpha}_1, \bar{\alpha}_2\}$. Consider the solution of this equation for real values of s and suppose for a certainty that $\alpha_2 > \alpha_1$. Then as it is shown at the Fig. 1 the following inequality holds

$$-\bar{\alpha}_2 < -\bar{\alpha}_1 < -\alpha_3.$$

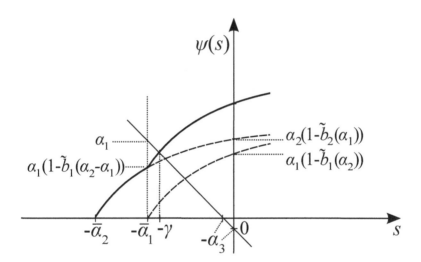

Fig. 1. Root of the equation $\psi(s) = -s - \alpha_3$.

For real values of s the functions $\tilde{b}_i(s + \alpha_{i*})$ are quite monotone ones (see Feller (1966), vol 2) and therefore are convex, thus the functions $1 - \tilde{b}_i(\cdot)$ are concave, as well as their linear combinations, which is the function $\psi(s)$. As it is possible to see from the Fig. 1 the value of the function $\psi(s)$ at point $s = -\bar{\alpha}_1$ equals to

$$\psi(-\bar{\alpha}_1) = \phi_1(-\bar{\alpha}_1 + \bar{\alpha}_2) = \alpha_1(1 - \tilde{b}_1(\alpha_2 - \alpha_1))$$

and less than α_1, which is the value of the straight line $f(s) = -s - \alpha_3$ at the point $s = \bar{\alpha}_1$. This means that the straight line $f(s) = -s - \alpha_3$ cross the curve

---

[2] Note that the maximal root of the function $\tilde{\pi}_3(s)$ is zero.

$\psi(s)$ at some point $s = -\gamma$, which is grater than $-\bar{\alpha}_1$. It shows that the equation (18) has a unique root $s = -\gamma$, which satisfy to the inequality $-\bar{\alpha}_1 < -\gamma < 0$ (see Fig. 1).

This argumentation show that the functions $\pi_i(t)$ and $R(t)$ have the forms

$$\pi_0(t) = A_0 e^{-\gamma t}(1 + \epsilon_0(t)),$$
$$\pi_i(t) = A_i e^{-\gamma t}(1 + \epsilon_i(t)) \ (i = 1, 2),$$
$$R(t) = A_R e^{-\gamma t}(1 + \epsilon_R(t)),$$

where functions $\epsilon_i(t)$ $(i = 1, 2, 3)$ and $\epsilon_R(t)$ are infinitely small with respect to $e^{-\gamma t}$ when $t \to \infty$.

These representations allow to calculate q.s.p.'s (16) as it shown in (17)

$$\bar{\pi}_i = \lim_{t \to \infty} \frac{\pi_i(t)}{R(t)} = \lim_{s \to -\gamma} \frac{\tilde{\pi}_i(s)}{\tilde{R}(s)} = \frac{A_i}{A_R} \ (i = 0, 1, 2), \tag{19}$$

where values A_i, A_R are residuals of the functions $\tilde{\pi}_i(s)$ $(i = 0, 1, 2)$ and $\tilde{R}(s)$ in the point $-\gamma$ that ends the proof

Corollary 2. *The explicit forms of the quasi-stationary probabilitied for the considered system are:*

$$\bar{\pi}_0 = \lim_{t \to \infty} \frac{\pi_0(t)}{R(t)} = \lim_{s \to -\gamma} \frac{\tilde{\pi}_0(s)}{\tilde{R}(s)} = \left[1 + \frac{\phi_1(-\gamma)}{\bar{\alpha}_2 - \gamma} + \frac{\phi_2(-\gamma))}{\bar{\alpha}_1 - \gamma}\right]^{-1},$$

$$\bar{\pi}_i = \lim_{t \to \infty} \frac{\pi_i(t)}{R(t)} = \lim_{s \to -\gamma} \frac{\tilde{\pi}_i(s)}{\tilde{R}(s)} = \frac{\phi_i(-\gamma))}{(\alpha_{i*} - \gamma)} \hat{\pi}_0 \ (i = 1, 2). \tag{20}$$

Proof. Substitution of the expressions (3) for the LT $\tilde{\pi}_i(s)$ and $\tilde{R}(s)$ of probabilities $\pi_i(t)$ $(i = 0, 1, 2)$ and of reliability function $R(t)$ into the formulas (19) after some calculations leads to the corollary statement.

6   Sensitivity Analysis

The above considerations show the evident dependence of the system reliability characteristics on the shape of repair time distributions in the form of their LTs. In this section it will be shown that this dependence becomes negligible in the case of rare failures of the components. For the considered model the rare failures should be understood as the slow intensity of failures with respect to the fixed repair times. Thus, fixing the minimal mean repair time as $c = \min\{b_1, b_2, b_3\}$ we will suppose that $q = \max\{\alpha_1, \alpha_2, \alpha_3\} \to 0$.

6.1   Asymptotic Insensitivity of the Reliability Function

The asymptotic insensitivity of the reliability function has been proved in Kozyrev a.o. (2018), but we remind it here without proof for completeness.

Naturally, analysis should be done with respect to a certain scale parameter. As such parameter the asymptotic mean system lifetime is considered. Using (2) and relations $\rho_i = \alpha_i b_i$ one can find that

$$\bar{\phi}_i(0) = \bar{\alpha}_i \alpha_i (1 - \tilde{b}_i(\bar{\alpha}_{i*})) \approx \rho_i \bar{\alpha}_i \bar{\alpha}_{i*} \quad (i = 1, 2) \tag{21}$$

as $q \to 0$, and therefore the system mean time to failure from (9) is calculated as follows:

$$m = \mathbf{E}[T] = \tilde{R}(0) = \frac{\bar{\alpha}_1 \bar{\alpha}_2 + \bar{\alpha}_1 \bar{\phi}_1(0) + \bar{\alpha}_2 \bar{\phi}_2(0)}{\bar{\alpha}_1 \bar{\alpha}_2 (\phi_1(0) + \phi_2(0) + \alpha_3)}$$

$$\approx \frac{1 + \rho_1 + \rho_2}{\bar{\alpha}_1 \rho_2 + \bar{\alpha}_2 \rho_1 + \alpha_3}. \tag{22}$$

Theorem 4. *Under rare components' failures the system reliability function becomes asymptotically insensitive to the shapes of their repair distributions and in scale of $m = \mathbf{E}[T]$ has an exponential form,*

$$\lim_{q \to 0} \mathbf{P} \left\{ \frac{T}{m} > t \right\} = e^{-t}.$$

6.2 Asymptotic Insensitivity of the Stationary Probabilities

Consider the asymptotic behavior of the stationary probabilities under the same conditions of rare failures of the components, which is modeled as $q = \max[\alpha_1, \alpha_2, \alpha_3] \to 0$.

Theorem 5. *Under the rare components' failures, as $q \to 0$, the stationary probabilities of the considered system take the following form:*

$$\pi_0 \approx [1 + \rho_1 + \rho_2 + \rho_3 + b_3(\bar{\alpha}_1 \rho_1 + \bar{\alpha}_2 \rho_2)]^{-1},$$
$$\pi_i \approx \rho_i \pi_0 \quad (i = 1, 2),$$
$$\pi_3 \approx [\rho_3 + b_3(\bar{\alpha}_1 \rho_1 + \bar{\alpha}_2 \rho_2)]\pi_0. \tag{23}$$

Proof. Applying Tailor expansion to formulas (11, 12) up to the second order of q, and taking into account that according to (21) $\phi_i(0) = \alpha_i(1 - \tilde{b}_i(\bar{\alpha}_{i*})) \approx \rho_i \bar{\alpha}_{i*}$ when $q \to 0$ one can find from (11)

$$\pi_i = \frac{\alpha_i(1 - \tilde{b}_i(\bar{\alpha}_{i*}))}{\bar{\alpha}_{i*}} \pi_0 \approx \frac{\alpha_i b_i \alpha_{i*}}{\bar{\alpha}_{i*}} \pi_0 = \rho_i \pi_0 \quad (i = 1, 2),$$
$$\pi_3 = [\alpha_1(1 - \tilde{b}_1(\bar{\alpha}_2)) + \alpha_2(1 - \tilde{b}_1(\bar{\alpha}_2)) + \alpha_3]b_3 \pi_0$$
$$\approx [(\rho_1 \bar{\alpha}_2 + \rho_2 \bar{\alpha}_1)b_3 + \rho_3]\pi_0,$$

while for π_0 from (12) it follows

$$\pi_0 \approx \left[1 + \rho_1 \bar{\alpha}_2 \left(b_3 + \frac{1}{\bar{\alpha}_2} \right) + \rho_2 \bar{\alpha}_1 \left(b_3 + \frac{1}{\bar{\alpha}_1} \right) + \rho_3 \right]^{-1}$$
$$= [1 + \rho_1 + \rho_2 + \rho_3 + (\rho_1 \bar{\alpha}_2 + \rho_2 \bar{\alpha}_1)b_3]^{-1}$$

that completes the proof of the theorem.

Remark 1. *One can notice that the part $(\bar{\alpha}_1\rho_1 + \bar{\alpha}_2\rho_2)b_3$ in probabilities π_0 and π_3 has the second order with respect to q, and therefore using only the first order of this value the above formulas can be rewritten as follows:*

$$\pi_0 \approx [1 + \rho_1 + \rho_2 + \rho_3]^{-1},$$

$$\pi_i \approx \frac{\rho_i}{1 + \rho_1 + \rho_2 + \rho_3} \quad (i = 1, 2, 3).$$

6.3   Asymptotic Insensitivity of the Quasi-stationary Probabilities

For q.s.p.'s with the help of Tailor expansion of formulas (20) as $q \to 0$ one can prove the following theorem.

Theorem 6. *Under the rare components' failures the quasi-stationary probabilities of the considered system take the form*

$$\bar{\pi}_0 \approx (1 + \rho_1 + \rho_2)^{-1},$$

$$\bar{\pi}_i \approx \frac{\rho_i}{1 + \rho_1 + \rho_2} \quad (i = 1, 2). \tag{24}$$

Proof. Indeed, by using Tailor expansion for $\hat{\pi}_0$ in formulas (20), taking into account that $\gamma < \min \alpha_i$ $(i = 1, 2)$, and that $\phi_i(-\gamma)) \approx \rho_i(\alpha_{i*} - \gamma)$ as $q \to 0$ one can find

$$\bar{\pi}_0 = \left[1 + \frac{\phi_1(-\gamma)}{\alpha_2 - \gamma} + \frac{\phi_2(-\gamma)}{\alpha_1 - \gamma}\right]^{-1} \approx (1 + \rho_1 + \rho_2)^{-1},$$

while for $\bar{\pi}_i$ $(i = 1, 2)$ the same argumentation leads to

$$\bar{\pi}_i = \frac{\phi_i(-\gamma)}{(\alpha_{i*} - \gamma)}\pi_0 \approx \frac{\rho_i}{1 + \rho_1 + \rho_2} \quad (i = 1, 2)$$

that proves the theorem.

The results of these theorems show the asymptotic insensitivity of the reliability function as well as stationary and quasi-stationary probabilities to the shapes of their components' repair distributions, but only to their mean values and the components' failure intensities.

Conclusions

Based on our previous papers Kozyrev et al. (2018) and Dimitrov and Rykov we prolong investigation of a heterogeneous double redundant hot-standby renewable system with Marshall-Olkin failure model. The stationary and quasi-stationary probability distributions of the system's states were calculated. It was shown that under rare failures the reliability characteristics are asymptotically insensitive to the shape of the components' repair time distributions up to their first moments.

Acknowledgments. The publication has been prepared with the support of the "RUDN University Program 5-100" (recipient V.V. Rykov, supervision). The reported study was funded by RFBR, project No. 17-01-00633 (recipient V.V. Rykov, formal analysis) and No. 17-07-00142 (recipient V.V. Rykov, mathematical model development).

References

Efrosinin, D., Rykov, V.: Sensitivity analysis of reliability characteristics to the shape of the life and repair time distributions. In: Dudin, A., Nazarov, A., Yakupov, R., Gortsev, A. (eds.) ITMM 2014. CCIS, vol. 487, pp. 101–112. Springer, Cham (2014). https://doi.org/10.1007/978-3-319-13671-4_13

Efrosinin, D., Rykov, V., Vishnevskiy, V.: Sensitivity of reliability models to the shape of life and repair time distributions. In: Proceedings of the 9-th International Conference on Availability, Reliability and Security (ARES 2014), pp. 430–437. IEEE (2014). Published in CD: 978-I-4799-4223-7/14. https://doi.org/10.1109/ARES 2014.65

Feller, W.: An Introduction to Probability Theory and Its Applications, vol. II. Willey, London (1966)

Gnedenko, B.V.: On cold double redundant system. Izv. AN SSSR. Texn. Cybern. **4**, 3–12 (1964a). (in Russian)

Gnedenko, B.V.: On cold double redundant system with restoration. Izv. AN SSSR. Texn. Cybern. **5**, 111–118 (1964b). (in Russian)

Kalashnikov, V.V.: Geometric Sums: Bounds for Rare Events with Applications: Risk Analysis, Reliability, Queueing, p. 256. Kluwer Academic Publishers, Dordrecht (1997)

Koenig, D., Rykov, V., Schtoyn, D.: Queueing Theory, p. 115. Gubkin University Press, Moscow (1979). (in Russian)

Kovalenko, I.N.: Investigations on Analysis of Complex Systems Reliability, p. 210. Naukova Dumka, Kiev (1976). (in Russian)

Kozyrev, D.V.: Analysis of asymptotic behavior of reliability properties of redundant systems under the fast recovery. Bull. Peoples' Friendship Univ. Russia. Ser. Math. Inf. Sci. Phys. **3**49–57 (2011). (in Russian)

Kozyrev, D.V., Rykov, V.V., Kolev, N.: Reliability function of renewable system under Marshall-Olkin failure model. In: Reliability: Theory & Applications, vol. 13, no. 1(48), pp. 39–46. Gnedenko Forum, San Diego (2018)

Marshall, A., Olkin, I.: A multivariate exponential distribution. J. Am. Stat. Assoc. **62**, 30–44 (1967)

Petrovsky, I.G.: Lectures on the theory of ordinary differential equations M.-L.: GITTL, p. 232 (1952). (in Russian)

Rykov, V.: Multidimensional alternative processes reliability models. In: Dudin, A., Klimenok, V., Tsarenkov, G., Dudin, S. (eds.) BWWQT 2013. CCIS, vol. 356, pp. 147–156. Springer, Heidelberg (2013). https://doi.org/10.1007/978-3-642-35980-4_17

Rykov, V.V., Kozyrev, D.V.: Analysis of renewable reliability systems by Markovization method. In: Rykov, V.V., Singpurwalla, N.D., Zubkov, A.M. (eds.) ACMPT 2017. LNCS, vol. 10684, pp. 210–220. Springer, Cham (2017). https://doi.org/10.1007/978-3-319-71504-9_19

Rykov, V., Kozyrev, D., Zaripova, E.: Modeling and simulation of reliability function of a homogeneous hot double redundant repairable system. In: Proceedings - 31st European Conference on Modelling and Simulation, ECMS 2017, pp. 701–705 (2017). https://doi.org/10.7148/2017-0701

Rykov, V., Ngia, T.A.: On sensitivity of systems reliability characteristics to the shape of their elements life and repair time distributions. Vestnik PFUR. Ser. Math. Inform. Phys. **3**, 65–77 (2014). (in Russian)

Rykov, V., Kozyrev, D.: On sensitivity of steady-state probabilities of a cold redundant system to the shapes of life and repair time distributions of its elements. In: Pilz, J., Rasch, D., Melas, V., Moder, K. (eds.) Statistics and Simulation. Springer Proceedings in Mathematics & Statistics, vol. 231, pp. 391–402. Springer, Cham (2018). https://doi.org/10.1007/978-3-319-76035-3_28

Rykov, V.: On reliability of renewable systems. In: Vonta, I., Ram, M. (eds.) Reliability Engineering. Theory and Applications, pp. 173–196. CRC Press (2018)

Sevast'yanov, B.A.: An ergodic theorem for Markov processes and its application to telephone systems with refusals. In: Theory of Probability and its Applications, vol. 2, no. 1 (1957)

Solov'ev, A.D.: On reservation with quick restoration. Izv. AN SSSR. Texn. Cybern. **1**, 56–71 (1970). (in Russian)

Unreliable Queueing System with Threshold Strategy of the Backup Server Connection

Valentina Klimenok[1]([⊠]), Chesoong Kim[2], V. M. Vishnevsky[3], and Alexander Dudin[1]

[1] Department of Applied Mathematics and Computer Science, Belarusian State University, Minsk 220030, Belarus
{klimenok,dudin}@bsu.by
[2] Sangji University, Wonju, Kangwon 26339, Republic of Korea
dowoo@sangji.ac.kr
[3] Institute of Control Sciences of Russian Academy of Sciences, Moscow, Russia
vishn@inbox.ru

Abstract. We consider an unreliable queuing system with an infinite buffer, the main unreliable server and absolutely reliable backup server. Customers arrive at the system according to a $BMAP$ (Batch Markovian Arrival Process). Breakdowns arrive at the main server in accordance with a MAP (Markovian Arrival Process). After a breakdown occurs, the main server fails and its repairing immediately begins. If at that moment the number of customers in the system exceeds a certain specified number (threshold), then the backup server is connected to the service of the customer whose service was interrupted by the breakdown arrival. Otherwise, the customer returns to the queue. Service and repair times have PH (Phase type) distribution. The system can be used to model a hybrid communication system consisting of FSO - Free Space Optics channel and radio wave channel. We derive the condition for the stable operation of the system, calculate the stationary distribution and the main performance measures of the system. We consider the optimization problem of choosing the threshold value minimizing the economic criterion of the quality of system operation.

Keywords: Unreliable queueing system · Backup server, threshold strategy · Stationary distribution · Performance measures · Optimization problem

1 Introduction

In recent years, the FSO - Free Space Optics technologies have become widespread due to their undoubted advantages. The main advantages of atmospheric optical (laser) communication link are high capacity and quality of communication. However, optical communication systems have also disadvantages, the main of which is the dependence of the communication channel on the

© Springer Nature Switzerland AG 2019
V. M. Vishnevskiy et al. (Eds.): DCCN 2019, LNCS 11965, pp. 249–262, 2019.
https://doi.org/10.1007/978-3-030-36614-8_19

weather condition. The unfavorable weather conditions such as rain, snow, fog, aerosols, smog can significantly reduce visibility and thus significantly reduce the effectiveness of atmospheric optical communication link.

As it is mentioned in [1], one of the main directions of creating the ultra-high speed (up to 10 Gbit/s) and reliable wireless means of communication is the development of hybrid communication systems based on laser and radio-wave technologies. Unlike the FSO channel, radio-wave IEEE802.11n channel is not sensitive to weather conditions and can be considered as absolutely reliable. However, it has a lower transmission speed compared with the FSO-channel. In hybrid communication system consisting of the FSO channel and the radio-wave IEEE802.11n channel the latter can be considered as a backup communication channel. Because of the high practical need for hybrid communication systems, a considerable amount of studies of this class of systems have appeared recently. Some results of these studies are presented in [1–7].

Papers from [2] are mainly focused on the study of stationary reliability characteristics, methods and algorithms for optimal channel switching in hybrid systems by means of simulation. The paper [3] deals with hybrid communication channel with so called *hot* redundancy, where the backup IEEE 802.11n channel continuously transmits data along with the FSO channel, but, unlike the latter, at low speed. In the paper [4], the hybrid communication system with *cold* redundancy is considered, where the radio-wave link is assumed to be absolutely reliable and backs up the atmospheric optical communication link only in cases when the latter interrupts its functioning because of the unfavorable weather conditions. The paper [1] is devoted to the study of a hybrid communication system where the millimeter-wave radio channel is used as a backup one. To model this system, the authors consider two-channel queueing system with unreliable heterogeneous servers which fail alternately. In further works [5–7], more complicated models of unreliable single-server queues are considered. They generalize models of [1,3,4], respectively, to the case of much more adequate processes describing the operation of the corresponding hybrid communication systems.

In the present paper, we consider queueing system suitable to model a hybrid communication channel with *cold* redundancy under more general, in comparison with papers [5–7], assumptions about strategy of switching between the main and backup server. The queue under consideration can be applied for modeling of hybrid communication system where the radio-wave link is assumed to be absolutely reliable (its work does not depend on the weather conditions) and backs up the atmospheric optical communication link only in cases when the latter interrupts its functioning because of the unfavorable weather conditions. Upon the occurrence of favorable weather conditions, the data packets begin transmission over the FSO channel. In order to save energy necessary for the backup server operation, a *threshold strategy* is used to connect or disconnect the backup server. The strategy is specified by a threshold which is a non-negative integer. If the number of customers in the system does not exceed the threshold,

the backup server does not connect to the service of customer even if the main
server is under repair. In this case the system does not serve customers at all.

We investigate the queue in steady state: derive the condition for the stable operation of the system, calculate the stationary distribution and the main
performance measures of the system. We consider the optimization problem of
choosing the threshold value minimizing the economic criterion of the quality of
system operation.

2 Model Description

We consider a queueing system consisting of two heterogeneous servers and infinite buffer. One of the servers (main server, server 1) is unreliable and the other
one (reserve server, server 2) is absolutely reliable. The latter is in the so-called
cold standby and connects to the service of a customer only in the case when
the main server is under repair. In order to save energy, a threshold strategy
for connecting the backup sever is used. It is defined by the threshold $j, j \geq 0$.
Under such a strategy the scenario of interaction of server 1 and server 2 is as
follows. If there is no customer in the system and the server 1 is fault-free at
an arrival epoch, this server immediately starts service of an arriving customer.
If the server 1 serves a customer at an arrival epoch, the arriving customer is
placed at the end of the queue in the buffer and is picked-up for service later on,
according to the FIFO discipline. If the server 1 is under repair and the number
i of customers in the system is greater than zero, an arriving customer is placed
in the buffer. If $i = 0$, then the arriving customer goes to the server 2 if $j > 0$
and is placed in the buffer if $j = 0$. If a breakdown arrives to the server 1 during
the service of a customer, the repair of the server 1 begins immediately. If at
the moment of the breakdown arrival the number i of customers in the system
is greater than j, the customer goes to the server 2 where it starts its service
anew. If $i \leq j$, the customer goes to the queue and waits for service. If the repair
period on the server 1 finishes but the service of a customer by the server 2 is
not completed, then the customer goes to the server 1 where it starts its service
anew.

Customers arrive into the system in accordance with a Batch Markovian
Arrival Process ($BMAP$). The $BMAP$ is defined by the underlying process
ν_t, $t \geq 0$, which is an irreducible continuous-time Markov chain with the finite
state space $\{0, \dots, W\}$, and the matrix generating function $D(z) = \sum_{k=0}^{\infty} D_k z^k$,
$|z| \leq 1$. The batches of customers enter the system only at the epochs of the
chain ν_t, $t \geq 0$, transitions. The $(W + 1) \times (W + 1)$ matrices D_k, $k \geq 1$, and
the non-diagonal entries of the matrix D_0 define the intensities of the process ν_t
transitions which are accompanied by generating the k-size batch of customers,
$k \geq 0$. The matrix $D(1)$ is an infinitesimal generator of the process ν_t. The
intensity (fundamental rate) of the $BMAP$ is defined as $\lambda = \boldsymbol{\theta} D'(1) \mathbf{e}$ where $\boldsymbol{\theta}$ is
the unique solution of the system $\boldsymbol{\theta} D(1) = \mathbf{0}$, $\boldsymbol{\theta} \mathbf{e} = 1$, and the intensity of batch
arrivals is defined as $\lambda_b = \boldsymbol{\theta}(-D_0) \mathbf{e}$. Here and in the sequel, $\mathbf{e}(\mathbf{0})$ is a column
(row) vector of appropriate size consisting of 1's (0's).

For more information about the $BMAP$, its special cases and properties see, e.g., [8], [9], [10].

Breakdowns arrive to the server 1 according to a MAP which is defined by the $(V+1) \times (V+1)$ matrices H_0 and H_1. The breakdowns fundamental rate is calculated as $h = \gamma H_1 \mathbf{e}$ where the row vector γ is the unique solution of the system $\gamma(H_0 + H_1) = \mathbf{0}$, $\gamma \mathbf{e} = 1$.

The service time of a customer on the kth server, $k = 1, 2$, has PH type distribution with an irreducible representation $(\beta^{(k)}, \mathbf{S}^{(k)})$. The service process is directed by the Markov chain $m_t^{(k)}$, $t \geq 0$, with the state space $\{1, \ldots, M^{(k)}, M^{(k)} + 1\}$ where the state $M^{(k)} + 1$ is an absorbing one. The intensities of transitions into the absorbing state are defined by the vector $\mathbf{S}_0^{(k)} = -\mathbf{S}^{(k)}\mathbf{e}$. The service rates are calculated as $\mu^{(k)} = -[\beta^{(k)}(\mathbf{S}^{(k)})^{-1}\mathbf{e}]^{-1}$, $k = 1, 2$. More detailed description of the PH distribution can be found in [11].

The repair period has PH type distribution with an irreducible representation (τ, T). The repair process is directed by the Markov chain ϑ_t, $t \geq 0$, with the state space $\{1, \ldots, R, R + 1\}$ where $R + 1$ is an absorbing state. The transitions rates into the absorbing state are defined by the vector $\mathbf{T}_0 = -T\mathbf{e}$. The repair rate is $\tau = -(\tau T^{-1}\mathbf{e})^{-1}$.

3   Process of the System States

Let, at time t,

- i_t be the number of customers in the system, $i_t \geq 0$;
- $n_t = 1$, if the main server is fault-free and $n_t = 2$, if the main server is under repair;
- $m_t^{(k)}$ is the state of underlying process of PH service time on the kth server, $m_t^{(k)} = \overline{1, M^{(k)}}$, $k = 1, 2$;
- ϑ_t is the state of underlying process of PH repair time, $\vartheta_t = \overline{1, R}$;
- ν_t and η_t are the states of underlying processes of $BMAP$ and MAP, respectively, $\nu_t = \overline{0, W}$, $\eta_t = \overline{0, V}$.

The process of the system states is described by a regular irreducible continuous time Markov chain ξ_t, $t \geq 0$, with state space

$$\Omega = \{(i, n, \nu, \eta), i = 0, n = 1, \nu = \overline{0, W}, \eta = \overline{0, V}\} \bigcup$$

$$\{(i, n, \nu, \eta, \vartheta), i = 0, n = 2, \nu = \overline{0, W}, \eta = \overline{0, V}, \vartheta = \overline{1, R}\} \bigcup$$

$$\{(i, n, \nu, \eta, m^{(1)}), i > 0, n = 1, \nu = \overline{0, W}, \eta = \overline{0, V}, m^{(1)} = \overline{1, M^{(1)}}\} \bigcup$$

$$\{(i, n, \nu, \eta, \vartheta), i = \overline{1, j}, n = 2, \nu = \overline{0, W}, \eta = \overline{0, V}, \vartheta = \overline{1, R}\} \bigcup$$

$$\{(i, n, \nu, \eta, m^{(2)}, \vartheta), i > j, n = 2, \nu = \overline{0, W}, \eta = \overline{0, V}, m^{(2)} = \overline{1, M^{(2)}}, \vartheta = \overline{1, R}\}.$$

Let us enumerate the states of the chain in the lexicographical order of its components. Denote by $Q_{i,l}$ the matrix of transition rates of the chain from the states corresponding to the value i of the first (countable) component to the states corresponding to the value l of this component, $i, l \geq 0$.

Lemma 1. *In the case $j > 0$ infinitesimal generator Q of the Markov chain ξ_t, $t \geq 0$, has the following block structure:*

$$Q = (Q_{i,l})_{i,l \geq 0}$$

$$= \begin{pmatrix}
Q_{0,0} & Q_{0,1} & Q_{0,2} & Q_{0,3} & Q_{0,4} & \cdots & Q_{0,j-1} & Q_{0,j} & | & Q_{0,j+1} & Q_{0,j+2} & \cdots \\
Q_{1,0} & \tilde{Q}_1 & \tilde{Q}_2 & \tilde{Q}_3 & \tilde{Q}_4 & \cdots & \tilde{Q}_{j-1} & \tilde{Q}_j & | & \hat{Q}_{j+1} & \hat{Q}_{j+2} & \cdots \\
O & \tilde{Q}_0 & \tilde{Q}_1 & \tilde{Q}_2 & \tilde{Q}_3 & \cdots & \tilde{Q}_{j-2} & \tilde{Q}_{j-1} & | & \hat{Q}_j & \hat{Q}_{j+1} & \cdots \\
\vdots & \vdots & \vdots & \vdots & \vdots & \ddots & & | & \vdots & \vdots & & \cdots \\
O & O & O & O & O & \cdots & \tilde{Q}_1 & \tilde{Q}_2 & | & \hat{Q}_3 & \hat{Q}_4 & \cdots \\
O & O & O & O & O & \cdots & \tilde{Q}_0 & \tilde{Q}_1 & | & \hat{Q}_2 & \hat{Q}_3 & \cdots \\
- & - & - & - & - & \cdots & - & - & - & - & - & \cdots \\
O & O & O & O & O & \cdots & O & Q_{j+1,j} & | & Q_1 & Q_2 & \cdots \\
O & O & O & O & O & \cdots & O & O & | & Q_0 & Q_1 & \cdots \\
O & O & O & O & O & \cdots & O & O & | & O & Q_0 & \cdots \\
\vdots & \vdots & \vdots & \vdots & \vdots & \vdots & & \vdots & \vdots & \vdots & & \ddots
\end{pmatrix},$$

where the non-zero blocks have the following form:

$$Q_{0,0} = \begin{pmatrix} D_0 \oplus H_0 \, I_{\bar{W}} \otimes H_1 \otimes \boldsymbol{\tau} \\ I_a \otimes T_0 \quad D_0 \oplus H \oplus T \end{pmatrix}, \quad Q_{0,k} = \begin{pmatrix} D_k \otimes I_{\bar{V}} \otimes \boldsymbol{\beta}^{(1)} & O \\ O & D_k \otimes I_{\bar{V}R} \end{pmatrix}, \, k = \overline{1, j},$$

$$Q_{0,k} = \begin{pmatrix} D_k \otimes I_{\bar{V}} \otimes \boldsymbol{\beta}^{(1)} & O \\ O & D_k \otimes I_{\bar{V}} \otimes \boldsymbol{\beta}^{(2)} \otimes I_R \end{pmatrix}, \, k > j,$$

$$Q_{1,0} = \begin{pmatrix} I_a \otimes \boldsymbol{S}_0^{(1)} & O \\ O & O_{aR} \end{pmatrix}, \quad \tilde{Q}_0 = \begin{pmatrix} I_a \otimes \boldsymbol{S}_0^{(1)} \boldsymbol{\beta}^{(1)} & O \\ O & O_{aR} \end{pmatrix},$$

$$\tilde{Q}_1 = \begin{pmatrix} D_0 \oplus H_0 \oplus S^{(1)} \, I_{\bar{W}} \otimes H_1 \otimes \mathbf{e}_{M^{(1)}} \otimes \boldsymbol{\tau} \\ I_a \otimes \boldsymbol{\beta}^{(1)} \otimes T_0^{(2)} \quad D_0 \oplus H \oplus T \end{pmatrix},$$

$$\tilde{Q}_k = \begin{pmatrix} D_{k-1} \otimes I_{\bar{V}M^{(1)}} & O \\ O & D_{k-1} \otimes I_{\bar{V}R} \end{pmatrix}, \, k \geq 2,$$

$$\hat{Q}_k = \begin{pmatrix} D_{k-1} \otimes I_{\bar{V}M^{(1)}} & O \\ O & D_{k-1} \otimes I_{\bar{V}} \otimes \boldsymbol{\beta}^{(2)} \otimes I_R \end{pmatrix}, \, k \geq 2,$$

$$Q_0 = \begin{pmatrix} I_a \otimes \boldsymbol{S}_0^{(1)} \boldsymbol{\beta}^{(1)} & O \\ O & I_a \otimes \boldsymbol{S}_0^{(2)} \boldsymbol{\beta}^{(2)} \otimes I_R \end{pmatrix},$$

$$Q_1 = \begin{pmatrix} D_0 \oplus H_0 \oplus S^{(1)} \quad I_{\bar{W}} \otimes H_1 \otimes \mathbf{e}_{M^{(1)}} \boldsymbol{\beta}^{(2)} \otimes \boldsymbol{\tau} \\ I_a \otimes \mathbf{e}_{M^{(2)}} \boldsymbol{\beta}^{(1)} \otimes T_0 \quad D_0 \oplus H \oplus S^{(2)} \oplus T \end{pmatrix},$$

$$Q_k = \begin{pmatrix} D_{k-1} \otimes I_{\bar{V}M^{(1)}} & O \\ O & D_{k-1} \otimes I_{\bar{V}M^{(2)}R} \end{pmatrix}, \, k \geq 2,$$

where $H = H_0 + H_1$, $\otimes$, $\oplus$ are symbols of Kronecker's sum and product respectively, see [12], $\bar{W} = W + 1$, $\bar{V} = V + 1$, $a = \bar{W}\bar{V}$.

The proof of the lemma is carried out by analyzing all possible transition probabilities of the chain in an infinitely small time interval.

Corollary 1. *In the case $j = 0$ the infinitesimal generator of the Markov chain ξ_t, $t \geq 0$, has the following block structure:*

$$Q = \begin{pmatrix} \bar{Q}_{0,0} & \bar{Q}_{0,1} & \bar{Q}_{0,2} & \bar{Q}_{0,3} & \bar{Q}_{0,4} & \cdots \\ \bar{Q}_{1,0} & Q_1 & Q_2 & Q_3 & Q_4 & \cdots \\ O & Q_0 & Q_1 & Q_2 & Q_3 & \cdots \\ O & O & Q_0 & Q_1 & Q_2 & \cdots \\ \vdots & \vdots & \vdots & \vdots & \vdots & \ddots \end{pmatrix},$$

where the matrix $\bar{Q}_{1,0}$ is of the form

$$\bar{Q}_{1,0} = \begin{pmatrix} I_a \otimes \boldsymbol{S}_0^{(1)} & O \\ O & I_a \otimes \boldsymbol{S}_0^{(2)} \otimes I_R \end{pmatrix}.$$

Corollary 2. *The Markov chain $\xi_t, t \geq 0$, belongs to the class of quasi-Toeplitz Markov chains with continuous time, see [13].*

Proof. The generator Q has block upper-Hessenberg structure and, for $i > j+1$, its non-zerow blocks $Q_{i,l}$ depend on the value i, l only via their difference $l - i$, precisely, $Q_{i,l} = Q_{l-i+1}$. Then, according the definition given in [13], the chain under consideration belongs to the class of quasi-Toeplitz Markov chains.

In the following, we will use expressions for the generating functions

$$\bar{Q}^{(j)} = \sum_{i=j+1}^{\infty} Q_{0,i} z^i, \quad Q(z) = \sum_{i=0}^{\infty} Q_i z^i, \quad |z| \leq 1.$$

These expressions are given by the following corollary.

Corollary 3. *The matrix generating functions $\bar{Q}(z)$, $Q(z)$ have the following forms:*

$$\bar{Q}^{(j)}(z) = \begin{pmatrix} \sum\limits_{i=j+1}^{\infty} D_i z^i \otimes I_{\bar{V}} \otimes \boldsymbol{\beta}^{(1)} & O \\ O & \sum\limits_{i=j+1}^{\infty} D_i z^i \otimes I_{\bar{V}} \otimes \boldsymbol{\beta}^{(2)} \otimes I_R \end{pmatrix}, \quad (1)$$

$$Q(z) = Q_0 + \mathcal{Q}z + z\,diag\{D(z) \otimes I_{\bar{V}M^{(1)}}, D(z) \otimes I_{\bar{V}M^{(2)}R}, \}, \quad (2)$$

where the matrix $\mathcal{Q}$ is of the form

$$\mathcal{Q} = \begin{pmatrix} I_{\bar{W}} \oplus H_0 \oplus S^{(1)} & I_{\bar{W}} \otimes H_1 \otimes \boldsymbol{e}_{M^{(1)}} \boldsymbol{\beta}^{(2)} \otimes \boldsymbol{\tau} \\ I_a \otimes \boldsymbol{e}_{M^{(2)}} \boldsymbol{\beta}^{(1)} \otimes \boldsymbol{T}_0 & I_{\bar{W}} \oplus H \oplus S^{(2)} \oplus T \end{pmatrix}.$$

4   Ergodicity Condition. Stationary Distribution

For the system under consideration, the condition for the existence of the system stationary distribution coincides with the ergodicity condition for the chain $\xi_t, t \geq 0$, which is given by the following theorem.

Theorem 1. *The Markov chain $\xi_t, t \geq 0$, is ergodic if and only if*

$$\lambda < \pi_1 S_0^{(1)} + \pi_2 S_0^{(2)}, \tag{3}$$

where

$$\pi_1 = \mathbf{x}_1(\mathbf{e}_{\bar{V}} \otimes I_{M^{(1)}}), \ \pi_2 = \mathbf{x}_2(\mathbf{e}_{\bar{V}} \otimes I_{M^{(2)}} \otimes \mathbf{e}_R), \tag{4}$$

and the vector $\mathbf{x} = (\mathbf{x}_1, \mathbf{x}_2)$ is the unique solution of the following system of linear algebraic equations:

$$\mathbf{x}\Gamma = 0, \ \mathbf{x}\mathbf{e} = 1, \tag{5}$$

where

$$\Gamma = \begin{pmatrix} H_0 \oplus S^{(1)} + I_{\bar{V}} \otimes S_0^{(1)}\beta^{(1)} & H_1 \otimes \mathbf{e}_{M^{(1)}}\beta^{(2)} \otimes \tau \\ I_{\bar{V}} \otimes \mathbf{e}_{M^{(2)}}\beta^{(1)} \otimes T_0 & H \oplus S^{(2)} \oplus T + I_{\bar{V}} \otimes S_0^{(2)}\beta^{(2)} \otimes I_R \end{pmatrix}.$$

Proof. By corollary 1, the process $\xi_t, t \geq 0$, is a quasi-Toeplitz Markov chain. Then, according [13], the necessary and sufficient condition for this chain to be ergodic is the fulfillment of the inequality

$$\mathbf{y}Q'(1)\mathbf{e} < 0, \tag{6}$$

where the vector $\mathbf{y}$ is the unique solution of the following system of linear algebraic equations:

$$\mathbf{y}Q(1) = \mathbf{0}, \tag{7}$$

$$\mathbf{y}\mathbf{e} = 1. \tag{8}$$

Represent the vector $\mathbf{y}$ in the form

$$\mathbf{y} = (\boldsymbol{\theta} \otimes \mathbf{x}_1, \boldsymbol{\theta} \otimes \mathbf{x}_2), \tag{9}$$

where $\mathbf{x} = (\mathbf{x}_1, \mathbf{x}_2)$ is a stochastic vector.

Then, taking into account the relation $\boldsymbol{\theta} \sum_{k=0}^{\infty} D_k = \mathbf{0}$, we reduce system (7)–(8) to the form (5). By substituting the vector $\mathbf{y}$ of form (9) in (6), expression for $Q'(1)$ derived using formula (2) taking into account that $\boldsymbol{\theta}D'(1)\mathbf{e} = \lambda$, we reduce inequality (6) to the following form:

$$\lambda + \mathbf{x}Q^-\mathbf{e} < 0, \tag{10}$$

where

$$Q^- = \begin{pmatrix} H_0 \oplus S^{(1)} & H_1 \otimes \mathbf{e}_{M^{(1)}}\beta^{(2)} \otimes \tau \\ I_{\bar{V}} \otimes \mathbf{e}_{M^{(2)}}\beta^{(1)} \otimes T_0 & H \oplus S^{(2)} \oplus T \end{pmatrix}.$$

Using the relations $He = (H_0+H_1)e = 0$, $Te+T_0 = 0$, we reduce inequality (10) to the form

$$\lambda < \mathbf{x}_1(e_{\bar{V}} \otimes I_{M^{(1)}})\mathbf{S}_0^{(1)} + \mathbf{x}_2(e_{\bar{V}} \otimes I_{M^{(2)}} \otimes e_R)\mathbf{S}_0^{(2)}.$$

After using the notation (4), this inequality takes the form (3).

Remark 1. Ergodicity condition (3) is easily interpreted if we take into account that this condition reflects the service process in the system under overload conditions and the vectors $\boldsymbol{\pi}_n$, $n = 1,2$, have the following meaning: the component $\pi_1(m^{(1)})$ of the vector $\boldsymbol{\pi}_1$ is the probability that server 1 is fault-free and serves a customer on the phase $m^{(1)}$, $m^{(1)} = \overline{1, M^{(1)}}$, the component $\pi_2(m^{(2)})$ of the vector $\boldsymbol{\pi}_2$ is the probability that server 1 is under repair and server 2 serves a customer on the phase $m^{(2)}$, $m^{(2)} = \overline{1, M^{(2)}}$. Then the right hand side of inequality (3) is a total service rate while the left hand side is an arrival rate. Obviously, for the existence of the stationary distribution, it is necessary and sufficient that the input rate be less than the total service rate.

Corollary 4. *In the case of stationary Poisson flow of breakdowns and exponential distributions of service and repair times, the ergodicity conditions (3)–(5) are reduced to the following inequality:*

$$\lambda < \frac{\tau}{\tau + h}\mu_1 + \frac{h}{\tau + h}\mu_2.$$

In the following, we assume that ergodicity condition (3) holds. Introduce the steady state probabilities of the chain:

$$p(0,1,\nu,\eta) = \lim_{t\to\infty} P\{i_t = 0, n_t = 1, \nu_t = \nu, \eta_t = \eta, \}, \ \nu = \overline{0,W}, \eta = \overline{0,V},$$

$$p(0,2,\nu,\eta,\vartheta) = \lim_{t\to\infty} P\{i_t = 0, n_t = 2, \nu_t = \nu, \eta_t = \eta, \vartheta_t = \vartheta\},$$

$$\nu = \overline{0,W}, \eta = \overline{0,V}, \vartheta = \overline{1,R},$$

$$p(i,1,\nu,\eta,m^{(1)}) = \lim_{t\to\infty} P\{i_t = i, n_t = 1, \nu_t = \nu, \eta_t = \eta, m_t^{(1)} = m^{(1)}\},$$

$$i > 0, \nu = \overline{0,W}, \eta = \overline{0,V}, m^{(1)} = \overline{1,M^{(1)}},$$

$$p(i,2,\nu,\eta,\vartheta) = \lim_{t\to\infty} P\{i_t = i, n_t = 2, \nu_t = \nu, \eta_t = \eta, \vartheta_t = \vartheta\},$$

$$i = \overline{1,j}, \ \nu = \overline{0,W}, \eta = \overline{0,V}, \vartheta = \overline{1,R},$$

$$p(i,2,\nu,\eta,m^{(2)},\vartheta) = \lim_{t\to\infty} P\{i_t = i, n_t = 2, \nu_t = \nu, \eta_t = \eta, m_t^{(2)} = m^{(2)}, \vartheta_t = \vartheta\},$$

$$i > j, n = 2, \ \nu = \overline{0,W}, \eta = \overline{0,V}, m^{(2)} = \overline{1,M^{(2)}}, \vartheta = \overline{1,R}.$$

Let us enumerate the steady state probabilities in the lexicographic order and form the row vectors $\mathbf{p}_i$ of these probabilities corresponding the value i of the first component, $i \geq 0$.

To calculate the vectors $\mathbf{p}_i$, we use the stable algorithm for calculation of the stationary distribution of quasi-Toeplitz Markov chain described in [13]. In what follows, we consider the vectors $\mathbf{p}_i$ $i \geq 0$, be known.

5 The Vector Generating Function. Performance Measures

Having the stationary probability vectors $\boldsymbol{p}_i$, $i \geq 0$, been calculated, we can find a number of performance measures of the system. When calculating the performance measures, the following result will be useful, especially in the case when the distribution $\boldsymbol{p}_i$, $i \geq 0$, is heavy tailed.

Lemma 2. *The generating function* $\mathbf{P}^{(j)}(z) = \sum\limits_{i=j+1}^{\infty} \boldsymbol{p}_i z^i$, $|z| \leq 1$, *satifies the following functional equation:*

$$\mathbf{P}^{(j)}(z)Q(z) = z^{j+1}\mathbf{p}_{j+1} - z\mathbf{p}_0 \bar{Q}^{(j)}(z) - \sum_{i=1}^{j} \mathbf{p}_i z^i \sum_{l=j-k+2}^{\infty} \hat{Q}_l z^l. \tag{11}$$

Taking into account that in practice the number of non-zero matrices D_k is finite, formula (11) can be used to calculate the values of the function $\mathbf{P}^{(j)}(z)$ and its derivatives at the point $z = 1$ without calculating infinite sums. The calculated values allow to find the moments of the number of customers in the system and a number of other characteristics of the system. Note that it is not possible to calculate directly the value of $\mathbf{P}^{(j)}(z)$ and its derivatives at the point $z = 1$ from Eq. (11), since the matrix $Q(1)$ is singular. This difficulty can be overcome by using the recursion formulas given below in Corollary 5.

Let $g^{(m)}(z)$ be the mth derivative of $g(z)$, $m \geq 1$, and $g^{(0)}(z) = g(z)$. Denote as $\mathcal{B}(z)$ the right hand side of Eq. (11).

Corollary 5. *The derivatives of the vector generating function* $\mathbf{P}^{(j)}(z)$, $|z| \leq 1$, *at the point $z = 1$ are calculated recursively as solutions of the following systems of linear algebraic equations:*

$$\begin{cases} \frac{d^m \mathbf{P}^{(j)}(z)}{dz^m}\big|_{z=1} Q(1) = \mathcal{B}^{(m)}(1) - \sum\limits_{l=0}^{m-1} C_m^l \frac{d^l \mathbf{P}^{(j)}(z)}{dz^l}\big|_{z=1} Q^{(m-l)}(1), \\ \frac{d^m \mathbf{P}^{(j)}(z)}{dz^m}\big|_{z=1} Q'(1)\mathbf{e} = \frac{1}{m+1}[\mathcal{B}^{(m+1)}(1) - \sum\limits_{l=0}^{m-1} C_{m+1}^l \frac{d^l \mathbf{P}^{(j)}(z)}{dz^l}\big|_{z=1} Q^{(m+1-l)}(1)]\mathbf{e}, \\ \qquad\qquad\qquad m \geq 0, \end{cases}$$

where the derivatives $\mathcal{B}^{(m)}(1)$ *and* $Q^{(m)}(1)$ *are calculated using formulas (1)–(2).*

The proof of the corollary is performed by analogy with the proof presented in [14].

Using the results obtained, we are able to calculate a number of performance measures of the system under consideration. Some of them are presented below.

(1) Throughput of the system (maximum rate of the input flow which can be passed through the system)

$$\varrho = \boldsymbol{\pi}_1 \mathbf{S}_0^{(1)} + \boldsymbol{\pi}_2 \mathbf{S}_0^{(2)}.$$

(2) Probability that there are i customers in the system

$$p_i = \mathbf{p}_i \mathbf{e}, i \geq 0.$$

(3) Mean number of customers in the system

$$L = \sum_{i=1}^{j} i p_i + \frac{d\mathbf{P}^{(j)}(z)}{dz}|_{z=1}\mathbf{e}.$$

(4) Variance of the number of customers in the system

$$V = \sum_{i=1}^{j} i^2 p_i + \frac{d^2\mathbf{P}^{(j)}(z)}{dz^2}|_{z=1} + L - L^2.$$

(5) Probability that server 1 is fault free (a share of time during which server 1 is available)

$$P^{(1)} = \mathbf{p}_0 diag\{I_a, O_{aR}\}\mathbf{e} + \sum_{i=1}^{j} \mathbf{p}_i diag\{I_{aM^{(1)}}, O_{aR}\}\mathbf{e}$$

$$+ \sum_{i=j+1}^{\infty} \mathbf{p}_i diag\{I_{aM^{(1)}}, O_{aM^{(2)}R}\}\mathbf{e}.$$

(6) Probability that server 1 is under repair and server 2 is not servicing customers (a share of time during which the system does not serve customers because server 1 is under repair and server 2 has not yet connected)

$$P^{(2-)} = \mathbf{p}_0 diag\{O_a, I_{aR}\}\mathbf{e} + \sum_{i=1}^{j} \mathbf{p}_i diag\{O_{aM^{(1)}}, I_{aR}\}\mathbf{e}.$$

(7) Probability that server 1 is under repair and server 2 serves customers (a share of time during which server 1 is under repair and server 2 serves customers)

$$P^{(2+)} = \mathbf{P}^{(j)}(1)diag\{O_{aM^{(1)}}, I_{aM^{(2)}R}\}\mathbf{e}.$$

(8) Probability that server 1 is under repair (a share of time during which server 1 is under repair)

$$P^{(2)} = P^{(2-)} + P^{(2+)}.$$

(9) Mean number of switchings from server to server per unit of time

$$\chi = \mathbf{p}_0 diag\{I_a, D^{(j+1)} \otimes I_{\bar{V}R}\}\mathbf{e} + \sum_{i=1}^{j} \mathbf{p}_i diag\{I_{aM^{(1)}}, D^{(j-i+1)} \otimes I_{\bar{V}R}\}\mathbf{e},$$

where

$$D^{(n)} = \sum_{k=n}^{\infty} D_k.$$

6   Optimization Problem

The optimization problem is formulated as the problem of choosing the optimal threshold j that minimizes the economic criterion of the quality of the system operation of the form

$$E = aL + c_1 P^{(2+)} + 2c_2 \chi, \tag{12}$$

where L is the mean number of customers in the system; a is the penalty per unit time for every customer staying in the system; $P^{(2+)}$ is a share of time during which server 1 is under repair and server 2 serves customers; c_1 is the cost per unit time of server 2 working, χ is the mean number of switchings from one server to another one per unit time, c_2 is the cost of one switch.

With this interpretation of the quantities occurred in (12), the criterion E represents an average charge per unit time under the steady-state operation of the system.

Below we present an example of numerical optimization. Our aim is to find numerically the optimal value j, $j \geq 0$, that provides the minimum value to cost criterion (12).

Consider the following input data. To define the $BMAP$ of customers we first introduce the matrices D_0 and D as

$$D_0 = \begin{pmatrix} -8.110725 & 0 \\ 0 & -0.26325 \end{pmatrix}, \quad D = \begin{pmatrix} 8.0568 & 0.053925 \\ 0.146625 & -0.26325 \end{pmatrix}.$$

Based on these matrices we construct the $BMAP$ of customers which is defined by the matrix D_0 and the matrices $D_k, k = \overline{1,3}$, which are calculates as $D_k = D q^{k-1}(1-q)/(1-q^3), k = \overline{1,3}$, where $q = 0.8$. Then we normalize the matrices $D_k, k = \overline{0,3}$, to get the $BMAP$ with fundamental rate $\lambda = 6.6$.

For this $BMAP$ $c_{cor} = 0,200504557$, $c_{var}^2 = 12,34004211$.

MAP of breakdowns are defined by the matrices

$$F_0 = \begin{pmatrix} -0.08110725 & 0 \\ 0 & -0.0026325 \end{pmatrix}, \quad F_1 = \begin{pmatrix} 0.080568 & 0.00053925 \\ 0.00146625 & 0.00116625 \end{pmatrix}$$

For this MAP $h = 1.8$, $c_{cor} = 0.200505$ and $c_{var}^2 = 12.340042$.

The PH service time distributions at server 1 and at server 2 are assumed to be the Erlangian of order 2 with parameters 20 and 15, respectively. The service rates $\mu_1 = 10, \mu_2 = 7.5$. The both squared coefficients of variation are $c_{var}^2 = 1/2$.

The PH repair time distribution is hyper-exponential one. It is defined by the vector τ and the matrix T, where

$$\tau = (0.05, 0.95), \quad T = \begin{pmatrix} -1.86075 & 0 \\ 0 & -146.9994 \end{pmatrix}.$$

For this PH, the rate of repair is $\tau = 15$ and the squared coefficient of variation of repair time is $c_{var}^2 = 25.07248$.

The cost criterion E as a function of the threshold j under different values of cost coefficients a, c_1, c_2, c_2 is presented in Fig. 1.

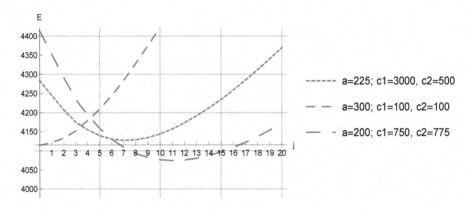

Fig. 1. Cost criterion E as a function of the threshold j

Figures 2, 3 and 4 show the behavior of performance measures $L, P^{(2+)}, \chi$ in this example as functions of threshold j.

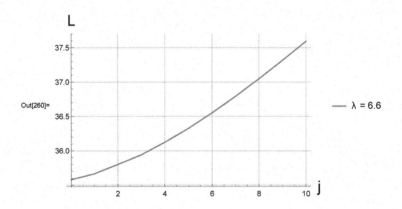

Fig. 2. Mean number of customers in the system, L, as a function of the threshold j

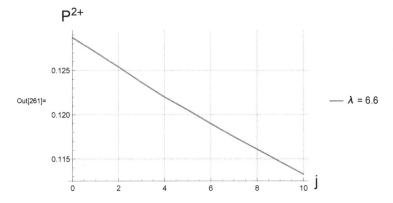

Fig. 3. A share of time during which server 2 serves customers, $P^{(2+)}$, as a function of the threshold j

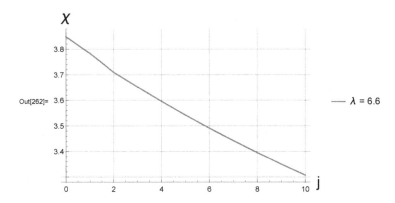

Fig. 4. Mean number of switching from server to server per unit of time, χ, as a function of the threshold j

7 Conclusion

In this paper, we investigate an unreliable single server queueing system with backup server and a Batch Markovian Arrival Process. Breakdowns arrive at the main server in accordance with a Markovian Arrival Process. During the repair period customers are served by the backup server if their number exceeds the value of threshold. Otherwise, the system does not serve the customers until the main server is recovered or the number of customers increases to the threshold. The process of the system states is described by the multidimensional Markov chain. The condition for existence of stationary distribution of this chain is derived and steady state probabilities are calculated. Expressions for the important performance measures of the system are obtained. The presented performance measures are used for numerical solving the problem of optimal choice of the threshold which minimizes the economic criterion for the quality of the system operation. The results can be exploited for performance evaluations and optimization of real world hybrid communication system consisting the atmospheric optical and radio-wave channels.

Acknowledgments. This research was supported by Basic Science Research Program through the National Research Foundation of Korea (NRF) funded by the Ministry of Education (Grant No. 2018K2A9A1A06072058) and by the joint grant of Belarusian Republican Foundation for Fundamental Research (No. F18R-136) and Russian Foundation for Fundamental Research (No. 18-57-00002).

References

1. Vishnevsky, V., Kozyrev, D., Semenova, O.V.: Redundant queueing system with unreliable servers. In: Proceedings of the 6th International Congress on Ultra Modern Telecommunications and Control Systems and Workshops (ICUMT), Moscow, pp. 383–386 (2014)
2. Arnon, S., Barry, J., Karagiannidis, G., Schober, R., Uysal, M.: Advanced Optical Wireless Communication Systems. Cambridge University Press, Cambridge (2012)
3. Vishnevsky, V.M., Semenova, O.V., Sharov, S.Y.: Modeling and analysis of a hybrid communication channel based on free-space optical and radio-frequency technologies. Autom. Remote Control **72**, 345–352 (2013)
4. Sharov, S.Y., Semenova, O.V.: Simulation model of wireless channel based on FSO and RF technologies. In: Distributed Computer and Communication Networks. Theory and Applications (DCCN 2010), pp. 368–374 (2010)
5. Dudin, A., Klimenok, V., Vishnevsky, V.: Analysis of unreliable single server queuing system with hot back-up server. Commun. Comput. Inf. Sci. **499**, 149–161 (2015)
6. Vishnevsky, V.M., Klimenok, V.I.: Unreliable queueing system with cold redundancy. Commun. Comput. Inf. Sci. **522**, 336–347 (2015)
7. Klimenok, V.I.: Two-server queueing system with unreliable servers and Markovian arrival process. Commun. Comput. Inf. Sci. **800**, 42–55 (2017)
8. Chakravarthy, S.R.: The batch Markovian arrival process: a review and future work. In: Krishnamoorthy, A., Raju, N., Ramaswami, V. (eds.) Advances in Probability Theory and Stochastic Processes, pp. 21–29. Notable Publications Inc., New Jersey (2001)
9. Lucantoni, D.: New results on the single server queue with a batch Markovian arrival process. Commun. Statist.-Stochastic Models **7**, 1–46 (1991)
10. Vishnevski, V.M., Dudin, A.N.: Queueing systems with correlated arrival flows and their applications to modeling telecommunication networks. Autom. Remote Control **78**, 1361–1403 (2017)
11. Neuts, M.: Matrix-Geometric Solutions in Stochastic Models. The Johns Hopkins University Press, Baltimore (1981)
12. Graham, A.: Kronecker Products and Matrix Calculus with Applications. Ellis Horwood, Cichester (1981)
13. Klimenok, V.I., Dudin, A.N.: Multi-dimensional asymptotically quasi-Toeplitz Markov chains and their application in queueing theory. Queueing Syst. **54**, 245–259 (2006)
14. Dudin, A., Klimenok, V., Lee, M.H.: Recursive formulas for the moments of queue length in the $BMAP/G/1$ queue. IEEE Commun. Lett. **13**, 351–353 (2009)

Detection and Detectability of Changes in a Multi-parameter Exponential Distribution

Igor Nikiforov$^{(\boxtimes)}$ (iD)

Université de Technologie de Troyes, UTT/ICD/LM2S, FRE CNRS,
12, rue Marie Curie CS 42060, 10004 Troyes Cedex, France
nikiforov@utt.fr

Abstract. The security of safety-critical systems requires special algorithmic tools detecting suddenly arriving faults, attacks, or intrusions. It is assumed that such anomalies lead to serious degradation of the system safety only if these anomalies are detected with the detection delay greater than the required time-to-alert. If the anomalies are detected with the delay smaller than or equal to the required time-to-alert, the monitored system can be reconfigurable/adaptable without compromising safety.

The goal of this paper is to study the reliable sequential detection of transient changes in a multi-parameter exponential distribution. The sequentially observed data are represented by a sequence of independent random vectors with the exponentially distributed components. The parameter vector consists of the expected values of exponentially distributed random variables (components of the vectors). This parameter vector changes at an unknown time (changepoint). It is necessary to reliably detect this changepoint. The considered optimality criterion minimizes the worst-case probability of missed detection provided that the worst-case probability of false alarm during a certain period is upper bounded. The statistical test discussed in the paper is optimal with respect to this criterion in a subclass of truncated sequential probability ratio tests.

Special attention is paid to the problem of change detectability. The maximum/minimum contrast vectors of post-change parameters are defined w.r.t. the vector of pre-change parameters by using a quadratic maximization/minimization problem. An application of the obtained results to the detection of spectral changes is also considered.

Keywords: Sequential tests · Multi-parameter exponential distribution · Transient changes · Finite moving average test · Detection of spectral changes

---

The author gratefully acknowledges the research and financial support of this work from the Commissariat à l'Energie Atomique et aux Energies Alternatives, France.

V. M. Vishnevskiy et al. (Eds.): DCCN 2019, LNCS 11965, pp. 263–275, 2019.
https://doi.org/10.1007/978-3-030-36614-8_20

1 Introduction

Sequential change-point detection deals with detecting changes in the properties of stochastic systems and processes [1–4]. The reliable detection of (transient) changes is motivated by two possible scenarios. The first scenario corresponds to the situation when the observed phenomena (say, underwater acoustic signal) is of short and maybe unknown (and random) duration Γ. It is called a transient change detection problem [5–9]. Sometimes even the "latent" detection (i.e., the detection after the end of transient change) is acceptable. The second scenario arises when the observed anomaly (say, aircraft equipment failure) leads to a serious degradation of the system safety when the change is detected with the delay greater than the required time-to-alert N, i.e., $T - t_0 + 1 > N$, where T is a stopping time and t_0 is the change-point. It is called a reliable detection of (transient) changes [10–13]. In the framework of this second scenario, the duration Γ of the transient change is assumed to be sufficient, at least $\Gamma \geq N$. If the transient change is detected with the detection delay greater than N, it is assumed to be missed. However, if the true duration Γ of the transient change is smaller than the required time-to-alert N, i.e., $\Gamma < N$, then such a transient change is considered as less dangerous because its impact on the system is limited or negligible. The present paper continuous further study of the reliable change detection considered in [10–13] for a special case of multi-parameter exponential distribution.

2 Reliable Change Detection

Let us formalize the reliable change detection problem. Consider a sequence $\{\xi_t\}_{t \geq 1}$ of independent random vectors $\xi_t \in \mathbb{R}^m$ with independent components $\xi_{1,t}, \ldots, \xi_{m,t}$. Let t_0 be the index of the first post-change observation (unknown and non-random) and the post-change period is of sufficient duration Γ such that $N \leq \Gamma$. Let us define the generative model of the transient change for the prescribed duration N:

$$\xi_t \sim \begin{cases} F_{\theta^1} & \text{if } t < t_0 \\ F_{\theta^2_{t-t_0+1}} & \text{if } t_0 \leq t \leq t_0 + N - 1, \end{cases} \tag{1}$$

where $F_{\theta^1} = \prod_{i=1}^m F_{i,\theta_i^1}$ is the pre-change joint cumulative distribution function (CDF), $F_{i,\theta_i^1} = F_{i,\theta_i^1}(x)$ is the pre-change marginal CDF of $\xi_{i,t}$ for $1 \leq t < t_0$, $F_{\theta_1^2}, \ldots, F_{\theta_N^2}$ are the post-change joint CDF's during the period N, $F_{\theta_j^2} = \prod_{i=1}^m F_{i,\theta_{i,j}^2}$ and $F_{i,\theta_{i,j}^2} = F_{i,\theta_{i,j}^2}(x)$ is the post-change marginal CDF of $\xi_{i,t}$ for $j = t - t_0 + 1$ and $t_0 \leq t \leq t_0 + N - 1$. It is worth noting that the transient change profile is defined only for N observations after change because all what happens after the time $t_0 + N - 1$ is considered as a missed detection.

Let $\mathcal{P}_{t_0}$ be the distribution of the sequence $\xi_1, \xi_2, \cdots, \xi_{t_0-1}, \xi_{t_0}, \xi_{t_0+1}, \cdots,$ ξ_{t_0+N-1} under which the observations $\xi_1, \xi_2, \cdots, \xi_{t_0-1}$ have the pre-change CDF F_{θ^1} and the observations $\xi_{t_0}, \cdots, \xi_{t_0+N-1}$ are the terms with CDF's

$F_{\theta_1^2}, \ldots, F_{\theta_N^2}$, respectively, when $t_0 < \infty$. Let $\mathcal{P}_0$ denote the distribution under which there is no change, i.e., all the observations $\xi_1, \xi_2, \cdots$ are i.i.d. with CDF F_{θ^1}. Let $\mathbb{E}_{t_0}$ (resp. $\mathbb{E}_0$) and $\mathbb{P}_{t_0}$ (resp. $\mathbb{P}_0$) be the expectation and probability w.r.t. the distribution $\mathcal{P}_{t_0}$ (resp. $\mathcal{P}_0$). The optimality criterion utilized in this paper is given by [11–13]:

$$\inf_{T \in C_\alpha} \left\{ \overline{\mathbb{P}}_{\mathrm{md}}(T) \stackrel{\mathrm{def}}{=} \sup_{t_0 \geq N} \mathbb{P}_{t_0}(T - t_0 + 1 > N \mid T \geq t_0) \right\}, \tag{2}$$

where $\overline{\mathbb{P}}_{\mathrm{md}}(T)$ is the worst-case probability of missed detection, among all stopping times $T \in C_\alpha$ satisfying

$$C_\alpha = \left\{ T : \overline{\mathbb{P}}_{\mathrm{fa}}(T; m_\alpha) \stackrel{\mathrm{def}}{=} \sup_{\ell \geq N} \mathbb{P}_0(\ell \leq T < \ell + m_\alpha) \leq \alpha \right\}, \tag{3}$$

where $\overline{\mathbb{P}}_{\mathrm{fa}}(T; m_\alpha)$ is the worst-case probability of false alarm during the reference period m_α measured in discrete time and α is the upper bound for $\overline{\mathbb{P}}_{\mathrm{fa}}(T; m_\alpha)$ used in the definition of the class C_α. Typically, the upper bound α for the probability of false alarm and the reference period m_α are usually defined by standards and norms. For instant, for some modes of civil aviation navigation, a typical value of α is of order $4 \cdot 10^{-6}$ per $m_\alpha = 15$ s (for the observation sampling period of 1 s).

3   Finite Moving Average Test

Following [13, Theorem 3], the optimization of the criterion (2)–(3) in a subclass of truncated sequential probability ratio tests leads to the finite moving average (FMA) test. The FMA test is organized as follows: the sequence $\{\xi_t\}_{t \geq 1}$ is observed sequentially and starting from the instant $t = N$, the sliding window of observations $\xi_{t-N+1}, \cdots, \xi_t$ is formed for each $t = N, N+1, N+2, \ldots$ and the decision function of the test is calculated by using these observations. The stopping time T_{FMA} of the FMA test adapted to the generative model (1) is given as follows

$$T_{\mathrm{FMA}} = \inf \left\{ n \geq N : \Lambda_{n-N+1}^n \geq h \right\}, \quad \Lambda_k^n = \sum_{j=k}^{n} \sum_{i=1}^{m} \log \frac{p_{\theta_{i,j-k+1}^2}(\xi_{i,j})}{p_{\theta_i^1}(\xi_{i,j})}, \tag{4}$$

where Λ_k^n is the log-likelihood ratio (LLR) calculated for the observations belonging to the sliding window $\xi_k, \ldots, \xi_n$. To get the upper bounds for the probabilities $\overline{\mathbb{P}}_{\mathrm{md}}(T_{\mathrm{FMA}})$ (2) and $\overline{\mathbb{P}}_{\mathrm{fa}}(T_{\mathrm{FMA}}; m_\alpha)$ (3), it is necessary to satisfy some technical conditions (see [12,13] for details).

In the present paper, it is assumed that the pre-change $p_{\theta_i^1}$ and post-change $p_{\theta_{i,j-k+1}^2}$ densities are perfectly known. In the case when the pre-change $p_{\theta_i^1}$ and post-change $p_{\theta_{i,j-k+1}^2}$ densities are unknown, a regularized estimation of the LLR Λ_k^n can be found by using two samples representing the pre-change and post-change periods. This theory and corresponding theorems establishing the convergence of the estimated LLR to the true LLR are obtained in [16].

Because the subsequent time windows of the FMA test are strongly overlapping, the most important condition is the assumption that the LLRs $\Lambda_N^N, \ldots, \Lambda_1^N, \Lambda_{N+1}^{N+1}, \ldots \Lambda_2^{N+1}, \ldots, \Lambda_{N+m_\alpha-1}^{N+m_\alpha-1}, \ldots, \Lambda_{m_\alpha}^{N+m_\alpha-1}$ are associated random variables (r.v.'s) in the pre-change mode (i.e., under the measure $\mathcal{P}_0$). The concept of associated r.v.'s and its application to different probabilistic/statistical problems have been introduced in [14,15].

Definition 1. ([14,15]). *The r.v.'s $\zeta_1, \ldots, \zeta_n$ are called associated if $\mathrm{cov}[f(\zeta_1, \ldots, \zeta_n), g(\zeta_1, \ldots, \zeta_n)] \geq 0$ for all coordinatewise nondecreasing functions f and g for which $\mathbb{E}[f(\zeta_1, \ldots, \zeta_n)]$, $\mathbb{E}[g(\zeta_1, \ldots, \zeta_n)]$, and $\mathbb{E}[f(\zeta_1, \ldots, \zeta_n)g(\zeta_1, \ldots, \zeta_n)]$ exist.*

Following [13, Corollary 2], the assumption that the LLRs are associated r.v.'s is satisfied for a family of distributions $\{F_\theta, \theta \in \mathbb{R}\}$ with a monotone likelihood ratio (LR). In contrast to such a one-parameter family, we consider now a multi-parameter family $\{F_\theta, \theta \in \mathbb{R}^m\}$. Hence, it is necessary to extend [13, Corollary 2] to a multi-parameter family of distributions. This extension is summarized in the following.

Lemma 1. *Let us consider a sequence $\{\xi_t\}_{t\geq 1}$ of independent random vectors $\xi_t \in \mathbb{R}^m$ with independent components $\xi_{1,t}, \ldots, \xi_{m,t}$. Let the marginal distributions F_i, $i = 1, \ldots, m$, of the components $\xi_{1,t}, \ldots, \xi_{m,t}$ be from the class of one-parameter distributions with a monotone (nondecreasing) LR for any $t \geq 1$. Let us define a parameter vector $\theta^1 = (\theta_1^1 \cdots \theta_m^1)^T \in \mathbb{R}^m$ and a set of parameter vectors $\theta_1^2, \ldots, \theta_N^2$, where $\theta_j^2 = (\theta_{1,j}^2 \cdots \theta_{m,j}^2)^T \in \mathbb{R}^m$ for $j = 1, \ldots, N$, such that the element-wise inequalities $\theta^1 \preceq \theta_j^2$ are satisfied for $j = 1, \ldots, N$. Then the LLRs $\Lambda_N^N, \ldots, \Lambda_1^N, \Lambda_{N+1}^{N+1}, \ldots \Lambda_2^{N+1}, \ldots, \Lambda_{N+m_\alpha-1}^{N+m_\alpha-1}, \ldots, \Lambda_{m_\alpha}^{N+m_\alpha-1}$ defined in (4) are associated r.v.'s under the measure $\mathcal{P}_0$.*

4   Multi-parameter Exponential Distribution

In many practical problems related to the queuing theory, reliability theory, reliability engineering, signal processing, the exponentially distributed variables are extensively used. This distribution is also used in physics, hydrology, econometrics, etc. For example, the service times of agents in a system are often modeled as exponentially distributed variables. Our interest in the exponential distribution is mainly motivated by some problems of signal processing.

Let us consider an exponential law $\mathrm{Exp}(\mu)$. Its probability density function (PDF) is given by

$$p_\mu(x) = \begin{cases} \frac{1}{\mu} e^{-\frac{x}{\mu}} & \text{if } x \geq 0, \\ 0 & \text{if } x < 0, \end{cases} \tag{5}$$

where μ is the scale parameter. In this case, the generative model of the transient change (1) is defined with the pre-change marginal CDF's $F_{i,\theta_i^1} = \mathrm{Exp}(\mu_i)$,

$i = 1, \ldots, m$ and the post-change marginal CDF's $F_{i,\theta_{i,j}^2} = \mathrm{Exp}(\mu_i + \vartheta_{i,j})$, $i = 1, \ldots, m$ and $j = 1, \ldots, N$. Following Lemma 1, it is assumed that all the transient change profiles are positive, i.e., $\vartheta_{i,j} > 0$, $i = 1, \ldots, m$ and $j = 1, \ldots, N$. Putting together (4) and (5), we get the LLR defined in (4)

$$\Lambda_k^n = \sum_{j=k}^{n} \sum_{i=1}^{m} \log \frac{\mu_i}{\mu_i + \vartheta_{i,j-k+1}} + \sum_{j=k}^{n} \sum_{i=1}^{m} \left(\frac{1}{\mu_i} - \frac{1}{\mu_i + \vartheta_{i,j-k+1}} \right) \xi_{i,j}. \qquad (6)$$

Following [13, Theorem 1], the worst-case probability of missed detection $\overline{\mathbb{P}}_{\mathrm{md}}(T_{\mathrm{FMA}})$ is upper bounded as follows

$$\overline{\mathbb{P}}_{\mathrm{md}}(T_{\mathrm{FMA}}) \leq G(h) \stackrel{\mathrm{def}}{=} \mathbb{P}_{t_0} \left(\Lambda_{t_0}^{N+t_0-1} < h \right), \ t_0 \geq N. \qquad (7)$$

Let us assume that the CDF of the LLR Λ_{n-N+1}^n

$$x \mapsto F(x) \stackrel{\mathrm{def}}{=} \mathbb{P}_0 \left(\Lambda_{n-N+1}^n \leq x \right), \ n \geq N \qquad (8)$$

is a continuous function under the measure $\mathcal{P}_0$. The worst-case probability of false alarm $\overline{\mathbb{P}}_{\mathrm{fa}}(T_{\mathrm{FMA}}; m_\alpha)$ for a given pre-changed period m_α is upper bounded [13, Theorems 2 and 3]

$$\overline{\mathbb{P}}_{\mathrm{fa}}(T_{\mathrm{FMA}}; m_\alpha) \leq H(h) \stackrel{\mathrm{def}}{=} 1 - [F(h)]^{m_\alpha}. \qquad (9)$$

The smallest value $\overline{\alpha}_1$ of the upper bound $G(h)$ provided that the upper bound for the worst-case probability of false alarm $H(h)$ is equal to a pre-assigned value $\overline{\alpha}_0$ and the optimal threshold h are given by [12,13]

$$\overline{\alpha}_1 = G \left[F^{-1} \left((1 - \overline{\alpha}_0)^{\frac{1}{m_\alpha}} \right) \right] \text{ and } h = F^{-1} \left((1 - \overline{\alpha}_0)^{\frac{1}{m_\alpha}} \right). \qquad (10)$$

The double sum $\sum_{j=k}^{n} \sum_{i=1}^{m} \left(\frac{1}{\mu_i} - \frac{1}{\mu_i + \vartheta_{i,j-k+1}} \right) \xi_{i,j}$ in the right hand side of equation (6) is distributed following a hypoexponential distribution (or a generalized Erlang distribution). The calculation of its PDF and CDF has been studied in [17–22]. This list is not exhaustive, but it is just a starting point to indicate some recent references on the topic. The PDF and CDF calculation is heavily based on the assumption about the following coefficients (expectations of the elements of the second double sum in (6))

$$\eta_{i,j} = 1 - \frac{\mu_i}{\mu_i + \vartheta_{i,j-k+1}} \text{ under the pre-change measure} \qquad (11)$$

and

$$\nu_{i,j} = \frac{\mu_i + \vartheta_{i,j-k+1}}{\mu_i} - 1 \text{ under the post-change measure.} \qquad (12)$$

The simplest case, where all the parameters are equal, i.e., $\eta_{i,j} = \eta$ (resp. $\nu_{i,j} = \nu$), leads to the Erlang distribution $Erl(Nm, \eta^{-1})$ (or to the Gamma distribution $\Gamma(Nm, \eta)$). The case, where the parameters $\eta_{i,j}$ (resp. $\nu_{i,j}$) are all distinct, is considered in [21]. The case, where these parameters are split into two groups (the first group includes the distributions with the equal values $\eta_{i,j}$ (resp. $\nu_{i,j}$) and the second one includes the distributions with distinct $\eta_{i,j}$) (resp. $\nu_{i,j}$) is discussed in [18]. Finally, the most general case, where the equal value parameters $\eta_{i,j}$ (resp. $\nu_{i,j}$) are grouped in an arbitrary number of distinct groups is considered in [17, 19]. The analysis of the hypoexponential distribution shows that the asymptotic Gaussian approximation for the CDF's $h \mapsto G(h)$ (7) and $x \mapsto F(x)$ (8) can be used for a preliminary calculation of $\overline{\alpha}_0$, $\overline{\alpha}_1$ and h in the case of large values of N and m.

5   Asymptotic Gaussian Approximation

As it follows from [17–22], the calculation of the hypoexponential distribution and its density (or their accurate approximations) for some combinations of the coefficients $\eta_{i,j}$ (resp. $\nu_{i,j}$) can lead to serious numerical difficulties. For this reason, it seems appealing to replace a complex calculation of the hypoexponential CDF by a simple Gaussian approximation, at least for a preliminary estimation of the relation between $\overline{\alpha}_1$ and $\overline{\alpha}_0$. Certainly, such an approximation can be more or less accurate only for large values of N and m but it is also interesting what happens for some moderate values of N and m.

Let us consider several examples to assess the accuracy of the asymptotic Gaussian approximation for the CDF's $h \mapsto G(h)$ (7) and $x \mapsto F(x)$ (8) and, also, to estimate the impact of this approximation on the relation between $\overline{\alpha}_1$ and $\overline{\alpha}_0$ given by Eq. (10).

Example 1. Let us begin with a simple example where $m = 2$ and $N = 1$ and the coefficients $\eta_{i,j} = \eta$ (resp. $\nu_{i,j} = \nu$) are different. The multi-parameter exponential distribution is defined as follows: $\mu_1 = 1$, $\mu_2 = 2$, $\vartheta_{i,1} = 3$ and $\vartheta_{i,1} = 5$. It is assumed that $m_\alpha = 5$. Obviously, the values of m and N are very small and hence the accuracy of the asymptotic Gaussian approximation should be very poor. The upper bound $\overline{\alpha}_1$ as a function of the upper bound $\overline{\alpha}_0$ is shown in Fig. 1. The function $\overline{\alpha}_0 \mapsto \overline{\alpha}_1(\overline{\alpha}_0)$ obtained by using the hypoexponential CDF is drawn in solid line and the same function obtained by the Gaussian approximation of the LLR Λ_k^n is drawn in dash-dotted line. It follows from Fig. 1 that the accuracy of such an approximation is quite poor.

Example 2. Let us consider another situation. Now, the multi-parameter exponential distribution is defined as follows: $\mu_i = \mu = 1$ for $i = 1, \ldots, m$ and $\vartheta_{i,j} = \vartheta = 0.5$ for $i = 1, \ldots, m$ and $j = 1, \ldots, N$, where $m = 10$ and $N = 10$. It is assumed that $m_\alpha = 100$. We begin with the accuracy of the asymptotic Gaussian approximation for the CDF $x \mapsto F(x) = Erl(Nm, \eta^{-1})$ (8) under the pre-change measure $\mathcal{P}_0$. The results of numerical calculation by using the standard MATLAB functions are shown in Fig. 2. The CDF of the Erlang distribution $Erl(Nm, \eta^{-1})$ (or Gamma distribution $\Gamma(Nm, \eta)$) is drawn in solid line. Seeking simplicity, the function $x \mapsto Erl(x; Nm, \eta^{-1})$ is drawn for $x \leq 33.23$

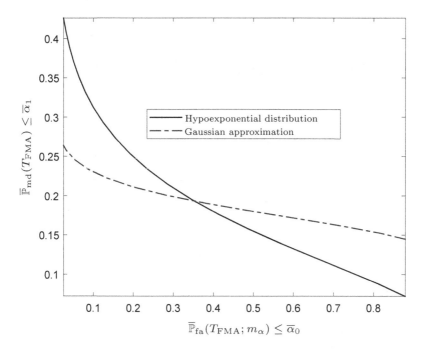

Fig. 1. The probability $\overline{\alpha}_1$ as a function of $\overline{\alpha}_0$.

and the function $x \mapsto 1 - Erl(x; Nm, \eta^{-1})$ is drawn for $x > 33.23$. As it follows from Fig. 2, the accuracy of the asymptotic Gaussian approximation is acceptable only for large values of probability $Erl(x; Nm, \eta^{-1})$ or $1 - Erl(x; Nm, \eta^{-1})$ lower bounded by 10^{-1}.

The upper bound $\overline{\alpha}_1$ as a function of the upper bound $\overline{\alpha}_0$ calculated by using the exact Erlang distribution and by using the asymptotic Gaussian approximation is shown in Fig. 3. The exact function $\overline{\alpha}_0 \mapsto \overline{\alpha}_1(\overline{\alpha}_0)$ is drawn in solid line and the Gaussian approximation of the LLR Λ_k^n is drawn in dash-dotted line. It follows from Fig. 3 that the accuracy of the asymptotic Gaussian approximation is better than the accuracy of such an approximation in Example 1.

Example 3. Let us consider the following case of multi-parameter exponential distribution: $\mu_i = \mu = 1$ for $i = 1, \ldots, m$ and $\vartheta_{i,j} = \vartheta = 0.1$ for $i = 1, \ldots, m$ and $j = 1, \ldots, N$, where $m = 100$ and $N = 100$. It is again assumed that $m_\alpha = 100$. We begin with the accuracy of the asymptotic Gaussian approximation for the CDF $x \mapsto F(x) = Erl(x; Nm, \eta^{-1})$ (8) under the pre-change measure $\mathcal{P}_0$. The results of numerical calculation by using the standard MATLAB functions are shown in Fig. 4. The CDF of the Erlang distribution $Erl(Nm, \eta^{-1})$ (or Gamma distribution $\Gamma(Nm, \eta)$) is drawn in solid line. Seeking simplicity, the function $x \mapsto Erl(x; Nm, \eta^{-1})$ is drawn for $x \leq 909$ and the function $x \mapsto 1 - Erl(x; Nm, \eta^{-1})$ is drawn for $x > 909$.

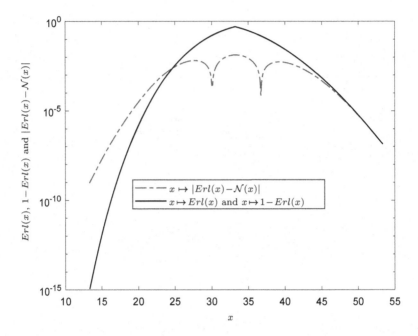

Fig. 2. CDF of Erlang distribution $x \mapsto Erl(x; Nm, \eta^{-1})$, $x \mapsto 1 - Erl(x; Nm, \eta^{-1})$ and the error of the asymptotic Gaussian approximation $|Erl(x; Nm, \eta^{-1}) - \mathcal{N}(x; Nm\eta, Nm\eta^2)|$.

The error $|Erl(x; Nm, \eta^{-1}) - \mathcal{N}(x; Nm\eta, Nm\eta^2)|$ of the asymptotic Gaussian approximation for the CDF $x \mapsto F(x) = Erl(x; Nm, \eta^{-1})$ is drawn in dash-dotted line. As it follows from Fig. 4, the accuracy of the asymptotic Gaussian approximation is acceptable, at least for values of probability $Erl(x; Nm, \eta^{-1})$ or $1 - Erl(x; Nm, \eta^{-1})$ which are lower bounded by 10^{-4}.

The upper bound $\overline{\alpha}_1$ as a function of the upper bound $\overline{\alpha}_0$ calculated by using the exact Erlang distribution and by using the asymptotic Gaussian approximation is shown in Fig. 5. The exact function $\overline{\alpha}_0 \mapsto \overline{\alpha}_1(\overline{\alpha}_0)$ is drawn in solid line and the Gaussian approximation of the LLR Λ_k^n is drawn in dash-dotted line.

6 Detectability of Changes in a Multi-parameter Exponential Distribution

In many practical problems, it is necessary to estimate the detectability of changes, i.e., to define the best/worst possible transient change profile $\vartheta_{i,j}$, $i = 1, \ldots, m$, $j = 1, \ldots, N$ for a given vector $(\mu_1 \cdots \mu_m)$. Seeking simplicity, the only case of *constant* transient change profile is considered now, i.e., it is assumed that $\vartheta_{i,j} = \vartheta_i$ for all $i = 1, \ldots, m$ and $j = 1, \ldots, N$. Let us assume that the post-change marginal CDF's are defined as follows $F_{i,\theta_{i,j}^2} = \mathrm{Exp}(\mu_i + \beta\vartheta_i)$, where the coefficient $\beta > 0$ plays the role of the signal-to-noise ratio (SNR).

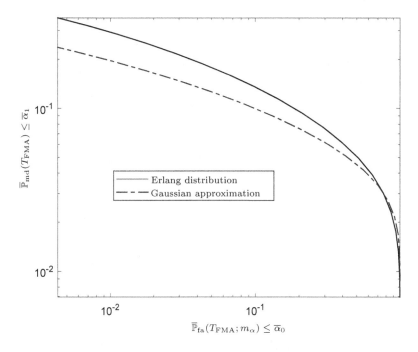

Fig. 3. The probability $\overline{\alpha}_1$ as a function of $\overline{\alpha}_0$.

Let $X = \left(\frac{\vartheta_1}{\mu_1} \cdots \frac{\vartheta_m}{\mu_m}\right)$ and $Y = \left(\frac{\vartheta_1}{\mu_1+\beta\vartheta_1} \cdots \frac{\vartheta_m}{\mu_m+\beta\vartheta_m}\right)$ be two vectors of size m. The expectations and variances of the LLR (6) (under the measures $\mathcal{P}_0$ and $\mathcal{P}_{t_0}$) are given as follows

$$\mathbb{E}_{t_0}(\Lambda_{t_0}^{N+t_0-1}) - \mathbb{E}_0(\Lambda_1^N) = N\beta^2(X, Y),$$
$$\mathrm{var}_0(\Lambda_1^N) = N\beta\|Y\|_2^2,$$
$$\mathrm{var}_{t_0}(\Lambda_{t_0}^{N+t_0-1}) = N\beta\|X\|_2^2, \tag{13}$$

where $(X, Y) = \sum_{i=1}^m x_i y_i$ is the inner product and $\|X\|_2$ is the Euclidean norm of the vector X.

We wish to find the best (resp. worst) profile $\vartheta_1, \ldots, \vartheta_m$ w.r.t. the pre-change mean vector $\mu_1, \ldots, \mu_m$ for calculating the LLR (6). The best profile minimizes $\overline{\alpha}_1$ for a given $\overline{\alpha}_0$ value and for a given (small) SNR β. The worst profile maximizes $\overline{\alpha}_1$ for a given $\overline{\alpha}_0$ values and for a given (small) SNR β. Taking into account the asymptotic Gaussian approximation for the CDF's $h \mapsto G(h)$ and $x \mapsto F(x)$, we get the following

Proposition 1. *Let us consider the FMA test* (4), (6) *in the case of weak signals, i.e., the SNR $\beta \to 0$ and $N \to \infty$ such that $\beta\sqrt{N} \to \varrho$, where ϱ is a given constant. Then*

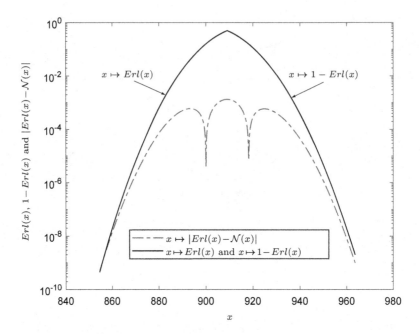

Fig. 4. CDF of Erlang distribution $x \mapsto Erl(x; Nm, \eta^{-1})$, $x \mapsto 1 - Erl(x; Nm, \eta^{-1})$ and the error of the asymptotic Gaussian approximation $|Erl(x; Nm, \eta^{-1}) - \mathcal{N}(x; Nm\eta, Nm\eta^2)|$.

1. *The finding of the best possible profile* $\vartheta_1, \ldots, \vartheta_m$ *is reduced to*

$$\underset{\vartheta_1,\ldots,\vartheta_m}{maximize}\{\|X(\vartheta_1,\ldots,\vartheta_m)\|_2\} \ subject \ to \sum_{i=1}^{m}\vartheta_i = \vartheta_0, \ \vartheta_1 \geq 0, \ldots, \vartheta_m \geq 0. \quad (14)$$

2. *The finding of the worst possible profile* $\vartheta_1, \ldots, \vartheta_m$ *is reduced to*

$$\underset{\vartheta_1,\ldots,\vartheta_m}{minimize}\{\|X(\vartheta_1,\ldots,\vartheta_m)\|_2\} \ subject \ to \sum_{i=1}^{m}\vartheta_i = \vartheta_0, \ \vartheta_1 \geq 0, \ldots, \vartheta_m \geq 0. \quad (15)$$

7 Detection of Spectral Changes

Let us consider a sequence of periodograms calculated by the Bartlett-Welch method [23–27]. The sequentially observed sample $(y_0, \ldots, y_{n-1})$ is subdivided into K smaller disjoint segments of size L, i.e., $n = KL$. The periodograms are calculated for the set of frequencies $f_k = \frac{k}{L}F_s$, $k = 0, 1, 2, \ldots, \frac{L}{2}$

$$\widehat{S}_y(f_k) = \frac{1}{K}\sum_{i=0}^{K-1}\widehat{S}_y^i(f_k), \ \widehat{S}_y^i(f_k) = \frac{1}{F_sL}\left|\sum_{n=0}^{L-1}y_{n+iL}e^{-jn\frac{2\pi k}{L}}\right|^2, \quad (16)$$

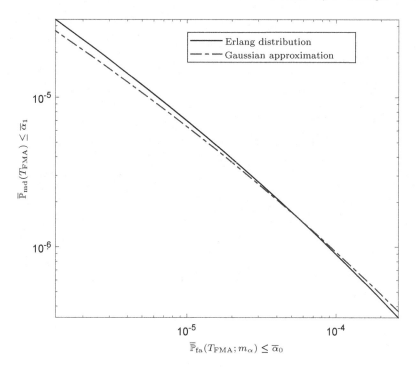

Fig. 5. The probability $\overline{\alpha}_1$ as a function of $\overline{\alpha}_0$.

where $S_y(f_k)$ is the signal power spectral density (PSD) and F_s is the sampling frequency measured in Hz. Following [24,25], the vector of estimations $\widehat{S}_y^i(f_k)$, where $0 < f_k < F_s/2$, converges in distribution to a vector of independent exponentially distributed r.v.'s. provided that $L \to \infty$ (see [25, Th. 10.3.2]). From the other hand, the estimations $\widehat{S}_y^i(f_k)$ and $\widehat{S}_y^j(f_k)$ calculated for the i-th and j-th segments of size L, $i \neq j$, are also asymptotically uncorrelated provided that $K \to \infty$ and $L \to \infty$, at least for a relatively flat spectrum. Assuming that $t_0 = Lk + 1$, where $k = 1, 2, \ldots$, the generative model (1) can be re-written for the sequence of periodograms calculated by segments:

$$\widehat{S}_y^i(f_k) \sim \begin{cases} \mathrm{Exp}\left(S_x(f_k)\right) & \text{if } i < i_0, \\ \mathrm{Exp}\left(S_{x+s}(f_k)\right) & \text{if } i_0 \leq i \leq i_0 + N - 1, \end{cases} \tag{17}$$

where i is the current segment number, i_0 is the change-point measured in segments, $S_x(f_k)$ is the noise (pre-change) PSD and $S_{x+s}(f_k) = S_x(f_k) + \beta S_s(f_k)$ is an additive sum of noise $S_x(f_k)$ and signal $S_s(f_k)$ (post-change) PSDs. Without loss of generality, it is assumed that

$$\mathrm{var}(x) = \int_0^{F_s/2} S_x(f)\,df = \mathrm{var}(s) = \int_0^{F_s/2} S_s(f)\,df. \tag{18}$$

Hence, β is the SNR. Following Proposition 1, the maximum and minimum contrast signal PSD's $f \to S_s(f)$ w.r.t. the noise PSD $f \to S_x(f)$ are defined from the optimization problem (14)–(15). The constant $\vartheta_0 = \mathrm{var}(x) = \mathrm{var}(s)$ is defined in (18).

8   Conclusion

The original contribution of this paper is threefold:

- the problem of sequential transient change detection is considered for a multi-parameter exponential distribution;
- the accuracy of the Gaussian approximation for the hypoexponential distribution and its impact on the relation between the bounds for the probabilities of missed detection and false alarm;
- the detectability of transient change profiles, i.e., the definition of the best/worst possible transient change profile with respect to the pre-change parameter vector.

The theoretical results are applied to the sequential detection of spectral changes by using a sequence of periodograms.

References

1. Basseville, M., Nikiforov, I.V.: Detection of Abrupt Changes: Theory and Application. Prentice Hall, Englewood Cliffs (1993). http://www.irisa.fr/sisthem/kniga
2. Lai, T.L.: Sequential analysis: some classical problems and new challenges (with discussion). Statistica Sinica **11**, 303–408 (2001)
3. Poor, H.V., Hadjiliadis, O.: Quickest Detection. Cambridge University Press, Cambridge (2009)
4. Tartakovsky, A., Nikiforov, I., Basseville, M.: Sequential Analysis: Hypothesis Testing and Changepoint Detection. CRC Press, Taylor & Francis Group, Boca Raton (2014)
5. Han, C., Willett, P.K., Chen, B., Abraham, D.A.: A detection optimal min-max test for transient signals. IEEE Trans. Inf. Theory **44**(2), 866–869 (1998)
6. Streit, R., Willett, P.: Detection of random transient signals via hyperparameter estimation. IEEE Trans. Signal Process. **47**(7), 1823–1834 (1999)
7. Han, C., Willett, P., Abraham, D.: Some methods to evaluate the performance of Page's test as used to detect transient signals. IEEE Trans. Signal Process. **47**(8), 2112–2127 (1999)
8. Wang, Z.J., Willett, P.: A performance study of some transient detectors. IEEE Trans. Signal Process. **48**(9), 2682–2685 (2000)
9. Wang, Z.J., Willett, P.: A variable threshold Page procedure for detection of transient signals. IEEE Trans. Signal Process. **53**(11), 4397–4402 (2005)
10. Bakhache, B., Nikiforov, I.: Reliable detection of faults in measurement systems. Int. J. Adapt. Control Signal Process. **14**(7), 683–700 (2000)
11. Guepie, B., Fillatre, L., Nikiforov, I.: Sequential monitoring of water distribution network. In: The Proceeding of the SYSID 2012, 16th IFAC Symposium on System Identification, Brussels, Belgium, pp. 392–397 (2012)
12. Guépié, B.K., Fillatre, L., Nikiforov, I.: Sequential detection of transient changes. Sequential Anal. **31**(4), 528–547 (2012)

13. Guépié, B.K., Fillatre, L., Nikiforov, I.: Detecting a suddenly arriving dynamic profile of finite duration. IEEE Trans. Inf. Theory **63**(5), 3039–3052 (2017)
14. Lehmann, E.L.: Some concepts of dependence. Ann. Math. Stat. **37**(5), 1137–1153 (1966)
15. Esary, J.D., Proschan, F., Walkup, D.W.: Association of random variables, with applications. Ann. Math. Stat. **38**(5), 1466–1474 (1967)
16. Stephaniuk, A.R.: Estimating the likelihood ratio function in the problem of "failure" of a stochastic process. Autom. Remote Control **47**(9), 1210–1216 (1986)
17. Amari, S.V., Misra, R.B.: Closed-form expressions for distribution of sum of exponential random variables. IEEE Trans. Reliab. **46**(4), 519–522 (1997)
18. Khuong, H.V., Kong, H.-Y.: General expression for pdf of a sum of independent exponential random variables. IEEE Commun. Lett. **10**(3), 159–161 (2006)
19. Nadarajah, S.: A review of results on sums of random variables. Acta Appl. Math. **103**(2), 131–140 (2008)
20. He, Q.-M., Zhang, H.-Q.: Coxian representations of generalized erlang distributions. Acta Mathematicae Applicatae Sinica, English Series **25**(3), 487–500 (2009)
21. Ross, S.M.: Introduction to Probability Models, 10th edn. Academic Press, San Diego (2011)
22. Smaili, K., Kadri, T., Kadry, S.: Hypoexponential distribution with different parameters. Appl. Math. **4**(4), 624–631 (2013)
23. Box, G., Jenkins, G.: Time Series Analysis: Forecasting and Control. Holden-Bay, San Francisco (1976)
24. Brillinger, D.R.: Time Series, Data Analysis and Theory. Rinehart & Winston, New York (1981)
25. Brockwell, P.J., Davis, R.A.: Time Series: Theory and Methods. Springer, New York (1987). https://doi.org/10.1007/978-1-4899-0004-3
26. Hayes, M.H.: Statistical Digital Signal Processing and Modeling. Wiley, New York (1996)
27. Manolakis, D.G., Ingle, V.K., Kogon, S.M.: Statistical and Adaptive Signal Processing. Artech House, Norwood (2005)

Distribution Parameters Estimation in Recurrent Synchronous Generalized Doubly Stochastic Flow of the Second Order

Lyudmila Nezhel'skaya[1(✉)], Michele Pagano[2], and Ekaterina Sidorova[1]

[1] National Research Tomsk State University, 36 Lenina ave., Tomsk 634050, Russia
ludne@mail.ru, katusha_sidorova@mail.ru
[2] University of Pisa, 16 Via Caruso, 56122 Pisa, Italy
m.pagano@iet.unipi.it

Abstract. We solve the estimation problem of the probability density parameters of the inter-event interval duration in a synchronous generalized flow of the second order, which can be used as a powerful mathematical model for the arrival processes in queuing systems and networks. The explicit form of the parameter estimates is determined by the method of moments on the basis of observations of the doubly stochastic flow under the recurrence conditions that are formulated in terms of the joint probability density of the durations of two adjacent inter-event intervals. The quality of the estimates is established by using the model, reproducing the flow behavior under conditions of complete observability.

Keywords: Recurrent synchronous generalized doubly stochastic event flow of the second order · Joint probability density · Flow recurrence conditions · Parameter estimation · Method of moments

1 Introduction

In the overwhelming majority of studies on queuing models, Poisson arrivals are used as incoming flows of events (messages, requests) [1] – this is primarily true for the fundamental works at the basis of queueing theory. The rapid evolution of computing technology, satellite, computer, wireless and mobile communication networks revealed the inadequacy of a stationary Poisson model for the actual information flows and the unsuitability of the corresponding queueing models for analyzing the processes occurring in modern telecommunication systems. Thus, the practice requirements stimulated the definition of new mathematical models of incoming flows in the form of doubly stochastic event flows with intensity represented by either a continuous [2] or piecewise constant [3,4] random process with a finite (arbitrary) number of states. Such models are widely used in many branches of science and technology, such as theory of communication networks and radio communications [5,6], study of intelligent systems and networks [7], statistical data processing [8] and others.

© Springer Nature Switzerland AG 2019
V. M. Vishnevskiy et al. (Eds.): DCCN 2019, LNCS 11965, pp. 276–288, 2019.
https://doi.org/10.1007/978-3-030-36614-8_21

Flows with step intensity function, commonly referred to as MC (Markov chain) flows [3] or MAP (Markovian Arrival Process) flows [9], are widely used as mathematical models of flows circulating in real telecommunication networks, in particular, in broadband wireless communication networks along the long transport highways [10–14]. This family of flows includes a synchronous generalized flow of events of the second order [15,16] (a generalization of a synchronous doubly stochastic flow [17]), which is considered in this paper.

The operation mode of a queuing system depends directly on the parameters of the MC (MAP) flow and the accompanying process state. In practice, these characteristics are quite often either only partially known, or unknown, or change randomly over the time. As a result, when implementing an adaptive control of a mathematical model of a real system, at first it is necessary to estimate the states of the incoming flow (filtering its intensity) at an arbitrary time [15,16,18] and (or) its parameters [17,19,20] on the basis of observations of the flow.

For the considered synchronous generalized flow of the second order, the problem of optimal estimation of the flow states was solved under conditions of observability of all its events in [15] and in the presence of unextendable dead time of fixed length (during which events of the original flow are lost) in [16]. In this paper, we solve the estimation problem of the probability density parameters of the inter-event interval duration in the recurrent event flow using the method of moments, which provides point estimates with *sufficiently good statistical properties* for large samples of observations.

2 Problem Statement

We consider a synchronous generalized event flow of the second order (flow), the accompanying process $\lambda(t)$ of which is a piecewise constant random process with two states S_1 and S_2; hereinafter, S_i is understood as ith state of $\lambda(t)$ and takes place when $\lambda(t) = \lambda_i$, $i = 1, 2$, $\lambda_1 > \lambda_2 \geq 0$.

The interval duration between flow events at the ith state is determined by a random variable $\eta_i = \min(\xi_i^{(1)}, \xi_i^{(2)})$, where random variables $\xi_i^{(1)}$ and $\xi_i^{(2)}$ are both independent and distributed according to the laws $F_i^{(1)}(t) = 1 - e^{-\lambda_i t}$ and $F_i^{(2)}(t) = 1 - e^{-\alpha_i t}$, respectively, $i = 1, 2$. At the moment when a flow event occurs, depending on the value η_i, the process $\lambda(t)$ either transits from the ith state to the jth state, $i \neq j$, or remains at the ith state, $i = j$, with probability $P_1^{(l)}(\lambda_j|\lambda_i)$, where superscript (l) indicates that the probability corresponds to the random variable $\xi_i^{(l)}$, $i, j = 1, 2$, $l = 1, 2$. Here $P_1^{(1)}(\lambda_j|\lambda_i) + P_1^{(1)}(\lambda_i|\lambda_i) = 1$, $P_1^{(2)}(\lambda_j|\lambda_i) + P_1^{(2)}(\lambda_i|\lambda_i) = 1$, $i, j = 1, 2$, $i \neq j$. Thus, the duration of the interval between flow events at the ith state of the process $\lambda(t)$ is a random variable with the distribution function $F_i(t) = 1 - e^{-(\lambda_i+\alpha_i)t}$, $i = 1, 2$.

Under the above assumptions, the accompanying piecewise constant random process $\lambda(t)$ for the considered doubly stochastic flow of events is a hidden Markov process [15], and infinitesimal characteristics matrices take the form

$$\mathbf{D}_0 = \left\| \begin{matrix} -(\lambda_1 + \alpha_1) & 0 \\ 0 & -(\lambda_2 + \alpha_2) \end{matrix} \right\|,$$

$$\mathbf{D}_1 = \left\| \begin{matrix} \lambda_1 P_1^{(1)}(\lambda_1|\lambda_1) + \alpha_1 P_1^{(2)}(\lambda_1|\lambda_1) & \lambda_1 P_1^{(1)}(\lambda_2|\lambda_1) + \alpha_1 P_1^{(2)}(\lambda_2|\lambda_1) \\ \lambda_2 P_1^{(1)}(\lambda_1|\lambda_2) + \alpha_2 P_1^{(2)}(\lambda_1|\lambda_2) & \lambda_2 P_1^{(1)}(\lambda_2|\lambda_2) + \alpha_2 P_1^{(2)}(\lambda_2|\lambda_2) \end{matrix} \right\|.$$

Elements of $\mathbf{D}_1$ are the intensities of the process $\lambda(t)$ transitions from state to state with a flow event occurrence. Diagonal elements of the matrix $\mathbf{D}_0$ are the intensities of the process $\lambda(t)$ output from its states, taken with the opposite sign. Off-diagonal elements of $\mathbf{D}_0$ represent the transition intensities without an event occurrence and identically equal to zero due to the flow feature.

We emphasize that a state change S_i to S_j of the process $\lambda(t)$, $i, j = 1, 2$, $i \neq j$, is possible only at the time moment when an event occurs, therefore, the flow under investigation is called synchronous.

We consider the steady-state (stationary) operation mode of the flow under study, i.e. observation begins when the flow is functioning infinitely long. At the same time the process $\lambda(t)$ is fundamentally unobservable on the interval (t_0, t), where t_0 and t represent the start and the end of observation, respectively; in other words, all the available information about the flow is provided by the sequence of observable time instants $t_1, t_2, \ldots, t_k, \ldots$ of occurrence of events. By virtue of the formulated prerequisites, the set $t_1, t_2, \ldots, t_k, \ldots$ forms a sequence $\{\lambda(t_k)\}$ that is an embedded Markov chain, i.e. the observed event flow has the Markov property, if its evolution is considered from the moment t_k, $k = 1, 2, \ldots$.

In the present study, on the basis of the observed sample $t_1, t_2, \ldots, t_k, \ldots$ we use the method of moments to estimate the probability density parameters of interval duration between neighboring events in the recurrent synchronous generalized flow of the second order; for this, the explicit form of the joint probability density of the durations of adjacent intervals in the flow of events is found, and the conditions for its recurrence are formulated.

3   Derivation of the Joint Probability Density

We denote the duration value of the kth interval between neighboring events of the observed flow t_k and t_{k+1}, $k = 1, 2, \ldots$, by $\tau_k = t_{k+1} - t_k$, $\tau_k \geq 0$. For the probability density of the values τ_k, due to the steady-state operation mode of the flow, the equality $p(\tau_k) = p(\tau)$, $\tau \geq 0$, is valid for any $k \geq 1$, which allows (without any loss of generality) to set the moment of occurrence of the event t_k equals to zero or, equivalently, the moment of the event occurrence is $\tau = 0$.

The location on the time axis of adjacent intervals (t_k, t_{k+1}) and (t_{k+1}, t_{k+2}) of durations $\tau_k = t_{k+1} - t_k$ and $\tau_{k+1} = t_{k+2} - t_{k+1}$, $\tau_k \geq 0$, $\tau_{k+1} \geq 0$, respectively, due to the flow stationarity is arbitrary, which makes it possible, by setting $k = 1$, to consider two adjacent intervals (t_1, t_2) and (t_2, t_3) with corresponding durations $\tau_1 = t_2 - t_1$ and $\tau_2 = t_3 - t_2$. In this case, $\tau_1 = 0$ corresponds to the moment t_1 of occurrence of the flow event, $\tau_2 = 0$ corresponds to the moment

t_2 of occurrence of the next flow event, and the joint probability density of the values τ_1 and τ_2 is $p(\tau_1, \tau_2)$, $\tau_1 \geq 0$, $\tau_2 \geq 0$.

For the studied flow, one can show the validity of Lemmas 1, 2, Theorem 1.

Lemma 1. *In a synchronous generalized flow of the second order the probability densities $\tilde{p}_{ij}(\tau)$, $i,j = 1,2$, that, without an event occurrence on the interval $(0, \tau)$ and an event occurrence at the moment τ, the process $\lambda(\tau)$ transits on this interval from state S_i to state S_j, $i,j = 1,2$, are determined by the formulas*

$$
\begin{aligned}
\tilde{p}_{11}(\tau) &= (\lambda_1 P_1^{(1)}(\lambda_1|\lambda_1) + \alpha_1 P_1^{(2)}(\lambda_1|\lambda_1))e^{-(\lambda_1+\alpha_1)\tau}, \\
\tilde{p}_{12}(\tau) &= (\lambda_1 P_1^{(1)}(\lambda_2|\lambda_1) + \alpha_1 P_1^{(2)}(\lambda_2|\lambda_1))e^{-(\lambda_1+\alpha_1)\tau}, \\
\tilde{p}_{21}(\tau) &= (\lambda_2 P_1^{(1)}(\lambda_1|\lambda_2) + \alpha_2 P_1^{(2)}(\lambda_1|\lambda_2))e^{-(\lambda_2+\alpha_2)\tau}, \\
\tilde{p}_{22}(\tau) &= (\lambda_2 P_1^{(1)}(\lambda_2|\lambda_2) + \alpha_2 P_1^{(2)}(\lambda_2|\lambda_2))e^{-(\lambda_2+\alpha_2)\tau}.
\end{aligned}
\tag{1}
$$

Let $\phi_1 = \lambda_1 P_1^{(1)}(\lambda_2|\lambda_1) + \alpha_1 P_1^{(2)}(\lambda_2|\lambda_1)$, $\phi_2 = \lambda_2 P_1^{(1)}(\lambda_1|\lambda_2) + \alpha_2 P_1^{(2)}(\lambda_1|\lambda_2)$.

Lemma 2. *In a synchronous generalized event flow of the second order the conditional final probabilities $\pi_i(0)$, $i = 1, 2$, that the process $\lambda(\tau)$ at the time moment $\tau = 0$ is at the ith state, $i = 1, 2$, provided that $\tau = 0$ is a moment of the flow event occurrence, are given by the expressions*

$$
\pi_1(0) = \frac{(\lambda_1 + \alpha_1)\phi_2}{(\lambda_1 + \alpha_1)\phi_2 + (\lambda_2 + \alpha_2)\phi_1}, \pi_2(0) = \frac{(\lambda_2 + \alpha_2)\phi_1}{(\lambda_1 + \alpha_1)\phi_2 + (\lambda_2 + \alpha_2)\phi_1}.
\tag{2}
$$

Theorem 1. *The one-dimensional probability density of the inter-event interval duration of a synchronous generalized flow of the second order has the form*

$$
p(\tau) = \gamma(\lambda_1 + \alpha_1)e^{-(\lambda_1+\alpha_1)\tau} + (1 - \gamma)(\lambda_2 + \alpha_2)e^{-(\lambda_2+\alpha_2)\tau}, \tau \geq 0,
\tag{3}
$$

where $\gamma = \pi_1(0)$, $\pi_1(0)$ is defined in (2).

Note that (3) is the density of the hyperexponential distribution with corresponding parameters; further we assume that $(\lambda_1 + \alpha_1) \neq (\lambda_2 + \alpha_2)$.

Lemmas 1, 2 allow us to formulate the following Theorem 2.

Theorem 2. *A synchronous generalized flow of the second order in the general case is a correlated flow and the joint probability density $p(\tau_1, \tau_2)$ has the form*

$$
\begin{aligned}
p(\tau_1, \tau_2) &= p(\tau_1)p(\tau_2) + [1 - \phi_1(\lambda_1 + \alpha_1)^{-1} - \phi_2(\lambda_2 + \alpha_2)^{-1} \\
&\times \gamma(1 - \gamma)[(\lambda_1 + \alpha_1)e^{-(\lambda_1+\alpha_1)\tau_1} - (\lambda_2 + \alpha_2)e^{-(\lambda_2+\alpha_2)\tau_1}] \\
&\times [(\lambda_1 + \alpha_1)e^{-(\lambda_1+\alpha_1)\tau_2} - (\lambda_2 + \alpha_2)e^{-(\lambda_2+\alpha_2)\tau_2}], \tau_1 \geq 0, \tau_2 \geq 0,
\end{aligned}
\tag{4}
$$

where $\gamma = \pi_1(0)$, $\pi_1(0)$ is defined in (2), $p(\tau_k)$ in (3) for $\tau = \tau_k$, $k = 1, 2$.

Proof. Since the sequence of instants of occurrence of the flow events generates an embedded Markov chain, the joint probability density of the durations of two adjacent intervals $p(\tau_1, \tau_2)$ is given by the formula

$$p(\tau_1, \tau_2) = \sum_{i=1}^{2} \pi_i(0) \sum_{j=1}^{2} \tilde{p}_{ij}(\tau_1) \sum_{k=1}^{2} \tilde{p}_{jk}(\tau_2), \tau_1 \geq 0, \tau_2 \geq 0. \tag{5}$$

Substituting in (5) first $p_{ij}(\tau_1)$ and $p_{jk}(\tau_2)$, calculated by (1) with $\tau = \tau_1$ and $\tau = \tau_2$, and then explicit expressions (2) for the probabilities $\pi_i(0)$, $i = 1, 2$, doing the necessary fairly laborious transformations, we arrive at (4).

Assuming $\alpha_1 = \alpha_2 = 0$ in (4), we obtain the joint probability density for a synchronous doubly stochastic flow of events [17].

4   Recurrence Conditions of the Event Flow

Let us consider the cases when the event flow under study becomes recurrent: from formula (4) for the joint probability density $p(\tau_1, \tau_2)$, we determine the conditions under which it can be factorized, i.e. $p(\tau_1, \tau_2) = p(\tau_1)p(\tau_2)$.

1. If $\phi_1(\lambda_1 + \alpha_1)^{-1} + \phi_2(\lambda_2 + \alpha_2)^{-1} = 1$ for $\phi_1 = \lambda_1 P_1^{(1)}(\lambda_2|\lambda_1) + \alpha_1 P_1^{(2)}(\lambda_2|\lambda_1)$, $\phi_2 = \lambda_2 P_1^{(1)}(\lambda_1|\lambda_2) + \alpha_2 P_1^{(2)}(\lambda_1|\lambda_2)$, then the joint density (4) factorizes, and from (3) as a result of the transformations it follows that

$$\gamma = 1 - \phi_1(\lambda_1 + \alpha_1)^{-1}, 1 - \gamma = 1 - \phi_2(\lambda_2 + \alpha_2)^{-1},$$
$$p(\tau_k) = (\lambda_1 P_1^{(1)}(\lambda_1|\lambda_1) + \alpha_1 P_1^{(2)}(\lambda_1|\lambda_1))e^{-(\lambda_1+\alpha_1)\tau_k}$$
$$+ (\lambda_2 P_1^{(1)}(\lambda_2|\lambda_2) + \alpha_2 P_1^{(2)}(\lambda_2|\lambda_2))e^{-(\lambda_2+\alpha_2)\tau_k},$$

$\tau_k \geq 0$, $k = 1, 2$, i.e.

$$p(\tau) = (\lambda_1 P_1^{(1)}(\lambda_1|\lambda_1) + \alpha_1 P_1^{(2)}(\lambda_1|\lambda_1))e^{-(\lambda_1+\alpha_1)\tau}$$
$$+ (\lambda_2 P_1^{(1)}(\lambda_2|\lambda_2) + \alpha_2 P_1^{(2)}(\lambda_2|\lambda_2))e^{-(\lambda_2+\alpha_2)\tau}, \tau \geq 0. \tag{6}$$

The next two conditions for factorizing $p(\tau_1, \tau_2)$ are obtained by analyzing the expression $\gamma(1 - \gamma) = (\lambda_1 + \alpha_1)\phi_2(\lambda_2 + \alpha_2)\phi_1[(\lambda_1 + \alpha_1)\phi_2 + (\lambda_2 + \alpha_2)\phi_1]^{-2}$.

2. If $\phi_2 = 0$, then $p(\tau_1, \tau_2) = p(\tau_1)p(\tau_2)$ under the condition that λ_2, α_2 are not equal to zero at the same time (otherwise, there is no flow in state S_2); by (3), $\gamma = 0$, $1 - \gamma = 1$, $p(\tau_k) = (\lambda_2 + \alpha_2)e^{-(\lambda_2+\alpha_2)\tau_k}$, $\tau_k \geq 0$, $k = 1, 2$, i.e.

$$p(\tau) = (\lambda_2 + \alpha_2)e^{-(\lambda_2+\alpha_2)\tau}, \tau \geq 0. \tag{7}$$

3. If $\phi_1 = 0$, then $p(\tau_1, \tau_2) = p(\tau_1)p(\tau_2)$, and according to (3), $\gamma = 1$, $1 - \gamma = 0$, $p(\tau_k) = (\lambda_1 + \alpha_1)e^{-(\lambda_1+\alpha_1)\tau_k}$, $\tau_k \geq 0$, $k = 1, 2$, i.e.

$$p(\tau) = (\lambda_1 + \alpha_1)e^{-(\lambda_1+\alpha_1)\tau}, \tau \geq 0. \tag{8}$$

If one of the listed conditions 1–3 is fulfilled and the one-dimensional probability density takes the form (6), (7) or (8), then the considered flow of events is a recurrent synchronous generalized flow of second order.

Remark 1. Probability densities of the form (7) and (8) coincide with the probability density of the values of the interval duration between neighboring events in the stationary Poisson flow with parameters $(\lambda_2 + \alpha_2)$ and $(\lambda_1 + \alpha_1)$, respectively.

5 Estimation of the Density Parameters in the Recurrent Flow Using the Method of Moments

It is not possible to estimate by the method of moments the twelve unknown parameters defining a synchronous generalized flow of the second order (or eight, if taking into account the restrictions on specifying transition probabilities), having only information about the form of $p(\tau)$, as shown below. As a result, we will estimate the parameters of the probability density of the values of the interval duration between adjacent events of the recurrent flow of the form (6).

Consider a sample $\tau_1, \tau_2, \ldots, \tau_n$ from the distribution $p(\tau|z_1, z_2, \beta_1, \beta_2) = \beta_1 e^{-z_1 \tau} + \beta_2 e^{-z_2 \tau}$, depending on the four unknowns $z_1 = \lambda_1 + \alpha_1$, $z_2 = \lambda_2 + \alpha_2$, $\beta_1 = \lambda_1 P_1^{(1)}(\lambda_1|\lambda_1) + \alpha_1 P_1^{(2)}(\lambda_1|\lambda_1)$, $\beta_2 = \lambda_2 P_1^{(1)}(\lambda_2|\lambda_2) + \alpha_2 P_1^{(2)}(\lambda_2|\lambda_2)$; here without any loss of generality, we assume $z_1 > z_2$. Due to the closeness of the theoretical and empirical distribution functions, i.e. unlimited convergence of their values, we should expect the closeness of their numerical characteristics, i.e. moments of the same order. The theoretical initial moment of the lth order $E[\tau^l] = \int_0^\infty \tau^l p(\tau|z_1, z_2, \beta_1, \beta_2) d\tau$, which is a function of the unknown parameters, is close to the corresponding sample moment – statistics $C_l = \frac{1}{n} \sum_{k=1}^n \tau_k{}^l$, where $\tau_k = t_{k+1} - t_k$ is the value of the interval duration between the moments t_k and t_{k+1} of occurrence of events in a synchronous generalized flow of the second order. Thus, the parameter estimates z_1, z_2, β_1, β_2 are defined as the solution of the system of moment equations $E[\tau^l] = C_l$, $l = \overline{1,4}$.

The theoretical initial moment of the lth order is determined by the formula

$$E[\tau^l] = \int_0^\infty \tau^l [\beta_1 e^{-z_1 \tau} + \beta_2 e^{-z_2 \tau}] d\tau = l! \frac{\beta_1}{z_1^{l+1}} + l! \frac{\beta_2}{z_2^{l+1}}, l = \overline{1,4},$$

on the basis of which we write the system of four equations for $\hat{z}_1$, $\hat{z}_2$, $\hat{\beta}_1$, $\hat{\beta}_2$:

$$\frac{\beta_1}{z_1^2} + \frac{\beta_2}{z_2^2} = C_1, 2\frac{\beta_1}{z_1^3} + 2\frac{\beta_2}{z_2^3} = C_2, 6\frac{\beta_1}{z_1^4} + 6\frac{\beta_2}{z_2^4} = C_3, 24\frac{\beta_1}{z_1^5} + 24\frac{\beta_2}{z_2^5} = C_4. \quad (9)$$

Making the necessary transformations for the system (9), we get

$$\frac{\beta_1}{z_1^2} + \frac{\beta_2}{z_2^2} = C_1, (z_1 + z_2)C_1 - z_1 z_2 C_2/2 - \left(\frac{\beta_1}{z_1} + \frac{\beta_2}{z_2}\right) = 0,$$

$$(z_1 + z_2)C_2 - z_1 z_2 C_3/3 = 2C_1, (z_1 + z_2)C_3 - z_1 z_2 C_4/4 = 3C_2. \quad (10)$$

From the third and fourth equations (10), applying the Kramer method, we find

$$\hat{z}_1 + \hat{z}_2 = \frac{12C_2C_3 - 6C_1C_4}{4C_3{}^2 - 3C_2C_4}, \hat{z}_1\hat{z}_2 = \frac{36C_2{}^2 - 24C_1C_3}{4C_3{}^2 - 3C_2C_4}.$$

The estimates $\hat{z}_1$, $\hat{z}_2$ of parameters z_1, z_2, according to the inverse Vieta's theorem, are roots of the quadratic equation $z - x_1 z + x_2 = 0$, where

$$x_1 = \hat{z}_1 + \hat{z}_2 = \frac{12C_2C_3 - 6C_1C_4}{4C_3{}^2 - 3C_2C_4}, x_2 = \hat{z}_1\hat{z}_2 = \frac{36C_2{}^2 - 24C_1C_3}{4C_3{}^2 - 3C_2C_4};$$

according to the condition $z_1 > z_2$, we have

$$\hat{z}_{1,2} = \frac{6C_2C_3 - 3C_1C_4}{4C_3{}^2 - 3C_2C_4} \pm \frac{1}{2}\sqrt{\left(\frac{12C_2C_3 - 6C_1C_4}{4C_3{}^2 - 3C_2C_4}\right)^2 - 4\frac{36C_2{}^2 - 24C_1C_3}{4C_3{}^2 - 3C_2C_4}}. \quad (11)$$

Estimates $\hat{\beta}_1$, $\hat{\beta}_2$ are determined uniquely from the first and second equations of the system (10) and have the form

$$\hat{\beta}_1 = \frac{2\hat{z}_1^3 C_1 - \hat{z}_1^3 \hat{z}_2 C_2}{2(\hat{z}_1 - \hat{z}_2)}, \hat{\beta}_2 = \frac{\hat{z}_1 \hat{z}_2^3 C_2 - 2\hat{z}_2^3 C_1}{2(\hat{z}_1 - \hat{z}_2)}. \quad (12)$$

Thus, system (10) has the unique solution (11), (12), which gives consistent estimates of the parameters $\hat{z}_1$, $\hat{z}_2$, $\hat{\beta}_1$, $\hat{\beta}_2$ [21].

Consider further $\tau_1, \tau_2, \ldots, \tau_n$ from $p(\tau | z_1, z_2, \beta_1{}^{(1)}, \beta_1{}^{(2)}, \beta_2) = (\beta_1{}^{(1)} + \beta_1{}^{(2)}) \times \times e^{-z_1 \tau} + \beta_2 e^{-z_2 \tau}$, depending on the unknowns $z_1 = \lambda_1 + \alpha_1$, $z_2 = \lambda_2 + \alpha_2$ $(z_1 > z_2)$, $\beta_1{}^{(1)} = \lambda_1 P_1^{(1)}(\lambda_1 | \lambda_1)$, $\beta_1{}^{(2)} = \alpha_1 P_1^{(2)}(\lambda_1 | \lambda_1)$ $(\beta_1 = \beta_1{}^{(1)} + \beta_1{}^{(2)})$ and $\beta_2 = \lambda_2 P_1^{(1)}(\lambda_2 | \lambda_2) + \alpha_2 P_1^{(2)}(\lambda_2 | \lambda_2)$. In this case, we write five moment equations

$$\frac{\beta_1{}^{(1)} + \beta_1{}^{(2)}}{z_1{}^2} + \frac{\beta_2}{z_2{}^2} = C_1,$$

$$(z_1 + z_2)C_1 - z_1 z_2 C_2 / 2 - \left(\frac{\beta_1{}^{(1)} + \beta_1{}^{(2)}}{z_1} + \frac{\beta_2}{z_2}\right) = 0, \quad (13)$$

$$(z_1 + z_2)C_2 - z_1 z_2 C_3 / 3 = 2C_1, (z_1 + z_2)C_3 - z_1 z_2 C_4 / 4 = 3C_2,$$

$$(z_1 + z_2)C_4 - z_1 z_2 C_5 / 5 = 4C_3.$$

Theorem 3. *The system of the moment equations* (13) *with respect to the unknown parameters* z_1, z_2, $\beta_1{}^{(1)}$, $\beta_1{}^{(2)}$, β_2 *of* $p(\tau)$ *of the form* (6) *is incompatible.*

Proof. We perform the variables change in the system (13):

$$z_1 + z_2 = x_1, z_1 z_2 = x_2, \frac{\beta_1{}^{(1)} + \beta_1{}^{(2)}}{z_1{}^2} + \frac{\beta_2}{z_2{}^2} = x_3, \frac{\beta_1{}^{(1)} + \beta_1{}^{(2)}}{z_1} + \frac{\beta_2}{z_2} = x_4,$$

which leads it to a linear form

$$x_3 = C_1, x_1 C_1 - x_2 C_2/2 - x_4 = 0, x_1 C_2 - x_2 C_3/3 = 2C_1,$$
$$x_1 C_3 - x_2 C_4/4 = 3C_2, x_1 C_4 - x_2 C_5/5 = 4C_3. \tag{14}$$

The ranks of the matrix and the extended matrix of the system of five linear inhomogeneous equations with respect to four unknowns x_l, $l = \overline{1,4}$, are respectively 4 and 5. Consequently, according to the corollary of the Kronecker–Capelli theorem, the system is incompatible, i.e. solution of (14) does not exist.

The latter theorem determines the impossibility of estimating by the method of moments more than four flow distribution parameters due to the insufficiency of the information contained in $p(\tau)$. In other words, the knowledge of the form (6) of the probability density makes it possible to obtain estimates of only four unknown parameters z_1, z_2, β_1, β_2; their consistency was proved in [21].

Remark 2. Estimation of the density parameters of the form (7), (8) in this work is not carried out by virtue of Remark 1, Theorem 3, according to which the only consistent estimate is defined as the estimate of the exponential distribution parameter $(z_2 = \lambda_2 + \alpha_2$ in (7), $z_1 = \lambda_1 + \alpha_1$ in (8)) and has the form C_1^{-1}.

6   Numerical Results of Estimation

For the numerical estimation of the distribution parameters of the recurrent event flow an ad-hoc program has been implemented as a graphical user interface using the object-oriented programming language C# in the integrated development environment Microsoft Visual Studio. The implementation is based on the simulation modeling of a synchronous generalized doubly stochastic flow of the second order in order to obtain statistics C_l, $l = \overline{1,4}$, followed by the calculation of the consistent estimates $\hat{z}_1$, $\hat{z}_2$, $\hat{\beta}_1$, $\hat{\beta}_2$ by formulas (11), (12).

The following statistical experiment was conducted to track the time interval needed for setting the stationary operation mode. The results of the simulation model work were obtained with the same values of the flow parameters reported in Table 1, for each modeling time value T_m (time of the flow observation) in $N = 100$ independent realizations of a synchronous generalized flow of the second order; for each realization, the estimates $\hat{\theta}^k$, $k = \overline{1,N}$, of the corresponding parameters θ were determined according to formulas (11), (12), on the basis of which the sample mean value $\hat{E}[\hat{\theta}] = \frac{1}{N}\sum_{k=1}^{N} \hat{\theta}^{(k)}$, displacement estimate $|\hat{E}[\hat{\theta}] - \hat{\theta}|$, sample square error $\sqrt{\hat{V}[\hat{\theta}]}$ as square root of the sample variance $\hat{V}[\hat{\theta}] = \frac{1}{N}\sum_{k=1}^{N}(\hat{\theta}^k - \theta)^2$, $\theta \in \{z_1, z_2, \beta_1, \beta_2\}$, $\hat{\theta} \in \{\hat{z}_1, \hat{z}_2, \hat{\beta}_1, \hat{\beta}_2\}$, were calculated.

The analysis of the current experiment results, including the numerical values of Table 2, highlights the strong dependence of the obtained parameter estimates $\hat{z}_1$, $\hat{z}_2$, $\hat{\beta}_1$, $\hat{\beta}_2$ on the modeling time. As the value T_m increases, the sample averages stabilize, and the quality of the estimates themselves improves in sense of decreasing the values of the sample quadratic error. The latter is quite natural

Table 1. Model parameters

$\lambda_1 = 6,1$	$P_1^{(1)}(\lambda_1|\lambda_1) = 0,9$	$P_1^{(1)}(\lambda_2|\lambda_1) = 0,1$
$\lambda_2 = 2,8$	$P_1^{(1)}(\lambda_2|\lambda_2) = 0,1$	$P_1^{(1)}(\lambda_1|\lambda_2) = 0,9$
$\alpha_1 = 3,5$	$P_1^{(2)}(\lambda_1|\lambda_1) = 0,9$	$P_1^{(2)}(\lambda_2|\lambda_1) = 0,1$
$\alpha_2 = 0,7$	$P_1^{(2)}(\lambda_2|\lambda_2) = 0,1$	$P_1^{(2)}(\lambda_1|\lambda_2) = 0,9$

and is explained first of all by the very concept of the method of moments, which is based on the values C_l, $l = \overline{1,4}$, required for the calculations. They contain information about the intervals between the observed moments given by the probability density $p(\tau)$ of the form (6), consequently, the greater their number in the realization, the more information will be collected in the statistics and, therefore, the accuracy of the density parameters estimation will be higher.

Table 2. Results of the statistical experiment

θ	T_m	400	500	600	700	800	900	1000
$z_1 = 9,6$	$\hat{E}[\hat{z}_1]$	9,820	9,382	9,974	9,908	9,857	9,579	9,684
	$|\hat{E}[\hat{z}_1] - z_1|$	0,220	0,218	0,374	0,308	0,257	0,021	0,084
	$\sqrt{\hat{V}[\hat{z}_1]}$	0,958	0,890	0,863	0,672	0,654	0,549	0,502
$z_2 = 3,5$	$\hat{E}[\hat{z}_2]$	2,767	2,704	2,849	2,825	2,960	2,837	2,988
	$|\hat{E}[\hat{z}_2] - z_2|$	0,733	0,796	0,651	0,675	0,540	0,663	0,512
	$\sqrt{\hat{V}[\hat{z}_2]}$	0,801	0,861	0,688	0,711	0,689	0,684	0,648
$\beta_1 = 8,64$	$\hat{E}[\hat{\beta}_1]$	8,875	8,396	8,886	8,855	8,765	8,647	8,707
	$|\hat{E}[\hat{\beta}_1] - \beta_1|$	0,235	0,244	0,246	0,215	0,125	0,007	0,067
	$\sqrt{\hat{V}[\hat{\beta}_1]}$	0,917	0,813	0,769	0,627	0,562	0,524	0,539
$\beta_2 = 0,35$	$\hat{E}[\hat{\beta}_2]$	0,358	0,262	0,324	0,311	0,328	0,267	0,312
	$|\hat{E}[\hat{\beta}_2] - \beta_2|$	0,008	0,088	0,026	0,039	0,022	0,083	0,038
	$\sqrt{\hat{V}[\hat{\beta}_2]}$	0,131	0,150	0,113	0,099	0,098	0,099	0,093

The most interesting are the results obtained using R programming language capabilities in the RStudio environment for the real traffic data collected by Leland and Wilson [22] over several Ethernet local area network at the Bellcore Morristown Research and Engineering Center and formed as sets, each of which contains one million packet arrivals. We considered the sets **BC-pAug89**, started at 11:25 on August 9, 1989 and lasting for about 3142,82 s (until one million packets were registered), and **BC-pOct89**, started at 11:00 on October 5, 1989 and lasting about 1759,62 s. Packet sizes were not taken into account; determining factor are the moments of their arrival, which form a real correlated information flow. In order to obtain a recurrent flow corresponding to [22], the shuffling technique described in detail in [23] and applied in [24], was used. The randomization of inter-arrival durations sequence leads to a partial destruction of the correlations between them, while maintaining the distribution.

From the sequences generated over each of **BC-pAug89** and **BC-pOct89** by this technique, according to the windowing method, 981 subsequences of 10000 elements were built consistently through steps of length 1000.

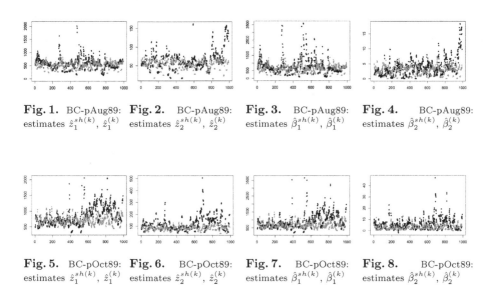

Fig. 1. BC-pAug89: estimates $\hat{z}_1^{sh(k)}$, $\hat{z}_1^{(k)}$ **Fig. 2.** BC-pAug89: estimates $\hat{z}_2^{sh(k)}$, $\hat{z}_2^{(k)}$ **Fig. 3.** BC-pAug89: estimates $\hat{\beta}_1^{sh(k)}$, $\hat{\beta}_1^{(k)}$ **Fig. 4.** BC-pAug89: estimates $\hat{\beta}_2^{sh(k)}$, $\hat{\beta}_2^{(k)}$

Fig. 5. BC-pOct89: estimates $\hat{z}_1^{sh(k)}$, $\hat{z}_1^{(k)}$ **Fig. 6.** BC-pOct89: estimates $\hat{z}_2^{sh(k)}$, $\hat{z}_2^{(k)}$ **Fig. 7.** BC-pOct89: estimates $\hat{\beta}_1^{sh(k)}$, $\hat{\beta}_1^{(k)}$ **Fig. 8.** BC-pOct89: estimates $\hat{\beta}_2^{sh(k)}$, $\hat{\beta}_2^{(k)}$

In Fig. 1, 2, 3, 4 and 5, 6, 7, 8, by light marker we indicate the parameter estimates $\hat{z}_1^{sh(k)}$, $\hat{z}_2^{sh(k)}$, $\hat{\beta}_1^{sh(k)}$, $\hat{\beta}_2^{sh(k)}$ calculated for them on the basis of the corresponding statistics C_l^k, $l = \overline{1,4}$, using formulas (11), (12); we compare them with the estimates $\hat{z}_1^{(k)}$, $\hat{z}_2^{(k)}$, $\hat{\beta}_1^{(k)}$, $\hat{\beta}_2^{(k)}$ (dark marker) obtained from samples of 10000 elements, formed by the windowing method from the original real data (before applying the shuffling procedure), $k = \overline{1,N}$, $N = 981$.

Note the strongly fluctuating nature of each of the considered parameter estimates $\hat{z}_1^{(k)}$, $\hat{z}_2^{(k)}$, $\hat{\beta}_1^{(k)}$, $\hat{\beta}_2^{(k)}$ (furthermore, some $\hat{z}_1^{(k)}$, $\hat{\beta}_1^{(k)}$ were out of the plotted intervals in Figs. 1 and 3), while for the estimates $\hat{z}_1^{sh(k)}$, $\hat{z}_2^{sh(k)}$, $\hat{\beta}_1^{sh(k)}$, $\hat{\beta}_2^{sh(k)}$ such tendency is not typical – there are relatively small deviations from the average values calculated for each sample, $k = \overline{1,N}$, $N = 981$.

For both the shuffled traces, among all the built subsequences with broken correlations, one was chosen for which the quality of approximation by a recurrent synchronous generalized doubly stochastic event flow of the second order was the best in the sense of the minimax criterion: $\delta^0 = \min_k \delta^k$, where δ^k denotes the relative error and is defined as $\delta^k = \max_\theta [|\theta - \hat{\theta}^{(k)}|/\theta]$, $k = \overline{1,N}$, $N = 981$, $\hat{\theta}^{(k)} \in \{\hat{z}_1^{(k)}, \hat{z}_2^{(k)}, \hat{\beta}_1^{(k)}, \hat{\beta}_2^{(k)}\}$, $\theta \in \{z_1, z_2, \beta_1, \beta_2\}$ are "true" values of the distribution parameters, set in accordance with the estimates obtained from the shuffled traces by the method of moments using formulas (11), (12). In the present experiment, the selected quality score achieved accuracy $\delta^0 = 0,005650$ for **BC-pAug89** and $\delta^0 = 0,004976$ for **BC-pOct89**.

Table 3 compares the values of the mean and variance calculated on the basis of the known probability density (6) with the sample mean and sample variance (statistical estimates of these probabilistic characteristics) of the random variable τ representing the interval duration between the flow events.

Table 3. Results of working with the real data

	$E[\tau]$	$\hat{E}[\tau]$	$\frac{|E[\tau]-\hat{E}[\tau]|}{E[\tau]}$	$D[\tau] \cdot 10^5$	$\hat{D}[\tau] \cdot 10^5$	$\frac{|D[\tau]-\hat{D}[\tau]|}{D[\tau]}$
pAug89	0,003103	0,003148	0,014502	3,333059	3,354130	0,006322
pOct89	0,001841	0,001821	0,010863	1,255121	1,254798	0,000257

The current analysis points out the correspondence of the recurrent synchronous generalized event flow of the second order to the arrival process in a certain time interval. In other words, the obtained results indicate that information flows in real systems can be described on some intervals by one doubly stochastic flow, on another by a similar flow (but with different input parameters) or in some cases by a different flow; it also indicates the noted earlier adequacy of the mathematical models of arrival processes of data packages in the form of doubly stochastic flows of events.

7   Conclusion

In this paper, we determined the explicit form of the joint probability density of the durations of adjacent inter-event intervals of a synchronous generalized flow of the second order (4), on the basis of which we formulated the recurrence conditions of the flow. By the method of moments, the parameter estimates of the defined by formula (6) distribution were found: $\hat{z}_1$, $\hat{z}_2$ in the form (11) and $\hat{\beta}_1$, $\hat{\beta}_2$ in the form (12). The numerical results of the statistical experiment conducted on the simulation model meet the physical interpretation and demonstrate the acceptable quality of the estimation in terms of the smallness of the selected quality indicators (the displacement estimate and sample quadratic error). The illustrated example with the real traffic data reflects the applicability of analytical estimation and simulation results to real information flows, the mathematical models of which are doubly stochastic flows.

References

1. Khinchin, A.Ya.: Mathematical methods of queuing theory. Proc. MIAS USSR **49**, 3–122 (1955)
2. Cox, D.R.: The analysis of non-Markovian stochastic processes by the inclusion of supplementary variables. Proc. Cambridge Philos. Soc. **51**(3), 433–441 (1955)
3. Basharin, G.P., Kokotushkin, V.A., Naumov, V.A.: On the equivalent substitutions method for computing fragments of communication networks. Part 1. Proc. USSR Acad. Sci. Tech. Cybern. **6**, 92–99 (1979)
4. Neuts, M.F.: A versatile Markov point process. J. Appl. Probab. **16**, 764–779 (1979)
5. Basharin, G.P., Gaidamaka, Y.V., Samouylov, K.E.: Mathematical theory of tele-traffic and its application to the analysis of multiservice communication of next generation networks. Autom. Control Comput. Sci. **47**(2), 62–69 (2013)
6. Dudin, A.N., Nazarov, A.A.: The MMAP/M/R/0 queueing system with reservation of servers operating in a random environment. Probl. Inf. Transm. **51**(3), 289–298 (2015)
7. Card, H.C.: Doubly stochastic Poisson processes in artificial neural learning. IEEE Trans. Neural Netw. **9**(1), 229–231 (1998)
8. Bouzas, P.R., Valderrama, M.J., Aguilera, A.M., Ruiz-Fuentes, N.: Modelling the mean of a doubly stochastic Poisson process by functional data analysis. Comput. Stat. Data Anal. **50**(10), 2655–2667 (2006)
9. Lucantoni, D.M., Neuts, M.F.: Some steady-state distributions for the MAP/SM/1 queue. Commun. Stat. Stoch. Models **10**, 575–598 (1994)
10. Klimenok, V., Dudin, A., Vishnevsky, V.: Tandem queueing system with correlated input and cross-traffic. Commun. Comput. Inf. Sci. **370**, 416–425 (2013)
11. Vishnevsky, V.M., Semenova, O.V.: Polling Systems: Theory and Applications for Broadband Wireless Networks. Academic Publishing, London (2012)
12. Vishnevsky, V.M., Dudin, A.N., Kozyrev, D.V., Larionov, A.A.: Performance evaluation of broadband wireless networks along the long transport routes. In: Proceedings of the Eighteenth International Scientific Conference on Distributed Computer and Communication Networks: Control, Computation, Communications (DCCN-2015), pp. 241–256. ICS RAS, Moscow (2015)
13. Farkhadov, M.P., Petukhova, N.V., Efrosinin, D.V., Semenova, O.V.: Two-phase model with unlimited queues to calculate the characteristics and optimize voice self-service portals. Control Sci. **6**, 53–57 (2010)
14. Pagano, M., Rykov, V.V., Khokhlov, Y.S.: Teletraffic Models. INFRA-M, Moscow (2019)
15. Nezhelskaya, L., Sidorova, E.: Optimal estimation of the states of synchronous generalized flow of events of the second order under its complete observability. Commun. Comput. Inf. Sci. **912**, 157–171 (2018)
16. Nezhelskaya, L.A., Sidorova, E.F.: Optimal estimate of the states of a generalized synchronous flow of second-order events under conditions of incomplete observability. Tomsk State Univ. J. Control Comput. Sci. **45**, 30–41 (2018)
17. Gortsev, A.M., Nezhelskaya, L.A.: Parameter estimation of synchronous twice-stochastic flow of events using the method of moments. Tomsk State Univ. J. **S1–1**, 24–29 (2002)
18. Gortsev, A.M., Zuevich, V.L.: Optimal estimation of states of the asynchronous doubly stochastic flow of events with arbitrary number of the states. Tomsk State Univ. J. Control Comput. Sci. **2**(11), 44–65 (2010)

19. Gortsev, A.M., Nissenbaum, O.V.: Estimation of the dead time and parameters of an asynchronous alternative flow of events with unextendable dead time period. Russ. Phys. J. **48**(10), 1039–1054 (2005)
20. Kalyagin, A.A., Nezhel'skaya, L.A.: The comparison of maximum likelihood method and moments method by estimation of dead time in a generalized semysynchronous flow of events. Tomsk State Univ. J. Control Comput. Sci. **3**(32), 23–32 (2015)
21. Malinkovsky, Y.V.: Probability Theory and Mathematical Statistics (Part 2. Mathematical Statistics). Francisk Skorina Gomel State University, Gomel (2004)
22. Leland, W.E., Wilson, D.V.: High time-resolution measurement and analysis of LAN traffic: implications for LAN interconnection. In: Proceedings of the IEEE INFOCOM 1991, Bal Harbour, FL, pp. 1360–1366 (1991)
23. Karagiannis, T., Faloutsos, M.: SELFIS: a tool for self-similarity and long-range dependence analysis. In: 1st Workshop on Fractals and Self-Similarity in Data Mining: Issues and Approaches (in KDD), Edmonton, Canada (2002)
24. Erramilli, A., Narayan, O., Willinger, W.: Experimental queueing analysis with long-range dependent packet traffic. IEEE/ACM Trans. Network. **4**(2), 209–223 (1996)

Queue with Partially Ignored Interruption in Markovian Environment

A. Krishnamoorthy[1,2(✉)] and S. Jaya[3]

[1] Centre for Research in Mathematics, C. M. S. College, Kottayam, India
achyuthacusat@gmail.com
[2] Department of Mathematics, CUSAT, Kochi, Kerala, India
[3] Department of Mathematics, Maharaja's College, Ernakulam, Kerala, India
jayasreelakam@gmail.com

Abstract. The paper carries out an analysis of the single server queueing system with Poisson arrival pattern, Erlang service pattern and partially ignored interruption. Interruptions are the outcome of the influence of various environmental factors on the service process. These environmental factors are states of a Markov chain. The replacement or the repair of the server is decided by the environmental factors. The resumption or restart of the interrupted service after the repair of the server is based on the realization of random clock or interruption clock started at the onset of interruption. Even though every interruption varies the service rate, the impact of interruption caused by a few environmental factors are indecisive. The paper ventures to establish important performance measures by means of Matrix a Analytic Method. The analytical approach is validated with the aid of illustrations and numerical results are computed.

Keywords: Partially ignored interruption · Environmental factor · Random clock · Interruption clock

1 Introduction

Queues with interruption have been studied extensively in the past and the same has been successfully used in various applied problems. In our day to day life we face different kinds of interruptions in banking, agriculture, networking, security etc. In spite of the improvement in facilities to cater to the growing demand for service, the interruptions continue to escalate.

In queueing system, preemptive priority, server vacation and server breakdown are some of the common sources of interruption. White and Christie were the pioneers of the concept of queues with interruption. The paper [15] deals with interruption due to preemptive priority. They assumed that at the end of each interruption the service of the interrupted customer is resumed. Jaiswal [6,7], Avi-Itzhak and Naor [1] and Thiruvengadam [13] analyse queueing models with general distribution for service and interruption time. In all the above mentioned papers the arrival of high priority customers caused interruption to the

© Springer Nature Switzerland AG 2019
V. M. Vishnevskiy et al. (Eds.): DCCN 2019, LNCS 11965, pp. 289–301, 2019.
https://doi.org/10.1007/978-3-030-36614-8_22

service of the low priority customers. A recent survey paper by Krishnamoorthy et al. [9] comprises a detailed description of research on queueing models with service interruption induced by server breakdown and customer induced service break. In [10], the discussion is about unacknowledged interruption for a short duration, in respect of the impact of various environmental factors.

To ensure that the efficiency of the system is maintained up to the desired level, the postponement of interruption is beneficial. The papers by Gaver, Jr. [4] and Blanc [5] takes up the idea of postponable interruption. Through [9] the author unfolds views on postponable interruption in which interruption is postponed until the termination of the current service. He also deals with preemptive repeat identical interruption and preemptive repeat different interruption. In the preemptive repeat identical interruption, instead of postponing the interruption, the server opts for repair at the onset of interruption. On completion of the repair process the service is resumed. In the case of preemptive repeat different interruption, on completion of repair of the server, new service starts to the same customer. An investigation of the effect of postponing interruption of the service in $M/G/1$ queueing system until the end of the service in progress is carried out in [5].

Vacation models are also considered as particular type of service interruption models. In classical vacation model the server opts for vacation at a service completion epoch. But in the queueing model analysed by Takagi and Leung [12], the server takes vacation when the service period exceeds a specified duration. Excellent surveys on the earlier works of vacation have been reported by Doshi [3], Takagi [11] and Tian and Zhang [14].

In almost all the papers on queue with service interruption the service is resumed or restarted on the removal of interruption. In the paper by Krishnamoorthy et al. [8] two random variables compete, on the onset of interruption to decide whether to repeat or resume the interrupted service. But in Fiems, Maertens and Bruneel, [2] the nature of service to be rendered after the completion of interruption is determined at the onset of interruption.

To the best of authors' knowledge, the study of queues with postponable interruption in the existing literature, focus is mainly on the interruption due to preemptive priority and server break down. This paper considers the effects of server continuing service with interruption and interruption induced by environmental factors. The model under consideration is an $M/E_a/1$ queueing system with interruption due to different environmental factors. The interruption to service might occur due to one of the $n + 1$ environmental factors. The environmental factors are labeled based on the severity of interruption caused by them. The interruption befalls according to Poisson Process and when the interruption due to any of the first n environmental factors arise, the service rate changes. Even if the service is interrupted as a result of interference of any one of the first n factors, it is ignored in the beginning and service is continued with interruption. On completion of interrupted service the server is taken for repair. The decision to repair or replace the server on the occurrence of interruption is taken based on the severity of interruption caused by the environmental factors. The interruption clock and random clock which are on track at the onset of interruption are pivotal in resolving the resumption or restart of the service of the interrupted customer.

As the service progresses with interruption, the severity of interruption may get accelerated. After some time, the cause of interruption may change from i^{th} factor to j^{th} factor, $(j > i)$ where the prior interruption is caused by the ith factor. The interruption due to $n + 1^{th}$ factor renders the server defunctive and server replacement is the alternative. In such cases the customer in service goes out of the system without awaiting service. The $n + 1$ environmental factors are the states of a Markov chain with initial probability vector $p_i, i = 1, 2, \ldots, n + 1$ and transition probability matrix $P = (p_{ij}), i, j = 1, 2, \ldots, n + 1$.

This model has application in all fields where there is a probability for piling up of a queue such as fault tolerant system, body area network, etc. During many a time in industries systems may have to continue to operate with interruption due to factors like consequences of system failure. The impact includes loss of property and time, financial loss, rise in product demand, safety issues etc. For instance consider a pumping system. A decided quantity of liquid (flammable or toxic in nature) has to be transferred from the storage point to another point of operation. There is a possibility for the occurrence of some of the interruptions like, pump developing a seal leak, increase in vibration, increase in power consumption, etc.

The operation may have to be continued ignoring the interruption considering the criticality of operation(e.g.: the transfer may be to a reactor where a definite quantity has to be transfered in a stipulated time). There is a pretty good chance that the service can be completed without crossing the threshold limit value of interruption duration. There is an equally good probability that the interruption may cross the threshold limit value of interruption duration and may stop in between for proceeding to repair.

The rest of the paper is organized as follows. Section 2 provides system description. The analysis of the service process is consolidated in Sect. 3. Analysis of the queueing model is discussed in Sect. 4. Section 5 is dedicated to important performance measures, a cost function is formulated in Sect. 6 and numerical illustration is provided in Sect. 7.

2   Model Description

Consider a single server queueing system in which arrival is characterized by Poisson Process with parameter λ. On arrival, if the customer finds the server busy, he joins the tail of the queue or else gets service immediately. The service is Erlang distributed with shape and scale parameters μ and a respectively. Here the assumption is that there are $n+1$ environmental factors causing interruption to the service. These factors are numbered 1 to $n + 1$ depending on the ascending order of severity of interruption caused by them. The interruption occurs according to a Poisson Process with parameter β. When the interruption befalls due to i^{th} factor the service rate changes from μ to μ_i. At the onset of interruption a random clock which is exponentially distributed with parameter α and an interruption clock which is Phase type distributed with representation (δ, U) of order m are started. Forward phase change alone is allowed for interruption clock $(U_{ij} = 0$ for $i > j)$. When the interruption occurs due to any of the first n factors, it is ignored for the time being and service is continued with interruption. Once the interruption clock realizes the server is taken for repair. The repair time is exponentially distributed with parameter η. After repair the service to the interrupted customer is resumed if the interruption clock is realized before the random clock; or else the service is restarted. There is a contingency for customer completing service with interruption. In that case the server goes for repair after the service completion. If the interruption is due to $n + 1^{th}$ factor the customer is lost and the server replacement takes place instantaneously. Once the interruption is rectified, both the clocks are reset to zero position.

As the duration of ignored interruption prolongs, the severity of interruption escalates. As time goes by, the cause of interruption varies from i^{th} factor to j^{th} factor, $(j > i)$ where prior interruption is caused by i^{th} factor. Then the service rate also changes from μ_i to μ_j. Again if the interruption stays abandoned further, the cause of interruption changes as well from j^{th} factor to k^{th} factor, $(k \geq j)$. The server gets replaced on being interrupted by the $n + 1^{th}$ factor.

The $n+1$ environmental factors are the states of a Markov chain with initial probability vector $p_i, i = 1, 2, \ldots, n+1$ and transition probability matrix $P = (p_{ij}), i, j = 1, 2, \ldots, n+1$. Graphical representation of the model is given in Fig. 1

3   Analysis of Service Process with Interruption

The service process $\{Y(t), t \geq 0\}$ where $Y(t) = (S(t), I_1(t), I_2(t), I_3(t), I_4(t))$ is a Markov chain with $(2+mn)a+1$ transient states given by $\{(0, j) \cup (1, i, j, l, 1) \cup (2, j, 1) \cup (2, 0)\}$ with $i = 1, 2, \ldots, n; j = 1, 2, \ldots, a; l = 1, 2, \ldots, m$; and one absorbing state which represents the customer moving out from the system either after service completion or due to interruption caused by $n + 1^{th}$ environmental factor. $S(t)$ denotes the status of the server at time t:

$$S(t) = \begin{cases} 0 & \text{the ongoing service is without interruption.} \\ 1 & \text{the ongoing service is with interruption.} \\ 2 & \text{server is under repair;} \end{cases}$$

$I_1(t)$ corresponds to the environmental factor that caused the current interruption to the service. In this model we consider $n + 1$ environmental factors; $I_2(t)$ denotes the phase of service. It varies from 1 to a;
$I_3(t)$ denotes the phase of interruption clock. It varies from 1 to m;
$I_4(t)$ denotes the phase of random clock. It takes the value 0 if clock is realized and 1 if it is functioning.

The infinitesimal generator of the process is given by $Q = \begin{bmatrix} W & W^0 \\ 0 & 0 \end{bmatrix}$

with initial probability vector

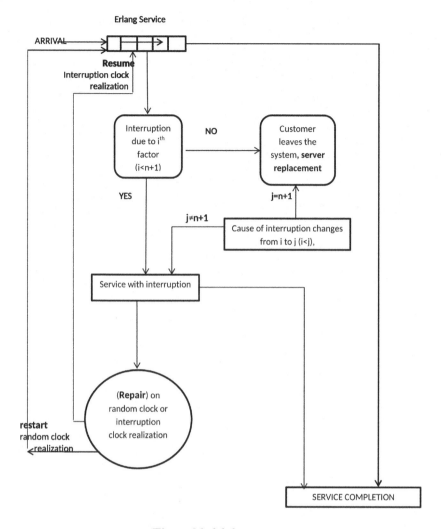

Fig. 1. Model description

$$\sigma = (1, \mathbf{0}, \mathbf{0}) \text{ where } W = \begin{bmatrix} D_0 & D_1 & 0 \\ 0 & D_2 & D_3 \\ D_4 & 0 & D_5 \end{bmatrix} \text{ and } W^0 = \begin{bmatrix} D_{01} \\ D_{11} \\ 0 \end{bmatrix}.$$

Let $\delta = (\delta_1, \delta_2, \ldots, \delta_m), I_a$ denotes identity matrix of order a and e_a denotes unit column matrix of order a.

then, $D_0(i, j) = \begin{cases} -\mu - \beta, & \text{if } i = j; \\ \mu, & \text{if } i = j - 1 \ i, j = 1, \ldots, a; \\ 0, & \text{otherwise}; \end{cases}$

$D_1 = \beta[p' \otimes (I_a \otimes \delta)]_{1 \times mna}$ with $p' = (p_1, p_2, \ldots, p_n)$;

$D_2 = I_{na} \otimes U - \overline{\mu} \otimes I_{ma} + \widehat{P} \otimes I_{am} - \alpha I_{mna} + J$;

where $\widehat{P} = \begin{bmatrix} P_{11} & P_{12} & \cdots & P_{1n} \\ & P_{22} & \cdots & P_{2n} \\ & & \ddots & \\ & & & P_{nn} \end{bmatrix}$ $\overline{\mu} = \begin{bmatrix} \mu_1 & & \\ & \mu_2 & \\ & & \ddots \\ & & & \mu_n \end{bmatrix}$;

and J is a square matrix of order mna.

$$J(i,j) = \begin{cases} \mu_r, & \text{for } j = i + m, i = (r-1)am + 1, \ldots, (ra-1)m; \\ & r = 1, \ldots, n; \\ 0, & \text{otherwise;} \end{cases}$$

$D_3 = I_n \otimes E$; with $E = \begin{bmatrix} E_1 & E_2 \end{bmatrix}$, where $E_1 = I_a \otimes U^0$, $M_2 = \alpha e_{ma}$.

D_4 is a matrix of order $((a+1) \times a)$ with

$$D_4(i,j) = \begin{cases} \eta, & \text{for } i = j = 1 \ \& \ j = i + 1; \\ 0, & \text{otherwise;} \end{cases}$$

$D_5 = -\eta I_{a+1}$.

D_{01} is a column vector, $D_{01}(i,1) = \begin{cases} \beta p_{n+1}, & \text{for } i = 1, 2, \ldots, a - 1; \\ \mu + \beta p_{n+1}, & \text{for } i = a; \end{cases}$

$$D_{11} = [K + \overline{P} \otimes e_{ma}]_{mna \times 1}; \text{ where } \overline{P} = \begin{bmatrix} P_{1n+1} \\ P_{2n+1} \\ \vdots \\ P_{nn+1} \end{bmatrix}.$$

K is an $(mna \times 1)$ matrix and

$$K(i,1) = \begin{cases} \mu_r, & \text{for } i = (ra-1)m + 1, \ldots, ram; r = 1, \ldots, n; \\ 0, & \text{otherwise;} \end{cases}$$

- Expected time for service completion, $E(ST) = \sigma(-W)^{-1}e$. Hence expected service rate $\mu_s = \frac{1}{E(ST)}$.
- Rate of replacement of server due to the interruption caused by $n + 1^{th}$ factor,

$$R_{replacement} = \sigma(-W)^{-1}W^{0'} \text{ where } W^{0'} = \begin{bmatrix} e_a \beta p_{n+1} \\ \overline{P} \otimes e_{ma} \\ 0 \end{bmatrix}.$$

3.1   Expected Number of Interruptions During the Service of a Customer

For calculating the expected number of interruptions during the service of a customer, consider the Markov chain $\left\{ \widehat{Y}(t), t \geq 0 \right\}$ where $\widehat{Y}(t) = (M(t), S(t), I_1(t), I_2(t), I_3(t), I_4(t))$, $M(t)$ is the number of interruptions occurred until time t. $S(t), I_1(t), I_2(t), I_3(t)$ and $I_4(t)$ are as mentioned earlier. The state space of $\widehat{Y}(t)$ is $\{(r, 0, j) \cup (r, 1, i, j, l, 1) \cup (r, 2, j, 1) \cup (r, 2, 0)\}$ with $r = 0, 1, 2, \ldots, \infty; i = 1, 2, \ldots, n; j = 1, 2, \ldots, a; l = 1, 2, \ldots, m;$ and one absorbing state which represents the customer moving out from the system either after service completion or due to interruption caused by $n + 1^{th}$ environmental factor. The infinitesimal generator matrix associated with $\widehat{Y}(t)$ is $\widehat{Q} = \begin{bmatrix} Y & Y^0 \\ 0 & 0 \end{bmatrix}$

where $Y = \begin{bmatrix} \widehat{B}_0 & \widehat{B}_1 & & & \\ & \widehat{A}_1 & \widehat{A}_0 & & \\ & & \widehat{A}_1 & \widehat{A}_0 & \\ & & & \ddots & \ddots \\ & & & & \ddots & \ddots \end{bmatrix}$ and $Y^0 = \begin{bmatrix} \widehat{B}_2 \\ \widehat{A}_2 \\ \widehat{A}_2 \\ \vdots \\ \vdots \end{bmatrix}$.

Here $\widehat{B}_0 = D_0$; $\widehat{B}_1 = D_1$; $\widehat{B}_2 = D_{01}$;

$$\widehat{A}_1 = \begin{bmatrix} D_2 & D_3 & 0 \\ 0 & D_5 & D_4 \\ 0 & 0 & D_0 \end{bmatrix}; \quad \widehat{A}_0 = \begin{bmatrix} 0 & 0 & 0 \\ 0 & 0 & 0 \\ D_1 & 0 & 0 \end{bmatrix}; \quad \widehat{A}_2 = \begin{bmatrix} D_{11} \\ 0 \\ D_{01} \end{bmatrix}.$$

- The probability for absorption after r interruptions$()a_r$, $r = 1, 2, \ldots, \infty)$, is
  $a_r = \sigma(-\widehat{B}_0^{-1}\widehat{B}_1)((\widehat{A}_1^{-1}\widehat{A}_0))^{r-1}(-\widehat{A}_1^{-1}\widehat{A}_2), a_0 = \sigma(-\widehat{B}_0^{-1}\widehat{B}_2)$,
- Expected number of interruptions before absorption,
  $E(NI) = \sum_{r=1}^{\infty} ra_r = \sigma(-\widehat{B}_0^{-1}\widehat{B}_1)(I - (\widehat{A}_1^{-1}\widehat{A}_0))^{-2}(-\widehat{A}_1^{-1}\widehat{A}_2)$

4   The Queueing Model

Let $N(t)$ be the number of customers in the system. Then $X = \{X(t), t \geq 0\}$, where $X(t) = (N(t), S(t), I_1(t), I_2(t), I_3(t), I_4(t))$, is a Continuous Time Markov Chain (CTMC) with state space $\{(r, 0, j) \cup (r, 1, i, j, l, 1) \cup (r, 2, j, 1) \cup (r, 2, 0)\}$ with $r = 1, 2, \ldots, \infty; i = 1, 2, \ldots, n; j = 1, 2, \ldots, a; l = 1, 2, \ldots, m$. $S(t), I_1(t), I_2(t), I_3(t)$ and $I_4(t)$ are as mentioned in Sect. 3. The infinitesimal generator matrix associated with the model

$$\widehat{Q} = \begin{bmatrix} B_0 & B_1 & & & \\ B_2 & A_1 & A_0 & & \\ & A_2 & A_1 & A_0 & \\ & & \ddots & \ddots & \ddots \\ & & & \ddots & \ddots & \ddots \end{bmatrix}$$ where $B_0 = [-\lambda]$; $B_1 = \begin{bmatrix} \lambda & 0 & 0 \end{bmatrix}$;

$B_2 = W^0$;
$A_0 = [\lambda I]$; $A_1 = W - \lambda I$; $A_2 = [W^0 \ \mathbf{0}]$ is a square matrix.

Theorem 1. *The queueing system is stable when $\lambda < \mu_s$.*

Proof. Let π denote the invariant probability vector of the generator $A_0 + A_1 + A_2$. That is, $\pi(A_0 + A_1 + A_2) = 0$; $\pi e = 1$. The Level Independent Quasi Birth Death (LIQBD) description of the model indicates that the queueing system is stable if and only if $\pi A_0 e < \pi A_2 e$. That is, the rate of drift to the left has to be higher than that to the right. Here,

$\pi A_0 e = \lambda,$
$\pi A_2 e = \pi W^0 = \frac{1}{\sigma(-W)^{-1}e} = \mu_s.$
Hence the queueing system is stable when $\lambda < \mu_s$.

4.1 Stationary Distribution

Let $\chi = (x_0, x_1, x_2, ...)$ be the steady state probability vector of the Markov chain $\{X(t), t \geq 0\}$. Each $x_i, i > 0$, are vectors with $(nm + 1)a$ elements. We assume that $x_2 = x_1 R$, and $x_i = x_1 R^{i-1}, i \geq 2$, where R is the minimal non-negative solution to the matrix quadratic equation $R^2 A_2 + R A_1 + A_0 = 0$. From $\chi Q = 0$ we get $x_0 B_0 + x_1 B_2 = 0$, $x_0 B_1 + x_1 (A_1 + R A_2) = 0$. Solving the above two equations we get x_0 and x_1 subject to the normalizing condition $x_0 e + x_1 (I - R)^{-1} e = 1$.

5 Performance Measures

After calculating the steady state probability vector we now calculate some important performance measures of the system to bring out the qualitative aspects of the model under study. These are listed below along with their formula for computation.

5.1 Expected Waiting Time

The expected waiting time of the customer who joins as the rth customer $(r > 0)$ in the queue can be obtained from the Markov chain $X'(t) = (M'(t), S(t), I_1(t), I_2(t), I_3(t), I_4(t))$ where $M'(t)$ is the rank of the tagged customer.

$S(t), I_1(t), I_2(t), I_3(t)$ and $I_4(t)$ are as defined in Sect. 3. The waiting time distribution of the tagged customer follows phase type distribution with representation (θ, S), and is given by

$$S = \begin{bmatrix} W & W^0\sigma & & \\ & W & W^0\sigma & \\ & & \cdots & \cdots \\ & & & W \end{bmatrix}, \quad S^0 = \begin{bmatrix} 0 \\ 0 \\ 0 \\ W^0 \end{bmatrix}.$$

The expected waiting time of the tagged customer is a column vector given by

$$E_w^r = -W^{-1}(I - (-W^0\sigma W^{-1}))^{(r-1)}(I + W^0\sigma W^{-1})^{-1}e.$$

Depending on the state of the server at the time of joining of the rth customer we get different values for expected waiting time of the tagged customer.

– The expected waiting time of a customer who waits in the queue is

$$E(W) = \sum_{r=1}^{\infty} X_r E_w^r.$$

5.2   Important Performance Measures

- Mean number of customers in the system, $E(s) = \sum_{i=1}^{\infty} i\mathbf{x}_i e$.

- Mean number of customers in the queue, $E(q) = \sum_{i=1}^{\infty} (i-1)\mathbf{x}_i e$.

- Probability that there is no customer in the system, $P_{idle} = \mathbf{x}_0$.

- Probability that the system is under interruption, $P(I) = \sum_{i=1}^{\infty} \mathbf{x}_{i1} e$.

- Probability that the system is under repair, $P(r) = \sum_{i=1}^{\infty} \mathbf{x}_{i2} e$.

- Effective interruption rate, $R_{int} = \beta \sum_{i=1}^{\infty} \mathbf{x}_{i0} e$.

- Effective repair rate, $R_{rep} = \eta \sum_{i=1}^{\infty} \mathbf{x}_{i1} e$.

- Effective rate of repetition of service, $R_{rpt} = \alpha \sum_{i=1}^{\infty} \mathbf{x}_{i1} e$.

- Rate at which service completion with interruption occurs before the random clock is realized is $\sum_{i=1}^{\infty} \sum_{j=1}^{n} \sum_{r=1}^{m} \mathbf{x}_{i1jar} \mu_j$.

- Probability of a customer completing service without any interruption, $P(s) = $ P(Service time $<$ exponentially distributed random variable with parameter β) $= \frac{\mu^a}{(\mu+\beta)^{a+1}}$.

- Probability that at least one interruption in service is $1 - \frac{\mu^a}{(\mu+\beta)^{a+1}}$.

- probability that the interruption is attended before the random clock is realized, $P_{ia} = $ P(interruption random variable $<$ random clock variable)$= \delta(\alpha I_m - U)^{-1} U^0$.

- Probability for restart of service is $1 - \delta(\alpha I_m - U)^{-1} U^0$.

6   Cost Function

To compute the expected cost, a cost function is constructed depending on important performance measures. The total cost for running the system,

$$E(C) = \mu a C_0 + R_{replacement} C_1 + E(q) C_2 + R_{rpt} C_3 + C_4 R_{rep},$$

where

- $C_0 -$ Unit time cost of service;
- $C_1 -$ Unit time cost for replacing the server;
- $C_2 -$ Holding cost per customer in the queue;
- $C_3 -$ Unit time cost for restarting the service;
- $C_4 -$ Unit time cost for repairing the server.

7   Numerical Examples

The performance of the queueing system is numerically illustrated in this section.

Let $n = m = a = 3$, $\lambda = 1$, $C_0 = \$100$, $C_1 = \$15000$, $C_2 = \$10$, $C_3 = \$200$, $C_4 = \$2000$, $\mu = 5, \mu_1 = 3, \mu_2 = 2, \mu_3 = 2$, $p' = (0.2, 0.2, 0.2)$, $p_{n+1} = 0.4$, $\delta = (0.3, 0.3, 0.4)$,

$$
P = \begin{bmatrix} 0.1 & 0.3 & 0.5 & 0.1 \\ 0 & 0.5 & 0.3 & 0.2 \\ 0 & 0 & 0.6 & 0.4 \\ 0 & 0 & 0 & 0 \end{bmatrix}; U = \begin{bmatrix} -30 & 15 & 5 \\ 0 & -20 & 10 \\ 0 & 0 & -10 \end{bmatrix}; U^0 = \begin{bmatrix} 10 \\ 10 \\ 10 \end{bmatrix}.
$$

By taking $\beta = 2$, $\eta = 4$ the numerical values of various performance measures are calculated here.

7.1   Effect of the Rate of Realization of Random Clock on Various Performance Measures

Table 1 illustrates that, as the rate of realization of random clock increases expected service rate decreases. When random clock realizes, the server goes for repair and after the repair of server the service of the customer in service restarts. This time lag reduces the service rate. Here the replacement rate decreases. This is because when random clock realizes the service is stopped and repair begins. So the progress in interruption is stopped. This reduces the chance of replacement of server. But the rate of restart of service increases. This is due to restart of service after every realization of random clock.

Table 1. Effect of α on various performance measures.

α	$E(s)$	$R_{replacement}$	μ_S	P_{idle}	P_{ia}
1	2.2246	1.0493	1.5736	0.2849	0.1609
2	2.2422	1.0458	1.5630	0.2851	0.8333
3	2.2472	1.0428	1.5546	0.2862	0.7692
4	2.2447	1.0401	1.5478	0.2876	0.7143
5	2.2379	1.0378	1.5423	0.2892	0.6667
6	2.2288	1.0356	1.5376	0.2909	0.6250

7.2   Effect of Rate of Realization of Random Clock and Rate of Interruption on $E(C)$

Table 2 illustrates that, as rate of realization of random clock increases, expected cost also increases. When rate of interruption increases expected cost also increases on expected lines.

Table 2. Effect of α and β on $E(C)$.

α	$E(C)$	β	$E(C)$
1	17262	1	1685
2	17210	2	17120
3	17165	3	17328
4	17124	4	17495
5	17089	5	17635
6	17057	6	17777

7.3 Effect of Rate of Interruption on Various Performance Measures

By taking $\mu = 5$, $\alpha = 4$, $\eta = 5$ the numerical values of various performance measures are calculated here. Inference from Table 3 is that, as occurrence of interruption increases expected number of customers in the system, rate of replacement of server increases and effective service rate decreases. This happens due to delay in service caused by repair of server or the reduced service rate of interrupted server. Increase in the occurrence of interruption reduces the probability for service completion without interruption. By taking $\alpha = 1$, $\beta = 1$ the numerical values of various performance measures are calculated here.

Table 3. Effect of β on various performance measures

β	$E(s)$	$R_{replacement}$	μ_S	P_{idle}	$P(s)$
1	1.4802	1.0226	1.6311	0.3651	0.3517
2	1.8160	1.0401	1.6182	0.3295	0.1609
3	2.2718	1.0537	1.6183	0.2894	0.0863
4	2.9943	1.0643	1.6259	0.2407	0.0514
5	4.4084	1.0727	1.6378	0.1786	0.0331
6	8.5368	1.0794	1.6521	0.0993	0.0225

7.4 Effect of Repair Rate on Various Performance Measures

Inference from Table 4 is that as the repair rate increases expected number of customers in the system and expected service rate decreases. But probability for idleness increases. The increased repair rate reduces the waiting time of customers due to which the queue length reduces. When queue length reduces the probability of idleness increases. As the repair rate increases the expected cost decreases. This is due to the decrease in waiting time of customer in the queue.

Table 4. Effect of η on various performance measures

η	$E(s)$	μ_S	P_{idle}	$E(C)$
2	32.8330	1.1848	0.0244	17104
3	5.9106	1.2793	0.1195	16769
4	3.9102	1.3324	0.1686	16749
5	3.2094	1.3664	0.1974	16742
6	2.8571	1.3901	0.2162	16738
7	2.6461	1.4075	0.2294	16736
8	2.5060	1.4208	0.2391	16735

7.5   Rate of Interruption Clock Realization & Efficiency of System

From Table 5 as the rate of realization of interruption clock increases rate of repair increases which causes reduction in the efficiency of the system.

Table 5. Rate of interruption clock realization & efficiency of system

Rate of interruption clock realization	Efficiency of system
0.0480	1.6307
0.0653	1.6058
0.0674	1.6031
0.0778	1.5894
0.0799	1.5863
0.0813	1.5848
0.1000	1.5630
0.1408	1.5195
0.1519	1.5081

8   Conclusion

In this paper $M/E_a/1$ queueing system with partially ignored interruption is considered. The $n + 1$ environmental factors which cause interruption to service are arranged in the ascending order of severity of interruption caused by them. These environmental factors are the states of a Markov chain. The interruption is ignored for the time being, provided it is due to first n environmental factors. The server is replaced if the cause of interruption is the $n + 1^{th}$ environmental factor. The ignored interruption may progress with time resulting in a change in service rate. At the onset of interruption, a random clock and an interruption clock start functioning. On completion of service with interruption the server is taken for repair. The realization of random clock or interruption clock during service with interruption will result in the repair of the server. Whether to repair or restart the service depends on the prior realization of random clock or interruption clock. The interruption caused by $n + 1^{th}$ environmental factor results in server replacement and customer loss. The system is studied using matrix analytic method and various performance measures like expected service rate, expected waiting time etc are obtained. The effect of various parameters on the profit is analysed by constructing a cost function.

References

1. Avi-Itzhak, B., Naor, P.: Some queuing problems with the service station subject to breakdown. Oper. Res. **11**(3), 303–320 (1963)
2. Fiems, D., Maertens, T., Bruneel, H.: Queueing systems with different types of server interruptions. Eur. J. Oper. Res. **188**(3), 838–845 (2008)
3. Doshi, B.T.: Queueing systems with vacations a survey. Queueing Syst. **1**, 29–66 (1986)
4. Gaver Jr., D.P.: A waiting line with interrupted service, including priorities. J. Roy. Stat. Socy. Ser. B (Methodol.) **24**, 73–90 (1962)
5. Blanc, H.(J.P.C.): M/G/1 queues with postponed interruptions. ISRN Probab. Stat. **2012**, 12 (2012)
6. Jaiswal, N.K.: Preemptive resume priority queue. Oper. Res. **9**(5), 732–742 (2012)
7. Jaiswal, N.K.: Time-dependent solution of the head-of-the-line priority queue. J. Roy. Stat. Soc. B **24**, 91–101 (1962)
8. Krishnamoorthy, A., Gopakumar, B, Narayanan, V.: A queueing model with interruption resumption/restart and reneging. Bull. Kerala Math. Assoc. Spec. Issue (2009)
9. Krishnamoorthy, A., Pramod, P.K., Chakravarthy, S.R.: Queues with interruption, a survey. Top **22**(1), 290–320 (2012)
10. Krishnamoorthy, A., Jaya, S., Lakshmy, B.: Queues with interruption in random environment. Ann. Oper. Res. **233**(1), 210–219 (2015)
11. Takagi, H.: Queueing Analysis: A Foundation of Performance Evaluation, vol. 1. Vacation and Priority System. Elsevier, Amsterdam (1991)
12. Takagi, H., Leung, K.K.: Analysis of a discrete-time queueing system with time-limited service. Queueing Syst. **18**(1), 183–197 (1994)
13. Thiruvengadam, K.: Queuing with breakdowns. Oper. Res. **11**(1), 62–71 (1963)
14. Tian, N.S., Zhang, Z.G.: Vacation Queueing Models: Theory and Applications. Springer, New York (2006). https://doi.org/10.1007/978-0-387-33723-4
15. White, H., Christie, L.: Queuing with preemptive priorities or with breakdown. Oper. Res. **6**(1), 79–95 (1958)

Modeling and Reliability Analysis of a Redundant Transport System in a Markovian Environment

Udo R. Krieger[1(✉)] and Natalia Markovich[2]

[1] Fakultät WIAI, Otto-Friedrich-Universität,
An der Weberei 5, 96047 Bamberg, Germany
udo.krieger@ieee.org
[2] V.A. Trapeznikov Institute of Control Sciences,
Russian Academy of Sciences,
Profsoyuznaya Street 65, Moscow 117997, Russia
markovic@ipu.rssi.ru

Abstract. We consider the multipath communication between a client and a server that is established at the transport and session layers of an SDN/NFV protocol stack in a fog computing scenario. We analyze the reliability function of an associated redundant transport system comprising two logical channels that are susceptible to failures. The failure processes of both channels are described by Markov-modulated failure times that are driven by the transitions of a common Markovian environment. We model the error-prone system with repair and independent phase-type distributed repair times by a continuous-time Markov chain. We identify its generator matrix in terms of Kronecker products of the underlying parameter matrices that are determined by the interarrival times driven by Markov-modulated failure processes and the independent phase-type distributed repair times. We show that the steady-state distribution of the restoration model can be effectively calculated by an iterative aggregation-disaggregation method for block matrices and compute the associated reliability function of the transport system by a uniformization method.

Keywords: SDN/NFV · Reliability modeling · Reliability function · Markov-modulated process · Phase-type distributed repair times · Kronecker matrices

1 Introduction

Currently, fog computing is applied to integrate new services based on modern multimedia and machine-to-machine communication in the evolving Internet-of-Things into a cloud computing infrastructure (cf. [1,2,7]). The architecture is based on software-defined networks (SDN) and network function virtualization (NFV) technologies (cf. [6,9,23]). In this context it has been realized that

© Springer Nature Switzerland AG 2019
V. M. Vishnevskiy et al. (Eds.): DCCN 2019, LNCS 11965, pp. 302–314, 2019.
https://doi.org/10.1007/978-3-030-36614-8_23

multipath communication established at the transport and session layers of the SDN/NFV protocol stack can provide means to improve the capacity and reliability of the interprocess communication of these new services. Protocols like SCTP, multipath TCP, or multipath QUIC support such concepts and offer a set of redundant transport paths between clients and servers that enable multihoming due to the use of multiple interfaces (see Fig. 1, cf. [3,5,8]). In future wireless networks triggered by the rapidly evolving 5G standard the virtualized server functionality will be shifted closer to the edge of the associated radio access network (RAN) to reduce the end-to-end delay and to improve the transport capacity. Diverse radio technologies may be integrated into the RAN environment that is attached to the SDN core network. The latter will support new cloud computing concepts like fog and mobile edge computing (see Fig. 2, cf. [7,14,20]).

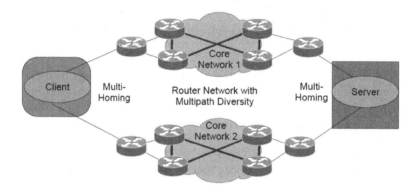

Fig. 1. Multipath diversity in modern SDN networks.

Regarding such a wireless access network, the modeling and analysis of the transport performance of single-path and multipath TCP communication associated with a client-server system that is embedded into a fog computing scenario provides numerous scientific challenges. The analysis of a simple TCP flow-control model has been a topic of our previous research (cf. [19]). In this performance study the related fog computing scenario will be considered from a different perspective.

In this paper we study a basic performance model describing two logical transport channels which provide a redundant communication system at the session layer between a client as sender and a server as receiver of a high-speed interprocess communication path in an SDN/NFV/RAN environment. The two system components modeling the transport channels are subject to errors that can decrease their throughputs below prescribed levels which is considered as a failure of the respective functional block. We model the failure times as Markov-modulated Poisson process (cf. [10]). The entire redundant system is managed by a scalable, virtualized management system applying Docker or Kubernetes container virtualization techniques (cf. [11,12]). It can instantiate one or two repair

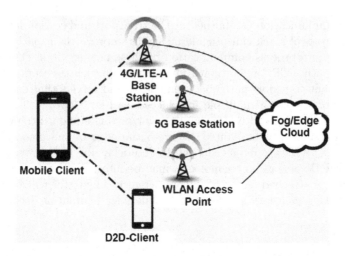

Fig. 2. Evolving radio access networks supporting fog and mobile edge computing.

functions as virtualized network functions and restore the original transport status after a generally distributed, nonnegative restoration period. The latter is approximated by a phase-type distribution (cf. [24]). We develop a Markovian reliability model of this redundant transport system in a random environment and compute its reliability function employing computational solution methods for finite Markov chains.

The paper is organized as follows. In Sect. 2 we describe the redundant transport system with two error-prone channels. In Sect. 3 we derive a finite Markov chain with its generator matrix and calculate its related steady-state vector. In Sect. 4 we compute the reliability function of the transport system. The exposition is finalized by some conclusions and an outlook on extensions of the reliability model.

2 Modeling a Redundant Transport System of Two Error-Prone Channels

We model the considered multipath transport system as a redundant technical system comprising two coupled logical (or physical) channels of identical structure that are able to transfer data packets (or more generally, energy or information) (see Figs. 1 and 2, see also [13,21,22]). The transfer function of each channel is subject to different types of failures that are driven by a common internal or external environment. The random transition behavior of the latter system is described by a Markov chain $\{Y(t), t \geq 0\}$ in continuous time with a finite state space $\Sigma_Y = \{1, \ldots, K\}$ and an irreducible generator matrix $Q \in \mathbb{R}^{K \times K}$. Its associated unique steady-state probability vector is denoted by $p \in \mathbb{R}^K$.

In the following, we formulate all algebraic relations of our model in terms of column-vectors to support a standard implementation as matrix-vector product by numerical methods of linear algebra. Furthermore, we use the order relation $0 \ll x$ for vectors $x \in \mathbb{R}^N$ indicating that all components $x_i > 0$ of a vector $x \in \mathbb{R}^N$ are positive. The order relation $0 < x$ means that $x \in \mathbb{R}^N$ is a nonnegative, non-zero vector, i.e. $0 \leq x_i$ for all $i \in \{1, \ldots N\}$ and at least one $0 < x_i$ occurs (cf. [4]).

We assume that for a given state $Y(t) = j \in \Sigma_Y$ the interarrival times of failures of channel 1 and 2 occur as independent exponentially distributed events with mean values $1/\lambda_{1j}$ and $1/\lambda_{2j}$, respectively. Let

$$0 \ll \lambda_1 = (\lambda_{11}, \ldots, \lambda_{1K})^T \in \mathbb{R}^K$$

and

$$0 \ll \lambda_2 = (\lambda_{21}, \ldots, \lambda_{2K})^T \in \mathbb{R}^K$$

denote the corresponding positive column vectors of the associated arrival rates and $\Lambda_1 = \mathrm{Diag}(\lambda_1) > 0, \Lambda_2 = \mathrm{Diag}(\lambda_2) > 0$ be the associated diagonal-positive diagonal matrices of these arrival rate vectors of the failures in the random Markovian environment Y. Let $\Lambda = \Lambda_1 + \Lambda_2$.

We suppose that each erroneous channel is immediately handled by independent repair activities. The associated independently distributed repair times R_1, R_2 of channel 1 and 2, respectively, are governed by general phase-type distributions

$$F_1(t) = \mathbb{P}\{R_1 \leq t\} = 1 - \beta^T \cdot \exp(T \cdot t) \cdot e, \tag{1}$$

and

$$F_2(t) = \mathbb{P}\{R_2 \leq t\} = 1 - \alpha^T \cdot \exp(S \cdot t) \cdot e \tag{2}$$

with the corresponding probability densities

$$f_1(t) = d\mathbb{P}\{R_1 \leq t\}/dt = \beta^T \cdot \exp(T \cdot t) \cdot T^0,$$
$$f_2(t) = d\mathbb{P}\{R_2 \leq t\}/dt = \alpha^T \cdot \exp(S \cdot t) \cdot S^0$$

for channels 1 and 2. Here e denotes the vector of all ones of corresponding dimension. It means that the two finite state phase-type representation matrices

$$(T, \beta), T \in \mathbb{R}^{n_1 \times n_1}, 0 < \beta \in \mathbb{R}^{n_1}, T^0 = -T \cdot e > 0$$

and

$$(S, \alpha), S \in \mathbb{R}^{n_2 \times n_2}, 0 < \alpha \in \mathbb{R}^{n_2}, S^0 = -S \cdot e > 0$$

with n_1 and n_2 states are used for channel 1 and 2, respectively. Then the associated mean repair times are given by

$$\mathbb{E}(R_1) = -\beta^T \cdot T^{-1} \cdot e, \tag{3}$$
$$\mathbb{E}(R_2) = -\alpha^T \cdot S^{-1} \cdot e \tag{4}$$

and their variances are determined by

$$\mathbb{V}(R_1) = \beta^T \cdot T^{-2} \cdot e - (\beta^T \cdot T^{-1} \cdot e)^2, \tag{5}$$
$$\mathbb{V}(R_2) = \alpha^T \cdot S^{-2} \cdot e - (\alpha^T \cdot S^{-1} \cdot e)^2. \tag{6}$$

3   Steady-State Analysis of the Reliable Transport System

3.1   A Markov Model of the Redundant Transport System

The redundant technical system comprising these two coupled logical (or physical) channels of identical logical structure with failure events whose parameters are driven by a common Markov-modulated environment Y and independent, phase-type distributed repair times R_1, R_2, respectively, can be described by a state-vector process

$$Z(t) = (X(t), H(t), Y(t)) = ((X_1(t), X_2(t)), (H_1(t), H_2(t)), Y(t)), \ t \geq 0 \quad (7)$$

on the finite state space Σ which can be embedded into the set $\{0,1\}^2 \times \{0,1,\ldots,n_1\} \times \{0,1,\ldots,n_2\} \times \{1,\ldots,K\}$. The tuple

$$X(t) = (X_1(t), X_2(t))$$

indicates by $X_1(t) = i_1 = 1$ or $X_2(t) = i_2 = 1$ that at time t a failure has occurred in channel 1 or 2, respectively, and the related channel of the system is under repair. A state $i_1 = 0$ or $i_2 = 0$ indicates a proper operation of the corresponding channel. The initial operational state is given by $X(t) = (0,0)$ and the error state by $X(t) = (1,1)$ where no further operation is possible until the repair of at least one channel has been executed successfully. The component

$$H(t) = (H_1(t), H_2(t))) = (j_1, j_2) \in \{0,1,\ldots,n_1\} \times \{0,1,\ldots,n_2\}$$

records the phases $h = (j_1, j_2) \geq 0$ of the running repair processes for a state $i_1 = 1$ or $i_2 = 1$ where a state $j_l = 0$ indicates an idle repair function for given $i_l = 0, l \in \{1,2\}$.

The state variable $Z(t)$ and its state space are arranged such that $X(t) = (X_1(t), X_2(t))$ is the leading system indicator variable of the continuous-time Markov chain (CTMC) with a subspace $\Sigma_X = \{0,1,2,3\}$ of four states that is arranged according to a lexicographical ordering, i.e.

$$0 \equiv (0,0), 1 \equiv (0,1), 2 \equiv (1,0), 3 \equiv (1,1).$$

$(H(t), Y(t)) \in \Sigma_{(H,Y)}$ with

$$
\begin{aligned}
\Sigma_{(H,Y)} = \quad & \{(0,0)\} \times \{1,\ldots,K\} \\
& \cup \{0\} \times \{1,\ldots,n_2\} \times \{1,\ldots,K\} \\
& \cup \{1,\ldots,n_1\} \times \{0\} \times \{1,\ldots,K\} \\
& \cup \{1,\ldots,n_1\} \times \{1,\ldots,n_2\} \times \{1,\ldots,K\}
\end{aligned}
$$

indicates the residual phase space.

The initial state

$$Z(t) = z = (x, h, y) \quad \text{with} \quad x = (0,0) \in \Sigma_X$$

consists of the $j_1 = K$ microstates $(x, h, y) \in \{((0,0), (0,0))\} \times \{1, \ldots, K\}$, whereas the final error state with $x = (1,1) \in \Sigma_X$ comprises the $j_4 = n_1 \cdot n_2 \cdot K$ microstates $\{(1,1)\} \times \{1, \ldots, n_1\} \times \{1, \ldots, n_2\} \times \{1, \ldots, K\}$. The two failure states with $x \in \{(0,1), (1,0)\} \subset \Sigma_X$ with one channel under repair consists of $j_1 = n_2 \cdot K$ and $j_2 = n_1 \cdot K$ microstates $\{(0,1)\} \times \{0\} \times \{1, \ldots, n_2\} \times \{1, \ldots, K\}$ and $\{(1,0)\} \times \{1, \ldots, n_1\} \times \{0\} \times \{1, \ldots, K\}$, respectively.

3.2 Generator Matrix of the Markov Model

The resulting generator matrix A of this finite CTMC $Z(t)$ has a block structure of the following form:

$$A = \begin{pmatrix} A_{00} & A_{01} & A_{02} & 0 \\ A_{10} & A_{11} & 0 & A_{13} \\ A_{20} & 0 & A_{22} & A_{23} \\ 0 & A_{31} & A_{32} & A_{33} \end{pmatrix} \in \mathbb{R}^{N \times N}, \tag{8}$$

Analyzing the transition behavior of the failure interarrivals in the Markovian environment Y driven by the irreducible generator matrix Q and the PH-type repairs governed by (T, β), (S, α) that run independently of each other, we define the corresponding blocks A_{ij} of the generator matrix A in terms of the Kronecker product and the Kronecker sum, i.e. $B \otimes C = (B_{ij} \cdot C)_{ij}$,

$$B \oplus C = B \otimes I_l + I_m \otimes C$$

for block matrices $B \in \mathbb{R}^{m \times m}, C \in \mathbb{R}^{l \times l}$ and identity matrices I_l, I_m of appropriate dimensions $l > 0, m > 0$:

$$A_{00} = 1 \otimes (Q - \Lambda) = Q - \Lambda$$
$$A_{01} = 1 \otimes \alpha^T \otimes \Lambda_2 = \alpha^T \otimes \Lambda_2$$
$$A_{02} = \beta^T \otimes 1 \otimes \Lambda_1 = \beta^T \otimes \Lambda_1$$
$$A_{10} = 1 \otimes S^0 \otimes I_K = S^0 \otimes I_K$$
$$A_{11} = 1 \otimes S \otimes I_K + 1 \otimes I_{n_2} \otimes (Q - \Lambda_1) = S \oplus (Q - \Lambda_1)$$
$$A_{13} = \beta^T \otimes I_{n_2} \otimes \Lambda_1$$
$$A_{20} = T^0 \otimes 1 \otimes I_K = T^0 \otimes I_K$$
$$A_{22} = T \otimes 1 \otimes I_K + I_{n_1} \otimes 1 \otimes (Q - \Lambda_2) = T \oplus (Q - \Lambda_2)$$
$$A_{23} = I_{n_1} \otimes \alpha^T \otimes \Lambda_2$$
$$A_{31} = T^0 \otimes I_{n_2} \otimes I_K$$
$$A_{32} = I_{n_1} \otimes S^0 \otimes I_K$$
$$A_{33} = T \otimes I_{n_2} \otimes I_K + I_{n_1} \otimes S \otimes I_K + I_{n_1} \otimes I_{n_2} \otimes Q = (T \oplus S) \oplus Q$$
$$A_{03} = A_{30} = A_{12} = A_{21} = 0$$

We define the part of the generator matrix A on the operational states

$$O = \{0, 1, 2\} \equiv \{(0,0), (0,1), (1,0)\} \subset \Sigma_X$$

excluding the failure state

$$F = \{3\} \equiv \{(1,1)\} \subset \Sigma_X$$

by the block matrix

$$A_O = \begin{pmatrix} A_{00} & A_{01} & A_{02} \\ A_{10} & A_{11} & 0 \\ A_{20} & 0 & A_{22} \end{pmatrix} \in \mathbb{R}^{M \times M} \tag{9}$$

with $M = K \cdot (1 + n_1 + n_2)$.

3.3   Calculation of the Steady-State Vector

Let $\Pi^T = (\Pi_0^T, \Pi_1^T, \Pi_2^T, \Pi_3^T) \gg 0$ denote the partitioned, unique steady-state row vector of the irreducible Markov chain $Z(t)$ supposing an irreducible Markovian environment matrix Q and two irreducible phase-type generators $T + T^0 \beta^T, S + S^0 \alpha^T$.

We may calculate Π by efficient numerical solution methods for ergodic, finite Markov chains such as direct or iterative solution techniques of the balance equations

$$\Pi^T \cdot A = 0, \quad \Pi^T \cdot e = 1, \tag{10}$$

for instance, aggregation-disaggregation methods such as an additive or multiplicative Schwarz decomposition method or another iterative scheme derived from an M-splitting (cf. [4, 15–18, 24]).

Let $\widetilde{A} = -A^T$ denote the irreducible M-matrix associated with the generator matrix A and

$$A = L + U - \Delta$$

be the block matrix decomposition into the diagonal block matrix

$$\Delta = -\mathrm{Diag}(A_{00}, A_{11}, A_{22}, A_{33}), \tag{11}$$

and lower- and upper-diagonal block matrices

$$L = \begin{pmatrix} 0 & 0 & 0 & 0 \\ A_{10} & 0 & 0 & 0 \\ A_{20} & 0 & 0 & 0 \\ 0 & A_{31} & A_{32} & 0 \end{pmatrix}, \quad U = \begin{pmatrix} 0 & A_{01} & A_{02} & 0 \\ 0 & 0 & 0 & A_{13} \\ 0 & 0 & 0 & A_{23} \\ 0 & 0 & 0 & 0 \end{pmatrix}, \tag{12}$$

respectively. Then we define the associated M-splitting

$$\widetilde{A} = -A^T = M - N$$

with the corresponding transposed matrices of the block matrix decomposition $M = \Delta^T - U^T, N = L^T$ and get the iteration matrix

$$J = M^{-1} \cdot N = [\Delta^T - U^T]^{-1} \cdot L^T$$

and the nonnegative matrix

$$\widetilde{T} = I - \widetilde{A} \cdot M^{-1} = N \cdot M^{-1} \tag{13}$$

with the property $M^{-1} \cdot \widetilde{T} \cdot M = J$. Then the column-stochastic matrix

$$T = I - \omega \widetilde{A} \cdot M^{-1} = (1 - \omega)I + \omega \widetilde{T} \tag{14}$$

is a semiconvergent, nonnegative matrix for any scaling $\omega \in (0,1)$ (cf. [4,15,24]).

Based on the block matrix decomposition of A in (8) we determine a partition $\Gamma = \{J_1, J_2, J_3, J_4\}$ into $m = 4$ subsets of the state space $\Sigma = \{1, \ldots, N\}$, $N = K + n_2 \cdot K + n_1 \cdot K + (n_1 \cdot n_2) \cdot K$, with the four disjoint subsets $J_1 = \{1, \ldots, K\}$, $J_2 = \{K+1, \ldots, (1+n_2) \cdot K\}$, $J_3 = \{(1+n_2) \cdot K + 1, \ldots, (1 + n_1 + n_2) \cdot K\}$, $J_4 = \{(1 + n_1 + n_2) \cdot K + 1, \ldots, (1 + n_1 + n_2 + n_1 \cdot n_2) \cdot K\}$.

Then the following iterative aggregation-disaggregation (IAD) algorithm is semiconvergent to the unique, normalized steady-state solution vector

$$\Pi^T = (\Pi_0^T, \Pi_1^T, \Pi_2^T, \Pi_3^T) \gg 0$$

with its components $\Pi_i^T \gg 0$ on the partition set J_{i+1} for each state $i \in \{0, 1, 2, 3\}$ (cf. [17,18]):

IAD-Algorithm for M-Matrices with Block Structure

1. We choose an initial probability vector

$$x^{(0)} \gg 0, e^T \cdot x^{(0)} = 1,$$

e.g., the uniform distribution, and real numbers $\epsilon, c_1, c_2 \in (0,1)$. We construct an aggregation matrix R and a prolongation matrix $P(x)$ for $0 < x \in \mathbb{R}^N$, $e^T \cdot x = 1$, by

$$R = \begin{pmatrix} e_{J_1}^T & 0 & 0 & 0 \\ 0 & e_{J_2}^T & 0 & 0 \\ 0 & 0 & e_{J_3}^T & 0 \\ 0 & 0 & 0 & e_{J_4}^T \end{pmatrix} \in \mathbb{R}^{4 \times N},$$

$$P(x) = \begin{pmatrix} y_1 & 0 & 0 & 0 \\ 0 & y_2 & 0 & 0 \\ 0 & 0 & y_3 & 0 \\ 0 & 0 & 0 & y_4 \end{pmatrix} \in \mathbb{R}^{N \times 4},$$

and $x = \begin{pmatrix} x_1 \\ x_2 \\ x_3 \\ x_4 \end{pmatrix} \in \mathbb{R}^N$, in terms of

$$[\alpha(x)]_j = e^T \cdot x_j$$

and
$$[y(x)]_j = x_j/[\alpha(x)]_j$$

provided that $x_j > 0$ holds and the uniform distribution in case of $x_j = 0$ for given $j \in \{1, \ldots, 4\}$. Using the iteration matrix T in (14) and $x \in \mathbb{R}^N$ we get the associated aggregated matrix $B(x) \in \mathbb{R}^{4\times 4}$ by

$$B(x) = R \cdot T \cdot P(x).$$

We set $k = 0$ and
$$r(x) = ||(I - T) \cdot x||_1$$

for the L_1-norm $||x||_1 = \sum_1^N |x_i|$ in $\mathbb{R}^N$.

2. We solve
$$B(x^{(k)}) \cdot \alpha(x^{(k)}) = \alpha(x^{(k)})$$

subject to
$$e^T \cdot \alpha(x^{(k)}) = 1, \alpha(x^{(k)}) > 0$$

and compute
$$\tilde{x} = P(x^{(k)}) \cdot \alpha(x^{(k)}).$$

3. We compute
$$x^{(k+1)} = T \cdot \tilde{x}.$$

4. If
$$r(\tilde{x}) \leq c_1 \cdot r(x^{(k)})$$

then go to step 5
else compute
$$x^{(k+1)} = T^h \cdot \tilde{x}$$

for $h > 1$ such that
$$r(x^{(k+1)}) \leq c_2 \cdot r(x^{(k)})$$

endif

5. If
$$||x^{(k+1)} - x^{(k)}||_1/||x^{(k)}||_1 < \epsilon$$

then step 6
else
$$k = k + 1,$$

and go to step 2
endif

6. At the end we perform a normalization after a successful convergence test:
$$\Pi = \frac{M^{-1} \cdot x^{(k+1)}}{e^T \cdot M^{-1} \cdot x^{(k+1)}}$$

Regarding the semiconvergence of this specific IAD-algorithm to the probability vector Π, we refer to the existing convergence theory related to numerical solution methods for finite Markov chains (cf. [17, 18, 24]).

4 Computing the Reliability Function of the Redundant Transport System

The reliability of the transport system is characterized by the sojourn time $S_T \geq 0$ in the set of the operational states $\widehat{O} = \{z = (x, h, y) \in \Sigma \mid x \in O \subset \Sigma_X\}$ of the overall state space Σ subject to the start in one of those states $z \in \widehat{O}$ in the steady-state regime with the steady-state row vector

$$\Pi_{\widehat{O}}^T = (\Pi_0^T, \Pi_1^T, \Pi_2^T) \gg 0$$

and its positive components $\Pi_i^T \gg 0$ associated with each non-failure state $i \in O = \{0, 1, 2\} \subset \Sigma_X$.

Then we can calculate the reliability function $F_R(t)$ as time-dependent probability of the Markov chain $Z(t)$ to reside in a state $z \in \widehat{O}$ up to time $t > 0$ given that the capturing in the absorbing states $\widehat{F} = \{z = (x, h, y) \in \Sigma \mid x \in F \subset \Sigma_X\}$ does not occur (see (9)):

$$F_R(t) = \mathbb{P}\{S_T > t \mid Z(0) \in \widehat{O}\} \cdot \mathbb{P}\{Z(0) \in \widehat{O}\}$$
$$= \mathbb{P}\{Z(t) \notin \widehat{F} \mid Z(0) \in \widehat{O}\} \cdot \mathbb{P}\{Z(0) \in \widehat{O}\} \tag{15}$$
$$= \Pi_{\widehat{O}}^T \cdot \exp(A_O t) \cdot e \tag{16}$$

The computation of the exponential matrix $\exp(A_O t)$ can be effectively performed by means of a uniformization approach (cf. [24]).

Let $D = \text{Diag}(D_{ii}) > 0$ denote the diagonal matrix determined by the positive diagonal elements $D_{ii} = -(A_O)_{ii} > 0, i \in \{1, \ldots, M\}$, of the M-matrix $-A_O$ in (9). We set

$$\gamma = \max_{1 \leq i \leq M} (D_{ii}) > 0$$

and define the sub-stochastic matrix $P_O = (P_{ij}), 1 \leq i \leq M, 1 \leq j \leq M$, by the transition probabilities

$$P_{ij} = \begin{cases} A_{ij}/\gamma \geq 0, & i \neq j \\ 1 + A_{ii}/\gamma \geq 0, & i = j \end{cases} \quad \text{or} \quad P_O = I_M + A_O/\gamma$$

at the embedded time epochs of events in the Markov chain $Z(t)$. Then it holds

$$A_O = \gamma \cdot (P_O - I_M)$$

with the identity matrix $I_M \in \mathbb{R}^{M \times M}$. It implies the representation

$$R_O(t) = \exp(A_o \cdot t) = \exp(\gamma t \cdot (P_O - I_M))$$
$$= \left[\sum_{n=0}^{\infty} \frac{(\gamma t)^n}{n!} e^{-\gamma t} \cdot ((P_O)^n)_{ij} \right]_{1 \leq i, j \leq M} \tag{17}$$

that enables a fast computation of the reliability function $F_R(t)$ in (16) as matrix-vector product

$$F_R(t) = \Pi_{\widehat{O}}^T \cdot R_O(t) \cdot e$$
$$= \sum_{n=0}^{\infty} \frac{(\gamma t)^n}{n!} e^{-\gamma t} \cdot \Pi_O^T \cdot (P_O)^n \cdot e \tag{18}$$

by means of a Poisson distribution with parameter γt and the n-th powers of the sub-stochastic matrix P_O.

5   Conclusions

In our study we have considered a multipath communication that is realized between clients and servers of new cloud computing concepts like fog and mobile edge computing at the transport and session layers of an underlying SDN/NFV network (cf. [7,14,20,23]). The latter network integrates diverse radio technologies into its attached RAN environments and supports the efficient interprocess communication of new services between the clients and servers.

We have modelled the reliability of the associated redundant transport system comprising two logical channels, that are susceptible to failures and restored with general phase-type distributed repair times, and computed its corresponding reliability function. The failure processes of both channels are described by Markov-modulated failure times that are driven by the transitions of a common Markovian environment.

First we have modelled this error-prone system with repair by a finite state, continuous-time Markov chain. We have identified its generator matrix in terms of associated Kronecker products of the underlying parameter matrices which are related to the Markov-modulated interarrival times of failures and the phase-type distributed repair times. We have shown that the steady-state distribution of the restoration model can be effectively computed by an iterative aggregation-disaggregation method and then computed the associated reliability function of the transport system by means of an appropriately defined finite, absorbing Markov chain.

Our further studies will elaborate on the extension of the reliability model to more general random environments. We will further consider its application to SDN/NFV networks with an integrated 5G RAN that can support fog and mobile edge computing in advanced IoT and machine-to-machine communication scenarios discussed in the literature (cf. [1,6,14,19]).

Acknowledgment. N.M. Markovich was partly supported by the Russian Foundation for Basic Research (grant 19-01-00090).

References

1. Aazam, M., Huh, E.-N.: Fog computing and smart gateway based communication for cloud of things. In: 2014 International Conference on Future Internet of Things and Cloud (FiCloud), Barcelona, Spain, 27–29 August, pp. 464–470 (2014)
2. Al-Fuqaha, A., et al.: Internet of things: a survey on enabling technologies, protocols, and applications. IEEE Commun. Surv. Tutor. **17**(4), 2347–2376 (2015)
3. Barré, S., Paasch, C., Bonaventure, O.: MultiPath TCP: from theory to practice. In: Domingo-Pascual, J., Manzoni, P., Palazzo, S., Pont, A., Scoglio, C. (eds.) NETWORKING 2011. LNCS, vol. 6640, pp. 444–457. Springer, Heidelberg (2011). https://doi.org/10.1007/978-3-642-20757-0_35

4. Berman, A., Plemmons, R.J.: Nonnegative Matrices in the Mathematical Sciences. Academic Press, Cambridge (1979)
5. Bonaventure, O., et al.: Multipath QUIC. In: CoNEXT 2017, Seoul/Incheon, South Korea, 12–15 December 2017. https://multipath-quic.org/
6. Cech, H., Großmann, M., Krieger, U.R.: A fog computing architecture to share sensor data by means of blockchain functionality. In: IEEE International Conference on Fog Computing, Prague, Czech Republic, 24–26 June 2019
7. Chiang, M., Zhang, T.: Fog and IoT: an overview of research opportunities. IEEE Internet Things J. **3**(6), 854–864 (2016)
8. Dreibholz, T., et al.: Stream control transmission protocol: past, current, and future standardization activities. IEEE Commun. Mag. **49**(4), 82–88 (2011)
9. Eiermann, A., Renner, M., Großmann, M., Krieger, U.R.: On a fog computing platform built on ARM architectures by Docker container technology. In: Eichler, G., Erfurth, C., Fahrnberger, G. (eds.) I4CS 2017. CCIS, vol. 717, pp. 71–86. Springer, Cham (2017). https://doi.org/10.1007/978-3-319-60447-3_6
10. Fischer, W., Meier-Hellstern, K.: The Markov-modulated Poisson process (MMPP) cookbook. Perform. Eval. **18**(2), 14–171 (1993)
11. Großmann, M., Ioannidis, C.: Continuous integration of applications for ONOS. In: 5th IEEE Conference on Network Softwarization, NetSoft 2019, Paris, France, 24–28 June 2019
12. Holla, S.: Orchestrating Docker. Packt Publishing Ltd., Birmingham (2015)
13. Kozyrev, D., Rykov, V., Kolev, N.: Reliability function of renewable system with Marshal-Olkin failure model. Reliab. Theory Appl. **13**(1(48)), 39–46 (2018)
14. Kozyrev, D., et al.: Mobility-centric analysis of communication offloading for heterogeneous Internet of Things devices. Wirel. Commun. Mob. Comput. **2018**, 11 (2018). Article ID 3761075
15. Krieger, U.R., Müller-Clostermann, B., Sczittnick, M.: Modeling and analysis of communication systems based on computational methods for Markov chains. IEEE J. Sel. Areas Commun. **8**(9), 1630–1648 (1990)
16. Krieger, U.R.: Analysis of a loss system with mutual overflow in a Markovian environment. In: Stewart, W. (ed.) Numerical Solution of Markov Chains, pp. 303–328. Marcel Dekker, New York (1990)
17. Krieger, U.R.: On a two-level multigrid solution method for finite Markov chains. Linear Algebra Appl. **223**(224), 415–438 (1995)
18. Krieger, U.R.: Numerical Solution of large finite markov chains by algebraic multigrid techniques. In: Stewart, W. (ed.) Computations with Markov Chains, pp. 403–424. Kluwer Academic Publishers, Boston (1995)
19. Krieger, U.R., Kumar, B.K.: Modeling the performance of ARQ error control in an LTE transmission system. In: German, R., Hielscher, K.-S., Krieger, U.R. (eds.) MMB 2018. LNCS, vol. 10740, pp. 202–217. Springer, Cham (2018). https://doi.org/10.1007/978-3-319-74947-1_14
20. Rimal, B.P., Van, D.P., Maier, M.: Mobile edge computing empowered fiber-wireless access networks in the 5G era. IEEE Commun. Mag. **55**, 192–200 (2017)
21. Rykov, V.V., Kozyrev, D.V.: Analysis of renewable reliability systems by Markovization method. In: Rykov, V.V., Singpurwalla, N.D., Zubkov, A.M. (eds.) ACMPT 2017. LNCS, vol. 10684, pp. 210–220. Springer, Cham (2017). https://doi.org/10.1007/978-3-319-71504-9_19

22. Rykov, V., Zaripova, E., Ivanova, N., Shorgin, S.: On sensitivity analysis of steady state probabilities of double redundant renewable system with Marshall-Olkin failure model. In: Vishnevskiy, V.M., Kozyrev, D.V. (eds.) DCCN 2018. CCIS, vol. 919, pp. 234–245. Springer, Cham (2018). https://doi.org/10.1007/978-3-319-99447-5_20
23. Stallings, W.: Foundations of Modern Networking: SDN, NFV, QoE, IoT, and Cloud. Pearson Education, London (2016)
24. Stewart, W.J.: Probability, Markov Chains, Queues, and Simulation. Princeton University Press, Princeton (2009)

Heterogeneous Queueing System MAP/GI$^{(n)}$/∞ with Random Customers' Capacities

Ekaterina Lisovskaya[1,2], Ekaterina Pankratova[3], Yuliya Gaidamaka[1,4],
Svetlana Moiseeva[2(✉)], and Michele Pagano[5]

[1] Peoples' Friendship University of Russia (RUDN University),
6 Miklukho-Maklaya Street, Moscow 117198, Russian Federation
lisovskaya-eyu@rudn.ru
[2] Tomsk State University, 36 Lenina Avenue, Tomsk 634050, Russian Federation
smoiseeva@mail.ru
[3] V. A. Trapeznikov Institute of Control Sciences of Russian Academy of Sciences,
65 Profsoyuznaya Street, Moscow 117997, Russian Federation
pankate@sibmail.com
[4] Federal Research Center "Computer Science and Control" of the Russian Academy
of Sciences (FRC CSC RAS), 44-2 Vavilov Street, Moscow 119333, Russian Federation
gaydamaka-yuv@rudn.ru
[5] Department of Information Engineering, University of Pisa,
Via Caruso 16, 56122 Pisa, Italy
michele.pagano@iet.unipi.it

Abstract. In this paper a model of a heterogeneous resource queueing system with a Markovian arrival process is considered. The customer accepted for servicing occupies random amount of resource with a given distribution function depending on the class of the customer and on the type of service it needs. At the end of the service, the customer leaves the system and releases the occupied resource. In this work, asymptotic formulas for calculating the main probability characteristics of the model, including the joint distribution functions of the customers number and the total resource amounts occupied by them, are obtained. Finally, the accuracy of the approximation is verified by using simulation.

Keywords: Resource queueing systems · Markovian arrival process · Asymptotic analysis · Gaussian approximation

The publication has been prepared with the support of the "RUDN University Program 5-100" (recipient E. Lisovskaya, mathematical model development). The reported study was funded by RFBR, project number 18-07-00576 (Y. Gaidamaka, simulation model development) and 17-07-00845 (Y. Gaidamaka, problem formulation and analysis) and the University of Pisa PRA 2018–2019 Research Project "CONCEPT Communication and Networking for vehicular Cyber-Physical systems" (recipient M. Pagano, simulation and numerical analysis).

© Springer Nature Switzerland AG 2019
V. M. Vishnevskiy et al. (Eds.): DCCN 2019, LNCS 11965, pp. 315–329, 2019.
https://doi.org/10.1007/978-3-030-36614-8_24

1 Introduction

Information networks have made a tremendous leap over the last century: from single–service telegraph and telephone networks to Integrated Services Digital Networks (ISDN), from Booadband ISDN (BISDN) to the ubiquitous Internet for the transmission of all types of information (speech, data, video, sensors measurements). The modern system of infocommunications is a complex of technical, software and organizational facilities, providing users with services related to the transfer, reception and processing of relevant data, as well as storage and issuance of accumulated information.

Organizationally, the infocommunication technologies have complex network structure, composed of a plurality of objects that perform specific functions for the transfer and processing of information, the development of the content and service consumers. The basis of this evolution is multi-channel information transmission systems based on electrical cables, optical fibers and radio links.

The development of new methods of asynchronous transfer of diverse information, multiplexing mechanisms and flow/congestion control, the creation of distributed networks, the use of combined segments of the radio spectrum for the transmission of radio signals provide high speed of information transmission, broadband access to data networks from anywhere in the world and while driving, high potential for integration and efficiency of global and national networks, including the Internet, reducing the cost of information and communication services. The use of heterogeneous queueing systems as mathematical models of infocommunication systems is relevant, as it is possible to take into account the ever growing transmission bandwidth as well as the heterogeneity of data and environment (see [15,16] and references therein).

In the last decade a growing interest is emerged towards queues with random customers' capacities, since they are useful for analysis and design issues in high-performance computer and communication systems, in which service time and customer volume are independent quantities (see [11,13]). For instance, in LTE (Long Term Evolution) networks the amount of required radio resources does not depend on the duration of the flow (i.e., the service time). In papers [4,20], resources for M2M communications are allocated in batches of fixed size; requests for them arrive according to a Markovian arrival process. Batch arrival with hysteretic overload control of the queue was investigated in [2]. For systems with a finite resource, besides queue overload control, systems with various adaptive admission control schemes [1] should be considered, including processor sharing for systems with unreliable server [17] and different types of resources for loss system with state-dependent arrival and service rates [12].

The article [3] presents an overview of the resource queuing systems used for modeling of a wide class of real systems with limited resources. The review considers the resource systems without waiting space with exponentially distributed service time. Under the same conditions in [18], the authors obtain the steady-state of the number of customers probability distribution.

Such queues are also important in modeling devices, where it is necessary to calculate a sufficient volume of buffer for data storing [14,19]. The analysis of the characteristics of the models will allow to establish to what extent the needs of the clientage in infocommunication services are met, to build a forecast of the development of the sphere of infocommunication services, and to assess the economic results and the quality of the multifunctional socio-economic systems. In the works [5,7,9,10], queues with renewal arrivals, Markovian Arrival processes (MAP) and Markov Modulated Poisson processes (MMPP) are studied under various asymptotic conditions.

Unlike above mentioned works, in this paper we consider a heterogeneous infinite-server queueing system, fed by MAP arrivals with random customers capacities and non-exponential service time distribution.

2 Problem Statement

2.1 Mathematical Model

Consider the queueing system (see Fig. 1) with unlimited number of servers of n different types [15, 16] and assume that each customer carries a random quantity of work (capacity of the customer) [6].

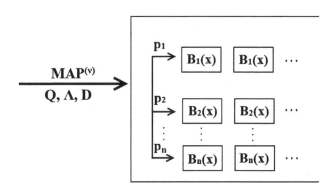

Fig. 1. Heterogeneous queue $MAP/GI^{(n)}/\infty$ with random customers' capacities

Customers arrive in the system according to a MAP, given by the underlying Markov chain $k(t)$ with a finite number of states K, the set of non-negative intensities λ_k and the probabilities of arrival when the Markov chain state changes $d_{\nu k}$ $(\nu, k = 1, \ldots, K)$. Note that $k(t)$ is determined by the infinitesimal generator matrix $\mathbf{Q} = |q_{\nu k}|$. At the time of occurrence of any event in the MAP-flow, only single customer arrives in the system and its type is randomly selected according to the set of probabilities p_i $(i = 1, \ldots, n)$. Then the customer goes to the appropriate device type, where its service is performed during a random time with distribution function $B_i(x)$ $(i = 1, \ldots, n)$ according to the type of the customer. Moreover, depending on its type $i = 1, \ldots, n$, each customer requires a random amount of resources $v_i > 0$, drawn from distribution function $G_i(y)$ $(i = 1, \ldots, n)$, which is independent of its service time. Denote by

$\{i_1(t), i_2(t), \ldots, i_n(t)\}$ and $\{V_1(t), V_2(t), \ldots, V_n(t)\}$ the number of each type's customers in the system at time t and their total capacities, respectively. We consider the $2n$-dimensional stochastic process $\{i_1(t), \ldots, i_n(t), V_1(t), \ldots, V_n(t)\}$; this process is non-Markovian, therefore, we use the dynamic screening method [9] for its investigation.

Let the system be empty at moment t_0, and let us fix some arbitrary moment T in the future as shown in Fig. 2. $S_1(t), S_2(t), \ldots, S_n(t)$ represent the probability that a customer arriving at time t is i-type and it will be served by the moment T, i.e. $S_i(t) = p_i(1 - B_i(T - t))$, $(i = 1, \ldots, n)$, for $t_0 \leq t \leq T$.

Denote by $\{m_1(t), m_2(t), \ldots, m_n(t)\}$ and $\{w_1(t), w_2(t), \ldots, w_n(t)\}$ the number of arrivals of each type screened before the moment t and their total capacities, respectively. As it is shown in [8], the probability distribution of the number of customers in the system at the moment T coincides with the probability distribution of the number of screened arrivals:

$$P\{i_1(T) = m_1, \ldots, i_n(T) = m_n\} = P\{m_1(T) = m_1, \ldots, m_n(T) = m_n\},$$

for $m_i = 0, 1, 2, \ldots$ It is easy to prove the same property for the extended stochastic process $\{i_1(t), \ldots, i_n(t), V_1(t), \ldots, V_n(t)\}$ for $m_i = 0, 1, 2, \ldots$; $w_i \geq 0$:

$$P\{i_1(T) = m_1, \ldots, i_n(T) = m_n, V_1(T) < w_1, \ldots, V_n(T) < w_n\} = $$
$$P\{m_1(T) = m_1, \ldots, m_n(T) = m_n, w_1(T) < w_1, \ldots, w_n(T) < w_n\}.$$

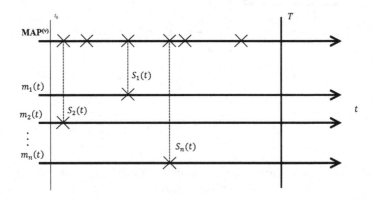

Fig. 2. Screening of the customers' arrivals

The above $2n$-dimensional process is non-Markovian, then we will add the state of Markov chain $k(t)$ and consider $(2n+1)$-dimensional Markovian process

$$\{k(t), m_1(t), \ldots, m_n(t), w_1(t), \ldots, w_n(t)\}. \tag{1}$$

2.2 Kolmogorov Integro-Differential Equations

For the (1) we can write the following system of Kolmogorov integro-differential equations:

$$
\frac{\partial P(k, m_1, \ldots, m_n, w_1, \ldots, w_n, t)}{\partial t} =
$$

$$
P(k, m_1, \ldots, m_n, w_1, \ldots, w_n, t)\lambda_k \left\{ \sum_{i=1}^{n} (1 - S_i(t)) - 1 \right\} +
$$

$$
\lambda_k S_1(t) \int_0^{w_1} P(k, m_1 - 1, \ldots, m_n, w_1 - y_1, \ldots, w_n, t)dG_1(y_1) + \cdots +
$$

$$
\lambda_k S_n(t) \int_0^{w_n} P(k, m_1, \ldots, m_n - 1, w_1, \ldots, w_n - y_n, t)dG_n(y_n) +
$$

$$
\sum_{\nu=1}^{K} \left\{ \left[1 - d_{\nu k} \sum_{i=1}^{n} S_i(t) \right] P(\nu, m_1, \ldots, m_n, w_1, \ldots, w_n, t) + \right.
$$

$$
d_{\nu k} \left[S_1(t) \int_0^{w_1} P(\nu, m_1 - 1, \ldots, m_n, w_1 - y_1, \ldots, w_n, t)dG_1(y_1) + \cdots + \right.
$$

$$
\left. \left. S_n(t) \int_0^{w_n} P(\nu, m_1, \ldots, m_n - 1, w_1, \ldots, w_n - y_n, t)dG_n(y_n) \right] \right\} q_{\nu k}.
$$

(2)

We define the initial conditions in the form

$$
P(k, m_1, \ldots, m_n, w_1, \ldots, w_n, t_0) = \begin{cases} r(k), \ m_i = w_i = 0, \ i = 1, \ldots, n, \\ 0, \text{otherwise}, \end{cases}
$$

(3)

where $r(k) = P\{k(t) = k\}$, $k = 1, \ldots, K$ is the stationary probability distribution of the underlying Markov chain $k(t)$. Denoting $P(k, m_1, \ldots, m_i - 1, \ldots, m_n, w_1, \ldots, w_i - y, \ldots, w_n, t) = \rho_i(y)$, the system of equations (2) takes the form

$$
\frac{\partial P(k, m_1, \ldots, m_n, w_1, \ldots, w_n, t)}{\partial t} =
$$

$$
\lambda_k \sum_{i=1}^{n} S_i(t) \left[\int_0^{w_i} \rho_i(y)dG_i(y) - P(k, m_1, \ldots, m_n, w_1, \ldots, w_n, t) \right] +
$$

$$
\sum_{\nu=1}^{K} \left\{ P(\nu, m_1, \ldots, m_n, w_1, \ldots, w_n, t) + \right.
$$

$$
\left. d_{\nu k} \left[\sum_{i=1}^{n} S_i(t) \left(\int_0^{w_i} \rho_i(y)dG_i(y) - P(\nu, m_1, \ldots, m_n, w_1, \ldots, w_n, t) \right) \right] \right\} q_{\nu k}.
$$

(4)

To solve (4), we introduce the partial characteristic function:

$$h(k, u_1, \ldots, u_n, z_1, \ldots, z_n, t) = \sum_{m_1=0}^{\infty} \cdots \sum_{m_n=0}^{\infty} e^{ju_1 m_1} \times \cdots \times$$

$$e^{ju_n m_n} \int_0^{\infty} \cdots \int_0^{\infty} e^{jz_1 w_1} \times \cdots \times e^{jz_n w_n} P(k, m_1, \ldots, m_n, dw_1, \ldots, dw_n, t),$$

where $j = \sqrt{-1}$ is the imaginary unit, and obtain the following system of equations:

$$\frac{\partial h(k, u_1, \ldots, u_n, z_1, \ldots, z_n, t)}{\partial t} =$$

$$\lambda_k h(k, u_1, \ldots, u_n, z_1, \ldots, z_n, t) \sum_{i=1}^{n} S_i(t) \left[e^{ju_i} G_i^*(z_i) - 1 \right] +$$

$$\sum_{\nu=1}^{K} h(\nu, u_1, \ldots, u_n, z_1, \ldots, z_n, t) q_{\nu k} +$$ (5)

$$\sum_{\nu=1}^{K} d_{\nu k} h(\nu, u_1, \ldots, u_n, z_1, \ldots, z_n, t) \sum_{i=1}^{n} S_i(t) \left[e^{ju_i} G_i^*(z_i) - 1 \right] q_{\nu k},$$

where

$$G_i^*(z_i) = \int_0^{\infty} e^{jz_i y} dG_i(y).$$

The solution $h(k, u_1, \ldots, u_n, z_1, \ldots, z_n, t)$ of system (5) satisfies condition

$$h(k, u_1, \ldots, u_n, z_1, \ldots, z_n, t_0) = r(k), \ (k = 1, \ldots, K).$$

Then we can write the vector-matrix equation

$$\frac{\partial \mathbf{h}(u_1, \ldots, u_n, z_1, \ldots, z_n, t)}{\partial t} =$$

$$\mathbf{h}(u_1, \ldots, u_n, z_1, \ldots, z_n, t) \left[\mathbf{Q} + \sum_{i=1}^{n} S_i(t) \left(e^{ju_i} G_i^*(z_i) - 1 \right) \mathbf{B} \right],$$ (6)

$$\mathbf{h}(u_1, \ldots, u_n, z_1, \ldots, z_n, t_0) = \mathbf{r},$$

here

$$\mathbf{h}(u_1, \ldots, u_n, z_1, \ldots, z_n, t) =$$

$$[h(1, u_1, \ldots, u_n, z_1, \ldots, z_n, t), \ldots, h(K, u_1, \ldots, u_n, z_1, \ldots, z_n, t)],$$

$$\mathbf{B} = \mathbf{\Lambda} + \mathbf{A}, \mathbf{\Lambda} = diag\{\lambda_k\}, \ \mathbf{A} = ||d_{\nu k} q_{\nu k}||, \ (\nu, k = 1, \ldots, K),$$

and $\mathbf{r} = [r(1), \ldots, r(K)]$ is the vector of the stationary probability distribution of the underlying Markov chain, defined by the following system of linear equations:

$$\begin{cases} \mathbf{rQ} = \mathbf{0}, \\ \mathbf{re} = 1, \end{cases}$$ (7)

where $\mathbf{e}$ is the column vector with all entries equal to 1.

The exact solution of the Eq. (6) is, in general, not available, but it is possible to get asymptotic results in case of heavy loads. To this aim we will use the condition that the service times of the different types of customers growth proportionally to each other.

3 Asymptotic Analysis

Denote

$$b_i = \int_0^\infty (1 - B_i(x))dx; \qquad \beta_i = \int_0^\infty (1 - B_i(x))^2 dx;$$

$$\beta_{ig} = \int_0^\infty (1 - B_i(x))(1 - B_g(x))dx, \ (i, g = 1, \ldots, n; \ i \neq g),$$

and we will solve the equation for the characteristic function (6) in the asymptotic condition that the service times of the different types of customers growth proportionally to each other, i.e. $b_i = bq_i$ for some real values q_i and $b \to \infty$ ($i = 1, \ldots, n$).

We state and prove the following theorem.

Theorem 1. *The asymptotic characteristic function of the probability distribution of the process* $\{m_1(t), \ldots, m_n(t), w_1(t), \ldots, w_n(t)\}$ *has the form*

$$h^{(2)}(u_1, \ldots, u_n, z_1, \ldots, z_n) =$$

$$\exp\left\{ \kappa_1 \sum_{i=1}^n p_i b_i \left[ju_i + jz_i a_i^{(1)} \right] + \frac{\kappa_1}{2} \sum_{i=1}^n p_i b_i \left[(ju_i)^2 + (jz_i)^2 a_i^{(2)} \right] + \right.$$

$$\kappa_1 \sum_{i=1}^n p_i b_i ju_i jz_i a_i^{(1)} + \frac{\kappa_2}{2} \sum_{i=1}^n p_i^2 \beta_i \left[ju_i + jz_i a_i^{(1)} \right]^2 +$$

$$\left. \frac{\kappa_2}{2} \sum_{i=1}^n \sum_{\substack{g=1, \\ g \neq i}}^n p_i p_g \beta_{ig} \left[ju_i + jz_i a_i^{(1)} \right] \left[ju_g + jz_g a_g^{(1)} \right] \right\},$$

where $\kappa_1 = \mathbf{r}\mathbf{B}\mathbf{e}$, $\kappa_2 = 2\mathbf{d}(\mathbf{B} - \kappa_1\mathbf{I})$ *and* $\mathbf{d}$ *is some row vector which satisfies the condition*

$$\begin{cases} \mathbf{d}\mathbf{Q} + \mathbf{r}(\mathbf{B} - \kappa_1\mathbf{I}) = 0, \\ \mathbf{d}\mathbf{e} = 0, \end{cases} \tag{8}$$

and $\mathbf{I}$ *is the identity matrix.*

Proof. At first, we prove an auxiliary statement.

Lemma 1. *The first-order asymptotic characteristic function of the probability distribution of the process* $\{m_1(t), \ldots, m_n(t), w_1(t), \ldots, w_n(t)\}$ *has the form*

$$h^{(1)}(u_1, \ldots, u_n, z_1, \ldots, z_n) \approx \exp\left\{ \kappa_1 \sum_{i=1}^{n} p_i \left(ju_i + jz_i a_i^{(1)} \right) b_i \right\},$$

where $a_i^{(1)} = \int\limits_{0}^{\infty} y \, dG_i(y)$ *are the means of customer's capacities.*

Proof. Put

$$\varepsilon = \frac{1}{q_i b}, \; S_i(t) = \tilde{S}_i(\tau), \; u_i = \varepsilon x_i, \; z_i = \varepsilon y_i, \; (i = 1, \ldots, n), \; t\varepsilon = \tau, \; t_0\varepsilon = \tau_0, \tag{9}$$
$$\mathbf{h}(u_1, \ldots, u_n, z_1, \ldots, z_n, t) = \mathbf{F_1}(x_1, \ldots, x_n, y_1, \ldots, y_n, \tau, \varepsilon).$$

Taking into account the above notation, we can write (6) as

$$\varepsilon \frac{\partial \mathbf{F_1}(x_1, \ldots, x_n, y_1, \ldots, y_n, \tau, \varepsilon)}{\partial \tau} =$$
$$\mathbf{F_1}(x_1, \ldots, x_n, y_1, \ldots, y_n, \tau, \varepsilon) \left[\mathbf{Q} + \sum_{i=1}^{n} \tilde{S}_i(\tau) \left(e^{j\varepsilon x_i} G_i^*(\varepsilon y_i) - 1 \right) \mathbf{B} \right]. \tag{10}$$

If $\varepsilon \to 0$ in (10), then obtain:

$$\lim_{\varepsilon \to 0} \mathbf{F_1}(x_1, \ldots, x_n, y_1, \ldots, y_n, \tau, \varepsilon) \mathbf{Q} = \mathbf{0}.$$

Since $\mathbf{rQ} = \mathbf{0}$, then we look for $\mathbf{F_1}(x_1, \ldots, x_n, y_1, \ldots, y_n, \tau)$ as

$$\mathbf{F_1}(x_1, \ldots, x_n, y_1, \ldots, y_n, \tau) = \mathbf{r}\Phi_1(x_1, \ldots, x_n, y_1, \ldots, y_n, \tau), \tag{11}$$

where $\Phi_1(x_1, \ldots, x_n, y_1, \ldots, y_n, \tau)$ is the desired scalar function.

Let us multiply (10) by the unit column vector $\mathbf{e}$:

$$\varepsilon \frac{\partial \mathbf{F_1}(x_1, \ldots, x_n, y_1, \ldots, y_n, \tau, \varepsilon)}{\partial \tau} \mathbf{e} =$$
$$\mathbf{F_1}(x_1, \ldots, x_n, y_1, \ldots, y_n, \tau, \varepsilon) \left[\mathbf{Qe} + \sum_{i=1}^{n} \tilde{S}_i(\tau) \left(e^{j\varepsilon x_i} G_i^*(\varepsilon y_i) - 1 \right) \mathbf{Be} \right].$$

Let us take into account that $\mathbf{Qe} = \mathbf{0}$, expand exponents in the previous equation into Taylor series, substitute into it the vector function $\mathbf{F_1}(x_1, \ldots, x_n, y_1, \ldots, y_n, \tau)$ in the form (11) and let $\varepsilon \to 0$, then:

$$\frac{\partial \mathbf{F_1}(x_1, \ldots, x_n, y_1, \ldots, y_n, \tau)}{\partial \tau} \mathbf{e} =$$
$$\mathbf{F_1}(x_1, \ldots, x_n, y_1, \ldots, y_n, \tau) \sum_{i=1}^{n} \tilde{S}_i(\tau) \left[jx_i + jy_i a_i^{(1)} \right] \mathbf{Be}. \tag{12}$$

Denoting $\kappa_1 = \mathbf{rBe}$ and substituting (11) into (12), we obtain the partial differential equation for the function $\Phi_1(x_1, \ldots, x_n, y_1, \ldots, y_n, \tau)$:

$$\frac{\partial \Phi_1(x_1, \ldots, x_n, y_1, \ldots, y_n, \tau)}{\partial \tau} =$$

$$\Phi_1(x_1, \ldots, x_n, y_1, \ldots, y_n, \tau) \kappa_1 \sum_{i=1}^{n} \tilde{S}_i(\tau) \left[j x_i + j y_i a_i^{(1)} \right].$$

Taking into account the initial condition $\Phi_1(x_1, \ldots, x_n, y_1, \ldots, y_n, \tau_0) = 1$, we obtain the following expression:

$$\Phi_1(x_1, \ldots, x_n, y_1, \ldots, y_n, \tau) = \exp \left\{ \kappa_1 \sum_{i=1}^{n} \left[j x_i + j y_i a_i^{(1)} \right] \int_{\tau_0}^{\tau} \tilde{S}_i(\xi) d\xi \right\}.$$

Thus, we can write the asymptotic ($\varepsilon \to 0$) approximate equality

$$\mathbf{h}(u_1, \ldots, u_n, z_1, \ldots, z_n, t) \approx \mathbf{r} \exp \left\{ \kappa_1 \sum_{i=1}^{n} \left[j u_i + j z_i a_i^{(1)} \right] \int_{t_0}^{t} S_i(\xi) d\xi \right\}.$$

Let $t_0 \to -\infty$ and $T = t$, we obtain the following expression for the asimptotic characteristic function $h(u_1, \ldots, u_n, z_1, \ldots, z_n)$:

$$h(u_1, \ldots, u_n, z_1, \ldots, z_n) = \mathbf{h}(u_1, \ldots, u_n, z_1, \ldots, z_n)\mathbf{e} =$$

$$\exp \left\{ \kappa_1 \sum_{i=1}^{n} p_i b_i \left[j u_i + j z_i a_i^{(1)} \right] \right\}.$$

$\square$

We seek the solution of the system (6) in the form

$$\mathbf{h}(u_1, \ldots, u_n, z_1, \ldots, z_n, t) =$$

$$\mathbf{h_2}(u_1, \ldots, u_n, z_1, \ldots, z_n, t) \exp \left\{ \kappa_1 \sum_{i=1}^{n} \left[j u_i + j z_i a_i^{(1)} \right] \int_{t_0}^{t} S_i(\xi) d\xi \right\}, \quad (13)$$

where $\mathbf{h_2}(u_1, \ldots, u_n, z_1, \ldots, z_n, t)$ is desired function.

After substituting (13) in (6), we obtain that $\mathbf{h_2}(u_1, \ldots, u_n, z_1, \ldots, z_n, t)$ is a solution of the differential equation:

$$\frac{\partial \mathbf{h_2}(u_1, \ldots, u_n, z_1, \ldots, z_n, t)}{\partial t} = \mathbf{h_2}(u_1, \ldots, u_n, z_1, \ldots, z_n, t) \times$$

$$\left[\mathbf{Q} + \sum_{i=1}^{n} S_i(t) \left\{ \left(e^{j u_i} G_i^*(z_i) - 1 \right) \mathbf{B} - \kappa_1 \left(j u_i + j z_i a_i^{(1)} \right) \mathbf{I} \right\} \right], \quad (14)$$

$$\mathbf{h_2}(u_1, \ldots, u_n, z_1, \ldots, z_n, t_0) = \mathbf{r}.$$

Let us denote $\varepsilon^2 = \dfrac{1}{q_i b}$, $i = 1, \ldots, n$, and make the following substitutions

$$t\varepsilon^2 = \tau, t_0 \varepsilon^2 = \tau_0, \mathbf{h_2}(u_1, \ldots, u_n, z_1, \ldots, z_n, t) = \mathbf{F_2}(x_1, \ldots, x_n, y_1, \ldots, y_n, \tau, \varepsilon).$$

Moreover, putting (9) into (14), we obtain:

$$\varepsilon^2 \frac{\partial \mathbf{F_2}(x_1, \ldots, x_n, y_1, \ldots, y_n, \tau, \varepsilon)}{\partial \tau} = \mathbf{F_2}(x_1, \ldots, x_n, y_1, \ldots, y_n, \tau, \varepsilon) \times$$
$$\left[\mathbf{Q} + \sum_{i=1}^{n} \tilde{S}_i(\tau) \left\{ \left(e^{j\varepsilon x_i} G_i^*(\varepsilon y_i) - 1 \right) \mathbf{B} - \kappa_1 \left(j\varepsilon x_i + j\varepsilon y_i a_i^{(1)} \right) \mathbf{I} \right\} \right], \tag{15}$$

with initial condition

$$\mathbf{F_2}(x_1, \ldots, x_n, y_1, \ldots, y_n, \tau_0, \varepsilon) = \mathbf{r}. \tag{16}$$

Let us find the asymptotic solution ($\varepsilon \to 0$) of the problem (15)–(16). We obtain

$$\begin{cases} \mathbf{F_2}(x_1, \ldots, x_n, y_1, \ldots, y_n, \tau)\mathbf{Q} = 0, \\ \mathbf{F_2}(x_1, \ldots, x_n, y_1, \ldots, y_n, \tau_0) = \mathbf{r}. \end{cases} \tag{17}$$

Taking into account (17), we look for solution of (15) of the form:

$$\mathbf{F_2}(x_1, \ldots, x_n, y_1, \ldots, y_n, \tau, \varepsilon) = \Phi_2(x_1, \ldots, x_n, y_1, \ldots, y_n, \tau) \{ \mathbf{r} +$$
$$\varepsilon \sum_{i=1}^{n} (jx_i + jy_i a_i^{(1)}) \tilde{S}_i(\tau) \mathbf{d} \} + O(\varepsilon^2). \tag{18}$$

By substituting (18) into (15), we obtain:

$$O(\varepsilon^2) = \Phi_2(x_1, \ldots, x_n, y_1, \ldots, y_n, \tau) \{ \mathbf{r}\mathbf{Q} +$$
$$\sum_{i=1}^{n} \tilde{S}_i(\tau) \left(j\varepsilon x_i + j\varepsilon y_i a_i^{(1)} \right) \mathbf{r}(\mathbf{B} - \kappa_1 \mathbf{I}) + \sum_{i=1}^{n} \tilde{S}_i(\tau) \left(j\varepsilon x_i + j\varepsilon y_i a_i^{(1)} \right) \mathbf{d}\mathbf{Q} \},$$

hence, taking into account $\mathbf{r}\mathbf{Q} = 0$, we can derive the following system of equations for the row vector $\mathbf{d}$ when $\varepsilon \to 0$:

$$\begin{cases} \mathbf{d}\mathbf{Q} + \mathbf{r}(\mathbf{B} - \kappa_1 \mathbf{I}) = 0, \\ \mathbf{d}\mathbf{e} = 0, \end{cases}$$

which coincides with (8).

We multiply equation (15) by vector $\mathbf{e}$ and perform the limit transition $\varepsilon \to 0$ in the obtained equality. The solution of the latter equation with the available initial condition $\Phi_2(x_1, \ldots, x_n, y_1, \ldots, y_n, \tau_0) = 1$ gives the expression:

$$\Phi_2(x_1, \ldots, x_n, y_1, \ldots, y_n, \tau) =$$
$$\exp \left\{ \frac{\kappa_1}{2} \sum_{i=1}^{n} \tilde{S}_i(\tau) \left[(jx_i)^2 + (jy_i)^2 a_i^{(2)} + 2jx_i jy_i a_i^{(1)} \right] + \right.$$
$$\frac{\kappa_2}{2} \sum_{i=1}^{n} \tilde{S}_i^2(\tau) \left[jx_i + jy_i a_i^{(1)} \right]^2 +$$
$$\left. \frac{\kappa_2}{2} \sum_{i=1}^{n} \sum_{\substack{g=1, \\ g \neq i}}^{n} \tilde{S}_i(\tau) \tilde{S}_g(\tau) \left[jx_i + jy_i a_i^{(1)} \right] \left[jx_g + jy_g a_g^{(1)} \right] \right\}.$$

Performing the inverse substitutions and putting $t_0 \to -\infty$, $T = t$, we obtain the following expression for the second-order asymptotic characteristic function of the process $\{i_1(t), \ldots, i_n(t), V_1(t), \ldots, V_n(t)\}$:

$$
\begin{aligned}
h(u_1, \ldots, u_n, z_1, \ldots, z_n) &= \mathbf{h}(u_1, \ldots, u_n, z_1, \ldots, z_n)\mathbf{e} = \\
\exp\Bigg\{ &\kappa_1 \sum_{i=1}^{n} p_i b_i \left[ju_i + jz_i a_i^{(1)} \right] + \frac{\kappa_1}{2} \sum_{i=1}^{n} p_i b_i \left[(ju_i)^2 + (jz_i)^2 a_i^{(2)} \right] + \\
&\kappa_1 \sum_{i=1}^{n} p_i b_i j u_i j z_i a_i^{(1)} + \frac{\kappa_2}{2} \sum_{i=1}^{n} p_i^2 \beta_i \left[ju_i + jz_i a_i^{(1)} \right]^2 + \\
&\frac{\kappa_2}{2} \sum_{i=1}^{n} \sum_{\substack{g=1, \\ g \neq i}}^{n} p_i p_g \beta_{ig} \left[ju_i + jz_i a_i^{(1)} \right] \left[ju_g + jz_g a_g^{(1)} \right] \Bigg\}.
\end{aligned}
\tag{19}
$$

$\square$

The structure of (19) implies that the $2n$-dimensional process $\{i_1(t), V_1(t),$ $\ldots, i_n(t), V_n(t)\}$ is asymptotically Gaussian with mean

$$
\mathbf{a} = \kappa_1 \left[\mathbf{a}_1 p_1 b_1 \ldots \mathbf{a}_n p_n b_n \right], \quad \mathbf{a}_i = \left[1 \; a_1^{(i)} \right],
$$

and covariance matrix

$$
\mathbf{K} = \kappa_1 \mathbf{K}^{(1)} + \kappa_2 \mathbf{K}^{(2)},
$$

where

$$
\mathbf{K}^{(1)} = \begin{bmatrix} \mathbf{K}_1^{(1)} p_1 b_1 & \cdots & 0 \\ \cdots & \cdots & \cdots \\ 0 & \cdots & \mathbf{K}_n^{(1)} p_n b_n \end{bmatrix}, \quad
\mathbf{K}^{(2)} = \begin{bmatrix} \mathbf{K}_{11}^{(2)} p_1^2 \beta_1 & \cdots & \mathbf{K}_{1n}^{(2)} p_1 p_n \beta_{1n} \\ \cdots & \cdots & \cdots \\ \mathbf{K}_{n1}^{(2)} p_n p_1 \beta_{n1} & \cdots & \mathbf{K}_{nn}^{(2)} p_n^2 \beta_n \end{bmatrix},
$$

$$
\mathbf{K}_i^{(1)} = \begin{bmatrix} 1 & a_1^{(i)} \\ a_1^{(i)} & a_2^{(i)} \end{bmatrix}, \quad
\mathbf{K}_{ig}^{(2)} = \begin{bmatrix} 1 & a_1^{(g)} \\ a_1^{(i)} & a_1^{(i)} a_1^{(g)} \end{bmatrix},
$$

4   Simulation and Numerical Examples

Let us consider as input a MAP with the following parameters:

$$
\mathbf{\Lambda} = \begin{bmatrix} 0.55 & 0 & 0 \\ 0 & 0.7 & 0 \\ 0 & 0 & 1 \end{bmatrix}, \quad
\mathbf{Q} = \begin{bmatrix} -0.6 & 0.3 & 0.3 \\ 0.2 & -0.4 & 0.2 \\ 0.4 & 0.4 & -0.8 \end{bmatrix}, \quad
\mathbf{D} = \begin{bmatrix} 0 & 0.9 & 0.1 \\ 0.3 & 0 & 0.7 \\ 0.4 & 0.6 & 0 \end{bmatrix},
$$

and $p_1 = 0.5$, $p_2 = 0.3$, $p_3 = 0.2$.

Table 1 shows the distribution laws of the random variables which characterize the amount of the occupied resource and the time required to service the customers.

Table 1. Types of customers and their distribution laws

Type	Distribution laws	
	Service time	Volume
First	Gamma $(0.5b, 0.5)$	Exponential (2)
Second	Gamma $(1.5b, 1.5)$	Exponential (1)
Third	Gamma $(2.5b, 2.5)$	Exponential (0.4)

By simulation, we obtained the marginal empirical probability distribution functions for the total amounts of resource occupied by each type of customers. Our goal is to compare empirical and asymptotic distribution laws. To this aim we will use the following definitions, to quantitatively verified the closeness of the corresponding distributions:

— Hellinger distance: $\Delta^H = \sqrt{\int\limits_0^\infty \left(\sqrt{f_{em}(x)} - \sqrt{f_{as}(x)} \right)^2 dx}$,

— Kullback-Leibler quasi-distance: $\Delta^{KL} = \int\limits_0^\infty f_{em}(x) \ln \left(\dfrac{f_{em}(x)}{f_{as}(x)} \right) dx$,

— areas distance: $\Delta^S = 1 - \int\limits_0^\infty \min \left(f_{em}(x), f_{as}(x) \right) dx$,

— Kolmogorov distance: $\Delta^C = \sup_x |F_{em}(x) - F_{as}(x)|$,

where subscript "em" and "as" denote the empirical and asymptotic densities (f) and distribution functions (F).

Tables 2, 3 and 4 show the values of these distances for the total volumes of the occupied resources of each type. The accuracy of the approximation increases with increasing mean of service time, as highlighted by Figs. 3 and 4.

Table 2. Metrics for the first type of resources

Metrics	b					
	10	20	50	100	200	500
Δ^H	0.141	0.118	0.089	0.063	0.045	0.032
Δ^{KL}	0.084	0.042	0.016	0.008	0.004	0.002
Δ^S	0.143	0.089	0.055	0.039	0.027	0.017
Δ^C	0.065	0.045	0.029	0.020	0.014	0.009

For sake of brevity we reported only the results for the occupied volume, but similar results have been obtained for the distribution of the number of busy servers in the system.

Table 3. Metrics for the second type of resources

Metrics	b					
	10	20	50	100	200	500
Δ^H	0.221	0.134	0.109	0.084	0.055	0.032
Δ^{KL}	0.181	0.082	0.027	0.013	0.006	0.003
Δ^S	0.219	0.123	0.072	0.049	0.035	0.022
Δ^C	0.085	0.059	0.037	0.026	0.019	0.012

Table 4. Metrics for the third type of resources

Metrics	b					
	10	20	50	100	200	500
Δ^H	0.257	0.161	0.118	0.100	0.071	0.044
Δ^{KL}	0.226	0.097	0.042	0.019	0.010	0.004
Δ^S	0.293	0.173	0.089	0.061	0.043	0.027
Δ^C	0.107	0.073	0.045	0.032	0.022	0.014

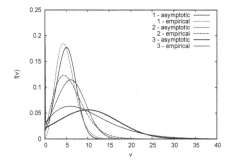

Fig. 3. Empirical and asymptotic distributions of the total amounts of the occupied resource of each type for $b = 20$

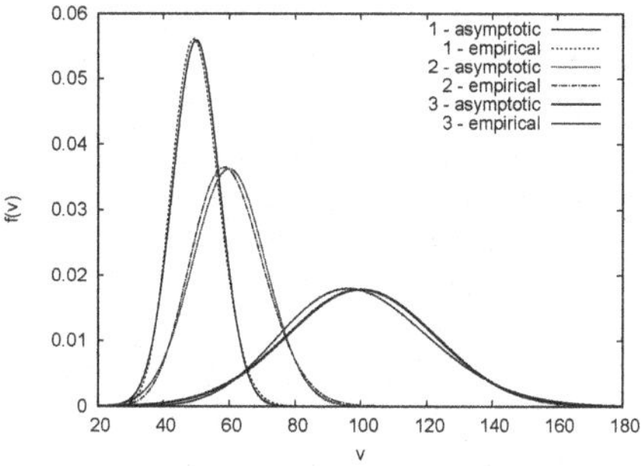

Fig. 4. Empirical and asymptotic distributions of the total amounts of the occupied resource of each type for $b = 200$

5   Conclusion

In this paper, we analyzed a resource queueing system with MAP arrivals, infinite servers numbers and heterogeneous customers, both in terms of service time and resource requirements. Taking advantage of the dynamic screening method, at first we wrote the corresponding Kolmogorov integrodifferential equations and then derived the first and second order asymptotic approximations (under the asymptotic condition of increasing service time) for the characteristic function of the multidimensional process, describing the number of busy servers and resource amounts occupied by customers of each type.

In more detail, we showed that the joint distribution of the considered process is multidimensional Gaussian and we obtained the parameters of this distribution. Finally, by means of discrete event simulations we verified the accuracy of the approximation both qualitatively (by visual comparison of the marginal probability density functions) and quantitatively (in terms of four well known distance metrics) as a function of the asymptotic parameter b.

References

1. Borodakiy, V.Y., Samouylov, K.E., Gudkova, I.A., Markova, E.V.: Analyzing meanbit rate of multicast video conference in LTE network with adaptive radio admission control scheme. J. Math. Sci. (U.S.) **218**(3), 257–268 (2016). https://doi.org/10.1007/s10958-016-3027-3

2. Gaidamaka, Y., Pechinkin, A., Razumchik, R., Samouylov, K., Sopin, E.: Analysis of an MG1R queue with batch arrivals and two hysteretic overload control policies. Int. J. Appl. Math. Comput. Sci. **24**(3), 519–534 (2014). https://doi.org/10.2478/amcs-2014-0038

3. Gorbunova, A.V., Naumov, V.A., Gaidamaka, Y.V., Samouylov, K.E.: Resource queuing systems as models of wireless communication systems. Informatika i ee Primeneniya **12**(3), 48–55 (2018). https://doi.org/10.14357/19922264180307

4. Gudkova, I., et al.: Analyzing impacts of coexistence between M2M and H2H communication on 3GPP LTE system. In: Mellouk, A., Fowler, S., Hoceini, S., Daachi, B. (eds.) WWIC 2014. LNCS, vol. 8458, pp. 162–174. Springer, Cham (2014). https://doi.org/10.1007/978-3-319-13174-0_13

5. Lisovskaya, E., Moiseeva, S., Pagano, M.: The total capacity of customers in the infinite-server queue with MMPP arrivals. Commun. Comput. Inf. Sci. **678**, 110–120 (2016). https://doi.org/10.1007/978-3-319-51917-3_11

6. Lisovskaya, E., Moiseeva, S., Pagano, M., Potatueva, V.: Study of the MMPP/GI/∞ queueing system with random customers' capacities. Informatika i ee Primeneniya **11**(4), 109–117 (2017). https://doi.org/10.14357/19922264170414

7. Melikov, A., Zadiranova, L., Moiseev, A.: Two asymptotic conditions in queue with MMPP arrivals and feedback. Commun. Comput. Inf. Sci. **678**, 231–240 (2016). https://doi.org/10.1007/978-3-319-51917-3_21

8. Moiseev, A., Moiseeva, S., Lisovskaya, E.: Infinite-server queueing tandem with MMPP arrivals and random capacity of customers. In: Proceedings - 31st European Conference on Modelling and Simulation, ECMS 2017, pp. 673–679 (2017). https://doi.org/10.7148/2017-0673

9. Moiseev, A., Nazarov, A.: Queueing network MAP - (GI/∞)K with high-rate arrivals. Eur. J. Oper. Res. **254**, 161–168 (2016). https://doi.org/10.1016/j.ejor.2016.04.011

10. Moiseev, A., Nazarov, A.: Tandem of infinite-server queues with Markovian arrival process. Commun. Comput. Inf. Sci. **601**, 323–333 (2016). https://doi.org/10.1007/978-3-319-30843-2_34

11. Morozov, E., Potakhina, L., Tikhonenko, O.: Regenerative analysis of a system with a random volume of customers. Commun. Comput. Inf. Sci. **638**, 261–272 (2016). https://doi.org/10.1007/978-3-319-44615-8_23

12. Naumov, V., Samouylov, K.: Analysis of multi-resource loss system with state-dependent arrival and service rates. Probab. Eng. Inf. Sci. **31**(4), 413–419 (2017). https://doi.org/10.1017/S0269964817000079

13. Naumov, V., Samouylov, K., Sopin, E., Andreev, S.: Two approaches to analyzing dynamic cellular networks with limited resources. In: International Congress on Ultra Modern Telecommunications and Control Systems and Workshops, vol. 2015-January, pp. 485–488 (2015). https://doi.org/10.1109/ICUMT.2014.7002149
14. Naumov, V.A., Samuilov, K.E., Samuilov, A.K.: On the total amount of resources occupied by serviced customers. Autom. Remote Control **77**(8), 1419–1427 (2016). https://doi.org/10.1134/S0005117916080087
15. Pankratova, E., Moiseeva, S.: Queueing system MAP/M/∞ with n types of customers. Commun. Comput. Inf. Sci. **487**, 356–366 (2014). https://doi.org/10.1007/978-3-319-13671-4_41
16. Pankratova, E., Moiseeva, S.: Queueing system with renewal arrival process and two types of customers. In: International Congress on Ultra Modern Telecommunications and Control Systems and Workshops, vol. 2015-January, pp. 514–517 (2015). https://doi.org/10.1109/ICUMT.2014.7002154
17. Samouylov, K., Naumov, V., Sopin, E., Gudkova, I., Shorgin, S.: Sojourn time analysis for processor sharing loss system with unreliable server. In: Wittevrongel, S., Phung-Duc, T. (eds.) ASMTA 2016. LNCS, vol. 9845, pp. 284–297. Springer, Cham (2016). https://doi.org/10.1007/978-3-319-43904-4_20
18. Tikhonenko, O., Zajac, P.: Queueing systems with demands of random space requirement and limited queueing or sojourn time. Commun. Comput. Inf. Sci. **718**, 380–391 (2017). https://doi.org/10.1007/978-3-319-59767-6_30
19. Tikhonenko, O.M., Kempa, W.: Queueing system with processor sharing and limited memory under control of the AQM mechanism. Autom. Remote Control **76**(10), 1784–1796 (2015). https://doi.org/10.1134/S0005117915100069
20. Vishnevsky, V.M., Samouylov, K.E., Naumov, V.A., Krishnamoorthy, A., Yarkina, N.: Multiservice queueing system with MAP arrivals for modelling LTE cell with H2H and M2M communications and M2M aggregation. Commun. Comput. Inf. Sci. **700**, 63–74 (2017). https://doi.org/10.1007/978-3-319-66836-9_6

Cluster Modeling of Lindley Process
with Application to Queuing

Natalia Markovich[1]([⊠])(iD) and Rostislav Razumchik[2,3](iD)

[1] V.A. Trapeznikov Institute of Control Sciences, Russian Academy of Sciences,
Profsoyuznaya Street 65, Moscow 117997, Russia
markovic@ipu.rssi.ru, nat.markovich@gmail.com
[2] Institute of Informatics Problems, FRC CSC RAS,
Vavilova Street 44-2, Moscow 119333, Russia
rrazumchik@ipiran.ru
[3] Peoples' Friendship University of Russia (RUDN University),
6 Miklukho-Maklaya Street, Moscow 117198, Russian Federation
razumchik-rv@rudn.ru

Abstract. We investigate clusters of extremes defined as subsequent exceedances of high thresholds in a Lindley process. The latter is usually used to model the waiting time or the length of a queue in queuing systems. Distributions of the cluster and inter-cluster sizes of the Lindley process are obtained for a given value of the threshold assuming that the process begins from the zero value. An example of a $M/M/1$ queue and the impact of service and arrival rates on the cluster and inter-cluster distributions are shown.

Keywords: Cluster of exceedances · Lindley process · Cluster and inter-cluster distributions · Extremal index · Queuing

1 Introduction

The cluster structure of stochastic processes is often observed in practice due to dependence and heavy-tailed distributions. We state the cluster as the number of consecutive exceedances of the process over a high threshold between two consecutive non-exceedances, see Fig. 1(a). Our aim is to consider such clusters of waiting times of customers of a queuing system and to investigate how arrival and service rates may impact on them. The main question is how does the relation between the service and arrival rates reflect on the cluster and inter-cluster distributions of the customer's waiting times.

Let start with an example. Let us observe the load of an escalator in the subway and the waiting times of passengers in front of it. Evidently, when the frequency of passengers is high, i.e. inter-arrival times between them is short, the escalator can be overloaded. Then the queue length in front of it can become long. Permanently, one can see clusters of passengers standing on the escalator and gaps between such clusters. During rush hours, the gaps are short or even absent, since the cluster length may became larger than the length of escalator.

© Springer Nature Switzerland AG 2019
V. M. Vishnevskiy et al. (Eds.): DCCN 2019, LNCS 11965, pp. 330–341, 2019.
https://doi.org/10.1007/978-3-030-36614-8_25

The length and the speed of the escalator determine its service capacity. Since the latter is upper bounded and fixed, the waiting times in front of the escalator can be long and exceed a high level. The waiting passengers form another cluster of exceedances with regard to their waiting times. Let the arrival rate be λ and the service rate be μ. The question is what could be the ratio λ/μ corresponding to shorter clusters of waiting times? In other words, how the distributions of integer-valued cluster and inter-cluster sizes of waiting times depend on these rates?

The Lindley process is accepted as a realistic Markov chain model for the waiting time and a $GI/GI/1$ queue is one of the examples of its application [2,13]. It is determined by

$$W_{n+1} = (W_n + D_n)^+ = \max(0, W_n + D_n), \ n = 0, 1, \dots,$$

where W_n denotes the waiting time of the nth customer until he is served. $D_n = B_n - A_n$ is the difference between the service time B_n of customer n and the inter-arrival time A_n between customer n and $n + 1$, if the quantity is positive, and 0 otherwise. $\{D_n\}_{n \geq 1}$ is assumed to be independent and identically distributed (i.i.d.) sequence of random variables (r.v.s) with common distribution function (d.f.) $H(x)$ and $ED_1 < 0$. The waiting time of the 0th customer W_0 is independent of $\{D_n\}$. Evidently, $W_{n+1} = 0$ implies the independence W_{n+2} and W_{n+1} and thus, $\{0\}$ build regenerative moments. Note, that regenerative moments or visits to 0 may happened only between clusters of exceedances over the threshold $u > 0$, see Fig. 1(a). The distribution of D_n depends on both distributions of the service and inter-arrival times that are mutually independent. Figures 1(b) and (c) show more solid clusters of exceedances when the service rate decreases.

Following [6,10,11] and [12] we seek to find distributions of the two r.v.s related to the local dependence of the Lindley process or its cluster structure. These are the inter-arrival times $T_1(u)$ between clusters and the inter-arrival times $T_2(u)$ within a cluster. Both are determined by the threshold u as follows

$$T_1(u) = \min\{j \geq 1 : M_{1,j} \leq u, W_{j+1} > u | W_1 > u\}, \tag{1}$$

$$T_2(u) = \min\{j \geq 1 : L_{1,j} > u, W_{j+1} \leq u | W_1 \leq u\}, \tag{2}$$

where $M_{1,j} = \max\{W_2, ..., W_j\}$, $M_{1,1} = -\infty$, $L_{1,j} = \min\{W_2, ..., W_j\}$, $L_{1,1} = \infty$. The cluster and inter-cluster sizes are then equal to $T_2(u) - 1$ and $T_1(u) - 1$, respectively. For practical applications, the case $j = 1$ is meaningless since $T_1(u) = 1$ and $T_2(u) = 1$ correspond to single inter-arrivals between consecutive nonexceedances and exceedances of $\{W_i\}$, respectively. Then the events $\{T_1(u) = j\}$ and $\{T_2(u) = j\}$ imply that the waiting times of $j - 1$ consecutive customers do not exceed or do exceed u, respectively. The practical implementation of such distributions may be connected to risks to lose customers with maximum waiting times (deadlines) exceeding the threshold u, see [8].

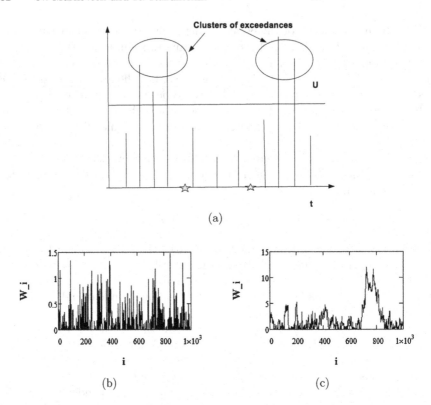

Fig. 1. Clusters of exceedances over the threshold u and the visits to zero marked by stars (Fig. 1(a)); Lindley time series for exponentially distributed service time B_n of customer n and the inter-arrival time A_n with the rates $(\mu, \lambda) = (5, 2)$ (Fig. 1(b)) and $(\mu, \lambda) = (2.5, 2)$ (Fig. 1(c)).

The asymptotic distributions of maxima of the Lindley process were proved both for non-lattice and lattice distributions in [1,13] and [3].

Our main objective is to find the distributions $P\{T_1(u) = j\}$ and $P\{T_2(u) = j\}$ for a fixed threshold u. From our best knowledge, this task was not solved yet. In fact, the arrival and service rates impact on the latter distributions. Since $D_n = B_n - A_n$ holds, we consider an example of known distributions of the vector (A_n, B_n) and find the distributions $P\{T_1(u) = j\}$ and $P\{T_2(u) = j\}$.

The paper is organized as follows. In Sect. 2 necessary definitions and results are reviewed. The distributions of the cluster and inter-cluster times for a fixed threshold u are obtained in Sect. 3. An example of the distribution for an $M/M/1$ queue is given in Sect. 4. We finalize the exposition with some conclusions in Sect. 5. Proofs are located in the Appendix.

2 Related Works

We recall that the extremal index (EI) is a dependence measure to study extremes distributions of stochastic processes.

Definition. *The stationary sequence $\{X_n\}_{n \geq 1}$ with distribution function $F(x)$ is said to have the extremal index $\theta \in [0,1]$ if for each $0 < \tau < \infty$ there is a sequence of real numbers $u_n = u_n(\tau)$ such that it holds*

$$\lim_{n \to \infty} n(1 - F(u_n)) = \tau, \tag{3}$$

$$\lim_{n \to \infty} P\{M_n \leq u_n\} = e^{-\tau\theta}, \tag{4}$$

where $M_n = \max\{X_1, ..., X_n\}$ ([9, p. 53]).

The EI may be determined as follows

$$\theta = \lim_{n \to \infty} P\{M_{1,p_n} \leq u_n | X_1 > u_n\},$$

where $\{p_n\}$ is an increasing sequence of positive integers, $p_n = o(n)$ as $n \to \infty$, $M_{1,p_n} = \max\{X_2, ..., X_{p_n}\}$ [4]. For independent r.v.s $\theta = 1$ holds. $\theta = 0$ implies a total dependence that corresponds to "very wide" clusters of exceedances when observations of the stochastic process likely never exceed a sufficiently high threshold u_n tending to infinity.

The example of the Lindley process with $\theta = 0$ generated by subexponential heavy-tailed noise D_n is given in [1]. In the Example 1 in [12] it is shown that this process satisfies so called j-dependence condition if a visit to 0 ($W_\tau = 0$) is located between W_1 and W_{j+1}. In other words, W_1 and W_{j+1} are independent if they belong to different regenerative cycles. Then for sufficiently large sample size n and the integer number j it was proved in [12, Theorem 2] that $|P\{T_1(x_{\rho_n}) = j\}/\rho_n - 1| < \varepsilon$ and $|P\{T_2(x_{\rho_n}) = j\}/q_n - 1| < \varepsilon$ for each $\varepsilon > 0$, where $\{x_{\rho_n}\}$ is a sequence of quantiles of the levels $\{1 - \rho_n\}^1$. It holds $\rho_n \sim \tau/n \to 0$ and $q_n = 1 - \rho_n \to 1$ as $n \to \infty$ due to (3). Then the latter result implies that the probability to have a huge cluster size tends to 1 and a huge inter-cluster size tends to 0 as a high quantile close to 100% is taken as the threshold u. This is in agreement with (4) since then $\lim_{n \to \infty} P\{M_n \leq u_n\} = 1$.

It is important for practice that $1/\theta$ approximates the mean cluster size. This allows one to estimate the EI by non-parametric methods like blocks, runs and intervals estimators [5,6].

In [2] it is found that if D_0 is non-lattice and there exists a $\gamma > 0$ with $E(\exp(\gamma D_1)) = 1$, $E(|D_1| \exp(\gamma D_1)) < \infty$, then $\{W_n\}$ has a stationary distribution P_π and $P_\pi\{W_0 > u\} \sim C \exp(-\gamma u)$, as $u \to \infty$, for the constant C. An explicit expression of the EI of a Lindley process,

$$\theta = \int_0^\infty P\{M \leq -x\} \gamma \exp(-\gamma x) dx, \tag{5}$$

---

[1] The latter theorem is valid for a process with $\theta = 0$ not necessarily the Lindley one irrespective on its stationary distribution and under specific mixing conditions.

where $M = \sup\{D_1, D_1 + D_2, ...\}$ is given in [13]. Another representation of θ is given in [3, Theorem 2].

Since the cluster distributions depend on the EI, it is important to know its precise value or, at least, to estimate it by nonparametric estimators.

3   Exact Cluster and Inter-cluster Distributions

Let us obtain the exact distributions of $T_1(u)$ and $T_2(u)$ for a fixed value u. From (1) and (2) we get

$$P\{T_1(u) = j\} = P\{M_{1,j} \leq u, W_{j+1} > u | W_1 > u\},$$
$$P\{T_2(u) = j\} = P\{L_{1,j} > u, W_{j+1} \leq u | W_1 \leq u\}.$$

Since $T_1(u)$ and $T_2(u)$ are integer-valued r.v.s and the case $j = 1$ is practically meaningless; thus we will assume that $j \geq 2$.

Remark 1. The events $\{T_1(u) = j\}$ may include the regenerative moments or visits to 0 between two consecutive exceedances over u. This implies that Proposition 2 below is valid not only within a single regenerative interval.

Let us begin with the probability $P\{T_2(u) = j\}$. Firstly we note that $P\{W_1 \leq u\} = P\{D_0 \leq u\} = H(u)$. Secondly, for $j \geq 3$ we can use the formula (31) from [12]:

$$
\begin{aligned}
P\{T_2(u) = j\}P\{W_1 \leq u\} &= P\{W_1 \leq u, W_{j+1} \leq u\} \\
&+ (-1)^{j-1}P\{W_1 \leq u, W_2 \leq u, ..., W_{j+1} \leq u\} \\
&+ \sum_{r=1}^{j-2}(-1)^r \sum_{i_1=2}^{j-r+1} \sum_{i_2=i_1+1}^{j-r+2} ... \sum_{i_r=i_{r-1}+1}^{j} P\{W_1 \leq u, W_{i_1} \leq u, ..., W_{i_r} \leq u, W_{j+1} \leq u\}.
\end{aligned}
\tag{6}
$$

In the following Propositions 1 and 2 we assume that $W_0 = 0$ holds, $H(x)$ is the distribution function of D_1 and its density $h(x) = H'(x)$ exists.

Proposition 1. *Regarding the Lindley process and $j = 2$ we have*

$$P\{T_2(u) = 2\} = H(u)^{-1} \int_{-\infty}^{u} h(t_0) \int_{b_{t_0,u}}^{\infty} h(t_1) \cdot H(b_{t_1,t_0,u}) dt_1 dt_0, \tag{7}$$

for $j \geq 3$ we have

$$P\{T_2(u) = j\}H(u) \tag{8}$$

$$
= \int_{-\infty}^{u} h(t_0) \int_{-\infty}^{\infty} h(t_1)... \int_{-\infty}^{\infty} h(t_{j-1})H(b_{t_{j-1},...,t_0,u}) dt_{j-1}...dt_0
$$

$$
+ (-1)^{j-1} \int_{-\infty}^{u} h(t_0) \int_{-\infty}^{b_{t_0,u}} h(t_1)... \int_{-\infty}^{b_{t_{j-2},...,t_0,u}} h(t_{j-1})H(b_{t_{j-1},...,t_0,u}) dt_{j-1}...dt_0
$$

$$
+ \sum_{r=1}^{j-2}(-1)^r \sum_{i_1=2}^{j-r+1} \sum_{i_2=i_1+1}^{j-r+2} ... \sum_{i_r=i_{r-1}+1}^{j}
$$

$$
\int_{-\infty}^{u} h(t_0) \int_{-\infty}^{b_{i_1}-1} h(t_{i_1-1})... \int_{-\infty}^{b_{i_r}-1} h(t_{i_r-1})H(b_{t_{j-1},...,t_0,u}) dt_{i_r-1}...dt_{i_1-1} dt_0,
$$

where we denote

$$b_{t_0,u} = \min(u, u - t_0), ..., b_{t_{j-1},...,t_0,u} = \min\left(u, u - t_{j-1}, ..., u - \sum_{i=0}^{j-1} t_i\right), \quad (9)$$

$$b_{i_m-1} = \min\left(u, u - t_{i_m-2}, ..., u - \sum_{i=0}^{i_m-2} t_i\right), \quad m = 1, ..., r, \quad i_m \geq 2.$$

Remark 2. The first term on the right-hand side of (8) can be rewritten shorter as

$$\int_{-\infty}^{u} h(t_0) E_{t_1,...,t_{j-1}} H(b_{t_{j-1},...,t_0,u}) dt_0,$$

where the expectation is taken over $t_1, ..., t_{j-1}$. For even $j \geq 4$ the first two terms in (8) can be combined in one integral.

Proof. Let us derive formula (7). For $j = 2$ we obtain

$$P\{T_2(u) = 2\} = P^{-1}\{W_1 \leq u\}$$
$$\times \left(P\{W_1 \leq u, W_3 \leq u\} - P\{W_1 \leq u, W_2 \leq u, W_3 \leq u\}\right).$$

We use essentially the Proposition 6.3 from [2]. It states

$$W_n = \max\{W_0 + S_n, S_n - S_1, ..., S_n - S_{n-1}, 0\},$$

where $S_0 = 0$, $S_k = D_0 + ... + D_{k-1}$. We use further the fact that $\{D_n\}$ are i.i.d. r.v.s. We get

$$P\{W_1 \leq u, W_3 \leq u\} = P\{S_1 \leq u, \max\{S_3, S_3 - S_1, S_3 - S_2, 0\} \leq u\} \quad (10)$$
$$= P\{D_0 \leq u, \max\{D_0 + D_1 + D_2, D_1 + D_2, D_2, 0\} \leq u\}$$
$$= \int_{-\infty}^{u} h(t_0) \int_{-\infty}^{\infty} h(t_1) \int_{-\infty}^{\min(u,u-t_1,u-(t_0+t_1))} h(t_2) dt_2 dt_1 dt_0.$$

Similarly, we have

$$P\{W_1 \leq u, W_2 \leq u, W_3 \leq u\}$$
$$= P\{S_1 \leq u, \max\{S_2, S_2 - S_1, 0\} \leq u, \max\{S_3, S_3 - S_1, S_3 - S_2, 0\} \leq u\}$$
$$= \int_{-\infty}^{u} h(t_0) \int_{-\infty}^{\min(u,u-t_0)} h(t_1) \int_{-\infty}^{\min(u,u-t_1,u-(t_0+t_1))} h(t_2) dt_2 dt_1 dt_0.$$

Hence, (7) follows.

Let us derive the formula (8). Turning back to (6) and generalizing (10) we obtain

$$P\{W_1 \leq u, W_{j+1} \leq u\}$$
$$= \int_{-\infty}^{u} h(t_0) \int_{-\infty}^{\infty} h(t_1)... \int_{-\infty}^{\infty} h(t_{j-1}) \int_{-\infty}^{b_{t_{j-1},...,t_0,u}} h(t_j) dt_j...dt_0.$$

Furthermore, we have

$$P\{W_1 \leq u, W_2 \leq u, ..., W_{j+1} \leq u\} \tag{11}$$

$$= \int_{-\infty}^{u} h(t_0) \int_{-\infty}^{b_{t_0,u}} h(t_1)... \int_{-\infty}^{b_{t_{j-2},...,t_0,u}} h(t_{j-1}) \int_{-\infty}^{b_{t_{j-1},...,t_0,u}} h(t_j) dt_j ... dt_0.$$

Similarly we obtain

$$P\{W_1 \leq u, W_{i_1} \leq u, ..., W_{i_r} \leq u, W_{j+1} \leq u\}$$

$$= \int_{-\infty}^{u} h(t_0) \int_{-\infty}^{b_{i_1-1}} h(t_{i_1-1})...$$

$$... \int_{-\infty}^{b_{i_r-1}} h(t_{i_r-1}) \int_{-\infty}^{b_{t_{j-1},...,t_0,u}} h(t_j) dt_j dt_{i_r-1}...dt_{i_1-1} dt_0.$$

We insert the last three formulae in (6) and obtain (8).

Proposition 2. *Regarding the Lindley process and $j \geq 2$ we have*

$$P\{T_1(u) = j\}\overline{H(u)} \tag{12}$$

$$= \int_{u}^{\infty} h(t_0) \int_{-\infty}^{u-t_0} h(t_1)... \int_{-\infty}^{b_{t_{j-2},...,t_0,u}} h(t_{j-1})\overline{H}(b_{t_{j-1},...,t_0,u}) dt_{j-1}...dt_0,$$

where $\overline{H}(x) = 1 - H(x)$.

Proof. Using (11) we obtain

$$P\{T_1(u) = j\}P\{W_1 > u\} = P\{W_1 > u, M_{1,j} \leq u\} - P\{W_1 > u, M_{1,j+1} \leq u\}$$

$$= \psi_j - \psi_{j+1},$$

where $\psi_j = P\{M_{1,j} \leq u\} - P\{M_j \leq u\}$ holds. It follows

$$\psi_j = \int_{u}^{\infty} h(t_0) \int_{-\infty}^{b_{t_0,u}} h(t_1)... \int_{-\infty}^{b_{t_{j-3},...,t_0,u}} h(t_{j-2})H(b_{t_{j-2},...,t_0,u}) dt_{j-2}...dt_0.$$

Here, we use the same notations as in (9). Since $t_0 \geq u > 0$ then $b_{t_0,u} = u - t_0$ holds. Then we get (12).

Remark 3. In Propositions 1 and 2 the distributions of the r.v.s B_n and A_n involved in the distribution H of $D_n = B_n - A_n$ are not specified. Therefore these statements do not allow one to analyse the impact of the arrival and service rates on the distributions of $T_1(u)$ and $T_2(u)$.

4 Special Case: $M/M/1$ Queue

Let us apply the results obtained above to the study of clusters in a $M/M/1$ queue. Then the r.v.s A_n and B_n introduced in the Introduction have the distribution functions $A(x) = 1 - e^{-\lambda x}$ and $B(x) = 1 - e^{-\mu x}$, respectively. Then the condition $\lambda < \mu$ is necessary and sufficient for the existence of the stationary regime. The distribution $H(x)$ of the r.v. D_1 is double exponential. Namely, we have (see, for example, [2, Example 6.1])

$$H(x) = P\{B_n - A_n \le x\} = P\{B_n \le x + A_n\}$$

$$= \begin{cases} \int_0^\infty \lambda e^{-\lambda u} \int_0^{x+u} \mu e^{-\mu y} du, \, x \ge 0, \\ \int_{-x}^\infty \lambda e^{-\lambda u} \int_0^{x+u} \mu e^{-\mu y} du, \, x < 0 \end{cases} = \begin{cases} 1 - \frac{\lambda}{\lambda + \mu} e^{-\mu x}, \, x \ge 0, \\ \frac{\mu}{\lambda + \mu} e^{\lambda x}, \qquad x < 0. \end{cases}$$

In other words, if λ and μ denote the arrival and the service rate in a $M/M/1$ queue, then

$$H(x) = \min(1, e^{\lambda x}) - \frac{\lambda}{\lambda + \mu} e^{\min(-\mu x, \lambda x)}, \quad -\infty < x < \infty. \tag{13}$$

Let us denote the system's load (utilization) by $\rho = \lambda/\mu$ and use the quantile x_ρ of the distribution $H(x)$ of level ρ, i.e. $H(x_\rho) = \rho$, as u. Taking into account that $H(u) = 1 - \frac{\lambda}{\lambda + \mu} e^{-\mu u}$, from the condition $H(x_\rho) = \rho$ we get that $e^{-\mu x_\rho} = \frac{1-\rho^2}{\rho}$ and $x_\rho = -\mu^{-1} \ln \left(\frac{1-\rho^2}{\rho} \right)$. Since x_ρ has to be positive one may only take such ρ which satisfy $(\sqrt{5} - 1)/2 < \rho < 1$. By plugging (13) into (7) we obtain

$$P(T_2(x_\rho) = 2) = \frac{1-\rho}{\rho(1+\rho)^3} \left(1 - \rho \ln \left(\frac{1-\rho^2}{\rho} \right) \right).$$

It is known that the EI, say θ, of the Lindley process $\{W_n, \, n \ge 1\}$ in a $M/M/1$ queue is related to ρ by $\theta = (1-\rho)^2$ (see [7]). Thus the probability $P(T_2(x_\rho) = 2)$ can be expressed solely in terms of the EI:

$$P(T_2(x_\rho) = 2) = \frac{\sqrt{\theta}}{(1 - \sqrt{\theta})(2 - \sqrt{\theta})^3} \left(1 - (1 - \sqrt{\theta}) \ln \left(\frac{1 - (1 - \sqrt{\theta})^2}{1 - \sqrt{\theta}} \right) \right).$$

In fact this holds for any value $P\{T_2(x_\rho) = j\}$ since the (unnormalized) distribution of the cluster size, which appears after the first customer arrived to the empty $M/M/1$ queue, has the form:

$$P\{T_2(x_\rho) = j\} = C_{j-1} \frac{\rho^{j-3}(1-\rho^2)}{(1+\rho)^{2j}} \left(1 - \rho \ln \left(\frac{1-\rho^2}{\rho} \right) \right), \, j \ge 2, \tag{14}$$

where $C_j = \binom{2j}{j}/(j+1)$, $j \ge 0$, are the Catalan numbers, $\binom{n}{k}$ is the binomial coefficient. Relation (14) is valid since $\rho < 1$. From (14) the moments of the cluster size are readily available. The mean cluster size, $E(T_2(x_\rho))$, can be expressed in the closed form as follows:

$$E(T_2(x_\rho)) = \frac{2}{\rho(1+\rho)} \left(1 - \rho \ln \left(\frac{1-\rho^2}{\rho} \right) \right). \tag{15}$$

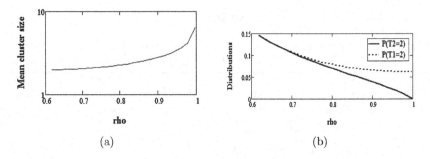

Fig. 2. Mean cluster size (Fig. 2(a)) and distributions $P\{T_1(x_\rho) = 2\}$ and $P\{T_2(x_\rho) = 2\}$ (Fig. 2(b)) as the function of the system's load ρ.

The proof of (15) is presented in the Appendix. As expected the growth of the utilization ρ leads to an increasing average cluster size, Fig. 2(a). This is in agreement with Fig. 1(c).

The inter-cluster distribution is much more difficult to obtain, even in this very special case. Probably this is due to the fact that the inter-cluster period may span several regenerative intervals. Particularly, the first three probabilities of the inter-cluster size distribution are equal to

$$P\{T_1(x_\rho) = 2\} = \frac{\rho^2}{(1+\rho)^4} + \frac{\rho}{(1+\rho)^4}\left(\frac{1-\rho^2}{\rho}\right)^{1+\rho},$$

$$P\{T_1(x_\rho) = 3\} = 2\frac{\rho^3}{(1+\rho)^6} + \frac{\rho}{(1+\rho)^5}\left(\frac{1+2\rho-\rho^2}{1+\rho} - 2\rho\ln\left(\frac{1-\rho^2}{\rho}\right)\right)\left(\frac{1-\rho^2}{\rho}\right)^{1+\rho},$$

$$P\{T_1(x_\rho) = 4\} = 5\frac{\rho^4}{(1+\rho)^8}$$
$$+ \frac{\rho}{(1+\rho)^8}\left(1+4\rho+3\rho^2-4\rho^3 - \rho\left(\frac{1-\rho^2}{\rho}\right)^{1+\rho}\right)\left(\frac{1-\rho^2}{\rho}\right)^{1+\rho}$$
$$+ \left(\frac{1-\rho^2}{\rho}\right)^{1+\rho}\ln\left(\frac{1-\rho^2}{\rho}\right)\frac{2\rho^2}{(1+\rho)^6}\left(\rho\ln\left(\frac{1-\rho^2}{\rho}\right) - \frac{1+3\rho-\rho^2}{1+\rho}\right).$$

The general expression for $P\{T_1(x_\rho) = j\}$, $j \geq 5$, remains an open issue.

5   Conclusions and Discussion

We have obtained the cluster and inter-cluster size distributions for the Lindley process that is considered as the model of the waiting times of the $GI/GI/1/\infty$ queue. Regarding the $M/M/1$ queue it was found that the distribution $P\{T_2(x_\rho) = j\}$ tends to zero as the utilization ρ tends to 1 but

$P\{T_1(x_\rho) = j\}$ does not, Fig. 2(b). Since the mean cluster size increases as the utilization increases, one may conclude that specifically for the $M/M/1$ queue there are likely groups of customers with long waiting times for the large utilization. For a $G/G/1/\infty$ queue this conclusion may be no longer valid.

The future research will be concerned with the cluster analysis in various specific $G/G/c/\infty$ queues. It is of special interest to investigate the cluster structure, if one switches to the Markov modulated environment (for example, to a $MAP/M/1$ or $MAP/MAP/1$ queue).

Although it was not shown here, the main ingredients of the formula (6), the probabilities $P\{W_1 \leq u, W_{i_1} \leq u, ..., W_{i_r} \leq u, W_{j+1} \leq u\}$, can be computed through the recursive procedure, which requires solutions of many integral equations not belonging to any special type (although very similar to Volterra equations of the second kind). The same concerns equations in Propositions 1 and 2. Thus the computational aspects remain the subject of further research.

Acknowledgment. The reported study was funded by RFBR, project number 19-01-00090 (recipient N. Markovich, conceptualization, mathematical model development, methodology development).The reported study was funded by RFBR, project number 17-07-00142 (recipient R. Razumchik, formal/numerical analysis, validation).

Appendix

Proof. The expression for $E(T_2(x_\rho))$ (formula (15)) follows from the application of the generating function method. Let us denote by $\mathcal{C}(z)$ — the probability generating function of Catalan numbers C_j, i.e. $\mathcal{C}(z) = \sum_{j=0}^{\infty} C_j z^j$, $0 < z < 1/4$. It is known that $\mathcal{C}(z) = \frac{1-\sqrt{1-4z}}{2z}$. Since $\mathcal{C}(z)$ satisfies the functional equation $\mathcal{C}(z) = 1 + z\mathcal{C}(z)^2$, the derivative of $\mathcal{C}(z)$ is equal to $\mathcal{C}'(z) = \frac{\mathcal{C}(z)^2}{1-2z\mathcal{C}(z)}$. Thus the value of $E(T_2(x_\rho))$ is equal to

$$E(T_2(x_\rho)) = \sum_{j=2}^{\infty} j P\{T_2(x_\rho) = j\} = \sum_{j=2}^{\infty} j C_{j-1} \frac{\rho^{j-3}(1-\rho^2)}{(1+\rho)^{2j}} \left(1 - \rho \ln\left(\frac{1-\rho^2}{\rho}\right)\right)$$

$$= \sum_{j=1}^{\infty} (j+1) C_j \frac{\rho^{j-2}(1-\rho^2)}{(1+\rho)^{2(j+1)}} \left(1 - \rho \ln\left(\frac{1-\rho^2}{\rho}\right)\right)$$

$$= \frac{(1-\rho^2)}{\rho^2(1+\rho)^2} \left(1 - \rho \ln\left(\frac{1-\rho^2}{\rho}\right)\right)$$

$$\times \left(\underbrace{\sum_{j=1}^{\infty} C_j [\frac{\rho}{(1+\rho)^2}]^j}_{=\mathcal{C}\left(\frac{\rho}{(1+\rho)^2}\right)-1} + \frac{\rho}{(1+\rho)^2} \underbrace{\sum_{j=1}^{\infty} j C_j [\frac{\rho}{(1+\rho)^2}]^{j-1}}_{=\mathcal{C}'\left(\frac{\rho}{(1+\rho)^2}\right)} \right).$$

Now we have

$$
\mathcal{C}\left(\frac{\rho}{(1+\rho)^2}\right) - 1 = \frac{1 - \sqrt{1 - 4\frac{\rho}{(1+\rho)^2}}}{2\frac{\rho}{(1+\rho)^2}} - 1 = \frac{1 - \frac{1}{1+\rho}\sqrt{1 + 2\rho + \rho^2 - 4\rho}}{2\frac{\rho}{(1+\rho)^2}} - 1
$$

$$
= \frac{(1+\rho)^2 - (1+\rho)\sqrt{(1-\rho)^2}}{2\rho} - 1 = \frac{2\rho^2 + 2\rho}{2\rho} - 1 = \rho.
$$

We also have

$$
\mathcal{C}'\left(\frac{\rho}{(1+\rho)^2}\right) = \frac{\mathcal{C}\left(\frac{\rho}{(1+\rho)^2}\right)^2}{1 - 2\frac{\rho}{(1+\rho)^2}\mathcal{C}\left(\frac{\rho}{(1+\rho)^2}\right)} = \frac{(1+\rho)^2}{1 - 2\frac{\rho}{(1+\rho)^2}(1+\rho)} = \frac{(1+\rho)^3}{1-\rho}.
$$

By putting everything together, we obtain

$$
E(T_2(x_\rho)) = \frac{(1-\rho^2)}{\rho^2(1+\rho)^2}\left(1 - \rho\ln\left(\frac{1-\rho^2}{\rho}\right)\right)\left[\rho + \frac{\rho}{(1+\rho)^2}\frac{(1+\rho)^3}{1-\rho}\right]
$$

$$
= \frac{(1-\rho^2)}{\rho(1+\rho)^2}\left(1 - \rho\ln\left(\frac{1-\rho^2}{\rho}\right)\right)\left[1 + \frac{1+\rho}{1-\rho}\right]
$$

$$
= \frac{2}{\rho(1+\rho)}\left(1 - \rho\ln\left(\frac{1-\rho^2}{\rho}\right)\right).
$$

References

1. Asmussen, S.: Subexponential asymptotics for stochastic processes: extremal behavior, stationary distributions and first passage probabilities. Ann. Appl. Probab. **8**, 354–374 (1998)
2. Asmussen, S.: Applied Probability and Queues, 2nd edn. Springer, New York (2003). https://doi.org/10.1007/b97236
3. Bacro, J.-N., Daudin, J.-J., Mercier, S., Robin, S.: Back to the local score in the logarithmic case: a direct and simple proof. Ann. Inst. Stat. Math. **54**(4), 748–757 (2002)
4. O'Brien, G.L.: Extreme values for stationary and Markov sequences. Ann. Probab. **15**(1), 281–291 (1987)
5. Beirlant, J., Goegebeur, Y., Teugels, J., Segers, J.: Statistics of Extremes: Theory and Applications. Wiley, Chichester (2004)
6. Ferro, C.A.T., Segers, J.: Inference for clusters of extreme values. J. Roy. Stat. Soc. Ser. B **65**, 545–556 (2003)
7. Hooghiemstra, G., Meester, L.E.: Computing the extremal index of special Markov chains and queues. Stoch. Process. Appl. **65**(2), 171–185 (1996)
8. Hyytiä, E., Righter, R., Virtamo, J.: Meeting soft deadlines in single-and multi-server systems. In: Proceedings of 28th International Teletraffic Congress (ITC 28), Würzburg, Germany, pp. 166–174 (2016). https://doi.org/10.1109/ITC-28.2016.129
9. Leadbetter, M.R., Lingren, G., Rootzén, H.: Extremes and Related Properties of Random Sequence and Processes. Springer, Heidelberg (1983). https://doi.org/10.1007/978-1-4612-5449-2

10. Markovich, N.M.: Modeling clusters of extreme values. Extremes **17**(1), 97–125 (2014)
11. Markovich, N.M.: Erratum to: modeling clusters of extreme values. Extremes **19**(1), 139–142 (2016)
12. Markovich, N.M.: Clusters of extremes: modeling and examples. Extremes **20**(3), 519–538 (2017)
13. Rootzén, H.: Maxima and exceedances of stationary Markov chains. Adv. Appl. Prob. **20**, 371–390 (1988)

Discrete-Time Insurance Models. Optimization of Their Performance by Reinsurance and Bank Loans

Ekaterina V. Bulinskaya$^{(\boxtimes)}$ [ID]

Lomonosov Moscow State University,
Leninskie Gory 1, Moscow 119991, Russia
ebulinsk@yandex.ru

Abstract. The popularity of discrete-time models in applied probability is explained as follows. They are more precise in some situations. In other cases they can be used as approximation of the corresponding continuous-time models. So, we consider two discrete-time insurance models and study the quality of their performance. The company reliability or the expected discounted costs incurred by its control can be chosen as an objective function (target or risk measure). It is possible to consider a finite or infinite planning horizon. The control includes reinsurance treaties and/or bank loans. The optimal control (maximizing the reliability or minimizing the costs) is established for the finite planning horizon. Its asymptotic behavior, as the horizon tends to infinity, is also investigated.

Keywords: Discrete time · Reinsurance · Bank loans · Optimization · Costs

1 Introduction

It is well known, see, e.g., [5], that almost all applied probability models arising in insurance, finance, queuing and reliability theory, dams and inventory theory, communications and population dynamics are of input-output type. Therefore such a model can be described by specifying the input, output, control, planning horizon $T \leq \infty$, as well as functional Ψ, reflecting the system configuration and operation mode, and objective function evaluating the quality of system performance. Input, output and control are some processes (deterministic or stochastic) defined for $t \leq T$, their dimensions are not necessarily the same. So, the system state (defined by application of functional Ψ to input, output and control) is also a process, maybe multi-dimensional. It is possible to control input, output or the system configuration and operation mode. An objective function (valuation criterium, target or risk measure) can be chosen in different

Supported by Russian Foundation for Basic Research according to the research project No. 17-01-00468.

ways. The most popular are reliability and cost approaches. In the first case, one wishes to minimize the ruin probability or maximize the uninterrupted working period. In the second case, the control is named optimal if it minimizes the (expected) costs associated with the system control or maximizes the (expected) profit obtained by the control application. For certainty, the further discussion will be conducted in terms of insurance models. However it is always possible to give another interpretation to the input and output processes to get a model arising in other applications. Thus, why insurance models were chosen? Insurance can be considered as risk management or decision making under uncertainty, see, e.g., [13,20]. It is necessary to protect the capital or other interests of citizens, enterprizes and organizations, i.e. individuals and legal entities. Moreover, insurance has the longest history among the above mentioned applied probability domains, see, e.g., [2]. In fact, methods for transferring or distributing risk were practiced by Chinese and Babylonian traders already 2–3 thousands years BC. Thus, Code of Hammurabi, c. 1750 BC, contained the laws for maritime insurance. Mutual societies, run by their members with no external shareholders to pay were first to appear. Next step is joint stock companies. A modern insurance company has a two-fold nature. Its primary task is indemnification of policyholders claims. The secondary, but very important, task is dividend payments to shareholders.

New problems have arisen in actuarial sciences during the last twenty years, see, e.g., [5]. This period is characterized by interplay of insurance and finance, unification of reliability and cost approaches, see, e.g., [6], as well as, consideration of complex systems. Sophisticated mathematical tools are used for analysis and optimization of insurance systems including dividends, reinsurance, capital injections and investment.

Nowadays, discrete-time models became popular in applied probability (see, e.g., [5]), since they are more precise in some situations. Thus, the reinsurance treaties are usually negotiated at the end of year, the decision on dividends payment is also based on the results of financial year. In other cases, the discrete-time models can be used as approximation of the corresponding continuous-time models when analytical results are unavailable, see, e.g., [12]. In such domains as inventories or population dynamics discrete-time models have arisen from the beginning. In insurance such models appeared later (see, e.g., [3] and [11]). The first review on discrete-time insurance models [14] was published in 2009. Below we develop the models proposed in [3–10]. Namely, we study the optimal policies for performance of insurance company. As control we choose reinsurance and bank loans.

2   Main Results

We consider the following discrete-time insurance model. Let $\{X_n\}_{n \geq 1}$ be a sequence of non-negative independent identically distributed (i.i.d.) random variables (r.v.'s) with a known distribution function F, possessing a density $\varphi(s) > 0$ for $s > 0$ and a finite expectation. Here X_k is the claim size during the k-th period. The initial capital (or surplus) of the company is equal to x. The premium paid at the beginning of the first period is included in the initial capital. The other premiums will be mentioned explicitly.

Our aim is to establish a strategy of bank loans minimizing the additional costs associated with claims indemnification. In the same time a company can use reinsurance. Below we suppose that the same reinsurance treaty is applied each period. The next step is to choose the optimal treaty or let it vary from one period to another.

2.1 Proportional Reinsurance

Suppose that we use a proportional reinsurance. More precisely, we apply a quota share treaty with coefficient $\beta > 0$. That means, the direct insurer retains $Z = \beta X$ if the initial demand for indemnification is X. Accordingly, the insurance company retains only part of premiums received from its clients. We denote this amount by M.

One-period Case. Let us begin by treating a one-period case. It is possible to get a bank loan at the beginning of the period, the interest rate being c, or after the claim arrival, with the interest rate $r > c$. Let $f_1(x)$ be the minimal expected costs associated with a loan.

Lemma 1. *The following statement is valid.*

$$f_1(x) = -cx + \min_{y \geq x} \underbrace{\left(cy + \frac{r}{\beta} \int\limits_{y}^{\infty} (s - y)\varphi\left(\frac{s}{\beta}\right) ds \right)}_{G_1(y)}. \tag{1}$$

Moreover, there exists a constant $\overline{y_1} = \beta F^{-1}(1 - \frac{c}{r})$ *such that*

$$f_1(x) = \begin{cases} -cx + G_1(\overline{y_1}) & \text{if } x \leq \overline{y_1}, \\ -cx + G_1(x) & \text{otherwise.} \end{cases} \tag{2}$$

Proof. Obviously, the interests for a loan up to level y at the beginning of the period are $c(y - x)$ and the company has to pay $rE(Z - y)^+ = \frac{r}{\beta} \int\limits_{y}^{\infty} (s - y)\varphi(\frac{s}{\beta}) ds$ for additional loan. So, relation (1) holds. Using the formula

$$\left(\int\limits_{\alpha(y)}^{\beta(y)} \gamma(y, x) dx \right)'_y = \int\limits_{\alpha(y)}^{\beta(y)} \gamma'_y(y, x) dx + \gamma(y, \beta(y))\beta'(y) - \gamma(y, \alpha(y))\alpha'(y),$$

we get immediately

$$G_1'(y) = c - r \int\limits_{\frac{y}{\beta}}^{\infty} \varphi(s) ds.$$

Now the equation $G_1'(y) = 0$ can be rewritten as $r - c = rF(\frac{y}{\beta})$, whence it follows that there exists a unique solution of this equation $\overline{y_1} = \beta F^{-1}(1 - \frac{c}{r})$. It is due to the fact that $G_1''(y) = \frac{r}{\beta}\varphi(\frac{y}{\beta}) > 0$. Thus, the optimal level for bank loan is

$$y_1(x) = \begin{cases} \overline{y_1} & \text{if } x \leq \overline{y_1}, \\ x & \text{otherwise,} \end{cases}$$

providing the desired result (2).   □

Multi-period Case. Now, turn to multi-period case. Let α be the discount factor and $f_k(x)$ denote the minimal expected discounted k-periods costs incurred by loans. The following result is valid.

Theorem 1. *The relation*

$$f_k(x) = -cx + \min_{y \geq x}(\underbrace{cy + \frac{r}{\beta}\int_y^\infty (s - y)\varphi(\frac{s}{\beta})\,ds + \alpha \int_0^\infty f_{k-1}(y + M - \beta s)\varphi(s)\,ds}_{G_k(y)})$$

takes place. Moreover, there exists a constant $\overline{y_k}$ such that the optimal bank loan $y_k(x)$ at the beginning of the k-step process, $k \geq 2$, is defined as follows

$$y_k(x) = \begin{cases} \overline{y_k} & \text{if } x \leq \overline{y_k}, \\ x & \text{otherwise.} \end{cases}$$

Hence,

$$f_k(x) = \begin{cases} -cx + G_k(\overline{y_k}) & \text{if } x \leq \overline{y_k}, \\ -cx + G_k(x) & \text{otherwise.} \end{cases} \tag{3}$$

The sequence $\{\overline{y_k}\}$ is non-decreasing in k.

Proof. We use the dynamic programming (see, e.g., [1]) and carry out the proof by induction. For $k = 2$ we easily obtain the relation

$$f_2(x) = -cx + \min_{y \geq x}(\underbrace{cy + \frac{r}{\beta}\int_y^\infty (s - y)\varphi(\frac{s}{\beta})\,ds + \alpha \int_0^\infty f_1(y + M - \beta s)\varphi(s)\,ds}_{G_2(y)})$$

called the Bellman equation. It can be rewritten as

$$f_2(x) = -cx + \min_{y \geq x}(G_1(y) + \alpha \int_0^\infty f_1(y + M - \beta s)\varphi(s)\,ds).$$

Hence, due to (2),

$$G_2'(y) = G_1'(y) + \alpha \int_0^{\frac{y+M-\overline{y_1}}{\beta}} G_1'(y + M - \beta s)\varphi(s)\, ds - \alpha c$$

and

$$G_2''(y) = G_1''(y) + \alpha \int_0^{\frac{y+M-\overline{y_1}}{\beta}} G_1''(y + M - \beta s)\varphi(s)\, ds > 0.$$

Since the function $G_2'(y)$ is increasing, we establish inequality $\overline{y_1} < \overline{y_2}$ verifying that $G_2'(\overline{y_1}) < 0$. In fact,

$$G_2'(\overline{y_1}) = \alpha \Big[\int_0^{\frac{M}{\beta}} (c - r \int_{\frac{\overline{y_1}+M-\beta s}{\beta}}^{\infty} \varphi(t)\, dt)\varphi(s)\, ds - c \Big].$$

The right-hand expression can be written in the form

$$\alpha[c(F(\frac{M}{\beta}) - 1) - r \int_0^{\frac{M}{\beta}} (1 - F(\frac{\overline{y_1} + M - \beta s}{\beta}))\varphi(s)\, ds],$$

where both terms are negative.

It is also clear that

$$\lim_{y \to \infty} G_2'(y) = c - \alpha c + \lim_{y \to \infty} (-r \int_{\frac{y}{\beta}}^{\infty} \varphi(s)\, ds)$$

$$+ \alpha \lim_{y \to \infty} \int_0^{\frac{y+M-\overline{y_1}}{\beta}} (c - r \int_{\frac{y+M-\beta s}{\beta}}^{\infty} \varphi(t)\, dt)\varphi(s)\, ds > 0$$

and equation $G_2'(y) = 0$ has a solution $\overline{y_2} < \infty$.

Thus,

$$f_2(x) = \begin{cases} -cx + G_2(\overline{y_2}) & \text{if } x \le \overline{y_2}, \\ -cx + G_2(x) & \text{otherwise,} \end{cases}$$

and the first step of induction is completed. In the same way, for $k > 2$,

$$f_k(x) = -cx + \underbrace{\min_{y \ge x}(cy + \frac{r}{\beta} \int_y^{\infty} (s - y)\varphi(\frac{s}{\beta})\, ds + \int_0^{\infty} f_{k-1}(y + M - \beta s)\varphi(s)\, ds)}_{G_k(y)}.$$

Assuming that for $k \leq n$ there exist $\overline{y_k}$ satisfying the relation $G'_k(y) = 0$ and

$$f_k(x) = \begin{cases} -cx + G_k(\overline{y_k}) & \text{if } x \leq \overline{y_k}, \\ -cx + G_k(x) & \text{otherwise,} \end{cases}$$

we have to prove that the same is true for $k = n + 1$ and $\overline{y_{n+1}} > \overline{y_n}$.
Clearly, for $k \leq n$,

$$f'_k(x) = \begin{cases} -c & \text{if } x \leq \overline{y_k}, \\ -c + G'_k(x) & \text{otherwise.} \end{cases}$$

Recall that $G'_n(y)$ has the form

$$G'_n(y) = c - r \int\limits_{\frac{y}{\beta}}^{\infty} \varphi(s)\,ds + \alpha \int\limits_0^{\infty} f'_{n-1}(y + M - \beta s)\varphi(s)\,ds \tag{4}$$

and

$$G''_n(y) = \frac{r}{\beta}\varphi\left(\frac{y}{\beta}\right) + \alpha \int\limits_0^{\infty} f''_{n-1}(y + M - \beta s)\varphi(s)\,ds$$

is nonnegative.
Obviously,

$$G'_{n+1}(y) = G'_n(y) + \alpha \int\limits_0^{\infty} (f'_n(y + M - \beta s) - f'_{n-1}(y + M - \beta s))\varphi(s)\,ds,$$

therefore $G'_{n+1}(\overline{y_n})$ is equal to

$$\alpha \int\limits_0^{\infty} (f'_n(\overline{y_n} + M - \beta s) - f'_{n-1}(\overline{y_n} + M - \beta s))\varphi(s)\,ds.$$

Now, we use the expression

$$f'_n(x) - f'_{n-1}(x) = \begin{cases} 0, & \text{if } x \leq \overline{y_{n-1}}, \\ -G'_{n-1}(x), & \text{if } x \in (\overline{y_{n-1}}, \overline{y_n}], \\ G'_n(x) - G'_{n-1}(x) & \text{otherwise.} \end{cases} \tag{5}$$

That means,

$$G'_{n+1}(\overline{y_n}) = \alpha \int\limits_0^{\frac{M}{\beta}} (G'_n(\overline{y_n} + M - \beta s) - G'_{n-1}(\overline{y_n} + M - \beta s))\varphi(s)\,ds$$

$$-\alpha \int_{\frac{M}{\beta}}^{\frac{\overline{y_n}-\overline{y_{n-1}}+M}{\beta}} G'_{n-1}(\overline{y_n}+M-\beta s)\varphi(s)\,ds. \tag{6}$$

Since $G'_{n-1}(y) > 0$ for $y > \overline{y_{n-1}}$, the second integral is positive (however it is taken with sign minus). The first integral can be transformed as follows

$$\int_0^{\frac{M}{\beta}} (G'_n(\overline{y_n}+M-\beta s) - G'_{n-1}(\overline{y_n}+M-\beta s))\varphi(s)\,ds$$

$$= \alpha \int_0^{\frac{M}{\beta}} (\int_0^{\infty} (f'_{n-1}(\overline{y_n}+2M-\beta(t+s)) - f'_{n-2}(\overline{y_n}+2M-\beta(t+s)))\varphi(t)\,dt)\varphi(s)\,ds.$$

Using the induction assumption the integral under consideration can be reduced, step by step, to the multiple integral with integrand depending on $G'_2(\cdot) - G'_1(\cdot)$ multiplied by α^{n-1} and the sum of negative terms similar to the second integral in (6) multiplied by α^k, $k = 2,\ldots,n-1$. We omit the explicit expression due to its bulkiness only mentioning that it is negative. Hence, it follows immediately that $f_{n+1}(x)$ has also the desired form. $\qquad \Box$

Theorem 1 enables us to treat infinite planning horizon and establish the optimal strategy of bank loans determined by one critical level $\overline{y}$ as shows

Corollary 1. *For $\alpha < 1$ there exists $\overline{y} = \lim\limits_{n\to\infty} \overline{y_n}$.*

Proof. It follows from (3) and (4) that

$$G'_n(y) = V(y) + \alpha \int_0^{\frac{y-\overline{y_{n-1}}+M}{\beta}} G'_{n-1}(y+M-\beta s)\varphi(s)\,ds$$

where $V(y) = G'_1(y) - \alpha c$.

Thus, $G'_n(y) = V(y)$ for $y + M \le \overline{y_{n-1}}$ and $G'_n(y) \ge V(y)$ otherwise, since $G'_{n-1}(y + M - \beta s) \ge 0$ if $0 \le s \le (y - \overline{y_{n-1}} + M)\beta^{-1}$.

Let $V(\overline{z}) = 0$, in other words, $\overline{F}(\overline{z}) = \frac{c}{r}(1-\alpha)$. Obviously, $G'_n(\overline{z}) \ge V(\overline{z}) = 0$, that is, $\overline{y_n} \le \overline{z} < \infty$. Hence, there exists $\overline{y} = \lim\limits_{n\to\infty} \overline{y_n}$. $\qquad \Box$

Theorem 2. *For $\alpha < 1$, functions $f_n(x)$ converge uniformly, as $n \to \infty$, to the solution of the functional equation*

$$f(x) = -cx + \min_{y \ge x}[G_1(y) + \alpha \int_0^{\infty} f(x+M-\beta s)\varphi(s)\,ds].$$

Proof. Let $y_n(x)$ be the optimal decision at the first step of the n-period process. Then

$$f_n(x) = -cx + \min_{y \geq x} G_n(y) = -cx + G_n(y_n(x)) \leq -cx + G_n(y_{n+1}(x)).$$

Therefore

$$G_{n+1}(y_{n+1}(x)) - G_n(y_{n+1}(x)) \leq f_{n+1}(x) - f_n(x) \leq G_{n+1}(y_n(x)) - G_n(y_n(x)).$$

That means,

$$|f_{n+1}(x) - f_n(x)| \leq \max(|G_{n+1}(y_n(x)) - G_n(y_n(x))|, |G_{n+1}(y_{n+1}(x)) - G_n(y_{n+1}(x))|).$$

The right-hand side of this inequality is bounded by $\max_y |G_{n+1}(y) - G_n(y)|$.

Put $u_n = \max_x |f_{n+1}(x) - f_n(x)|$. Since

$$G_{n+1}(y) - G_n(y) = \alpha \int_0^\infty (f_n(y + M - \beta s) - f_{n-1}(y + M - \beta s))\varphi(s)\, ds,$$

one gets immediately

$$u_n \leq \alpha u_{n-1} \leq \alpha^2 u_{n-2} \leq \dots \leq \alpha^{n-1} u_1.$$

Thus, if we show that $u_1 < \infty$, then the uniform convergence of $f_n(x)$ to a limit $f(x)$ is obvious. It is due to the fact that, for any x, the function $|f_n(x) - f_1(x)| \leq |f_2(x) - f_1(x)| + |f_3(x) - f_2(x)| + \dots + |f_n(x) - f_{n-1}(x)|$ is majorized by the partial sum of a geometric progression $\{u_1\alpha^k\}_{k \geq 1}$.

Recall that using (5) we have

$$u_1 \leq \max[G_2(\overline{y_2}) + G_1(\overline{y_2}), \alpha \cdot \max_{x \geq \overline{y_1}} \int_0^\infty (f_1(x + M - \beta s)\varphi(s)\, ds].$$

Clearly, $G_2(\overline{y_2}) + G_1(\overline{y_2}) < \infty$. Accordingly to (2) it is possible to write

$$\int_0^\infty f_1(x + M - \beta s)\varphi(s)\, ds = \beta r \int_0^{\frac{x+M-\overline{y_1}}{\beta}} \varphi(s) \int_{\frac{x+M-\beta s}{\beta}}^\infty s_1\varphi(s_1)\, ds_1\, ds$$

$$-r \int_0^{\frac{x+M-\overline{y_1}}{\beta}} (x + M - \beta s)\overline{F}(\frac{x + M - \beta s}{\beta})\varphi(s)\, ds + G_1(\overline{y_1})\overline{F}(\frac{x + M - \overline{y_1}}{\beta})$$

$$-c(x + M)\overline{F}(\frac{x + M - \overline{y_1}}{\beta}) + c\beta \int_{\frac{x+M-\overline{y_1}}{\beta}}^\infty s\varphi(s)\, ds.$$

Since we treat only the domain $\{x \geq \overline{y_1}\}$, all the summands of the above equality are bounded. This is due to existence of $EX_1 = \mu < \infty$.

Thus, uniform convergence of $f_n(x)$ to $f(x)$ is established. Obviously, $f(x)$ satisfies the functional equation stated in the theorem.   □

2.2   Non-proportional Reinsurance

Now we turn to the case of non-proportional reinsurance, namely, suppose that a stop-loss treaty with retention a is applied each period. That means, instead of the claim X_k, $k \geq 1$, insurer has to pay $Z_k = \min(X_k, a)$ during the k-th period. Since the claims are supposed to be i.i.d. r.v.'s, the premiums obtained by the insurance company are equal to

$$M = (1 + \gamma_1)EX - (1 + \gamma_2)E(X - a)^+.$$

Here X has the same distribution function F as all X_k, $k \geq 1$, whereas γ_1 and γ_2 are the safety loadings of insurer and reinsurer, respectively. We assume that F has a density $\varphi(x) > 0$, for $x > 0$, and a finite mean value.

We would like to establish the optimal policy of bank loans minimizing the associated expected costs.

One-period Case. As previously, we begin by consideration of one-step process establishing the form of $f_1(x)$. If insurer decides to take a loan to raise the surplus up to level y, the expected costs are equal to $c(y - x) + rE(Z_1 - y)^+$. It is clear that

$$f_1(x) = -cx + \min_{y \geq x} G_1(y)$$

where

$$G_1(y) = cy + D(y) \quad \text{with } D(y) = r \int_0^\infty [\min(s, a) - y]^+ \, dF(s).$$

Now we can prove the following result.

Lemma 2. *Optimal loan level is given by $\overline{y_1} = \min(a, F^{-1}(1 - \frac{c}{r}))$.*

Proof. It is clear that under the stop-loss treaty with retention a the insurer never has to pay more than a. Thus, it is unreasonable to take a loan up to level greater than a.

Rewriting $G_1(y)$, for $y \leq a$, in the form

$$G_1(y) = cy + r \int_y^a (s - y)\varphi(s) \, ds + r(a - y)P(X > a)]$$

we get

$$G_1'(y) = c - r\overline{F}(y) \quad \text{and } G_1''(y) = r\varphi(y) > 0. \tag{7}$$

Therefore $G_1'(y) < 0$ for $y < F^{-1}(1 - \frac{c}{r})$. Hence, the desired result is obvious. $\square$

Multi-period Case. For the multi-step case we have the following Bellman equation

$$f_k(x) = -cx + \min_{y \ge x} G_k(y)$$

with $G_k(y)$ given by

$$G_1(y) + \alpha \int_0^\infty f_{k-1}(y + M - \min(s,a))\, dF(s).$$

The following result is valid.

Theorem 3. *The optimal levels $\overline{y_k}$, $k \ge 1$, for the bank loans strategy at the beginning of k-step process form an increasing sequence.*

Proof. We begin by treating the case $\overline{F}(a) \le \frac{c}{r}$, in other words, $\overline{y_1} \le a$. Thus,

$$f_1(x) = -cx + \begin{cases} G_1(\overline{y_1}) & \text{if } x \le \overline{y_1}, \\ G_1(x) & \text{otherwise.} \end{cases}$$

Therefore,

$$f_1'(x) = -c + \begin{cases} 0 & \text{if } x \le \overline{y_1}, \\ G_1'(x) & \text{otherwise} \end{cases}$$

and, according to (7), $f_1'(x) \le 0$ for all x. It follows immediately, that

$$G_2'(y) = G_1'(y) + \alpha \int_0^\infty f_1'(y + M - \min(s,a))\, dF(s) \le G_1'(y).$$

Clearly,

$$G_2'(\overline{y_1}) \le G_1'(\overline{y_1}) = 0.$$

Hence, $\overline{y_2} \ge \overline{y_1}$ and initial step of induction is proved.

Now suppose that for $k \le n$ we have established that there exist the critical levels $\overline{y_k}$ given by $G_k'(\overline{y_k}) = 0$. They form a non-decreasing sequence and

$$f_k'(x) = -c + \begin{cases} 0 & \text{if } x \le \overline{y_k}, \\ G_k'(x) & \text{otherwise.} \end{cases}$$

Moreover, we assume that $G_k'(x) - G_{k-1}'(x) \le 0$ for all x.

Obviously,

$$f_k'(x) - f_{k-1}'(x) = \begin{cases} 0 & \text{if } x \le \overline{y_{k-1}}, \\ -G_{k-1}'(x) & \text{if } x \in (\overline{y_{k-1}}, \overline{y_k}), \\ G_k'(x) - G_{k-1}'(x) & \text{if } x \ge \overline{y_k}. \end{cases} \tag{8}$$

Consider

$$G'_{n+1}(x) = G'_n(x) + \alpha \int_0^\infty (f'_n(y + M - \min(s, a)) - f'_{n-1}(y + M - \min(s, a)))\, dF(s).$$

Due to (8) and other induction assumptions, $f'_n(x) - f'_{n-1}(x) \le 0$ for all x. That means $G'_{n+1}(x) \le G'_n(x)$ for all x. Taking $x = \overline{y}_n$ we get the desired result $\overline{y}_n \le \overline{y}_{n+1}$. The case $\overline{F}(a) < \frac{c}{r}$ can be treated similarly.   □

It is possible to consider the infinite planning horizon proving

Corollary 2. *A functional equation*

$$f(x) = -cx + \min_{y \ge x}[cy + \int_y^a r(s - y)\varphi(s)\, ds + r(a - y)P(X > a)$$

$$+ \alpha \int_0^\infty f(y + M - \min(s, a))\, dF(s)]$$

has a unique solution for $\alpha < 1$.

The proof is omitted because the methods employed are similar to those used in Theorem 2.

3   Conclusion

We considered two discrete-time models with bank loans and proportional (or non-proportional) reinsurance for a finite planning horizon. It is established that the optimal loans policy is determined by a sequence of critical levels $\overline{y}_n$, $n \ge 1$. That means, if the surplus at the beginning of the n-step process $x < \overline{y}_n$, $n \ge 1$, then optimal decision is to raise it up to level $\overline{y}_n$, otherwise the loan is not necessary. The sequence of critical levels $\{\overline{y}_n\}$ is bounded non-decreasing, so it converges, as $n \to \infty$, to the limit $\overline{y}$ if $\alpha < 1$. Moreover, it is proved that the minimal costs $f_n(x)$ converge uniformly in x to a function $f(x)$ which is the unique solution of a functional equation.

For $\alpha = 1$ we have to choose another risk measure introducing the notion of asymptotically optimal policy. It is possible as well to calculate the company ruin probability for any α under the optimal loans strategy.

The next step is investigation of models stability with respect to small fluctuations of system parameters and perturbations of the underlying distributions. The books [16,17] and [18] are useful for this purpose, as well as the results obtained in [8,15,19].

References

1. Bellman, R.: Dynamic Programming. Princeton University Press, Princeton (1957)
2. Bernstein, P.L.: Against the Gods: The Remarkable Story of Risk. Wiley, New York (1996)
3. Bulinskaya, E.: On the cost approach in insurance. Rev. Appl. Ind. Math. **10**(2), 276–286 (2003). (in Russian)
4. Bulinskaya, E.: Asymptotic analysis of insurance models with bank loans. In: Bozeman, J.R., Girardin, V., Skiadas, C.H. (eds.) New Perspectives on Stochastic Modeling and Data Analysis, pp. 255–270. ISAST, Athens (2014)
5. Bulinskaya, E.: New research directions in modern actuarial sciences. In: Panov, V. (ed.) MPSAS 2016. PROMS, vol. 208, pp. 349–408. Springer, Cham, Switzerland (2017). https://doi.org/10.1007/978-3-319-65313-6_15
6. Bulinskaya, E.: Cost approach versus reliability. In: Vishnevskiy, V. (ed.) Proceedings of International Conference DCCN-2017, 25–29 September 2017, pp. 382–389. Technosphera, Moscow (2017)
7. Bulinskaya, E.: Asymptotic analysis and optimization of some insurance models. Appl. Stoch. Models Bus. Ind. **34**(6), 762–773 (2018)
8. Bulinskaya, E., Gusak, J.: Optimal control and sensitivity analysis for two risk models. Commun. Stat. - Simul. Comput. **45**(5), 1451–1466 (2016)
9. Bulinskaya, E., Kolesnik, A.: Reliability of a discrete-time system with investment. In: Vishnevskiy, V.M., Kozyrev, D.V. (eds.) DCCN 2018. CCIS, vol. 919, pp. 365–376. Springer, Cham (2018). https://doi.org/10.1007/978-3-319-99447-5_31
10. Bulinskaya, E., Gusak, J., Muromskaya, A.: Discrete-time insurance model with capital injections and reinsurance. Methodol. Comput. Appl. Probab. **17**(4), 899–914 (2015)
11. De Finetti, B.: Su un'impostazione alternativa della teoria collettiva del rischio. Trans. XV-th Int. Congr. Actuaries **2**, 433–443 (1957)
12. Dickson, D.C.M., Waters, H.R.: Some optimal dividends problems. ASTIN Bull. **34**, 49–74 (2004)
13. Dionne, G. (ed.): Handbook of Insurance, 2nd edn. Springer, New York (2013). https://doi.org/10.1007/978-1-4614-0155-1
14. Li, S., Lu, Y., Garrido, J.: A review of discrete-time risk models. Rev. R. Acad. Cien. Serie A. Mat. **103**, 321–337 (2009)
15. Oakley, J.E., O'Hagan, A.: Probabilistic sensitivity analysis of complex models: a Bayesian approach. J. Roy. Stat. Soc. B. **66**(Part 3), 751–769 (2004)
16. Rachev, S.T., Klebanov, L., Stoyanov, S.V., Fabozzi, F.: The Methods of Distances in the Theory of Probability and Statistics. Springer, New York (2013). https://doi.org/10.1007/978-1-4614-4869-3
17. Rachev, S.T., Stoyanov, S.V., Fabozzi, F.J.: Advanced Stochastic Models, Risk Assessment, Portfolio Optimization. Wiley, Hoboken (2008)
18. Saltelli, A., et al.: Global Sensitiivity Analysis. The Primer. Wiley, Chichester (2008)
19. Sobol', I.M., et al.: Estimating the approximation error when fixing unessential factors in global sensitivity analysis. Reliab. Eng. Syst. Saf. **92**, 957–960 (2007)
20. Williams, A., Heins, M.H.: Risk Management and Insurance, 2nd edn. McGraw - Hill, New York (1995)

Hidden Markov Model of Information System with Component-Wise Storage Devices

Yuriy E. Obzherin$^{(\boxtimes)}$ ⓘ, Stanislav M. Sidorov ⓘ, and Mikhail M. Nikitin ⓘ

Sevastopol State University, Universitetskaya Street 33, 299053 Sevastopol, Russia
objsev@mail.ru, xaevec@mail.ru, m.nikitin.1979@gmail.com

Abstract. In the paper, using the stationary phase merging algorithm, a merged semi-Markov model was constructed, describing the operation of a two-component information system with component-wise storage devices. On the basis of the merged semi-Markov model, a hidden Markov model is built, in which the hidden states are the states of the embedded Markov chain of the merged model. The main tasks of the hidden Markov models theory are considered, which allow to evaluate the characteristics of the embedded Markov chain of the merged model and predict its states basing on given vector of signals.

Keywords: Information system · Component-wise storage devices · Semi-Markov model · Phase merging algorithm · Hidden Markov model · Vector of signals · Characteristics evaluation · State prediction

1 Introduction

Semi-Markov processes [1–14], hidden Markov and semi-Markov models [15–24] are widely used for constructing models of systems for various purposes. When constructing a semi-Markov model, it is necessary to introduce the phase space of the states of the system. In a number of cases, it is sufficient to use a finite or countable set of states that reflect the physical states of the system [4,13,14]. In other cases, when constructing a semi-Markov model of a system, in the phase states of the system need to introduce additional continuous components that are necessary for the correct construction of the model [1,8,10,11]. These components can be the time elapsed since the beginning of the operation of the system element; time remaining until the restore of the system element; time remaining until the system is monitored; the amount of the remaining time reserve, etc. Note that these additional continuous components carry important information about the functioning of the system. In this case, it is necessary to use the discrete-continuous state phase space [1,8,10,11], and to construct

The research was carried out within the state assignment of the Minobrnauki of Russia (No. 1.10513.2018/11.12), with financial support by RFBR (project No. 18-01-00392a).

V. M. Vishnevskiy et al. (Eds.): DCCN 2019, LNCS 11965, pp. 354–364, 2019.
https://doi.org/10.1007/978-3-030-36614-8_27

a semi-Markov model and system analysis it is necessary to use the theory of semi-Markov processes with a common phase space of states [1–3,9,10].

An important component of the semi-Markov process is an embedded Markov chain (EMC), which is responsible for transitions between states of the system. The phase space of the states of the EMC coincides with the phase space of states of the semi-Markov process. When using systems for which a semi-Markov model is constructed, it is not always possible to obtain the information contained in the state encoding when changing its states, but only the possibility to obtain a certain signal (information) associated with the states of the EMC (semi-Markov process). For example, in the phase state of a semi-Markov process, for each element of the system, it is indicated whether it is in working order or on recovery, and when using the system, it is possible to receive a signal only about the number of operable elements. For queuing systems, for example, data can only be obtained on the number of free instruments, and not on the state of each device, contained in phase states. When using systems, it is difficult or impossible to obtain the values of additional continuous components, which, as noted, carry useful information about the functioning of the system. In these cases, the states of the EMC can be considered hidden (unobservable). Therefore, the problem arises of finding estimates of the characteristics of the EMC and the semi-Markov process on the basis of the observed signal vector.

This can be done by hidden Markov models (HMM) [17,19–24] and hidden semi-Markov models (HsMM) [15,16,18], which are widely used to solve a wide range of scientific and engineering problems, including speech recognition, decoding and modeling of channels in digital communication , modeling information systems, communication networks and many others.

In [8], a semi-Markov model of a two-component system with a component-wise time reserve was constructed, the stationary distribution of EMC and the stationary characteristics of the system reliability were found. In this paper, the merged semi-Markov model of this system is built, and then the hidden Markov model constructed on the basis of the merged model is considered. For the resulting hidden Markov model on the basis of given vector of signals, the problems of estimating the characteristics of the EMC of merged model and predicting its states are solved.

2   Construction of a Merged Semi-Markov Model

Let us consider the system [8], consisting of two components, times to failure of which are random variables (RVs) α_i with the distribution functions (DFs) $F_i(t)$, a restoration times are RVs β_i with DFs $G_i(t)$. Each component of the system has a random instantly replenished time reserve τ_i with DF $R_i(t)$. RVs $\alpha_i, \beta_i, \tau_i$ are assumed to be independent in aggregate, having finite mathematical expectations; DFs $F_i(t)$, $G_i(t)$, $R_i(t)$ have distribution densities $f_i(t)$, $g_i(t)$, $r_i(t)$.

The time reserve starts to be used at the time the component begins to recover. The failure of system occurs when both elements are restored and the time reserve for each element is completely spent. It continues until the restoration of one of the failed components, while the time reserve of the restored element is instantly replenished to the level τ_i.This system is a model of a two-component information system with component-wise information storage devices.

In [8], to build a semi-Markov model of the system S, a semi-Markov process with a discrete-continuous phase space of states of the form $\xi(t)$ is used:

$$E = \left\{ i\bar{d}x : \bar{d} = (d_1, d_2), x > 0 \right\}, \tag{1}$$

where $i = 1, 2$ is the number of the component in which the state change occurred. Component d_k of the vector $\bar{d}$ describes the physical state of the element with the number k:

$$
d_k = \begin{cases}
0, & \text{if } k-th \text{ component is in failure,} \\[2mm]
1, & \text{if } k-th \text{ component is operational,} \\[2mm]
\bar{1}, & \text{if } k-th \text{ component is restored and} \\[2mm]
& \text{operates due to time reserve,}
\end{cases}
$$

the continuous component x indicates the elapsed time since the last change in the system state.

In [8] the stationary distribution of EMC and the stationary reliability characteristics of the system under consideration were found.

In order to simplify the system model, we have merged the semi-Markov model constructed in [8], using the stationary phase merging algorithm proposed in [1,9].

The phase space of states E of the original model is split into $N = 9$ classes:

$$E_{00} = \{100x, 200x\}, \quad E_{11} = \{1, 111x, 211x\}, \quad E_{1\bar{1}} = \{1\bar{1}1x, 2\bar{1}1x\},$$

$$E_{\bar{1}1} = \{1\bar{1}1x, 2\bar{1}1x\}, E_{1\bar{1}} = \{11\bar{1}x, 21\bar{1}x\}, E_{10} = \{110x, 210x\},$$

$$E_{01} = \{101x, 201x\}, E_{0\bar{1}} = \{10\bar{1}x, 20\bar{1}x\}, E_{\bar{1}0} = \{1\bar{1}0x, 2\bar{1}0x\},$$

each of which is "glued" into one state of the merged model.

The phase space of states of the merged model is:

$$\widehat{E} = \{00, 11, 1\bar{1}, \bar{1}1, \bar{1}\bar{1}, 10, 01, \bar{1}0, 0\bar{1}\}.$$

The physical meaning of the introduced state classes is:

$E_{00}-$ both components are in failure, the time reserve of each component is completely consumed;

E_{11} – both components are operational;

$E_{\bar{1}\bar{1}}$ – both components are on recovery and operate at the expense of the time reserve;

$E_{\bar{1}1}$ – the first component is on recovery and operates at the expense of the time reserve, the second is operational;

$E_{1\bar{1}}$ – the second component is on recovery and operates at the expense of the time reserve, the first is operational;

E_{10} – the first component is operational, the second is in failure;

E_{01} – the second component is operational, the first is in failure;

$E_{0\bar{1}}$ – the first component is in failure, the second is on recovery and operates at the expense of the time reserve;

$E_{\bar{1}0}$ – the second component is in failure, the first is on recovery and operates at the expense of the time reserve.

We define the transition probabilities $\widehat{p}_{kr}$ of the EMC and the average sojourn times $\widehat{m}_k$ in the states of the merged model, which according to [1,9] are found by the formulas:

$$\hat{p}_{kr} = \frac{\int\limits_{E_k} \rho(de)P(e, E_r)}{\rho(E_k)}, k, r = \overline{1, N}, \qquad (2)$$

$$\hat{m}_k = E\theta_k = \frac{\int\limits_{E_k} \rho(de)m(e)}{\rho(E_k)}, k = \overline{1, N}, \qquad (3)$$

where $\rho(de)$ – stationary distribution of original EMC, $P(e, E_r)$ – probabilities of transition of EMC of the merged model, $m(e)$ – average times of being in the states of the model being merged.

Using the formula (2) let us find the EMC transition probabilities of the merged model, which will be used in constructing of the hidden Markov model:

$$P_{00}^{01} = \frac{E\left([\beta_1 - \tau_1]^+\right)}{E\left([\beta_1 - \tau_1]^+\right) + E\left([\beta_2 - \tau_2]^+\right)}, P_{00}^{10} = \frac{E\left([\beta_2 - \tau_2]^+\right)}{E\left([\beta_1 - \tau_1]^+\right) + E\left([\beta_2 - \tau_2]^+\right)},$$

$$P_{11}^{\bar{1}1} = \frac{E\alpha_2}{E\alpha_1 + E\alpha_2}, \quad P_{11}^{1\bar{1}} = \frac{E\alpha_1}{E\alpha_1 + E\alpha_2}, \quad P_{\bar{1}\bar{1}}^{\bar{1}0} = \frac{p_2 E(\beta_1 \wedge \tau_1)}{E(\beta_1 \wedge \tau_1) + E(\beta_2 \wedge \tau_2)},$$

$$P_{\bar{1}\bar{1}}^{0\bar{1}} = \frac{p_1 E(\beta_2 \wedge \tau_2)}{E(\beta_1 \wedge \tau_1) + E(\beta_2 \wedge \tau_2)}, \quad P_{\bar{1}1}^{11} = \frac{(1 - p_2)E\alpha_1}{E\alpha_1 + E(\beta_2 \wedge \tau_2)},$$

$$P_{\bar{1}\bar{1}}^{\bar{1}1} = \frac{(1 - p_2)E(\beta_1 \wedge \tau_1)}{E(\beta_1 \wedge \tau_1) + E(\beta_2 \wedge \tau_2)}, \quad P_{\bar{1}\bar{1}}^{1\bar{1}} = \frac{(1 - p_1)E(\beta_2 \wedge \tau_2)}{E(\beta_1 \wedge \tau_1) + E(\beta_2 \wedge \tau_2)},$$

$$P_{1\bar{1}}^{11} = \frac{(1 - p_1)E\alpha_2}{E\alpha_2 + E(\beta_1 \wedge \tau_1)}, \quad P_{1\bar{1}}^{\bar{1}\bar{1}} = \frac{E(\beta_1 \wedge \tau_1)}{E\alpha_2 + E(\beta_1 \wedge \tau_1)}, \quad P_{1\bar{1}}^{01} = \frac{p_1 E\alpha_2}{E\alpha_2 + E(\beta_1 \wedge \tau_1)},$$

$$P_{\bar{1}1}^{1\bar{1}} = \frac{E(\beta_2 \wedge \tau_2)}{E\alpha_1 + E(\beta_2 \wedge \tau_2)}, \quad P_{\bar{1}1}^{10} = \frac{p_2 E\alpha_1}{E\alpha_1 + E(\beta_2 \wedge \tau_2)},$$

$$P_{10}^{11} = \frac{E\alpha_1}{E\alpha_1 + E\left([\beta_2 - \tau_2]^+\right)}, \qquad P_{10}^{\bar{1}0} = \frac{E\left([\beta_2 - \tau_2]^+\right)}{E\alpha_1 + E\left([\beta_2 - \tau_2]^+\right)},$$

$$P_{01}^{11} = \frac{E\alpha_2}{E\alpha_2 + E\left([\beta_1 - \tau_1]^+\right)}, \qquad P_{01}^{0\bar{1}} = \frac{E\left([\beta_1 - \tau_1]^+\right)}{E\alpha_2 + E\left([\beta_1 - \tau_1]^+\right)},$$

$$P_{01}^{00} = \frac{p_2 E\left([\beta_1 - \tau_1]^+\right)}{E(\beta_2 \wedge \tau_2) + E\left([\beta_1 - \tau_1]^+\right)}, \qquad P_{01}^{1\bar{1}} = \frac{E(\beta_2 \wedge \tau_2)}{E(\beta_2 \wedge \tau_2) + E\left([\beta_1 - \tau_1]^+\right)},$$

$$P_{01}^{01} = \frac{(1 - p_2)E\left([\beta_1 - \tau_1]^+\right)}{E(\beta_2 \wedge \tau_2) + E\left([\beta_1 - \tau_1]^+\right)}, \qquad P_{10}^{00} = \frac{p_1 E\left([\beta_2 - \tau_2]^+\right)}{E(\beta_1 \wedge \tau_1) + E\left([\beta_2 - \tau_2]^+\right)},$$

$$P_{10}^{\bar{1}1} = \frac{E(\beta_1 \wedge \tau_1)}{E(\beta_1 \wedge \tau_1) + E\left([\beta_2 - \tau_2]^+\right)}, \qquad P_{10}^{10} = \frac{(1 - p_1)E\left([\beta_2 - \tau_2]^+\right)}{E(\beta_1 \wedge \tau_1) + E\left([\beta_2 - \tau_2]^+\right)},$$

$$\tag{4}$$

where $[\beta_i - \tau_i]^+$ is a RV with CDF

$$P\{[\beta_i - \tau_i]^+ > t\} = \frac{\int\limits_0^\infty r_i(z)\bar{G}_i(t + z)dz}{P(\beta_i > \tau_i)},$$

$\wedge$ is the minimum sign,

$$\bar{G}_i(t) = 1 - G_i(t), p_i = P(\beta_i > \tau_i) = \int\limits_0^\infty \bar{G}_i(t)r_i(t)dt,$$

$$E(\beta_i \wedge \tau_i) = \int\limits_0^\infty \bar{G}_i(t)\bar{R}_i(t)dt, \qquad E([\beta_i - \tau_i]^+) = \frac{\int\limits_0^\infty \bar{G}_i(t)R_i(t)dt}{P(\beta_i > \tau_i)}.$$

Using the formula (3) let us find the average sojourn times in the states of the merged model:

$$m_{00} = \frac{E\left([\beta_1 - \tau_1]^+\right)E\left([\beta_2 - \tau_2]^+\right)}{E\left([\beta_1 - \tau_1]^+\right) + E\left([\beta_2 - \tau_2]^+\right)}, \qquad m_{\bar{1}\bar{1}} = \frac{E(\beta_1 \wedge \tau_1)E(\beta_2 \wedge \tau_2)}{E(\beta_1 \wedge \tau_1) + E(\beta_2 \wedge \tau_2)},$$

$$m_{\bar{1}1} = \frac{E\alpha_2 E(\beta_1 \wedge \tau_1)}{E\alpha_2 + E(\beta_1 \wedge \tau_1)}, \qquad m_{11} = \frac{E\alpha_1 E\alpha_2}{E\alpha_1 + E\alpha_2}, \qquad m_{1\bar{1}} = \frac{E\alpha_1 E(\beta_2 \wedge \tau_2)}{E\alpha_1 + E(\beta_2 \wedge \tau_2)},$$

$$m_{10} = \frac{E\alpha_1 E\left([\beta_2 - \tau_2]^+\right)}{E\alpha_1 + E\left([\beta_2 - \tau_2]^+\right)}, \qquad m_{01} = \frac{E\alpha_2 E\left([\beta_1 - \tau_1]^+\right)}{E\alpha_2 + E\left([\beta_1 - \tau_1]^+\right)},$$

$$m_{0\bar{1}} = \frac{E(\beta_2 \wedge \tau_2)E\left([\beta_1 - \tau_1]^+\right)}{E(\beta_2 \wedge \tau_2) + E\left([\beta_1 - \tau_1]^+\right)}, \qquad m_{\bar{1}0} = \frac{E(\beta_1 \wedge \tau_1)E\left([\beta_2 - \tau_2]^+\right)}{E(\beta_1 \wedge \tau_1) + E\left([\beta_2 - \tau_2]^+\right)}. \tag{5}$$

Note that formulas (4)–(5) are invariant with respect to the laws of distribution of the RVs α_1, α_2.

Below the merged semi-Markov model is used to construct the hidden Markov model of the system under consideration.

3 Hidden Markov Model of a Merged Semi-Markov Model

Let $\{X_n, n = 1, 2, ...\}$ be the merged model EMC with transitions probabilities defined by the formulas (4).

Let us suppose that when the system S is operating, the merged model EMC states are not observable (hidden states), and only the number of operable elements of the system can be observed at the moment of the system transition to a new state.

Hence, the set of signals has the form:

$$J = \{0, 1, 2\}.$$

Let us consider the interconnection between the EMC states and the signals, i.e. define the emission function $R(s|x)$ [20].

$$R(s|x) = P(S_n = s | X_n = x), x \in \hat{E}, s \in J, \sum_{s \in J} R(s|x) = 1, \qquad (6)$$

where S_n is the n-th signal.

Emission function $R(s|x)$ connecting the EMC states with probabilities of the signals obtained is presented in the Table 1.

Table 1. Emission function $R(s|x)$ of communication of the EMC states with signals

Signals, s	States, x								
	00	11	$\bar{1}\bar{1}$	$\bar{1}1$	$1\bar{1}$	10	01	$0\bar{1}$	$\bar{1}0$
$s = 0$	1	0	0	0	0	0	0	0	0
$s = 1$	0	0	0	0	0	1	1	1	1
$s = 2$	0	1	1	1	1	0	0	0	0

4 Solution of the Problems of Hidden Markov Models Theory

Let us consider the main problems of the hidden Markov models theory, following [20, 21], as applied to the hidden Markov model constructed. Let $\bar{S}^n = (S_1, S_2, ..., S_n)]$ be a random vector of the first n signals. For a given vector of signals $\bar{s}_n = (s_1, s_2, ..., s_n)$ let it be $\bar{s}_k = (s_1, s_2, ..., s_k)$, $k < n$.

The problem is to estimate merged (hidden) model EMC characteristics based on given vector of signals $\bar{s}_n$. It is assumed that at the initial moment of time the system is in state 11.

Let us introduce the functions $F_k(i)$ [20]:

$$F_k(i) = P(\bar{S}^k = \bar{s}_k, X_k = i), k = 1, 2, ..., n, \qquad (7)$$

that are called forward variables [20,21]. For these functions the next recurrent formula [20] is valid

$$F_k(i) = R(s_k|i) \sum_j F_{k-1}(j) P_j^i, \ F_1(i) = R(s_1|i) p_i, \tag{8}$$

where P_j^i are transition probabilities for merged model EMC defined by the formulas (4), p_i is EMC initial state distribution.

Using the formula (8) let us find the first three functions $F_k(i)$:

$$F_1(j) = \begin{cases} 0, j \neq 11, \\ R(s_1|11), j = 11, \end{cases}$$

$$F_2(j) = \begin{cases} 0, j \neq \bar{1}1, j \neq 1\bar{1}, \\ R(s_2|1\bar{1}) R(s_1|11) P_{11}^{1\bar{1}}, j = 1\bar{1}, \\ R(s_2|\bar{1}1) R(s_1|11) P_{11}^{\bar{1}1}, j = \bar{1}1, \end{cases}$$

$$F_3(j) = \begin{cases} R(s_3|11) \left[R(s_2|1\bar{1}) R(s_1|11) P_{11}^{1\bar{1}} P_{1\bar{1}}^{11} + R(s_2|\bar{1}1) R(s_1|11) P_{11}^{\bar{1}1} P_{\bar{1}1}^{11} \right], j = 11, \\ R(s_3|10) R(s_2|1\bar{1}) R(s_1|11) P_{11}^{1\bar{1}} P_{1\bar{1}}^{10}, j = 10, \\ R(s_3|01) R(s_2|\bar{1}1) R(s_1|11) P_{11}^{\bar{1}1} P_{\bar{1}1}^{01}, j = 01, \\ R(s_3|\bar{1}\bar{1}) \left[R(s_2|1\bar{1}) R(s_1|11) P_{11}^{1\bar{1}} P_{1\bar{1}}^{\bar{1}\bar{1}} + R(s_2|\bar{1}1) R(s_1|11) P_{11}^{\bar{1}1} P_{\bar{1}1}^{\bar{1}\bar{1}} \right], j = \bar{1}\bar{1}, \\ 0, \text{for other states.} \end{cases}$$

Other functions used while finding of hidden model characteristics estimates are the functions $B_k(i)$ called backward variables [20,21]:

$$B_k(i) = \sum_j R(s_{k+1}|j) B_{k+1}(j) P_i^j, \ B_{n-1}(i) = \sum_j P_i^j R(s_n|j). \tag{9}$$

For the probability $P(\bar{S}^n = \bar{s}_n)$ the next formulas [15] are valid

$$P(\bar{S}^n = \bar{s}_n) = \sum_i F_n(i) = \sum_i R(s_1|i) B_1(i) p_i \tag{10}$$

as well as [20]

$$P(\bar{S}^n = \bar{s}_n) = \sum_i F_k(i) B_k(i) \tag{11}$$

with any fixed k.

As an example of hidden merged model characteristics estimates let us consider the system S with average times to failure of components K_1 and K_2 are equal to $E\alpha_1 = 8\,\text{h}$ and $E\alpha_2 = 6\,\text{h}$ respectively; RVs β_1 and β_2 have 5-th order Erlang distribution, $E\beta_1 = 0.71\,\text{h}$, $E\beta_2 = 0.83\,\text{h}$. The component K_i has non-random time reserve equal to h_i (i.e. $R_i(t) = 1(t - h_i)$), $h_1 = 0.5\,\text{h}$, $h_2 = 0.4\,\text{h}$.

Let there be defined the vector of signals $\{2, 2, 1, 1, 0, 1, 1\}$ ($n = 7$) . Let us consider the next problems on hidden model characteristics estimations:

1. Let us determine the probabilities of hidden model states at the moment of 7th signal emission. Let us use the formula [20]:

$$P\{X_n = j | \bar{S}^n = \bar{s}_n\} = \frac{F_n(j)}{\sum_i F_n(i)}. \tag{12}$$

Then, at step 7, the system was in a state $0\bar{1}$ with a probability of 0.3991 and in a state $\bar{1}0$ with a probability of 0.6009. For other states this probability is zero.

2. Let us determine the probabilities with which the system will make the transition to the appropriate state in the next step. To do this let us use the formula [15]:

$$P(X_{n+1} = j | \bar{s}_n) = \sum_i P(X_n = i | \bar{s}_n) P_i^j, \tag{13}$$

with the formula (12).

We are obtaining the next transition probabilities of hidden model at the 8-th step:

to the state 00 with probability 0.3924; to the state $11 - 0$; to the state $\bar{1}\bar{1} - 0$; to the state $\bar{1}1 - 0.2928$; to the state $1\bar{1} - 0.2123$; to the state $10 - 0.0846$; to the state $01 - 0.0179$; to the state $0\bar{1} - 0$; to the state $\bar{1}0 - 0$.

3. Determine the probability of occurrence of signal j in the next step for a given chain of signals.

Solution:

The probability of occurrence at step 8 of signal 2 is 0.5050; signal 1 is equal to 0.1025; signal 0 is equal to 0.3925.

4. To determine the probability of occurrence (emission) of a given chain of signals.

Solution:

Let us find the probability of the chain $\{2, 2, 1, 1, 0, 1, 1\}$, using formula (4). In this case, it is equal to 0.000896.

The probability of occurrence of a given chain of signals, with different values of the constant time reserve, is presented in Table 2.

You can also determine the same probability by the formula [20]:

$$P\{\bar{S}^n = \bar{s}_n\} = \sum_j F_k(j) B_k(j). \tag{14}$$

Using the formula (8) we get 0.000896.

Table 2. The probability of signals chain occurrence with different time reserve

h_1	0	0.1	0.1	0.2	0.3	0.4	0.5	0.7
h_2	0	0.4	0.6	0.2	0.5	0.4	0.4	0.8
S	0.0092	0.0032	0.0020	0.0039	0.0013	0.0012	0.0009	0.0001

5. Prediction of system states for given signals.

Suppose that the first n = 7 observed signals are $\{2,2,1,1,0,1,1\}$, and from them we want to predict (restore) the first n = 7 system states.

For $k \leq n$ the forecast is carried out according to the formula [20]:

$$P\{X_k = j | \bar{S}^n = \bar{s}_n\} = \frac{F_k(j)B_k(j)}{\sum_j F_k(j)B_k(j)}. \tag{15}$$

For the 7th signal, this forecast was given in problem 1.

Thus, the optimal prediction for system states will be the values of j, which maximize $F_k(j)B_k(j)$.

Then, the optimal vector of the first 7 states of the system corresponding to the given signals will be $\{11, 1\bar{1}, 10, \bar{1}0, 00, 10, \bar{1}0\}$.

The maximum values for the system state classes will be: 11 with probability 1; $1\bar{1}$ with probability 0.6139; 10 with probability 0.6139; $\bar{1}0$ with probability 0.6139; 00 with probability 1; 10 with probability 0.6009; $\bar{1}0$ with probability 0.6009.

Trellis diagram of merged model operation is depicted in the Fig. 1 using for simplicity the next notation: $1 \leftrightarrow$ state 00, $2 \leftrightarrow 11$, $3 \leftrightarrow \bar{1}\bar{1}$, $4 \leftrightarrow \bar{1}1$, $5 \leftrightarrow 1\bar{1}$, $6 \leftrightarrow 10$, $7 \leftrightarrow 01$, $8 \leftrightarrow 0\bar{1}$, $9 \leftrightarrow \bar{1}0$. The thick line shows the most likely transitions.

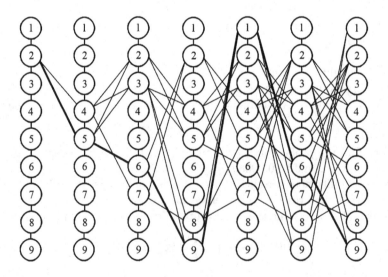

Fig. 1. The trellis of maximized transition probabilities for considered chain of signals.

To search for the most likely estimate of the hidden Markov model parameters for a given set of observations, we apply the Baum–Welch algorithm.

5 Conclusion

A large number of systems for various purposes allow the construction of a semi-Markov model. When constructing a semi-Markov model, it is necessary to build a rather complex phase space of states that reflect the physical states of the system, and ensure the correctness of the construction of the semi-Markov model. An important component of the semi-Markov model is the embedded Markov chain, responsible for the transitions of the system from one state to another. The phase space of the embedded Markov chain coincides with the phase space of the semi-Markov model.

Using a system for which a semi-Markov model is constructed, in some cases, when changing the states of the system, it is not possible to obtain all the information contained in the phase states of the semi-Markov model, but it is possible to obtain only some information (signal) associated with the phase states. In this case, the phase states of the embedded Markov chain (the semi-Markov model) can be considered hidden, so the problem arises of estimating the characteristics of the embedded Markov chain and the semi-Markov model on the basis of the obtained vector of signals. Hidden Markov chains allow solving this problem.

In this paper, a hidden Markov model of an information system is built on the basis of an merged semi-Markov model of a system with component-wise storages.The probability of occurrence of a given chain of signals is determined using the forward–backward algorithm.The Viterbi algorithm was used to find the most likely sequence of states corresponding to a given chain of signals.The HMM was trained using the Baum-Welsh algorithm.

In the future, it is planned to build and use hidden Markov and hidden semi-Markov models for forecasting states and analyzing the reliability of information systems.

References

1. Korolyuk, V.S., Turbin, A.F.: Markovian Renewal Processes in the Problems of System Reliability. Naukova Dumka, Kiev (1982)
2. Jansen, J., Limnios, N. (eds.): Semi-Markov Models and Applications. Kluwer Academic Publishers, Amsterdam (1999)
3. Limnios, N., Oprisan, G.: Semi-Markov Processes and Reliability. Springer, Boston (2001). https://doi.org/10.1007/978-1-4612-0161-8
4. Grabski, F.: Semi-Markov Processes: Applications in System Reliability and Maintenance. Elsevier Science, Amsterdam (2014)
5. Korolyuk, V., Swishchuk, A.: Semi-Markov Random Evolutions. Kluwer Academic Publishers, Amsterdam (1995)
6. Korolyuk, V.S., Limnios, N.: Stochastic Systems in Merging Phase Space. World Scientific, Imperial College Press, London (2005)

7. Silvestrov, D., Silvestrov, S.: Nonlinearly Perturbed Semi-Markov Processes. Springer, Cham (2017). https://doi.org/10.1007/978-3-319-60988-1
8. Obzherin, Yu.E., Sidorov, S.M., Nikitin, M.M.: Analysis of reliability of systems with component-wise storages. In: E3S Web of Conferences, vol. 58, art. no. 02024 (2018). https://doi.org/10.1051/e3sconf/20185802024
9. Korolyuk, V.S., Korolyuk, V.V.: Stochastic Models of Systems. Springer, Dordrecht (1999). https://doi.org/10.1007/978-94-011-4625-8
10. Obzherin, Y.E., Boyko, E.G.: Semi-Markov Models: Control of Restorable Systems with Latent Failures. Elsevier Academic Press, London (2015)
11. Kopp, V.Y., Obzherin, Y.E., Peschanskiy, A.I.: Stochastic Models of Automized Systems With Time Reservation. Publishing of Sevastopol State Technical University, Sevastopol (2000)
12. Limnios, N., Nikulin, M.: Recent Advances in Reliability Theory: Methodology, Practice and Inference. Springer, Boston (2000). https://doi.org/10.1007/978-1-4612-1384-0
13. Korolyuk, V.S., Turbin, A.F.: Semi-Markov Processes and Their Application. Naukova Dumka, Kiev (1976)
14. Feller, W.: An Introduction to Probability Theory and Its Applications, vol. 2, 2nd edn. Wiley, Hoboken (1971)
15. Yu, S.-Z.: Hidden Semi-Markov Models: Theory. Elsevier Science, Algorithms and Applications (2015)
16. Barbu, V.S., Limnios, N.: Semi-Markov Chains and Hidden Semi-Markov Models Towards Applications: Their Use in Reliability and DNA Analysis. Springer, New York (2008). https://doi.org/10.1007/978-0-387-73173-5
17. Elliott, R., Aggoun, L., Moore, J.: Hidden Markov Models, Estimation and Control. Springer, New York (1995). https://doi.org/10.1007/978-0-387-84854-9
18. Elliott, R., Limnios, N., Swishchuk, A.: Filtering hidden semi-Markov chains. Stat. Probab. Lett. **83**(9), 2007–2014 (2013)
19. Cappé, O., Moulines, E., Rydeén, T.: Inference in Hidden Markov Models. Springer, New York (2005). https://doi.org/10.1007/0-387-28982-8
20. Ross, S.M.: Introduction to Probability Models, 9th edn. Elsevier Academic Press, Cambridge (2006)
21. Rabiner, L.R.: A tutorial on hidden Markov models and selected applications in speech recognition. Proc. IEEE **77**(2), 257–286 (1989)
22. Churchill, G.: Hidden Markov chains and the analysis of genome structure. Comput. Chem. **16**(2), 107–115 (1992)
23. Durand, J.-B., Gaudoin, O.: Software reliability modelling and prediction with hidden Markov chains. Stat. Model. - An Int. J. **5**(1), 75–93 (2005)
24. Archer, G.-E.B., Titterington, D.M.: Parameter estimation for hidden Markov chains. Stat. Plann. Infer. **108**(1–2), 365–390 (2002)

Statistical Distributions of Partial Correlators of Network Traffic Aggregated Packets for Distinguishing DDoS Attacks

Andrey Evgenievich Krasnov$^{(\boxtimes)}$ and Dmitrii Nikolaevich Nikol'skii

Russian State Social University, Moscow, Russia
krasnovmgutu@yandex.ru, nikolskydn@mail.ru

Abstract. The limitations of classical spectral-correlation methods of network traffic analyzing for detecting and classifying its anomalous states are considered, and the perspective of using phase-portrait statistics and statistics in the form of partial correlators of network traffic aggregates suitable for both stationary and unsteady signals analysis is shown. These statistics are entered on the basis of an indirect analogy between the flow of aggregates of network traffic and the flow of wave packets of a coherent electromagnetic field. To do this, using the number of packets entering the aggregates and having various flags in their headers, a set of analytical signals are formed, the real parts of which correspond to the generalized coordinates of the aggregates, and the imaginary parts conjugated by the Hilbert transform correspond to their generalized velocities. Analytical signals modulate a coherent electromagnetic field, forming its envelopes, the statistics of which in the form of phase portraits and distributions of the values of the partial correlators of the wave packets of the field allow us to describe normal and abnormal states of network traffic. Partial correlators are formed by averaging the traffic evolution operator over the flag states of its aggregates, which makes it possible to describe non-stationary fields using distributions of correlator values. The effectiveness of using partial correlators for distinguishing of complex network attacks (TCP Connection Flood, Slow Loris, HTTP Get Flood) is confirmed by a computational experiment using an example of states analysis of the real network traffic of a TCP streaming protocol.

Keywords: Network traffic · Phase portrait · Partial correlator

The article was prepared as part of research work on the topic "Automated Intelligent Management Information System (AIMIS) at a digital university. 'I. Digital Faculty'", funded by Russian State Social University.

© Springer Nature Switzerland AG 2019
V. M. Vishnevskiy et al. (Eds.): DCCN 2019, LNCS 11965, pp. 365–378, 2019.
https://doi.org/10.1007/978-3-030-36614-8_28

1 Introduction

1.1 Objective

Over the past two decades, a large number of works have appeared on the analysis of network traffic, due to the importance of the problem of protecting network resources from various attacks. Relatively general methods of analyzing network traffic are divided into signature and behavioral. The former are designed to detect known, clearly described attacks, and are based on reconciling sequences of characters and events with a database of attack signatures. The disadvantage of signature methods is the inability to detect new types of attacks and their modifications. Behavioral methods are designed to detect unknown attacks and are based on the identification of anomalies or deviations from the normal functioning mode. Their advantage lies in the possibility of analyzing the dynamics of processes to identify new types of attacks. Behavioral methods are used to detect anomalies or deviations from the normal mode of operation of telecommunication systems for packet data transmission. Of these, the most common are: classical spectral-correlation, fractal (multifractal), wavelet methods, as well as combined analysis methods. However, these methods often give a detection error, and, moreover, are not intended for a more subtle analysis of network traffic associated, for example, with the classification of attack types. Therefore, some systematic generalization of the results of studies on the analysis of network traffic by classical spectral-correlation methods, as well as the authors approach related to them, which allows not only to detect deviations in the normal states of traffic, but also to distinguish its anomalous states, is of considerable interest. In this paper, we consider the limitations of classical spectral-correlation methods for analyzing network traffic for detecting and classifying its anomalous states, and also show the prospect of using so-called phase-portrait statistics and statistics in the form of partial correlators of network traffic aggregates suitable for analysis as stationary and non-stationary signals.

1.2 Disadvantages of Classical Spectral-Correlation Analysis for Behavioral Methods of Network Traffic Analysis

An analysis of the temporal structure or behavioral nature of network traffic is one of the widespread approaches to the detection and identification (classification) of its anomalous conditions associated with DDOS attacks, leading to failure of the computer network [1]. At the same time, to increase the analysis speed, signals generated on the basis of service information collected from the address and load characteristics of the headers of data packets are used [2,3].

Numerous studies have shown that network traffic signals are in the nature of fluctuation processes, whose autocorrelation functions and spectral densities vary according to power laws [4]. Moreover, at different time scales, the spectral densities are not smooth, like the spectra of "white" noise, but have fluctuation outliers, leading to unpredictability of network traffic [5].

In radiophysics, a similar behavior of spectral densities in the frequency ranges (10^{-1} Hz $< \omega < 10^4$ Hz), which violates one of the stationary conditions of the processes, is known as the flicker effect or flicker noise [6].

Therefore, using only one correlation function or spectral density of network traffic signals, it is practically impossible to judge their differences, especially at short time intervals, which is required to detect attacks [7]. Indeed, it was shown in [8] that the main energy of DoS and DDoS attacks is distributed at high and low frequencies, respectively, whereas in ordinary TCP traffic, energy is distributed evenly over the entire frequency range.

It was also shown (see, for example, [9]) that network traffic is self-similar in nature, the degree of fracture of which is integrally described by the fractal dimension using the indicator or Hurst index [10,11].

However, the Hurst index allows judging only about a certain correspondence of the signal corresponding to traffic with a noise-like process. Relatively recent publications, for example, [12] have confirmed the impossibility of using Hurst index to distinguish attacks, although they have shown that along with other statistics (for example, wavelet representation coefficients [13]), Hurst index can be used as their indicator.

The reason for the rather rough estimates of the signals states generated using Hurst index lies in the direct connection of this indicator with their autocorrelation functions [9].

Thus, studies of the applicability of spectral-correlation analysis and its derivative methods to estimating the statistical properties of network traffic once again confirmed the importance of the results of earlier fundamental studies in the analysis of stationary fluctuating signals [6,14,15]. So, for example, in [6] it was shown that a serious drawback of classical spectral-correlation descriptions is the loss of phase information—causal relationships of the sequence between individual signal components. As a result, degeneracy is observed: identical descriptions correspond to signals of different structure; when "shuffling" a components of signals, their descriptions do not change.

Similar restrictions are inherent in the descriptions of non-stationary signals using various dynamic "window" representations [16,17] and hybrid models based on them [18].

We also note that the fractal nature of many physical processes and their thermodynamic description (average description) using spectral-correlation and fractal methods are used in the stochastic dynamics of natural objects [19,20]. From this it is completely clear that correlation functions, as well as the related power spectra and fractal characteristics are very general behavioral statistics of fluctuating processes.

To form a non-degenerate description of network traffic, it is necessary to apply statistics that take into account the phase relationships between components of corresponding signals. For example, one of the approaches to accounting for the sequence of components of network traffic is based on recording successive values of the sizes of network data packets and the intervals between them [21].

Below we present another approach developed on the basis of indirect analogies of the network data packets flow with the wave packets flow in electrodynamics.

1.3   A Statistical Description of Structural Signals Taking into Account Phase Relations

In [22,23], it was suggested that the observed (measured) signal x_t of an oscillatory nature associated with any process of a real physical object can be considered stationary as a first approximation and its spectral representation based on the complex amplitudes $C(\omega_n)$ of the harmonic components of the signal can be used. These components correspond to the cyclic frequencies of transitions between different stationary energy levels of the physical system generating the signal. The average spectral density (power spectrum for energy processes) $\left\langle |C(\omega_n)|^2 \right\rangle$ of the signal x_t, determined through its second-order moment by the well-known Einstein–Wiener–Khinchin dependence, is a description of the overall signal structure in the spectral representation [6,15]. The averaging is performed over an ensemble of different initial conditions for the signal measurement procedure using the multidimensional probability distribution density $p\left[\{|C(\omega_n)|\}\right]$ depending only on the absolute value $|C(\omega_n)|$ of spectral amplitudes.

In the next approximation, it is necessary to take into account the interaction between the system in question and other parts of the object, including between the system and the measuring device. In this case, the stationary states of the system are perturbed, and the signal becomes unsteady and is described by the corresponding probability distribution density $p\left[\{C(\omega_n)\}\right]$ of the set $\{C(\omega_n)\}$ of complex values of spectral amplitudes [6]. This shows that the most general description of the structure of both stationary and non-stationary signals in the spectral representation are the statistical probability distribution densities of the spectral amplitudes of the signals themselves.

The decomposition of the real signal x_t into complex harmonic oscillations contains two complex components: positive-frequency x_t^+ with complex spectral amplitudes $\{C(\omega_n)\}$; negative-frequency x_t^- with complex spectral amplitudes $\{C^*(\omega_n)\}$. On the real frequency axis $(-\infty < \omega < +\infty)$, the signal x_t is completely determined by its "half-spectrum", for example, with frequencies $(-\infty \leq \omega \leq +\infty)$, since the other half of the spectrum with frequencies $(-\infty < \omega < 0)$ is formally determined through the operation complex conjugation and carries the same information about the signal, since $C(-\omega_n) = C^*(\omega_n)$.

Thus, there is a one-to-one correspondence between the signal x_t and the analytical signal x_t^+ [6]. Therefore, the statistical description of the signal x_t is equivalent to the description of the analytical signal x_t^+.

The causal behavior of the system is associated with the dynamic operator $\partial/\partial t$, which describes the rate of change of the system state. In closed systems, time is uniform, therefore the law of the energy conservation is fulfilled. Due to this, the behavior of the system remains causal during time reversal: $t \rightarrow -t$. However, when registering a signal, the system is not closed, and the causal behavior of the system corresponds to the real flow of time: $t > 0$. Such time flow correspond the positive frequencies. It is these frequencies that form the complex signal x_t^+. Thus, the transition from the signal x_t to its positive-frequency component x_t^+ means taking into account the causality principle.

Mathematically, the causal relationship of the components (time samples) of the analytical signal x_t^+ is expressed as follows. Since this signal contains only positive frequencies ($x_t^- = 0$ for $\omega_n < 0$), its real $\operatorname{Re} x_t^+$ and imaginary $\operatorname{Im} x_t^+$ parts are related by Hilbert relations, and $\operatorname{Re} x_t^+ = x_t/2$ [24]. Thus, $x_t^+ = (x_t + iy_t)/2$, where $y_t = x_t \otimes P/(\pi t)$ is the Hilbert-image of x_t obtained by convolution ($\otimes$) of the measured signal x_t with the core P/t (P is the main value generalized function $1/t$). The length of the P/t core gives a nonlocal causal relationship between the samples of the analytical signal x_t^+, taking into account such phase relationships as its past, present and future.

In practical terms, the Hilbert image y_t of the signal is most easily generated using a discrete filter h_t with a finite impulse response, for which $h_t \equiv 0$ for all even discrete samples t, and for odd ones $h_t = 1/(\pi t)$ [24].

For a statistical description of a causal analytical signal x_t^+ of vibrational nature [22], as well as envelopes of electromagnetic signals [23], we used the results of [25], based on which the distribution $w[x, y]$, which is a reduced estimate of the multidimensional probability distribution density $p\left[\{C(\omega_n)\}\right]$ and called phase portrait (FP).

FP $w[x, y]$ plays a role analogous to the probability distribution density $p\left[\{C(\omega_n)\}\right]$, but already in the two-dimensional space generated by the values of the measured signal and its Hilbert-image. Its main advantage over the initial description given in the space of values of the spectral amplitudes of the signal x_t^+ is a significant decrease in dimension. At the same time, both statistical descriptions are equivalent, since they give the same average value $< F >$ of any random physical quantity—the real function $F[x, y] = F\left[\{C(\omega_n)\}\right]$, which depends on the signal x_t^+ through the set $\{C(\omega_n)\}$ of its complex spectral amplitudes [23].

Theoretically, distribution $w[x, y]$ corresponds to the probability density w of joint events, consisting in the fact that the signal and its Hilbert-image take specific values of x and y in the "elementary" region $(x, x + dx; y, y + dy)_{dx, dy \to 0}$. Therefore, $w[x, y]$ can be easily estimated experimentally by measuring all possible combinations of the joint values x, y of the time samples of the signals x_t and y_t falling into the elementary n, m – regions with coordinates x_n, y_m and then constructing a normalized two-dimensional histogram or scatter diagram $w_{emp}[x_n, y_m]$ $(\sum_n \sum_m w_{emp}[x_n, y_m] = 1)$.

When evaluating $w[x, y]$, it must be taken into account that all implementations of random signals x_t and y_t are practically inaccessible. Therefore, it was proposed to estimate $w[x, y]$ by "blurring" the empirical distribution $w_{emp}[x_n, y_m]$, formed by realizing the signal x_t with a sufficiently large observation time T, using a smooth two-dimensional function.

To simplify the process of using FP $w[x, y]$ to detect and classify abnormal states of network traffic, it was proposed to reduce two-dimensional empirical distributions $w_{emp}[x_n, y_m]$ into one-dimensional [26].

The considered phase portrait statistics were used to detect and classify various types of DDoS attacks using Wald's method of sequential statistical analysis [26]. Nevertheless, the use of the considered phase-portrait statistics is an extremely laborious task. Therefore, we will consider another approach below, also based on the analogy of the flow of network traffic aggregates with the flow of wave packets of a coherent electromagnetic field, for which the apparatus of the so-called field correlators is used in physics [25].

2   Partial Correlators of Aggregates of Network Traffic

2.1   Statistical and Dynamic Models of the Flow of Aggregates of Network Traffic

The study used the format of the streaming data transfer protocol (TCP) to extract the values of the control binary fields (flags) of the headers of the transmitted data packets [28,29]. The binary sequence of flags forms the binary code of the possible flag states of the packet. All packets observed on the ΔT interval were grouped by flag states, and for each j-th flag state the number $N_{\Delta T}^{(j)}$ of packets and their total information capacity $I_{\Delta T}^{(j)}$ were calculated $(j = 1, 2, \ldots, J)$.

A stream of discrete-time sequences $t_k = \Delta k$ $(k = 1, 2, \ldots, K)$ with grouped packets or aggregate of network traffic was generated, where each k-th unit is associated with its generalized dynamic coordinate

$$X_{\Delta T}^{(j)}(t_k) = \sqrt{N_{\Delta T}^{(j)}(t_k)}. \tag{1}$$

Using the Hilbert transform considered above, the values $Y_{\Delta T}^{(j)}(t_k)$ of aggregates generalized velocities were also calculated from the values $X_{\Delta T}^{(j)}(t_k)$ of their generalized coordinates.

A discrete statistical model of the network traffic aggregates flow was formed as a set of analytical signals:

$$Z_{\Delta T}^{(j)}(t_k) = X_{\Delta T}^{(j)}(t_k) + iY_{\Delta T}^{(j)}(t_k), \quad k = 1, 2, \ldots, K \tag{2}$$

and phase-portrait statistics $w[X_{\Delta T}, Y_{\Delta T}]$ $(j = 1, 2, \ldots, J)$ considered in Sect. 1.3.

If we assume that the signals $X_{\Delta T}^{(j)}(t_k)$ are independent for different flag states, then the resulting phase transition can be represented as $\prod_j^J w\left[X_{\Delta T}^{(j)}, Y_{\Delta T}^{(j)}\right]$. However, for simplicity, we will use the empirical distribution in the form

$$w_{emp}\left[X_{\Delta T}^{(j)}, Y_{\Delta T}^{(j)}\right],$$

where:

$$X_{\Delta T} = \sum_{j=1}^{J} q_j(t_k) X_{\Delta T}^{(j)}(t_k), \quad Y_{\Delta T} = \sum_{j=1}^{J} q_j(t_k) Y_{\Delta T}^{(j)}(t_k), \tag{3}$$

and $q_j(t_k) = I_{\Delta T}^{(j)}(t_k) / \sum_{j=1}^{J} I_{\Delta T}^{(j)}(t_k)$ are statistical weights or significance of aggregates with j-th flag states $(j = 1, 2, \ldots, J)$.

In describing the dynamic model of the flow of network traffic aggregates, we will use an approach based on the evolution operator [29, 30], but with reference not to the input stream of analytical signals (2), but to their normalized images, which are described by the complex phase vector $z_{\Delta T}(t_k)$:

$$\mathbf{z}_{\Delta T}(t_k) = \begin{vmatrix} \mathbf{x}_{\Delta T}(t_k) \\ i\mathbf{y}_{\Delta T}(t_k) \end{vmatrix}, \quad \mathbf{z}_{\Delta T}^{+}(t_k) = \mathbf{x}_{\Delta T}(t_k) - i\mathbf{y}_{\Delta T}(t_k),$$

$$\|\mathbf{z}_{\Delta T}(t_k)\| = \sqrt{\mathbf{x}_{\Delta T}^{T}(t_k)\mathbf{x}_{\Delta T}(t_k) + \mathbf{y}_{\Delta T}^{T}(t_k)\mathbf{y}_{\Delta T}(t_k)} = 1, \tag{4}$$

$$\mathbf{x}_{\Delta T}(t_k) = \frac{X_{\Delta T}(t_k)}{\sqrt{\sum_{j=1}^{J} \left[\left(X_{\Delta T}^{(j)}(t_k) \right)^2 + \left(Y_{\Delta T}^{(j)}(t_k) \right)^2 \right]}},$$

$$\mathbf{y}_{\Delta T}(t_k) = \frac{Y_{\Delta T}(t_k)}{\sqrt{\sum_{j=1}^{J} \left[\left(X_{\Delta T}^{(j)}(t_k) \right)^2 + \left(Y_{\Delta T}^{(j)}(t_k) \right)^2 \right]}},$$

$$\mathbf{X}_{\Delta T}(t_k) = \left[X_{\Delta T}^{(1)}(t_k), X_{\Delta T}^{(2)}(t_k), \ldots X_{\Delta T}^{(J)}(t_k) \right],$$

$$\mathbf{Y}_{\Delta T}(t_k) = \left[Y_{\Delta T}^{(1)}(t_k), Y_{\Delta T}^{(2)}(t_k), \ldots Y_{\Delta T}^{(J)}(t_k) \right], \quad k = 0, 1, \ldots, K.$$

Then the dynamic model of the change in the phase vector moving in accordance with (4) along the unit hypersphere is described by the evolution operator:

$$\mathbf{z}_{\Delta T}(t_s) = \mathbf{S}(t_s, t_p)\mathbf{z}_{\Delta T}(t_p), \quad p < s, \quad s = 1, 2, \ldots, K. \tag{5}$$

Using the results of [30, 31], we write the expression for the non-Hermitian evolution operator from (5) for each t_s, t_p-th discrete time instants in the form:

$$\mathbf{S}(t_s, t_p) = \mathbf{z}_{\Delta T}(t_s)\mathbf{z}_{\Delta T}^{+}(t_p), \quad p < s, \quad s = 1, 2, \ldots, K. \tag{6}$$

In the future, we will use only the real part of the evolution operator:

$$\mathrm{Re}\,\mathbf{S}(t_s, t_p) = \mathrm{Re}\,\mathbf{z}_{\Delta T}(t_s)\mathbf{z}_{\Delta T}^{+}(t_p), \quad p < s, \quad s = 1, 2, \ldots, K. \tag{7}$$

2.2 An Analogy of the Flow of Network Traffic Aggregates with the Flow of Wave Packets of a Coherent Electromagnetic Field

Using signals $Z_{\Delta T}^{(j)}(t_k)$ $(j = 1, 2, \ldots, J)$, we perform the external spatial modulation [32] of the coherent electromagnetic field of a single source having a complex amplitude $\tilde{E}_0$ of electric intensity with a given unit polarization vector $\mathbf{e}$ and a cyclic optical frequency ω. In this case, to each j-th flag index of the aggregate we associate the j-th spatial coordinate or field mode. As a result, we obtain the flow of wave packets of the coherent electromagnetic field:

$$\tilde{\mathbf{E}}_{\Delta T}^{+}(t) = \mathbf{e} \left[\tilde{E}_{\Delta T}^{(1)}(t), \tilde{E}_{\Delta T}^{(2)}(t), \ldots, \tilde{E}_{\Delta T}^{(J)}(t) \right], \tag{8}$$

each j-th spatial mode of which is described as:

$$\tilde{E}_{\Delta T}^{(j)}(t) = \tilde{E}_0(t) \exp\left(i\omega t\right) \sum_{k=0}^{K} Z_{\Delta T}^{(j)}(t_k) d_{\Delta T}(t - t_k), \quad j = 1, 2, \ldots, J, \tag{9}$$

where $d_{\Delta T}(t - t_k)$ is a single pulse with a duration shifted in time by a value t_k.

The complex envelope of the field, slowly changing in comparison with the frequency ω, has in each of its j-th modes of each k-th wave packet the amplitude:

$$\alpha_{\Delta T}^{(j)} = \left| \tilde{E}_0 \right| \sqrt{N_{\Delta T}^{(j)}(t_k)} \sqrt{1 + \left(\frac{Y_{\Delta T}^{(j)}(t_k)}{X_{\Delta T}^{(j)}(t_k)} \right)^2} \tag{10}$$

This shows that the dependence $X_{\Delta T}^{(j)}(t_k) = \sqrt{N_{\Delta T}^{(j)}(t_k)}$ of the generalized dynamic coordinate of the network traffic aggregate was not chosen randomly, but was consistent with the electrodynamic dependence $E_0 = \sqrt{\omega \hbar N^{photon}}$, where $\hbar$ is the Planck constant, and N^{photon} is the number of photons emitted by a coherent source with a power of $PW \sim \left| \tilde{E}_0 \right|^2$ Thus, for every j-th mode of the field of each k-th wave packet there will be $N_{\Delta T}^{photon} N_{\Delta T}^{(j)}(t_k) / \sum_j N_{\Delta T}^{(j)}(t_k)$ photons proportional to the number $N_{\Delta T}^{(j)}(t_k)$ of packets in the j-th flag state for the k-th aggregate, where $N_{\Delta T}^{photon} = P\Delta T / (\omega \hbar)$ is the number of photons falling on the wave packet.

For example, for the optical range ($\omega \sim 10^{14}$ Hz), the number of photons emitted by a cw laser with a power of $PW = 10$ mW in the unit time interval $\Delta T = 50$ ms will be of the order of $N_{\Delta T}^{photon} \sim 10^{15}$ photon. Therefore, even for insignificant weights $N_{\Delta T}^{(j)}(t_k) / \sum_{j=1}^{J} \sim 1/J$ of flag states for example, for $J = 64$, the number of photons in each mode of the field will be large ($\sim 10^{13}$). Such a flow of wave packets from (10) can be considered as a semiclassical field, and the statistics of its shells can be described both by the phase-portrait statistics considered in Sect. 1.3 [23] and by the classical field correlators [25].

2.3   Partial Correlators of a Network Traffic Aggregates Flow

In electrodynamics, the contrast of the interference pattern formed by electromagnetic waves with amplitudes $\tilde{E}^{(j)}(t_s) = E^{(j)}(t_s) \exp\left(-i\omega t_s\right)$ and $\tilde{E}^{(j)}(t_p) =$

$E^{(j)}(t_s)\exp(-i\omega t_p)$ observed at two space-time points j, t_s and $m.t_p$ is described by a normalized correlator or a complex first-order coherence function [25]:

$$\gamma^{(1)}(j, t_s, m, t_p) = \frac{\langle E^{(j)}(t_s)E^{*(m)}(t_p)\rangle \exp(-i\omega(t_s - t_p))}{\sqrt{\langle |E^{(j)}(t_s)|^2 \rangle \langle |E^{(m)}(t_p)|^2 \rangle}}, \tag{11}$$

whose module $|\gamma^{(1)}(j, t_s, m, t_p)|$ determines the contrast of j, t_s, m, t_p-th interference gratings $\langle E^{(j)}(t_s)E^{*(m)}(t_p)\rangle$.

It is easy to see that the evolution operator $\mathbf{S}(t_s, t_p) = \operatorname{Re}\mathbf{z}_{\Delta T}(t_s)\mathbf{z}_{\Delta T}^+(t_p)$ considered in (7) is almost similar to $|\gamma^{(1)}(j, t_s, m, t_p)|$, however, it determines the contrast not of a separate interference grating, but of the whole interference structure $\mathbf{z}_{\Delta T}(t_s)\mathbf{z}_{\Delta T}^+(t_p)$ formed by a set $(J \times J)$ of gratings. In almost all experiments, the averaging $\langle \cdot \rangle$ given in (11) is carried out by temporary accumulation. Due to this, a high contrast $(|\gamma^{(1)}| \equiv 1)$ of the interference bands of coherent signals appears, since for them [33]

$$\left\langle E^{(j)}(t_s)E^{*(m)}(t_p)\right\rangle \sim E^{(j)}(t_s)E^{*(m)}(t_p).$$

However, in our case, this averaging is not suitable for fluctuating signals, since the formation of an averaged structure $\langle \mathbf{z}_{\Delta T}(t_s)\mathbf{z}_{\Delta T}^+(t_p)\rangle$ will lead to a decrease in its contrast.

Therefore, we will use the real part of evolution operator (7) corresponding to adjacent aggregates of network traffic, and for the quantitative description of their connections we introduce partial (not time-averaged, but averaged over flag indices) first-order correlators:

$$h(t_k, t_{k-1}) = \operatorname{tr}\left[\mathbf{Q}(t_k, t_{k-1})\operatorname{Re}\mathbf{S}(t_k, t_{k-1})\right] =$$

$$\sum_{j}^{J}\sum_{m}^{J} q_{jm}(t_k, t_{k-1})\left[x_{\Delta T}^{(j)}(t_k)x_{\Delta T}^{(m)}(t_{k-1}) + y_{\Delta T}^{(j)}(t_k)y_{\Delta T}^{(m)}(t_{k-1})\right], \tag{12}$$

$$q_{jm}(t_k, t_{k-1}) = \begin{cases} \dfrac{I_{\Delta T}^{(j)}(t_k)I_{\Delta T}^{(m)}(t_{k-1})}{\sum\limits_{j}^{J} I_{\Delta T}^{(j)}(t_k)\sum\limits_{m}^{J} I_{\Delta T}^{(m)}(t_{k-1})}, & i \neq j, \\[4mm] \dfrac{I_{\Delta T}^{(j)}(t_k)I_{\Delta T}^{(j)}(t_{k-1})}{\sum\limits_{j}^{J} I_{\Delta T}^{(j)}(t_k)I_{\Delta T}^{(j)}(t_{k-1})}, & \operatorname{tr}\mathbf{Q}(t_k, t_{k-1})\sum\limits_{j}^{J} q_{ij}(t_k, t_{k-1}) = 1. \end{cases} \tag{13}$$

The diagonal matrix elements $q_{ij}(t_k, t_{k-1})$ from (13) correspond to the weights or values of the same flag indices of adjacent aggregates. The offdiagonal matrix elements determine the statistical weights of the bonds of their heterogeneous flag indices. Compared with the correlators introduced in [30] and used in [31], normalized partial correlators as average (by flag indices) values of a physical quantity corresponding to the evolution operator take into account cross-links not only of the generalized coordinates of adjacent aggregates of network traffic, but also their generalized speeds, and also contain information on the statistical weights of the flag indices of adjacent aggregates.

3   Computational Experiment

3.1   Partial Correlators of Flows of Network Traffic Aggregates

In a computational experiment, empirical distributions $w_{emp}[X_{\Delta T}, Y_{\Delta T}|r]$ and $w_{emp}[h|r]$ were studied to describe various states of network traffic: normal state—normal ($r = 0$), tcpcon attack states—TCP Connection Flood ($r = 1$), slowloris—Slow Loris ($r = 2$), httpget—HTTP Get Flood ($r = 3$). These attacks relate to application-level attacks of the Open Systems Interaction Model (OSI) and, according to the classification of [1], are difficult to detect. Below are the results of a study of some empirical distributions of phase-portrait statistics and partial correlators for different states of network traffic.

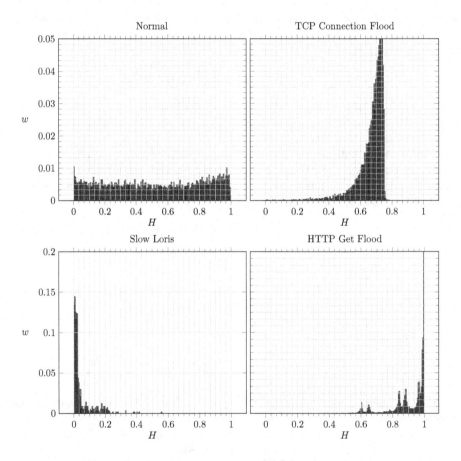

Fig. 1. Empirical historgam $w_{emp}[h|r]$ for $\Delta T = 50\,\text{ms}$.

Figure 1 shows the distribution $w_{emp}[h|r]$ of the values of the partial correlators from (12) for $T = 50\,\text{ms}$. The significant difference in the obtained distributions is clearly visible. Additional numerical experiments showed that this time is minimal at which the criterion:

$$D = \frac{1}{6} \sum_{r=0}^{2} \sum_{r \neq r'} \frac{2 \sum_h w_{emp}(h|r) w_{emp}(h|r')}{\sum_h \left[w_{emp}^2(h|r) + w_{emp}^2(h|r') \right]} \tag{14}$$

for the difference in distributions satisfies the relation $D > 0{,}9$.

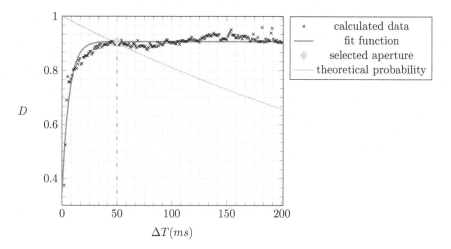

Fig. 2. Dependences of measure D (14) of difference of distributions $w_{emp}\,[h|r]$ and the probability $p\,(\Delta T | \Delta T_{eff})$ (15) of pure states of aggregates on their temporal aperture ΔT.

Figure 2 shows the experimental (calculated data) and smoothed (fit function) data. With an increase of ΔT over 50 ms (selected aperture), criterion D remains almost unchanged, but the theoretical probability:

$$\rho(\Delta T | \Delta T_{eff}) = \exp\left(-\frac{\Delta T}{\Delta T_{eff}} \right) \tag{15}$$

of pure states of traffic aggregates decreases: the probability of including packages related to various types of attacks increases in its aggregates. The probability of pure states shown in Fig. 2 corresponds to the effective time $\Delta T_{eff} = 475$ ms. The increase of ΔT will entail a situation in which aggregates will consist of packets belonging to different types of attacks, which will lead to an error when evaluating traffic states.

3.2   Conclusion

In classical electrodynamics, the so-called field correlators of coherent electromagnetic fields are widely used for statistical analysis of their stationary states.

When forming correlators for semiclassical fields, the products are averaged over time for different orders of field amplitudes, which is unacceptable for the analysis of non-stationary fields.

In the present work, it is shown that for the analysis of such fields it is possible to use both their phase-portrait statistics and partial correlators, which are also formed by the products of their amplitudes, but averaged only over their different spatial modes.

The main practical result of the study is the construction of statistical and dynamic models of the flow of network traffic aggregates based on the introduction of analytical signals that depend on the number of data packets included in the aggregates with different flag indices. By drawing a conditional analogy between the flow of network traffic aggregates and the envelope of the flow of wave packets of a coherent electromagnetic field, phase portraits and partial correlators of the first order of adjacent aggregates of network traffic are considered.

In a computational experiment, it was shown that empirical one-dimensional distributions of partial correlator values can be used to describe various normal and abnormal states of network traffic caused by complex DDoS attacks in order to distinguish them.

A computational experiment based on the TCP protocol showed that the use of partial correlators allows the formation of statistical distributions of their values, which significantly differ for a number of traffic states (Normal, TCP Connection Flood, Slow Loris, HTTP Get Flood), which is an undoubted contribution to the development behavioral methods of statistical analysis of network traffic.

References

1. Bhattacharyya, D., Kalita, J.: DDoS Attacks: Evolution, Detection, Prevention, Reaction, and Tolerance. CRC Press Taylor & Francis Group, Boca Raton (2016)
2. Shui, Y., Zhou, W., Jia, W., Xiang, Y., Tang, F.: Discriminating DDoS atacks from flash crowds flow correlation coefficient. IEEE Trans. Parallel Distrib. Syst. **23**(6), 412–425 (2012)
3. Zhou, L., Liao, M., Yuan, C., Zhang, H.: Low-rate DDoS attack detection using expectation of packet size. Secur. Commun. Netw. **2017** (2017). https://www.hindawi.com/journals/scn/2017/3691629/. Accessed 15 Sept 2019
4. Cebrail, C., Ali, G., Abdullah, T.: Packet traffic features of IPv6 and IPv4 protocol traffic. Turk. J. Electr. Eng. Comput. Sci. **20**(5), 727–749 (2012)
5. Zhongda T.: Chaotic characteristic analysis of network traffic time series at different time scales. Chaos Solitons Fractals **130**, 1–16 (2020). https://doi.org/10.1016/j.chaos.2019.109412. Accessed 15 Sept 2019
6. Rytov, S.M., Kravtsov, Y.A., Tatarskii, V.I.: Principles of Statistical Radiophysics 1. Elements of Random Process Theory. Springer, Berlin (1987)
7. Cheng, C.-M., Kung, H.T., Tan, K.-S.: Use of spectral analysis in defense against DoS attacks. In: Proceedings of IEEE GLOBECOM, pp. 2143–2148. Institute of Electrical and Electronics Engineers, Taiwan (2002)
8. Fouladi, R., Seifpoor, T., Anarim, E.: Frequency characteristics of DoS and DDoS attacks. In: 21st Signal Processing and Communications Applications Conference, SIU, pp. 1–4. IEEE, Turkey (2013)

9. Kettani, H., Gubner, J.A.: A novel approach to the estimation of the Hurst parameter in self-similar traffic. In: Proceedings of IEEE Conference on Local Computer Networks, Florida, pp. 160–165 (2002)
10. Hurst, H.E.: Long-term storage of reservoirs: an experimental study. Trans. Am. Soc. Civ. Eng. **116**, 770–799 (1951)
11. Mandelbrot, B.B.: The Fractal Geometry of Nature. W. H. Freeman and Co., New York (1983)
12. Gezer, A.: Identification of abnormal DNS traffic via Hurst parameter. Balk. J. Electr. Comput. Eng. **6**(3), 46–52 (2018)
13. Abry, P., Veitch, D.: Wavelet analysis of long-range dependent traffic. IEEE Trans. Inf. Theory **44**(1), 2–15 (1998)
14. Robinson, E.A.: A historical perspective of spectrum estimation. Proc. IEEE **70**, 885–907 (1982)
15. Yaglom, A.M.: Correlation Theory of Stationary and Related Random Functions. Springer, New York (1987). https://doi.org/10.1007/978-1-4612-4628-2
16. Claasen, T.A.C.M., Mecklenbmker, W.F.G.: The Wigner distribution—a tool for time-frequency signal analysis. Philips J. Res. **35**(6), 372–389 (1980)
17. Xiaoyan, M.A., Hongguang, L.I.: An approach to dynamic estimation for Hurst index of network traffic. Int. J. Commun. Netw. Syst. Sci. **3**, 167–172 (2010)
18. Kanarachos, S., Mathew, J., Chroneos, A., Fitzpatrick, M.E.: Anomaly detection in time series data using a combination of wavelets, neural networks and Hilbert transform. In: 6th IEEE International Conference on Information, Intelligence, Systems and Applications (IISA), pp. 1–6. IEEE, Greece (2015)
19. Shelton, D.P.: Long-range orientation correlation in liquids. J. Chem. Phys. **136**(4), 044503-1–044503-5 (2012)
20. Zhongda, T., Shujiang, L., Yanhong, W., Yi, S.: A prediction method based on wavelet transform and multiple models fusion for chaotic time series. Chaos Solitons Fractals **98**, 158–172 (2017)
21. Crotti, M., Dusi, M., Gringoli, F., Salgarelli, L.: Traffic classification through simple statistical fingerprinting. ACM SIGCOMM Comput. Commun. Rev. **37**(1), 5–16 (2007)
22. Krasnov, A.E.: Use of Hilbert filtering of an electromagnetic signal to identify invariant characteristics of its spatial structure. Optoelectron. Instrum. Data Process. **5**, 106–108 (1987)
23. Krasnov, A.E.: Envelopes phase portraits of coherent electromagnetic field on the plane: using the phase portraits for the optimal discerning of field states. Radiotekhnika (2), 49–54 (1997)
24. Titchmarsh, E.: Introduction to the Theory of Fourier Integrals. Oxford University Press, England (1948)
25. Glauber R.: Optical Coherence and Photon Statistics (In book Quantum Theory of Optical Coherence: Selected Papers and Lectures). Wiley-VCH Verlag GmbH & Co. KGaA, Weinheim (2007)
26. Krasnov, A.E., Nadezhdin, E.N., Galayev, V.S., Zykova, E.A., Nikol'skii, D.N., Repin, D.S.: DDoS attack detection based on network traffic phase coordinates analysis. Int. J. Appl. Eng. Res. **13**(8), 5647–5654 (2018)
27. Krasnov, A.E., Nikol'skii, D.N., Repin, D.S., Galyaev, V.S., Zykova, E.A.: Detecting DDoS attacks using the analysis of network traffic as dynamical system. In: International Scientific and Technical Conference Modern Computer Network Technologies (MoNeTeC), pp. 1–7. IEEE, Moscow (2018)
28. RFC 792 – Internet control message protocol. https://tools.ietf.org/html/rfc792. Accessed 15 Sept 2019

29. RFC 793 – Transmission Control Protocol. https://tools.ietf.org/html/rfc793. Accessed 15 Sept 2019
30. Krasnov, A.E., Nadezhdin, E.N., Nikolsky, D.N., Galyaev, V.S.: Direct and reverse problems of reconstruction of evolution operators in the analysis of the dynamics of multidimensional processes. Chebyshevskii Sb. **19**(2), 217–233 (2018). (in Russian)
31. Krasnov, A.E., Nadezhdin, E.N., Nikol'skii, D.N., Repin, D.S., Galyaev, V.S.: Detecting DDoS attacks by analyzing the dynamics and interrelation of network traffic characteristics. Vestn. Udmurtsk. Univ. Mat. Mekh. Komp. Nauki **28**(3), 407–418 (2018)
32. Agrawal, G.: Fiber-Optic Communications Systems. Wiley, New York (2002)
33. Klauder, J.R., Sudarshan, E.C.G.: Fundamentals of Quantum Optics. Dover Publications, New York (2006)

Approximate Product Form Solution for Performance Analysis of Wireless Network with Dynamic Power Control Policy

Yves Adou[1] , Ekaterina Markova[1(✉)] , and Irina Gudkova[1,2,3]

[1] Department of Applied Probability and Informatics,
Peoples' Friendship University of Russia (RUDN University),
6 Miklukho-Maklaya St., Moscow 117198, Russian Federation
{1032135491,markova-ev,gudkova-ia}@rudn.ru
[2] Institute of Informatics Problems, Federal Research Center
"Computer Science and Control" of the Russian Academy of Sciences,
44-2 Vavilov St., Moscow 119333, Russian Federation
[3] Department of Telecommunication, Brno University of Technology,
Brno, Czech Republic

Abstract. As the number of mobile devices increases and new networks technologies emerge, radio spectrum is becoming a rare resource. This stands as the major problem in the development of modern wireless technologies. In recent years, several methods and technologies were implemented to solve this problem, for example the licensed assisted-access (LAA) and the licensed shared access (LSA) frameworks allowing to use more efficiently the available radio resources. At the same time, radio resource management (RRM) researches showed that mechanisms (i.e. downlink power reduction, user service interruption) using efficiently the radio spectrum can also be applied. In this paper, we aim to propose different methods for the analysis of possible admission control scheme models to access the radio resources of a wireless network with the implementation of downlink power policy and user service interruption mechanisms. Such models could be described by a queuing system with unreliable servers within a random environment (RE).

Keywords: Wireless network · Radio resource management · Limit power policy · Service interruption · Queuing system · Random environment · Recursive algorithm · Approximate method

The publication has been prepared with the support of the "RUDN University Program 5-100" (Gudkova I.A., mathematical model development). The reported study was funded by RFBR, project number 18-37-00231 and 18-00-01555(18-00-01685) (Markova E.V., numerical analysis). This article is based as well upon support of international mobility project MeMoV, No. CZ.02.2.69/0.0/0.0/16_027/00083710 funded by European Union, Ministry of Education, Youth and Sports, Czech Republic and Brno University of Technology.

© Springer Nature Switzerland AG 2019
V. M. Vishnevskiy et al. (Eds.): DCCN 2019, LNCS 11965, pp. 379–390, 2019.
https://doi.org/10.1007/978-3-030-36614-8_29

1   Introduction

The recent years have seen a rapid growth in modern cellular networks traffic, contributed by billions of mobile devices and brand new technologies [1,2]. This growth of traffic affects the availability of radio resource necessary for serving users with the required level of quality of service (QoS) [3,4]. The main technologies developed to solve this problem are orthogonal frequency division multiple access (OFDM) [5,6], multiple input multiple output (MIMO) [7–9], multimedia broadcast multicast service (MBMS) [10], licensed assisted-access framework (LAA) [11] and licensed shared access framework (LSA) [12,13]. Within these technologies, researchers propose different radio resource management (RRM) mechanisms, for example transmission power limitation [14,15] and user service interruption [16], that allow using more efficiently available radio spectrum. Models implementing such mechanisms could be described by a queuing system model with unreliable servers [17]. One type of these systems are systems operating in a random environment [18,19].

Random environment (RE) is a significant obstacle to the efficient diffusion and use of telecommunication systems. It can strongly affect the system performance measures, for example the blocking probability and the achievable bit rate.

The aim of this paper is to propose methods for analyzing performance measures of one of the possible admission control schemes models for wireless network, described by a queuing system with unreliable servers within a RE.

The rest of the paper is organized as follows. In Sect. 2, we propose a general description of the model operating in RE, state of which can vary. In Sect. 3, we propose an accurate method to calculate the stationary probability distribution of the model. In Sect. 4 is proposed an approximate method that significantly reduces the complexity of the model. Section 5 describes and analyzes the main performance measures of the model. Conclusions are drawn in Sect. 6.

2   Mathematical Model

Let us consider a single cell of a wireless mono-service network with radius R within a RE that can impact on the QoS level of users requests. We assume that user requests arrive in the system according to the Poisson law with rate λ and are serviced according to the exponential law with rate μ. We suppose that the distance between all the users and the Base Station (BS) is constant and equal to d. Each request processed in the system is serviced with the guaranteed bit rate (GBR) equaling r_0.

We assume that the RE can change its state. Unlike [20], in this paper the RE changes state from 0 to 2. Such limitation allows to obtain an approximate solution which significantly reduces the complexity of calculations. Transition from state s to $s - 1$ is made with rate $\alpha_s, s = 1, 2$ and from s to $s + 1$ with rate $\beta_s, s = 0, 1$. Note that the rates α_s and β_s are exponentially distributed.

Access to radio resource of the system is implemented in such a way that downlink power can vary under RE influence. It decreases from p_s to $p_{s-1}, s = 1, 2$ with rate $\alpha_s, s = 1, 2$, when transition to state $s - 1$ occurs. Recovery to

$p_s, s = 1, 2$ is made with rate $\beta_s, s = 0, 1$. Note that the rates α_s and β_s are exponentially distributed.

Changing downlink power affects the QoS level of users in the system, due primarily to the deterioration of such parameters as the achievable bit rate r_s and the maximum number of requests $N_s, s = \overline{0,2}$ that can be serviced. Transition to state $s - 1, s = 1, 2$ leads to a decrease of that maximum number from N_s to $N_{s-1}, s = 1, 2$. According to Shannon formula and system current state, the value of $N_s, s = \overline{0,2}$ (1) can be defined as ratio of achievable bit rate $r_s, s = \overline{0,2}$ to GBR r_0.

$$N_s = \left\lfloor \frac{\omega \ln\left(1 + \frac{Gp_s}{d^\kappa N_0}\right)}{r_0} \right\rfloor, s = \overline{0,2}, \tag{1}$$

where ω is the bandwidth of uplink channel, G - the propagation constant, κ - the propagation exponent and N_0 - the noise power.

We describe the behavior of the system using a two-dimensional vector (n, s) over the state space $\mathbf{X} = \{(n, s) \colon 0 \leq n \leq N_s, s = \overline{0,2}\}$, where n represents the current number of requests in the system and s - the current state of the RE. Data transfer is performed with the minimum downlink power p_0 in state $s = 0$ and with the maximum p_2 in state $s = 2$. It should be noted that under RE influence, the services of $N_s - N_{s-1}, s = 1, 2$ requests are interrupted, when transition occurs from state s to $s - 1, s = 1, 2$, since $N_2 > N_1 > N_0$. The corresponding state transition and central state transition diagrams are shown respectively in Figs. 1 and 2.

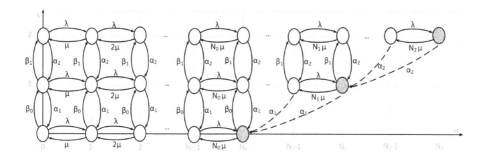

Fig. 1. State transition diagram.

In accordance with the central state transition diagram shown in Fig. 2, the corresponding Markov process is described by the following system of equilibrium equations

$$\left[\lambda I(n < N_s, s = \overline{0,2}) + n\mu I(n > 0) + \alpha_s I(n = \overline{0, N_s}, s = 1, 2) + \beta_s\right.$$
$$\left. \times I(n = \overline{0, N_s}, s = 0, 1)\right] \cdot p(n, s) = \lambda I(n > 0) \cdot p(n - 1, s) + (n + 1)\mu I$$
$$(n < N_s) \cdot p(n + 1, s) + \alpha_{s+1} I(s < 2) \cdot p(n, s + 1) + \beta_{s-1} I(s > 0) \cdot p(n, s - 1), \tag{2}$$

where $p(n, s), (n, s) \in \mathbf{X}$ represents the stationary probability distribution and I - the indicator function equaling 1 when the condition is met and 0 otherwise.

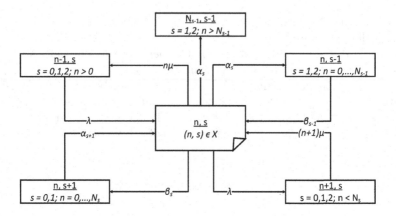

Fig. 2. Central state transition diagram.

3 Accurate Methods for Calculating Probability Distribution

3.1 Numerical Solution

The process describing the considered system is not a reversible Markov process. Therefore, due to the implementation of the user service interruption mechanism, the system stationary probability distribution $p(n,s)_{(n,s)\in\mathbf{X}} = \mathbf{p}$ can be computed, first of all using a numerical solution of the system of equilibrium equations $\mathbf{p}\cdot\mathbf{A} = \mathbf{0}, \mathbf{p}\cdot\mathbf{1}^T = 1$, where $\mathbf{A}$ is the infinitesimal generator of Markov process, elements $a((n,s)(n',s'))$ of which are defined as follows

$$
\begin{cases}
\alpha_s, & \text{if } n' = n, s' = s-1, n = \overline{0,N_{s-1}}, s = 1,2, \\
& \text{or } n' = N_{s-1}, s' = s-1, n \geq N_{s-1}, s = 1,2; \\
\beta_s, & \text{if } n' = n, s' = s+1, n = \overline{0,N_s}, s = 0,1; \\
\lambda, & \text{if } n' = n+1, s' = s, n < N_s, s = \overline{0,2}; \\
n\mu, & \text{if } n' = n-1, s' = s, n > 0, s = \overline{0,2}; \\
*, & \text{if } n' = n, s' = s; \\
0, & \text{otherwise,}
\end{cases}
\tag{3}
$$

where $* = -(\alpha\cdot I(s \neq 0) + \beta\cdot I(s \neq S, n \leq N_s) + \lambda\cdot I(n < N_s, s = \overline{0,2}) + n\mu\cdot I(n > 0))$, $(n,s) \in \mathbf{X}$.

3.2 Recursive Algorithm

The system stationary probability distribution $p(n,s), (n,s) \in \mathbf{X}$ can also be computed using a recursive algorithm that significantly reduces the complexity of the calculations.

Let us consider the unnormalized probabilities $q(n,s), (n,s) \in \mathbf{X}$. To calculate these probabilities, we use the algorithm described below.

Step 1. The unnormalized probabilities $q(n, s)$ are defined by formulas

$$q(0,0) = 1; \tag{4}$$

$$q(0,1) = x; \tag{5}$$

$$q(n, s) = \gamma_{ns} + \delta_{ns}x + \theta_{ns}y, (n, s) \in \mathbf{X} : n > 0; \tag{6}$$

$$x = (P \cdot Q - R \cdot L) / (P \cdot T - U \cdot L); \tag{7}$$

$$y = (R \cdot T - U \cdot Q) / (P \cdot T - U \cdot L), \tag{8}$$

where

$$P = \lambda\theta_{N_1-1,1} - \alpha_2 \sum_{n=N_1}^{N_2} \theta_{n2} + (\alpha_1 + N_1\mu + \beta_1)\theta_{N_1,1};$$

$$Q = \lambda\gamma_{N_2-1,2} - (\alpha_2 + N_2\mu)\gamma_{N_2,2};$$

$$R = \alpha_2 \sum_{n=N_1}^{N_2} \gamma_{n2} + \lambda\gamma_{N_1-1,1} - (\alpha_1 + N_1\mu + \beta_1)\gamma_{N_1,1};$$

$$L = (\alpha_2 + N_2\mu)\theta_{N_2,2} - \lambda\theta_{N_2-1,2};$$

$$T = (\alpha_2 + N_2\mu)\delta_{N_2,2} - \lambda\delta_{N_2-1,2};$$

$$U = (\alpha_1 + N_1\mu + \beta_1)\delta_{N_1,1} - \alpha_2 \sum_{n=N_1}^{N_2} \delta_{n2} + \lambda\delta_{N_1-1,1}.$$

Step 2. The coefficients γ_{ns}, δ_{ns} and θ_{ns} are defined by recursive formulas

$$\gamma_{00} = 1, \delta_{00} = 0, \theta_{00} = 0; \quad \gamma_{01} = 0, \delta_{01} = 1, \theta_{01} = 0; \quad \gamma_{02} = 0, \delta_{02} = 0, \theta_{02} = 1;$$

$$\gamma_{10} = \frac{\lambda + \beta_0}{\mu}, \delta_{10} = -\frac{\alpha_1}{\mu}, \theta_{10} = 0; \quad \gamma_{11} = -\frac{\beta_0}{\mu}, \delta_{11} = \frac{\lambda + \alpha_1 + \beta_1}{\mu}, \theta_{11} = -\frac{\alpha_2}{\mu};$$

$$\gamma_{12} = 0, \delta_{12} = -\frac{\beta_1}{\mu}, \theta_{12} = \frac{\lambda + \alpha_2}{\mu};$$

$$n = \overline{2, N_0 - 1},$$

$$\gamma_{n0} = [(\gamma_{10} + (n+1))\gamma_{n-1,0} + \delta_{10}\gamma_{n-1,1} - \rho\gamma_{n-2,0}] \cdot n^{-1},$$

$$\delta_{n0} = [(\gamma_{10} + (n+1))\delta_{n-1,0} + \delta_{10}\delta_{n-1,1} - \rho\delta_{n-2,0}] \cdot n^{-1},$$

$$\theta_{n0} = [(\gamma_{10} + (n+1))\theta_{n-1,0} + \delta_{10}\theta_{n-1,1} - \rho\theta_{n-2,0}] \cdot n^{-1};$$

$$n = \overline{2, N_0},$$

$$\gamma_{n1} = [(\rho + (n-1) - \delta_{10} - \delta_{12})\gamma_{n-1,1} - \rho\gamma_{n-2,1} + \theta_{11}\gamma_{n-1,2} + \gamma_{11}\gamma_{n-1,0}] \cdot n^{-1},$$

$$\delta_{n1} = [(\rho + (n-1) - \delta_{10} - \delta_{12})\delta_{n-1,1} - \rho\delta_{n-2,1} + \theta_{11}\delta_{n-1,2} + \gamma_{11}\delta_{n-1,0}] \cdot n^{-1},$$

$$\theta_{n1} = [(\rho + (n-1) - \delta_{10} - \delta_{12})\theta_{n-1,1} - \rho\theta_{n-2,1} + \theta_{11}\theta_{n-1,2} + \gamma_{11}\theta_{n-1,0}] \cdot n^{-1};$$

$$n = \overline{2, N_1},$$

$$\gamma_{n2} = [(\rho + (n-1) - \theta_{11})\gamma_{n-1,2} - \rho\gamma_{n-2,2} + \delta_{12}\gamma_{n-1,1}] \cdot n^{-1},$$

$$\delta_{n2} = [(\rho + (n-1) - \theta_{11})\delta_{n-1,2} - \rho\delta_{n-2,2} + \delta_{12}\delta_{n-1,1}] \cdot n^{-1},$$

$$\theta_{n2} = [(\rho + (n-1) - \theta_{11})\theta_{n-1,2} - \rho\theta_{n-2,2} + \delta_{12}\theta_{n-1,1}] \cdot n^{-1};$$
$$n = \overline{N_0 + 1, N_1 - 1},$$
$$\gamma_{n1} = [(\rho + (n-1) - \delta_{10} - \delta_{12})\gamma_{n-1,1} - \rho\gamma_{n-2,1} + \theta_{11}\gamma_{n-1,2}] \cdot n^{-1},$$
$$\delta_{n1} = [(\rho + (n-1) - \delta_{10} - \delta_{12})\delta_{n-1,1} - \rho\delta_{n-2,1} + \theta_{11}\delta_{n-1,2}] \cdot n^{-1},$$
$$\theta_{n1} = [(\rho + (n-1) - \delta_{10} - \delta_{12})\theta_{n-1,1} - \rho\theta_{n-2,1} + \theta_{11}\theta_{n-1,2}] \cdot n^{-1};$$
$$n = \overline{N_1 + 1, N_2 - 1},$$
$$\gamma_{n2} = [(\rho + (n-1) - \theta_{11})\gamma_{n-1,2} - \rho\gamma_{n-2,2}] \cdot n^{-1},$$
$$\delta_{n2} = [(\rho + (n-1) - \theta_{11})\delta_{n-1,2} - \rho\delta_{n-2,2}] \cdot n^{-1},$$
$$\theta_{n2} = [(\rho + (n-1) - \theta_{11})\theta_{n-1,2} - \rho\theta_{n-2,2}] \cdot n^{-1}.$$

Finding the unnormalized probabilities $q(n,s), (n,s) \in \mathbf{X}$, one can compute the stationary probability distribution of the system as follows:

$$p(n,s) = \frac{q(n,s)}{\sum_{(i,j)\in\mathbf{X}} q(i,j)}, (n,s) \in \mathbf{X}. \tag{9}$$

4   Approximate Method

According to the model considered in the previous sections, the stationary probability distribution $p(n,s)_{(n,s)\in\mathbf{X}} = \mathbf{p}$ is not of product form.

To determine that form we simplify the model. We suppose that under RE influence, downlink power reduction occurs only when the current number of requests n in state s, $s = 1, 2$ is equal to maximum number of serviced requests in state $s - 1$, i.e. $n = N_{s-1}, s = 1, 2$. In other words user service is not interrupted. In Figs. 3 and 4 are shown respectively the state transition and the central state transition diagrams of the new model.

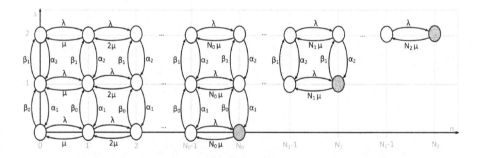

Fig. 3. State transition diagram of model without service interruption.

The behavior of the system is described using a two-dimensional vector (n,s) over the state space $\mathbf{X} = \{(n,s): 0 \leq n \leq N_s, s = 0, 1, 2\}$, where n represents the current number of requests in the system and s - the system current state.

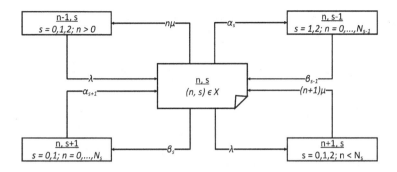

Fig. 4. Central state transition diagram of model without service interruption.

The discussed Markov process is described by the following system of equilibrium equations:

$$(\lambda + \beta_0)p(0,0) = \mu p(1,0) + \alpha_1 p(0,1);$$
$$(\lambda + \alpha_1 + \beta_1)p(0,1) = \mu p(1,1) + \alpha_2 p(0,2) + \beta_0 p(0,0);$$
$$(\lambda + \alpha_2)p(0,2) = \mu p(1,2) + \beta_1 p(0,1);$$
$$n = \overline{1, N_0 - 1}$$
$$(\lambda + n\mu + \beta_0)p(n,0) = \lambda p(n-1,0) + (n+1)\mu p(n+1,0) + \alpha_1 p(n,1);$$
$$n = \overline{1, N_0}$$
$$(\lambda + n\mu + \alpha_1 + \beta_1)p(n,1) = \lambda p(n-1,1) + (n+1)\mu p(n+1,1) + \alpha_2 p(n,2) +$$
$$+ \beta_0 p(n,0);$$
$$n = \overline{1, N_1}$$
$$(\lambda + n\mu + \alpha_2)p(n,2) = \lambda p(n-1,2) + (n+1)\mu p(n+1,2) + \beta_1 p(n,1);$$
$$n = \overline{N_0 + 1, N_1 - 1}$$
$$(\lambda + n\mu + \beta_1)p(n,1) = \lambda p(n-1,1) + (n+1)\mu p(n+1,1) + \alpha_2 p(n,2);$$
$$n = \overline{N_1 + 1, N_2 - 1}$$
$$(\lambda + n\mu)p(n,2) = \lambda p(n-1,2) + (n+1)\mu p(n+1,2);$$
$$(N_0\mu + \beta_0)p(N_0,0) = \lambda p(N_0 - 1,0) + \alpha_1 p(N_0,1);$$
$$(N_1\mu + \beta_1)p(N_1,1) = \lambda p(N_1 - 1,1) + \alpha_2 p(N_1,2);$$
$$N_2\mu p(N_2,2) = \lambda p(N_2 - 1,2),$$

where $p(n,s), (n,s) \in \mathbf{X}$ is the stationary probability distribution.

After simplification, the process describing the behavior of the system becomes a reversible Markov process. This is verifiable using Kolmogorov criterion. Then, the stationary probability distribution $p(n,s), (n,s) \in \mathbf{X}$ could be calculated using a product form (10), obtained throughout solution of system partial balance equations

$$p(n,s) = \left(\sum_{(i,j) \in \mathbf{X} \setminus \{(0,0)\}} \frac{\rho^i}{i!} \cdot f(j) + 1 \right)^{-1} \cdot \frac{\rho^n}{n!} \cdot f(s), \quad (n,s) \in \mathbf{X} \setminus \{(0,0)\}, \quad (10)$$

where $f(z)$ is the function equaling $\prod_{k=1}^{z} \beta_{k-1}/\alpha_k$ when argument is strictly positive and 1 when null.

Note that, the stationary probability distribution $p(n, s), (n, s) \in \mathbf{X}$ could also be calculated using a recursive algorithm like in Subsect. 3.2.

5   Performance Measures Analysis Using Both Methods

5.1   Performance Measures

Having found the stationary probability distribution $p(n, s), (n, s) \in \mathbf{X}$, one can calculate the main performance measures of the considered model such as

– Blocking probability B

$$B = \sum_{s=0}^{2} p(N_s, s); \tag{11}$$

– Average number of serviced requests $\overline{N}$

$$\overline{N} = \sum_{s=0}^{2} \overline{N}_s = \sum_{s=0}^{2} \sum_{n=1}^{N_s} n p(n, s); \tag{12}$$

– Mean bit rate $\overline{r}$

$$\overline{r} = \frac{\sum_{s=0}^{2} r_s \cdot \sum_{n=1}^{N_s} p(n, s)}{1 - \sum_{s=0}^{2} p(0, s)}; \tag{13}$$

– Interruption probability Π

$$\Pi = \sum_{n=N_0+1}^{N_1-1} \left(\frac{\alpha_1}{\alpha_1 + \beta_1 + n\mu + \lambda} \cdot \frac{n - N_0}{n} \cdot p(n, 1) \right) + \frac{\alpha_1}{\alpha_1 + \beta_1 + N_1\mu}$$

$$\times \frac{N_1 - N_0}{N_1} \cdot p(N_1, 1) + \sum_{n=N_1+1}^{N_2-1} \left(\frac{\alpha_2}{\alpha_2 + n\mu + \lambda} \cdot \frac{n - N_1}{n} \cdot p(n, 2) \right) \tag{14}$$

$$+ \frac{\alpha_2}{\alpha_2 + N_2\mu} \cdot \frac{N_2 - N_1}{N_2} \cdot p(N_2, 2).$$

5.2   Comparative Analysis of Results

We evaluate how effective is the developed approximate method with a comparative analysis of results obtained after the performance measures calculation, i.e. using the recursive algorithm described in Subsect. 3.2 and formula (10).

To understand the physical interpretation of the RE, let us consider the operation of the system within the LSA framework [11]. LSA facilitates access for additional licensees in bands which are already in use by one or more incumbents allowing to dynamically share the frequency band, whenever and wherever it is unused by the incumbent users, but only on the basis of an individual authorization, i.e. licensed.

We consider one frequency band in use by one mobile operator (licensee) who rent it from another mobile operator (incumbent). Whenever needed by the incumbent, the band is returned leading to a downlink power reduction and consequently, to a QoS degradation in the licensee network. We suppose that the incumbent does not use the frequency band permanently, i.e. every 30 min (urban use of the LSA system). Downlink power reduction occurs only when the data transfer quality in the incumbent network is not satisfying. We assume that the licensee requests to use the LSA band every 30 s. When access is granted, the downlink power increases to level p_1. At that level, the licensee sends a request every 60 s to reach the maximum downlink power p_2. We summarize all the parameters of the comparative analysis in Tables 1 and 2 [11].

Table 1. Constant parameters of the comparative analysis.

α_1^{-1}	α_2^{-1}	β_0^{-1}	β_1^{-1}	N_0	ω	λ	μ^{-1}	r_0	G
1200 s	1800 s	30 s	60 s	−60 dBm	10 MHz	5	0.1 s	1 Mbps	197.53

Table 2. Variable parameters of the comparative analysis.

-	d	p_2	p_1	p_0
Scenario 1	200–300 m	35, 39 dBm	$p_2 \cdot 2/3$	$p_2 \cdot 1/3$
Scenario 2	200, 250 m	33–42 dBm		

Let us consider several scenarios for the analysis of the system blocking probability.

Scenario 1. We illustrate the behavior of the blocking probability B depending on the distance d between the users and the BS (Fig. 5). The results presented in the graph show that at low downlink power (i.e. 35 dBm), the approximate method is suitable only for small cells. With the increase of downlink power, approximate method becomes more accurate and can then be used for big cells.

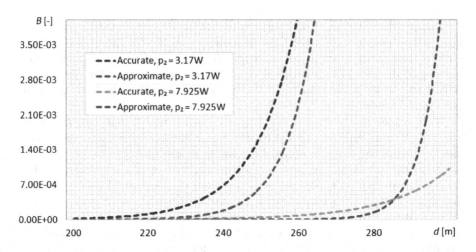

Fig. 5. Blocking probability B depending on distance d to BS.

Scenario 2. We illustrate the behavior of the blocking probability B depending on the maximum downlink power p_2 (Fig. 6). The results show that the higher maximum downlink power gets, the more accurate approximate method becomes.

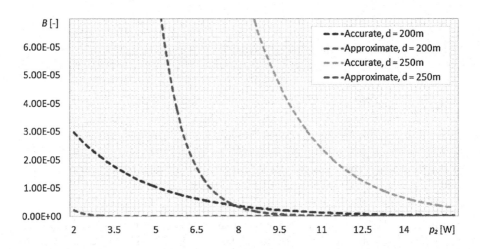

Fig. 6. Blocking probability B depending on maximum downlink power p_2.

6   Conclusion

In this paper, a mathematical model for accessing radio resources of wireless communication networks, i.e. using downlink power reduction and user service interruption mechanisms, is constructed and investigated. The model is described as a queuing system operating in RE. An accurate method (recursive algorithm) is proposed to calculate the stationary probability distribution of the model. An approximate method based on assumption that simplifies the model is also given. Formulas are proposed to calculate the main performance measures of the model - blocking probability, average number of serviced requests, mean bit rate and interruption probability. A comparative analysis of the blocking probability calculation using both methods showed that the approximation is quite precise at high downlink power and suitable for small cells only at low downlink power.

References

1. Ericsson: Ericsson mobility report, November 2018
2. Cisco Visual Networking Index: Global Mobile Data Traffic Forecast Update, 2017–2022, White Paper (2019)
3. 3GPP TS 22.105: Services and service capabilities: Release 14. – 3GPP. – 2017-03
4. Basaure, A., Sridhar, V., Hammainen, H.: Adoption of dynamic spectrum access technologies: a system dynamics approach. Telecommun. Syst. **63**(2), 169–190 (2016)
5. 3GPP TS 36.300: Evolved Universal Terrestrial Radio Access (E-UTRA) and Evolved Universal Terrestrial Radio Access Network (E-UTRAN); Overall description; Stage 2: Release 14. – 3GPP. – 2017-07
6. Acar, Y., Aldirmaz, S., Basar, E.: Channel estimation for OFDM-IM systems. Turk. J. Electr. Eng. Comput. Sci. **27**, 1908–1921 (2019). https://doi.org/10.3906/elk-1803-101
7. Ghallab, R., Shokair, M., Abouelazm, A., Sakr, A.A., Saad, W., Naguib, A.: Performance enhancement using MIMO electronic relay in massive MIMO cellular networks. IET Netw. (2019). https://doi.org/10.1049/iet-net.2018.5023
8. Garcia-Rodriguez, A., Geraci, G., Galati, L., Bonfante, A., Ding, M., Lopez-Perez, D.: Massive MIMO unlicensed: a new approach to dynamic spectrum access. IEEE Commun. Mag. **56**, 186–192 (2017). https://doi.org/10.1109/MCOM.2017.1700533
9. Ouyang, F.: Massive MIMO for dynamic spectrum access, pp. 9–12 (2017). https://doi.org/10.1109/ICCE.2017.7889210
10. 3GPP TS 23.246: Multimedia Broadcast/Multicast Service (MBMS); Architecture and functional description: Release 15. – 3GPP. – 2017-12
11. Markova, E., et al.: Flexible spectrum management in a smart city within licensed shared access framework. IEEE Access **5**, 22252–22261 (2017)
12. Maule, M., Moltchanov, D., Kustarev, P., Komarov, M., Andreev, S., Koucheryavy, Y.: Delivering fairness and QoS guarantees for LTE/Wi-Fi coexistence under LAA operation. IEEE Access **6**, 7359–7373 (2018)
13. Markova, E., Moltchanov, D., Gudkova, I., Samouylov, K., Koucharyavy, Y.: Performance assessment of QoS-aware LTE sessions offloading onto LAA/WiFi systems. IEEE Access **7**, 36300–36311 (2019)
14. Borodakiy, V., Gudkova, I., Markova, E., Samouylov, K.: Modelling and performance analysis of pre-emption based radio admission control scheme for video conferencing over LTE. In: Proceedings of the ITU Kaleidoscope Academic Conference, pp. 53–59. ITU, Geneva (2014)

15. Basharin, G.P., Samouylov, K.E., Yarkina, N.V., Gudkova, I.A.: A new stage in mathematical teletraffic theory. Autom. Remote Control **70**(12), 1954–1964 (2009)
16. Vishnevsky, V., Kozyrev, D., Rykov, V.: On the reliability of hybrid system information transmission evaluation. In: Proceedings of the Belarusian Winter Workshops in Queueing Theory BWWQT 2013, Minsk, Belarus, 28–31 January 2013, pp. 192–202 (2013)
17. Ali, A., Shah, G., Arshad, M.: Energy efficient resource allocation for M2M devices in 5G. Sensors **19**, 1830 (2019). https://doi.org/10.3390/s19081830
18. Chen, S., Ma, R., Chen, H.-H., Zhang, H., Meng, W., Liu, J.: Machine-to-machine communications in ultra-dense networks - a survey. IEEE Commun. Surv. Tutor. **19**, 1478–1503 (2017). https://doi.org/10.1109/COMST.2017.2678518
19. Su, J., Xu, H., Xin, N., Cao, G., Zhou, X.: Resource allocation in wireless powered iot system: a mean field stackelberg game-based approach. Sensors **18**, 3173 (2018). https://doi.org/10.3390/s18103173
20. Adou, Y., Markova, E.V., Gudkova, I.: Performance measures analysis of admission control scheme model for wireless network, described by a queuing system operating in random environment. In: Proceedings of the 10th International Congress on Ultra Modern Telecommunications and Control Systems ICUMT-2018, Moscow, Russia, 5–9 November 2018, pp. 262–267. IEEE, Piscataway (2018)

The Modeling of Call Center Functioning in Case of Overload

Sergey N. Stepanov$^{(\boxtimes)}$ (ID), Mikhail S. Stepanov (ID), and Hanna M. Zhurko (ID)

Department of Communication Networks and Commutation Systems,
Moscow Technical University of Communication and Informatics,
8A, Aviamotornaya str., Moscow 111024, Russia
stpnvsrg@gmail.com, mihstep@yandex.ru, hazhurko@gmail.com

Abstract. The mathematical model of call center functioning in case of overload is constructed and analyzed. In the model multi-skilled routing based on usage of one group of operators capable of serving simple requests and several groups of experts (consultants) handling more advanced topics is taken into account together with the possibility of request repetition in case of blocking or unsuccessful waiting time. Markov process that describes model functioning is constructed. Main performance measures of interest are defined through the values of stationary probabilities of model's states. Algorithm of characteristics estimation is suggested based on solving the system of state equations. Expressions that relates introduced performance measures in form of conservation laws are derived. It is shown how to use found relations for quantitative and qualitative analysis of the model functioning. The usage of the model for elimination of call center overload based on reducing of the input flow is considered.

Keywords: Call center · System of state equations · Multi-skilled routing · Repeated attempts

1 Introduction

Call centers are playing a significant role in creating the communication between customers and companies and between citizens and public organizations. The term *call center* corresponds to the case when a customer communicate with service system with help of fixed or mobile phone. Another option is the possibility of using multiple communication channels such as email, chat, social net. Such type of service system is called *contact center*. Throughout the paper the term call center will be used because it is supposed that the telephone type of communication is applied.

The most valuable resource of call center is the operators responsible for serving the coming requests from customers. Their salaries constitutes round 60–80

The publication has been prepared with the support of the Russian Foundation for Basic Research, project No. 16-29-09497ofi-m.

V. M. Vishnevskiy et al. (Eds.): DCCN 2019, LNCS 11965, pp. 391–406, 2019.
https://doi.org/10.1007/978-3-030-36614-8_30

% of total call center operating costs. This statement motivates the researchers to create the models that allow to study the process of serving the arriving requests with taking into account the peculiarities of call center functioning. Numerous references describing mathematical [1–9] and engineering [10–13] backgrounds of call center modeling can be found in the literature. The construction and analysis of such models meet some problems especially in case of overload when customer with some probability repeats the unsuccessful request [8–10]. In this situation a customer usually demonstrates a high level of persistency in contacts with call center. As result of such activity the intensity of retrials can be several times higher than the intensity of primary requests.

There are two approachers to overcome arising problems. If the increasing of input traffic has random character and caused by happening of some events like holidays, disasters and so on then it is necessary to reduce number of excepted requests. If the increasing of the input traffic caused by appearing of additional customers attracted for example by new services then it is required to estimate the intensity of primary requests and then calculate the right number of operators, consultants and waiting positions. In both cases to find an answer it is necessary to construct the queueing model of call center with possibility for blocked request to repeat the attempt. It was done for some simple models of call center functioning in [10]. In the paper this problem will be solved for the model with taking into account the multi-skilled routing and the possibility of retrial after blocking or unsuccessful waiting time.

In Sect. 2, we introduce main parameters and define the process of requests coming and serving. In Sect. 3 we construct a Markov process that describes the model functioning and give formal definitions for the main performance measures. Values of characteristics of requests servicing can be found after solving the system of state equations by Gauss-Seidel iterative algorithm. This approach will be described in Sects. 4. In Sect. 5 the constructed model is used to derive the conservation laws that relate introduced performance measures. It is shown how to use found relations for theoretical analysis of the constructed model, in particular, for choice of the borders of truncated space of states and for indirect measurement of the intensity of primary requests and other characteristics of call center functioning. In Sect. 6 we consider numerical results that illustrate the usage of the model for decreasing the influence of repetitions in case of overload.

2   Model Description

Requests for servicing from customers enter the call center via telephone access lines and further are serviced by IVR (Interactive Voice Response) and if needed by operators or by consultants from selected group. Let us denote by v the number of operators, and by v_k the number of consultants in the kth group, $k = 1, \ldots, n$. If operators or selected group of consultants are occupied than customer can wait the beginning of servicing. The number of waiting positions for the beginning of servicing is limited. Let us denote by w the number of waiting positions for operators, and by w_k the number of waiting positions for kth group of consultants, $k = 1, \ldots, n$.

Primary requests for servicing arrive according to a Poisson model with intensity λ. It is supposed that the model describes the call center functioning in case of overload. It means that apart of primary requests, the call center serves repeated requests. There are two main reasons for call repetition. The first reason, when there are no available operators, consultants in selected group, and places where to wait the beginning of servicing. Another reason is related with procedure of waiting. It is supposed that maximum allowed time of waiting is restricted and after it finishing customer can repeat the request. In both cases, a customer with probability H repeats the request after random time having exponential distribution with parameter ν and with additional probability $1 - H$ the customer stops his attempts for servicing and leaves the system. It is supposed that maximum allowed time of waiting the beginning of operator service and time of waiting the beginning of consultant service for group of consultant number k, $k = 1, \ldots, n$, has exponential distribution with parameter σ and σ_k correspondingly.

The process of servicing of a customer may include three phases: getting a recorded message from the IVR, receiving the information of general character from an operator, and getting specialized information from a consultant of selected group number k, $k = 1, \ldots, n$. We suppose that durations of the last two stages have exponential distribution with parameters μ and μ_k correspondingly. The selection of phases is described by fixed probabilities depending on the type of the request: primary or repeated. With probability q_p for primary requests and with probability q_r for retrials after getting service from IVR a customer with probability a_k is trying to get service at kth group of consultant, $k = 1, \ldots, n$ and with probability $a = 1 - \sum_{k=1}^{n} a_k$ is trying to get servicing at group of operators. With additional probability $1 - q_p$ for primary requests and with probability $1 - q_r$ for retrials a customer stops his attempts to get the information service and leaves the system unserved. After finishing of operator service a customer with probability cc_k is trying to continue service at kth group of consultant, $k = 1, \ldots, n$ and with probability $1 - c$ a customer stops his attempts to get the service and leaves the system unserved. Model's main parameters and the process of requests coming and serving are shown on Fig. 1.

3 Markov Process and Main Performance Measures

Let us denote the state of the system by vector $(j, i, i_1, \ldots, i_n)$ where j is the number of customers repeating a request for servicing, i is the number of occupied operators and waiting positions and i_k is the number of occupied consultants and waiting positions of kth group of consultants, $k = 1, \ldots, n$. The components of the vector varies as follows

$$j = 0, 1, \ldots,; \quad i = 0, 1, \ldots, v + w; \tag{1}$$

$$i_k = 0, 1, \ldots, v_k + w_k; \quad k = 1, \ldots, n.$$

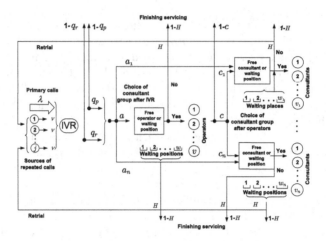

Fig. 1. The structure of call center mathematical model.

Let us denote by $j(t)$ the number of customers that at time t are in the state of repeating a request for servicing, by $i(t)$ we denote the number of operators and waiting positions occupied at time t; by $i_k(t)$ we denote the number of consultants and waiting positions of kth group of consultants occupied at time t, $k = 1, \ldots, n$.

The dynamics of model states changing is described by Markov process

$$r(t) = (j(t), i(t), i_1(t), \ldots, i_n(t)),$$

defined on the space of states S given by relations (1). The Markov character of $r(t)$ is due to the fact that durations of model staying at each state has exponential distribution and independent from each other just like transition probabilities from state to state. We consider the model functioning in stationary mode. In order for such a mode to exist it is sufficient to require that $H < 1$ and $\nu > 0$. If these inequalities are valid than in case of overload a customer leaves the system with probability tending to one after some time having bounded average.

Let us denote by $p(j, i, i_1, \ldots, i_n)$ probability of stationary states of the model $(j, i, i_1, \ldots, i_n) \in S$. The values $p(j, i, i_1, \ldots, i_n)$ can be interpreted as the portion of time the model stands in the state $(j, i, i_1, \ldots, i_n)$. Using this interpretation, we can introduce formal definitions for main performance measures of the analyzed model.

Mean Values of Components of the Model State. Let us denote by M_r the mean number of customers repeating a request for servicing, by M_i and M_w we denote the mean number of occupied operators and waiting positions correspondingly. Their formal definitions look as follows:

$$M_r = \sum_{\{(j,i,i_1,\ldots,i_n)\in S|j>0\}} p(j, i, i_1, \ldots, i_n)j; \tag{2}$$

$$M_i = \sum_{\{(j,i,i_1,...,i_n) \in S | i \leq v\}} p(j,i,i_1,...,i_n)i$$

$$+ \sum_{\{(j,i,i_1,...,i_n) \in S | i > v\}} p(j,i,i_1,...,i_n)v;$$

$$M_w = \sum_{\{(j,i,i_1,...,i_n) \in S | i > v\}} p(j,i,i_1,...,i_n)(i-v).$$

In the same way we denote by $M_{i,k}$ and $M_{w,k}$ the mean number of occupied consultants and waiting positions for kth group of consultants correspondingly, $k = 1,...,n$. Their formal definitions can be written as:

$$M_{i,k} = \sum_{\{(j,i,i_1,...,i_n) \in S | i_k \leq v_k\}} p(j,i,i_1,...,i_n)i_k \qquad (3)$$

$$+ \sum_{\{(j,i,i_1,...,i_n) \in S | i_k > v_k\}} p(j,i,i_1,...,i_n)v_k;$$

$$M_{w,k} = \sum_{\{(j,i,i_1,...,i_n) \in S | i_k > v_k\}} p(j,i,i_1,...,i_n)(i_k - v_k).$$

Intensities of Blocked Requests According to the Phase of Servicing. Let us denote by I_b and by $I_{b,k}$ the intensities of primary and repeated requests lost in attempt to get service from operators and from kth group of consultants correspondingly, $k = 1,...,n$. By $I_{b,g}$ we denote the intensity of requests blocked when entering a call-center. Formal definition of introduced characteristics looks as follows:

$$I_b = \sum_{\{(j,i,i_1,...,i_n) \in S | i = v + w\}} p(j,i,i_1,...,i_n)(\lambda q_p + j\nu q_r)a; \qquad (4)$$

$$I_{b,k} = \sum_{\{(j,i,i_1,...,i_n) \in S | i_k = v_k + w_k\}} p(j,i,i_1,...,i_n)(\lambda q_p + j\nu q_r)a_k$$

$$+ \sum_{\{(j,i,i_1,...,i_n) \in S | i \leq v, i_k = v_k + w_k\}} p(j,i,i_1,...,i_n)i\mu cc_k$$

$$+ \sum_{\{(j,i,i_1,...,i_n) \in S | i > v, i_k = v_k + w_k\}} p(j,i,i_1,...,i_n)v\mu cc_k;$$

$$I_{b,g} = \sum_{\{(j,i,i_1,i_2) \in S | i = v + w\}} p(j,i,i_1,i_2)(\lambda q_p + j\nu q_r)a$$

$$+ \sum_{k=1}^{n} \sum_{\{(j,i,i_1,...,i_n) \in S | i_k = v_k + w_k\}} p(j,i,i_1,...,i_n)(\lambda q_p + j\nu q_r)a_k.$$

Intensities of Arrived Requests According to the Phase of Servicing. Let us denote by I_t and by $I_{t,k}$ the intensities of primary and repeated requests coming to get service from operators and from kth group of consultants correspondingly, $k = 1, \ldots, n$. By $I_{t,g}$ we denote the intensity of requests coming to get service from operators or consultants. The values of defined characteristics can be written in the following way:

$$I_t = \lambda q_p a + \sum_{\{(j,i,i_1,\ldots,i_n) \in S\}} p(j,i,i_1,\ldots,i_n) j \nu q_r a; \tag{5}$$

$$I_{t,k} = \lambda q_p a_k + \sum_{\{(j,i,i_1,\ldots,i_n) \in S\}} p(j,i,i_1,\ldots,i_n) j \nu q_r a_k$$

$$+ \sum_{\{(j,i,i_1,\ldots,i_n) \in S | i \leq v\}} p(j,i,i_1,\ldots,i_n) i \mu c c_k$$

$$+ \sum_{\{(j,i,i_1,\ldots,i_n) \in S | i > v\}} p(j,i,i_1,\ldots,i_n) v \mu c c_k;$$

$$I_{t,g} = \lambda q_p + \sum_{\{(j,i,i_1,\ldots,i_n) \in S\}} p(j,i,i_1,\ldots,i_n) j \nu q_r.$$

Losses According to the Phase of Servicing. Let us denote by π_t and by $\pi_{t,k}$ the portions of time when all operators and waiting positions are occupied and when all consultants and waiting positions of kth group are occupied correspondingly, $k = 1, \ldots, n$. Their definitions can be written in the following way:

$$\pi_t = \sum_{\{(j,i,i_1,\ldots,i_n) \in S | i = v + w\}} p(j,i,i_1,\ldots,i_n); \tag{6}$$

$$\pi_{t,k} = \sum_{\{(j,i,i_1,\ldots,i_n) \in S | i_k = v_k + w_k\}} p(j,i,i_1,\ldots,i_n).$$

Let us denote by π_c and by $\pi_{c,k}$ the ratios of lost primary and repeated requests coming to get service from operators and from kth group of consultants correspondingly, $k = 1, \ldots, n$. Their formal definitions looks as follows:

$$\pi_c = \frac{I_b + M_w \sigma}{I_t}; \quad \pi_{c,k} = \frac{I_{b,k} + M_{w,k} \sigma_k}{I_{t,k}}. \tag{7}$$

Mean Time of Waiting or Servicing According to the Phase of Servicing. Let us denote by t_w and by $t_{w,k}$ the mean times of waiting or servicing of requests coming to get service from operators and from kth group of consultants correspondingly, $k = 1, \ldots, n$. The formulas defining the introduced characteristics can be written in the following way:

$$t_w = \frac{M_i + M_w}{I_t - I_b}; \quad t_{w,k} = \frac{M_{i,k} + M_{w,k}}{I_{t,k} - I_{b,k}}. \tag{8}$$

Global Measures of Requests Servicing at Analyzed Model of Call Center. Let us denote by $\pi_{c,t}$ the ratio of lost primary and repeated requests coming to get service at call center; by $t_{w,t}$ we define the mean time of waiting or servicing of requests coming to get service at call center; by $M_{w,t}$ we define the mean number of requests waiting the beginning of service and by $M_{i,t}$ define the mean number of requests servicing at call center, correspondingly. The formal definitions of introduced performance measures looks as follows:

$$\pi_{c,t} = \frac{I_b + M_w\sigma + \sum_{k=1}^{n}(I_{b,k} + M_{w,k}\sigma_k)}{I_{t,g}}; \tag{9}$$

$$t_{w,t} = \frac{M_i + M_w + \sum_{k=1}^{n}(M_{i,k} + M_{w,k})}{I_{t,g} - I_{b,g}};$$

$$M_{w,t} = M_w + \sum_{k=1}^{n} M_{w,k}; \quad M_{i,t} = M_i + \sum_{k=1}^{n} M_{i,k}.$$

4   System of State Equations

The model performance measures are expressed through values of $p(j, i, i_1, \ldots, i_n)$ that can be found from the solution of the system of state equations. Let us denote by $P(j, i, i_1, \ldots, i_n)$ the unnormalized values of $p(j, i, i_1, \ldots, i_n)$. After equating the intensity of leaving arbitrary model state $(j, i, i_1, \ldots, i_n) \in S$ to the intensity of entering the state $(j, i, i_1, \ldots, i_n)$ we obtain the following system of linear equations:

$$P(j, i, i_1, \ldots, i_n) \tag{10}$$

$$\times \left\{ j\nu(1 - q_r) + \left(\lambda q_p + j\nu q_r\right)\left(aI(i < v + w) + \sum_{k=1}^{n} a_k I(i_k < v_k + w_k)\right) \right.$$

$$+ \left(\lambda q_p H + j\nu q_r(1 - H)\right)\left(aI(i = v + w) + \sum_{k=1}^{n} a_k I(i_k = v_k + w_k)\right)$$

$$+ i\mu I(i \le v) + v\mu I(i > v) + \sum_{k=1}^{n}\left(i_k\mu_k I(i_k \le v_k) + v_k\mu_k I(i_k > v_k)\right)$$

$$\left. + (i - v)\sigma I(i > v) + \sum_{k=1}^{n}(i_k - v_k)\sigma_k I(i_k > v_k)\right\}$$

$$= P(j, i - 1, i_1, \ldots, i_n)\lambda q_p aI(i > 0)$$

$$+ \sum_{k=1}^{n} P(j, i, i_1, \ldots, i_k - 1, \ldots, i_n)\lambda q_p a_k I(i_k > 0)$$

$$+ P(j + 1, i - 1, i_1, \ldots, i_n)(j + 1)\nu q_r aI(i > 0)$$

$$+ \sum_{k=1}^{n} P(j + 1, i, i_1, \ldots, i_k - 1, \ldots, i_n)(j + 1)\nu q_r a_k I(i_k > 0)$$

$$+P(j-1,i,i_1,\ldots,i_n)\lambda q_p H I(j>0)$$

$$\times\left(aI(i=v+w)+\sum_{k=1}^{n}a_k I(i_k=v_k+w_k)\right)$$

$$+P(j+1,i,i_1,\ldots,i_n)(j+1)\nu q_r(1-H)$$

$$\times\left(aI(i=v+w)+\sum_{k=1}^{n}a_k I(i_k=v_k+w_k)\right)$$

$$+P(j+1,i,i_1,\ldots,i_n)(j+1)\nu(1-q_r)$$

$$+P(j,i+1,i_1,\ldots,i_n)\mu(1-c)$$

$$\times\left((i+1)I(i+1\le v)+vI(v<i+1\le v+w)\right)$$

$$+\sum_{k=1}^{n}P(j,i+1,i_1,\ldots,i_k-1,\ldots,i_n)\mu cc_k I(i_k>0)$$

$$\times\left((i+1)I(i+1\le v)+vI(v<i+1\le v+w)\right)$$

$$+\sum_{k=1}^{n}P(j,i+1,i_1,\ldots,i_n)\mu cc_k(1-H)I(i_k=v_k+w_k)$$

$$\times\left((i+1)I(i+1\le v)+vI(v<i+1\le v+w)\right)$$

$$+\sum_{k=1}^{n}P(j-1,i+1,i_1,\ldots,i_n)\mu cc_k H I(i_k=v_k+w_k,j>0)$$

$$\times\left((i+1)I(i+1\le v)+vI(v<i+1\le v+w)\right)$$

$$+\sum_{k=1}^{n}P(j,i,i_1,\ldots,i_k+1,\ldots,i_n)\times\left((i_k+1)\mu_k I(i_k+1\le v_k)\right.$$

$$+(v_k\mu_k+(i_k+1-v_k)\sigma_k(1-H))I(v_k<i_k+1\le v_k+w_k)\Big)$$

$$+P(j,i+1,i_1,\ldots,i_n)(i+1-v)\sigma(1-H)I(v<i+1\le v+w)$$

$$+P(j-1,i+1,i_1,\ldots,i_n)(i+1-v)\sigma H I(v<i+1\le v+w,j>0)$$

$$+\sum_{k=1}^{n}P(j-1,i,i_1,\ldots,i_k+1,\ldots,i_n)$$

$$\times(i_k+1-v_k)\sigma_k H I(v_k<i_k+1\le v_k+w_k,j>0),$$

$$(j,i,i_1,\ldots,i_n)\in S.$$

By $I(\cdot)$ in (10) the indicator function is defined

$$
I(\cdot) = \begin{cases} 1, & \text{if condition formulated in brackets is fulfilled,} \\ 0, & \text{if this condition isn't fulfilled.} \end{cases}
$$

Values of $P(j, i, i_1, \ldots, i_n)$ should be normalized.

In order to solve the system by standard algorithms of linear algebra it is necessarily to limit the number of unknowns in the system of state equations. For this purpose, it sufficient to restrict the number of customers repeating the attempt by some integer j_m. The value j_m can be found through numerical experiments (see details in Sect. 5). When calculating the performance measures it will be supposed further that in (1) the value of j varies in interval $j = 0, 1, \ldots, j_m$. System of state equation can be easily rewritten for the case $j_m < \infty$ but looks too complicated and because of this reason is omitted here. Almost all elements of the matrix of the system of state Eq. (10) are zeros. In this case the optimal approach to solve (10) consist in using Gauss-Seidel iterative algorithm [8,9,14] or by other approachers [15,16].

It is possible to construct a counterexample that shows that convergence of Gauss-Zeidel algorithm for singular system (10) is not guaranteed. In order to guarantee the convergence it is sufficient to put one of unknown probabilities $P(j, i, i_1, \ldots, i_n)$ in (10) to one, remove from (10) equation corresponding to the chosen probability and after implement Gauss-Seidel algorithm to obtained non-singular system of linear equations. The convergence of Gauss-Seidel and Jacobi algorithms in this case is guaranteed because the matrix of the constructed system of linear equations belongs to the class of irreducibly diagonally dominant matrices [14]. It is worth to mention that convergence of Gauss-Seidel algorithm for singular system (10) is often much faster than for considered above nonsingular case.

5   Conservation Laws and Their Applications

Derivation of Conservation Laws. Model performance measures introduced in Sect. 3 are related by conservation laws that can be obtained after multiplication of the system of state Eq. (10) consequently by $j, i, i_1, \ldots, i_n$. After collecting the terms we get $(n+2)$ relations according to the number of components in the vector of state $(j, i, i_1, \ldots, i_n)$:

$$
M_r \nu = \left(I_b + M_w \sigma + \sum_{k=1}^{n} (I_{b,k} + M_{w,k} \sigma_k) \right) H; \tag{11}
$$

$$
I_t = I_b + M_w \sigma + M_i \mu;
$$

$$
I_{t,k} = I_{b,k} + M_{w,k} \sigma_k + M_{i,k} \mu_k, \quad k = 1, \ldots, n.
$$

Relations (11) can be also proved with help of Little's formula. Let us show it. The mean time for a customer to repeat the request is $\frac{1}{\nu}$. According to the Little formula $\frac{1}{\nu}$ equals to the mean number of customers M_r that are in the state of request repetition divided to the intensity of requests that are going to repeat an attempt $\left(I_b + M_w\sigma + \sum_{k=1}^{n}(I_{b,k} + M_{w,k}\sigma_k)\right)H$. As result we have relation

$$\frac{1}{\nu} = \frac{M_r}{\left((I_b + M_w\sigma + \sum_{k=1}^{n}(I_{b,k} + M_{w,k}\sigma_k)\right)H}$$

that is equivalent to the first conservation law in (11). It has very simple explanation: intensity of retrials equals to the intensity of events that caused repeated attempts. In the same way it is possible to proof other laws in (11).

The Choice of Truncation Level. In order to solve (10) by standard algorithms of linear algebra it is necessarily to limit the number of unknowns in the system of state equations. For this purpose, it is sufficient to restrict the number of customers repeating the attempt by some integer j_m. If a customer gets refusal in establishing the connection with operator or consultant and call center has already j_m customers repeating a request for servicing then blocked customer with probability equals to one stops his attempts to get servicing. Let us denote the characteristics of truncated model by the same symbols as was used for model without restriction on j only with superscript $*$. Let us find an empirical formula for estimation of the error caused by truncation of the used space of states. We solve the formulated problems with help of conservation laws.

By the same way as it was done for the initial model it is possible to prove the validity of conservation laws for truncated model. They are looking as follows

$$M_r^*\nu = \left(I_b^* + M_w^*\sigma + \sum_{k=1}^{n}(I_{b,k}^* + M_{w,k}^*\sigma_k)\right)H - \gamma; \qquad (12)$$

$$I_t^* = I_b^* + M_w^*\sigma + M_i^*\mu;$$

$$I_{t,k}^* = I_{b,k}^* + M_{w,k}^*\sigma_k + M_{i,k}^*\mu_k, \quad k = 1,\ldots,n,$$

where γ is calculating as follows

$$\gamma = \sum_{\{(j,i,i_1,\ldots,i_n)\in S|i=v+w\}} p(j_m,i,i_1,\ldots,i_n)\lambda q_p a H \qquad (13)$$

$$+ \sum_{\{(j,i,i_1,\ldots,i_n)\in S|v<i+1\leq v+w\}} p(j_m,i+1,i_1,\ldots,i_n)(i+1-v)\sigma H$$

$$+ \sum_{k=1}^{n}\left\{ \sum_{\{(j,i,i_1,\ldots,i_n)\in S|i_k=v_k+w_k\}} p(j_m,i,i_1,\ldots,i_n)\lambda q_p a_k H + \right.$$

$$+ \sum_{\{(j,i,i_1,\ldots,i_n)\in S|i+1\leq v,i_k=v_k+w_k\}} p(j_m,i+1,i_1,\ldots,i_n)(i+1)\mu c c_k H$$

$$+ \sum_{\{(j,i,i_1,\ldots,i_n)\in S|v<i+1\leq v+w,i_k=v_k+w_k\}} p(j_m,i+1,i_1,\ldots,i_n)v\mu c c_k H$$

$$+ \sum_{\{(j,i,i_1,\ldots,i_n)\in S|v_k<i_k+1\leq v_k+w_k\}} p(j_m,i,i_1,\ldots,i_k+1,\ldots,i_n)(i_k+1-v_k)\sigma_k H \Bigg\}.$$

Let us denote the difference between characteristics of initial and truncated models by symbol Δ. For example, $\Delta M_r = M_r - M_r^*$. Subtracting from (11) the relations (12) and using the definitions of characteristics (2)–(5) it is easily to derive the following expression

$$\Delta M_r \nu(1-q_r H) = \gamma - \Delta M_i \mu(1-c)H - \sum_{k=1}^{n} \Delta M_{i,k} \mu_k H. \qquad (14)$$

By restricting the maximum allowed number of repeating customers we decrease in the truncated model the total intensity of coming requests for servicing. Because of this property we can expect that characteristics of initial model defined by relations (2)–(5) are upper estimates for corresponding characteristics of truncated model. This result is strictly proved for the basic model with repeated calls [15] and is supported by numerical experiments for introduced model of call center. From (14) follows empirical upper estimate for absolute error caused by using M_r^* instead of M_r. It looks as follows

$$\Delta M_r \leq \frac{\gamma}{\nu(1-q_r H)}. \qquad (15)$$

Similar estimation can be found for other characteristics. It is worth to mention that this error is expressed in terms of performance measures of truncated model and the error is proportional to the γ.

Let us consider a numerical example that illustrates the accuracy of estimation (15). Model input parameters are as follows: $\lambda = 30$; $q_p = 0,7$; $q_r = 0,9$; $n = 2$; $a_1 = 0,25$; $a_2 = 0,35$; $a = 1 - a_1 - a_2 = 0,4$; $H = 0,7$; $\nu = 5$; $j_m = 10$; $\mu = 1$; $\mu_1 = 0,5$; $\mu_2 = 0,25$; $c = 0,6$; $c_1 = 0,7$; $c_2 = 0,3$; $\sigma = 1$; $\sigma_1 = 0,5$; $\sigma_2 = 0,25$; $v = 20$; $w = 10$; $v_1 = 8$; $w_1 = 4$; $v_2 = 6$; $w_2 = 3$, $\varepsilon = 10^{-15}$. The value of j_m varies from 2 to 40. The values of characteristics found for $j_m = 40$ are excepted as exact values corresponding to the case of unlimited interval of changing j_m. As a time unit was chosen the mean time of servicing a request by operator. In the Table 1 are presented the values of j_m and depending on j_m the values of M_r, ΔM_r, the upper estimation of ΔM_r found from (15) (we denote it as $\Delta^* M_r$) and γ.

As it is seen from the content of the table the estimation of the error (15) is quite close to the exact value of the error for all chosen values of j_m. It means that suggested approach can be used for choice of the truncation level when making calculation of the performance measures of the studied model of call center. More detailed investigation of the problem of truncation of the used state space for models with taking into account repeated calls can be found in [15].

Table 1. The dependence on j_m of the estimation of the error (15) caused by truncation of the used space of states.

j_m	M_r	ΔM_r	$\Delta^* M_r$	γ
2	1,3497738094	$2,290 \cdot 10^0$	$2,880 \cdot 10^0$	$5,329 \cdot 10^0$
4	2,4235447417	$1,216 \cdot 10^0$	$1,518 \cdot 10^0$	$2,808 \cdot 10^0$
6	3,1347875935	$5,051 \cdot 10^{-1}$	$6,236 \cdot 10^{-1}$	$1,154 \cdot 10^0$
8	3,4877652669	$1,521 \cdot 10^{-1}$	$1,855 \cdot 10^{-1}$	$3,432 \cdot 10^{-1}$
10	3,6075970046	$3,228 \cdot 10^{-2}$	$3,888 \cdot 10^{-2}$	$7,193 \cdot 10^{-2}$
12	3,6349125744	$4,968 \cdot 10^{-3}$	$5,905 \cdot 10^{-3}$	$1,093 \cdot 10^{-2}$
14	3,6392970316	$5,831 \cdot 10^{-4}$	$6,837 \cdot 10^{-4}$	$1,265 \cdot 10^{-3}$
16	3,6398254497	$5,469 \cdot 10^{-5}$	$6,319 \cdot 10^{-5}$	$1,169 \cdot 10^{-4}$
18	3,6398758872	$4,257 \cdot 10^{-6}$	$4,843 \cdot 10^{-6}$	$8,959 \cdot 10^{-6}$
20	3,6398798602	$2,836 \cdot 10^{-7}$	$3,176 \cdot 10^{-7}$	$5,876 \cdot 10^{-7}$
22	3,6398801272	$1,662 \cdot 10^{-8}$	$1,832 \cdot 10^{-8}$	$3,390 \cdot 10^{-8}$
24	3,6398801430	$8,754 \cdot 10^{-10}$	$9,507 \cdot 10^{-10}$	$1,759 \cdot 10^{-9}$
40	3,6398801438	$0,000 \cdot 10^0$	$2,878 \cdot 10^{-21}$	$5,325 \cdot 10^{-21}$

Indirect Measurements of λ, H and Other Characteristics. The constructed model can be used for estimation of necessary amount of operators, consultants and waiting positions to avoid of call center overload. The accuracy of estimation depends on the accuracy of model input parameters. Main among them are λ the intensity of primary requests for servicing and H the probability of repetition. It is difficult to estimate the values of these parameters because we should divide the total flow of coming requests into parts corresponding to primary and repeated components. It is hard to do so for big call centers because we should keep the history of contacts of each customer with call center. However we can solve this problem indirectly by relations derived from conservation laws (11).

By using standard measurement equipment we can easily estimate the value of $\Lambda = I_{t,g}$ the total intensity of primary and repeated requests entering call center to get servicing from operators or consultants and the value of $\pi_{c,t}$ the ratio of lost primary and repeated requests coming to get service at call center. In measuring Λ and $\pi_{c,t}$ we do not have to distinguish primary and repeated requests. Adding to the left- and right-hand sides of the first relation (11) the value of λq_p and using the definition of $\pi_{c,t}$, we get

$$\lambda q_p = \Lambda(1 - \pi_{c,t} H q_r). \tag{16}$$

From (11) we find expressions than can be used for indirect measurement of λ and H

$$\lambda = \Lambda(1 - \pi_{c,t} H q_r)\frac{1}{q_p}, \qquad H = \frac{\Lambda - \lambda q_p}{\Lambda \pi_{c,t} q_r}. \tag{17}$$

In a similar way we can derive formulas for indirect estimation of other characteristics of call center functioning that are difficult to measure by straightforward methods. Let us define by $M_{r,p}$ the mean number of retrials per one primary request. Then from (11) and (16) we obtain

$$M_{r,p} = \frac{M_r \nu}{\lambda} = \frac{\pi_{c,t} H q_p}{1 - \pi_{c,t} H q_r}. \tag{18}$$

6 The Usage of the Model for Elimination of Call Center Overload

The process of normal functioning of call center may be disturbed by overload caused by increasing of the intensity of coming requests for servicing. In the first scenario this raise is random and caused by happening of some events like holidays, disasters and so on. In the considered case to keep call center functioning in normal order it is necessary to reduce number of excepted requests. After a short period the raise of total flow is stabilized. In another scenario the increasing of input traffic can be caused by appearing of additional customers attracted for example by new services. In this case it is necessary to estimate the intensity of primary requests and then calculate the required number of operators, consultants and waiting positions. In both cases the elaborated model can be used for solving the arising problems. Let us consider the first scenario. The analysis of the second scenario will be done in the separate paper.

Let us consider a numerical example that illustrates the usage of the call center model to study the process of reducing of incoming flow of requests in case of overload. Model input parameters are as follows: $\lambda = 50$; $q_p = 1$; $q_r = 1$; $n = 2$; $a_1 = 0{,}2$; $a_2 = 0{,}1$; $a = 1 - a_1 - a_2 = 0{,}7$; $H = 0{,}9$; $\nu = 10$; $j_m = 60$; $\mu = 1$; $\mu_1 = 0{,}75$; $\mu_2 = 0{,}5$; $c = 0{,}2$; $c_1 = 0{,}6$; $c_2 = 0{,}4$; $\sigma = 1$; $\sigma_1 = 0{,}75$; $\sigma_2 = 0{,}5$; $v = 18$; $w = 9$; $v_1 = 10$; $w_1 = 6$; $v_2 = 8$; $w_2 = 5$; $\varepsilon = 10^{-15}$. The intensity of input flow is reduced by decreasing the probabilities of entering call center $q = q_p = q_r$ from 1 (this choice corresponds to the overload) to 0,25 with step 0,025. The rate of decreasing $\pi_{c,t}$ and π_c with decreasing of the probability of entering call center $q = q_p = q_r$ is shown on Fig. 2. If the goal of call center management is to reduce the ratio of lost requests below 0,1 then it is enough to choose $q = q_p = q_r \leq 0{,}47$. The behavior of M_r and $M_{w,t}$ with decreasing of the probability of entering call center $q = q_p = q_r$ is shown on Fig. 3.

The presented results clearly show that by decreasing the probability of entering call center for coming requests it is possible to eliminate the negative consequences of overload. Among them are heavy losses of coming requests, large number of customers repeating the attempt of connection and big number of customers waiting the beginning of servicing. The constructed model allows to study the process of forming and servicing requests in case of overload and choose the right values of parameters that can be used for control the call center functioning.

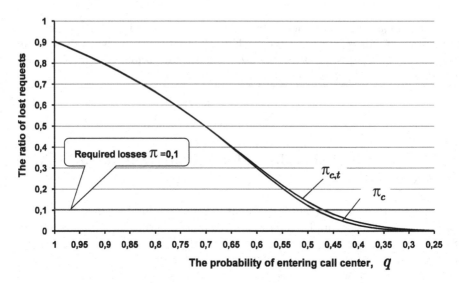

Fig. 2. The rate of decreasing the ratio of lost requests with decreasing of the probability of entering call center.

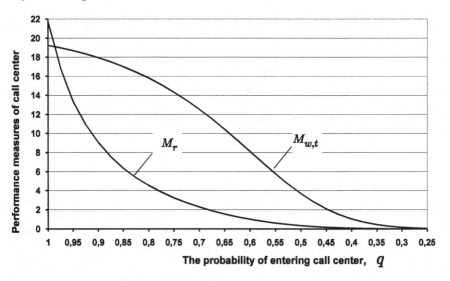

Fig. 3. The rate of decreasing M_r and $M_{w,t}$ with decreasing of the probability of entering call center.

7   Conclusion

In this paper the mathematical model of call center functioning in case of overload is constructed and analyzed. In the model multi-skilled routing based on usage of one group of operators capable of serving simple requests and several groups of experts (consultants) handling more advanced topics is taken into account together with the possibility of call repetition in case of blocking or unsuccessful waiting time. Primary and repeated requests for servicing are coming after exponentially distributed time intervals. It is supposed that service times at operators and consultants groups also have exponential distribution with parameters depending on chosen group. Markov process that describes model functioning is constructed. In the framework of the proposed model the definitions of main performance measures of interest are formulated through values of probabilities of model's stationary states. Algorithm of performance measures estimation is suggested based on solving the system of state equations by iterative algorithm. Expressions that relates introduced performance measures in form of conservation laws are derived. It is shown how to use found relations for indirect measurement of the intensity of primary requests and other characteristics of call center functioning.

The model and derived algorithms of performance measures estimation can be used to produce the quantitative and qualitative analysis of the dependence of model's performance measures on the values of input parameters with taking into account the customer behavior in case of overload. The constructed analytical framework additionally offers the possibility to find the numbers of operators, consultants and waiting positions required for serving coming requests with given values of performance indicators and to find the size of buffers and maximum allowed time for requests to be in the buffers that permits to serve traffic flows with acceptable value of delay instead of increasing the number of operators or consultants. Proposed model can be further developed to include the possibility of non exponential distribution of time interval between retrials, service times and maximum allowed waiting times.

References

1. Gans, N., Koole, M., Mandelbaum, A.: Telephone call-centers: tutorial, review and research prospects. Manuf. Serv. Manag. **5**, 79–141 (2003)
2. Stolletz, R., Helber, S.: Performance analysis of an inbound call-center with skills-based routing. OR Spectr. **26**, 331–352 (2004)
3. Borst, S., Mandelbaum, A., Reiman, M.I.: Dimensioning large call centers. Oper. Res. **52**(1), 17–34 (2004)
4. Koole, G., Mandelbaum, A.: Queueing models of call centers: an introduction. Ann. Oper. Res. **113**(4), 41–59 (2002)
5. Cezik, M.T., L'Ecuyer, P.: Staffing multiskill call centers via linear programming and simulation. Manag. Sci. **54**(2), 310–323 (2008)
6. Mandelbaum, A., Zeltyn, S.: Staffing many-server queues with impatient customers: constraint satisfaction in call centers. Oper. Res. **57**(5), 1189–1205 (2009)
7. Bhulai, S., Koole, G.: A queueing model for call blending in call centers. IEEE Trans. Autom. Control **48**(8), 1434–1438 (2003)
8. Stepanov, S.N., Stepanov, M.S.: Construction and analysis of a generalized contact center model. Autom. Remote Control **75**(11), 1936–1947 (2014)
9. Stepanov, S.N., Stepanov, M.S.: Algorithms for estimating throughput characteristics in a generalized call center model. Autom. Remote Control **77**(7), 1195–1207 (2016)

10. Aguir, S., Karaesmen, F., Aksin, O.Z., Chauvet, F.: The impact of retrials on call center performance. OR Spectr. **26**(3), 353–376 (2004)

11. Aksin, Z., Armony, M., Mehrotra, A.: The modern call center: a multi- disciplinary perspective on operations management research. Prod. Oper. Manag. **16**(6), 665–688 (2007)

12. Whitt, W.: Engineering solution of a basic call-center model. Manag. Sci. **51**(2), 221–235 (2005)

13. Whitt, W.: Staffing a call center with uncertain arrival rate and absenteeism. Prod. Oper. Manag. **15**(1), 88–102 (2006)

14. Barker, V.A.: Numerical solution of sparse singular systems of equations arising from ergodic Markov chains. Stoch. Models **5**(3), 335–381 (1989)

15. Stepanov, S.N.: Markov models with retrials: the calculation of stationary performance measures based on the concept of truncation. Math. Comput. Model. **30**, 207–228 (1999)

16. Stepanov, S.N.: Generalized model with retrials in case of extreme load. Queueing Syst. **27**, 131–151 (1998)

Evaluation of Information Transmission Resource While Processing Heterogeneous Traffic in Data Networks

Veronika M. Antonova[1,2](✉) ⓘ, Natalia A. Grechishkina[1] ⓘ,
Ludmila Yu. Zhilyakova[3] ⓘ, and Nickolay A. Kuznetsov[1] ⓘ

[1] Kotelnikov Institute of Radioengineering and Electronics (IRE)
of Russian Academy of Sciences, Moscow, Russia
xarti@mail.ru
[2] Bauman Moscow State Technical University, Moscow, Russia
[3] V. A. Trapeznikov Institute of Control Sciences of RAS, Moscow, Russia
zhilyakova.ludmila@gmail.com

Abstract. The paper briefly describes a mathematical model of multiservice traffic processing in a data network. Methods for the model's quality indicator assessment based on working out a system of equilibrium equations and its subsequent solving by means of the iterative Gauss-Seidel method are proposed. The work presents numerical results and their usage for planning the values of the model's structural parameters such as evaluation results of the rate in the cell providing the service of specified request flows of real time and non-real time traffic with the predetermined amount of maximum losses as well as results of the rate estimation in the cell, which provides the service of the specified requests flows of elastic traffic and real time services traffic, with the specified values of the maximum losses and the average elastic traffic transfer time.

Keywords: Real-time traffic · Elastic traffic · QoS · M2M

1 Introduction

Currently, an increasing number of enterprise security systems, like alarm systems, control of physical access to facilities, monitoring of various detectors, and telemetry, use cellular networks for data transmission. Due to the decrease in revenue from traditional communication services, mobile operators begin to actively develop the market segment of machine-to-machine (M2M) communication equipment. However, increasing in traffic induced by machine-to-machine communication should not affect sensitive real-time services such as voice connections, video conferencing, and etc. To meet this requirement the article proposes applying guaranteed thresholds for elastic traffic.

The work is partially supported by the Russian Foundation for Basic Research (project No. 19-07-00525 A – Developing flow-based models of routing problems in telecommunications networks).

2   The Model of Servicing Heterogeneous Traffic in a Cell

We denote by C the rate of information transmission that is expressed in bits/s and provided by the technical capabilities of a separate cell in a data network. The model considers the process of receiving two flows of requests: the first one is for real time traffic and the second one is for elastic/non-real time data traffic. Requests for real time traffic transfer are accepted according to Poisson law with the intensity λ_r and the rate equal to c_r bps. The time for processing a request t_r has an exponential distribution with an average of $\frac{1}{\mu_r}$, where μ_r is the parameter of exponential distribution [2,3].

The receipt of new requests for elastic traffic transferring also obeys the Poisson law with the intensity λ_d and the rate of a request being serviced varying in the range of $c_1 \leq c_d \leq c_2$. We will assume that the size of the transmitted non-real time request has an exponential distribution with an average value F expressed in bits. The transmission time of elastic traffic using only the minimal c_1 or only the maximal c_2 speeds has an exponential distribution with the corresponding parameters $\mu_{d,1} = \frac{c_1}{F}$ and $\mu_{d,2} = \frac{c_2}{F}$. This is shown in Fig. 1.

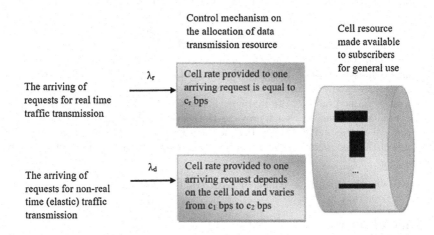

Fig. 1. An operation scheme of a network cell model

A new request can only be accepted to be serviced if every previously accepted request is able to be processed at an acceptable transfer rate. Requests for real-time services traffic transfer have a relative priority in using the information transfer resource. Denote by i_r and i_d the number of real-time requests and elastic traffic ones being serviced, respectively.

Thus, when the transmission rate increases or decreases, the mean value of residual time for data file transfer decreases or increases proportionally. The data transfer rate changes dynamically in accordance with the cell loading.

The process of request arriving is shown in Fig. 2. Hatched cells show real-time traffic transfer, squared cells mean elastic traffic transmission.

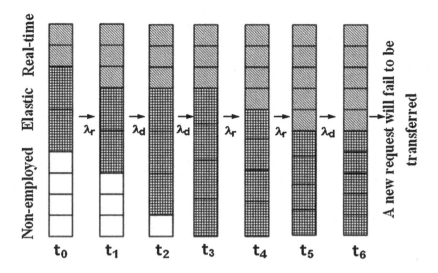

Fig. 2. The example of resource allocation for the researched network cell model

If the cell has sufficient free capacity, which means that the inequality

$$i_r c_r + i_d c_2 + c_r \leq C$$

is valid, then the request is accepted for service and configured with a resource in the amount of c_r bps. When the given value is not available, but the inequality

$$i_r c_r + i_d c_1 + c_r \leq C$$

is valid, the transmission rate of all non-real time requests being serviced decreases from the value $\frac{C - i_r c_r}{i_d}$ to the value of $\frac{C - (i_r + 1) c_r}{i_d}$, this new value of elastic traffic transmission rate being greater than c_1.

If the inequality

$$i_r c_r + i_d c_1 + c_r > C$$

is valid, the received request for transmitting real-time traffic is denied since in this situation, either there is no available resource in the amount of c_r or it's impossible to obtain the specified amount of the resource by reducing the transmission rate of the non-real time requests i_d that are being processed because the transmission rate c_1 is a minimum.

It is assumed that when a request arrives, the system has requests for real-time traffic i_r and requests elastic traffic i_d, respectively. If the cell has a sufficient amount of free resources, i.e.

$$i_r c_r + i_d c_2 + c_2 \leq C$$

is valid, the request is accepted for servicing and the maximum amount of resources equal to c_2 bps is allocated for it. When there isn't the specified resource in the system, but the ratio

$$i_r c_r + i_d c_1 + c_1 \leq C$$

is performed, the transfer rate of all non-real time requests is reduced from $\frac{C-i_r c_r}{i_d}$ to $\frac{C-i_r c_r}{i_d+1} > c_1$. When the inequality

$$i_r c_r + i_d c_1 + c_1 > C$$

is valid, the received request for elastic traffic transfer is denied.

3   Mathematical Description of the Resource Allocation Model

To estimate the portion of lost requests for real-time and elastic traffic transferring [4], the volume of the occupied resource and the time of non-real time request transmission, it is enough to know the time period when the cell is in a state with a fixed number of requests of all types that are currently being serviced. The total number of requests being processed specifies the type of the state and the components of the random process that will be used for the evaluation of the introduced indicators. Let $i_r(t)$ be the number of requests for real-time traffic transfer that are being serviced at the moment of time t, and let $i_d(t)$ be the number of elastic traffic requests that are being serviced at the same time t. The dynamics of changes in the total number of requests processed is described by a two-dimensional Markov process $r(t) = (i_r(t), i_d(t))$ defined on the finite state space S. It consists of vectors (i_r, i_d), with the components

$$i_r = 0, 1, \ldots, \left[\frac{C}{c_r}\right]; \; i_d = 0, 1, \ldots, \left[\frac{C - i_r c_r}{c_1}\right].$$

Let us denote by $p(i_r, i_d)$ the values of stationary probabilities of $(i_r, i_d) \in S$. They have the interpretation of the time period when the cell is in the state (i_r, i_d), and can be used to evaluate main indicators for the simultaneous service of incoming requests.

The portion of real-time traffic requests lost due to the lack of free transfer resource of the cell π_r is defined as the residence time period when the process $r(t)$ is in the states satisfying the inequality $i_r c_r + i_d c_1 + c_r > C$:

$$\pi_r = \sum_{\{((i_r, i_d) \in S | i_r c_r + i_d c_1 + c_r > C)\}} p(i_r, i_d).$$

The portion of requests for non-real time traffic transfer lost due to the lack of free transfer cell resource π_d is defined as the residence time period when the process $r(t)$ is in the states that satisfy the inequality $i_r c_r + i_d c_1 + c_1 > C$:

$$\pi_d = \sum_{\{((i_r, i_d) \in S | i_r c_r + i_d c_1 + c_1 > C)\}} p(i_r, i_d).$$

The average value c_d of the cell data transmission resource is used for processing one elastic traffic request and calculated from the ratio

$$c_d = \frac{s_d}{m_d}.$$

The average time T_d for processing this request is calculated from Little's law

$$T_d = \frac{m_d}{\lambda_d(1-\pi_d)}$$

Little's law is also used to define the ratios linking service parameters for real time traffic

$$\lambda_r(1-\pi_r) = m_r\mu_r$$

and for elastic traffic

$$\lambda_d(1-\pi_d) = \frac{s_d}{F}.$$

The given ratios are in the nature of conservation laws of flow intensities both for arrived and processed requests in the considered system. The laws are utilized further to assess the convergence of iterative methods for solving a system of equilibrium equations and to establish some correlations between introduced parameters of request processing. The ratio for assessing T_d can be defined as

$$T_d = \frac{m_d}{\lambda_d(1-\pi_d)} = \frac{m_d F}{s_d}.$$

4   A System of Equilibrium Equations

In order to take advantage of the introduced definitions, it is necessary to work out and solve a system of statistical equilibrium equations linking the values of stationary probabilities $p(i_r, i_d)$. Applying the fundamental results of the theory of Markov processes to formulate the system of equilibrium equations we should obtain and sum the intensities of all analyzed in the model event instances that cause the process $r(t)$ to leave the arbitrary state $(i_r, i_d) \in S$ (the left-hand member of the system of equilibrium equations), and then equate them to the cumulative intensity of $r(t)$ transition into the state (i_r, i_d) (the right-hand member of the system of equilibrium equations). The feasibility of each event instance depends on the ratio between the number of real time traffic requests and the number of non-real time requests being processed by the system. Consider step-by-step application of the formulated provision.

The state (i_r, i_d) changes as the result of the following event instances:

1. When the number of real time traffic requests that are being serviced increases. The intensity of occurrence for the given events is equal to λ_r providing the cell resource employed for processing real time and elastic types of traffic in this state is less than or equal to $C - c_r$.

2. When the number of non-real time traffic requests that are being serviced increases. The intensity of occurrence for the given events is equal to λ_d providing the cell resource employed for processing real time and elastic types of traffic in this state is less than or equal to $C-c_1$.
3. When processing one of the requests i_r for real time traffic transmission is completed. The intensity of occurrence for the given events is equal to $i_r\mu_r$ providing that the requests of this type are being serviced.
4. When processing one of the requests i_d for non-real time traffic transmission is completed. The intensity of occurrence for the given events is equal to $\min\left\{\frac{i_d c_2}{F}, \frac{C-i_r c_r}{F}\right\}$ providing that the requests of this type are being serviced.

Let the right-hand member of the proposed system of equilibrium equations have the summands that indicate what initial states, what intensity and what conditions enable the transition of the process $r(t)$ to the state (i_r, i_d). Consider these summands and the conditions of their emergence. The transition of the process $r(t)$ to the state (i_r, i_d) is possible in the following cases:

1. From the state (i_r-1, i_d) with the intensity λ_r if a real time traffic request arrives to be processed. The validity of the inequality $i_r>0$ is a required condition for the event instance.
2. From the state (i_r, i_d-1) with the intensity λ_d if a non-real time traffic request arrives to be processed. The validity of the inequality $i_d>0$ is a required condition for the event instance.
3. From the state (i_r+1, i_d) with the intensity $(i_r+1)\mu_R$ if the processing of a real time traffic request has been completed. The event instance necessitates that the components of the state (i_r+1, i_d) satisfy the ratio $i_r c_r + i_d c_1 + c_r \leq C$.
4. From the state (i_r, i_d+1) with the intensity $\min\left\{\frac{(i_d+1)c_2}{F}, \frac{C-i_r c_r}{F}\right\}$ if the processing of a non-real time traffic request has been completed. The event instance necessitates that the components of the state (i_r, i_d+1) satisfy the ratio $i_r c_r + i_d c_1 + c_1 \leq C$.

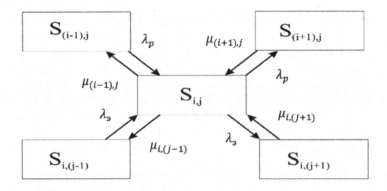

Fig. 3. Graph of transitions to acceptable adjacent states

Further, in accordance with the general rule it's necessary to set the intensity of leaving the arbitrary state $(i_r, i_d) \in S$ equal to the transition intensity of $r(t)$ into the state (i_r, i_d). As a result, we can get the equation for the state $(i_r, i_d) \in S$ from the system of statistical equilibrium equations. To write down the conditions for the event instances that change the state of the model we use a function indicator $I(\cdot)$ expressed as the ratio (Fig. 3).

$$I(\cdot) = \begin{cases} 1, & \text{if the condition in brackets is satisfied} \\ 0, & \text{if the condition is not satisfied} \end{cases}$$

Some necessary transformations give us the following finite system of linear equations

$$p(i_r, i_d) [\lambda_r I (i_r c_r + i_d c_1 + c_r \leq C) + \lambda_d I (i_r c_r + i_d c_1 + c_1 \leq C)$$

$$+ \lambda_r \mu_r I (i_r > 0) + \min \left\{ \frac{i_d c_2}{F}, \frac{C - i_r c_r}{F} \right\} I (i_d > 0)$$

$$= P (i_r - 1, i_d) \lambda_r I \{i > 0\} + + p (i_r, i_d - 1) \lambda_d I \{i_d > 0\}$$

$$+ p (i_r + 1, i_d) (i_r + 1) \mu_r I (i_r c_r + i_d c_1 + c_r) \leq C$$

$$+ p (i_r, i_d + 1) \min \left\{ \frac{(i_d + 1) c_2}{F}, \frac{C - i_r c_r}{F} \right\} I \{i_r c_r + i_d c_1 + c_1 \leq C\}.$$

Normalizing condition: $\sum_{(i_r, i_d) \in S} P(i_r, i_d) = 1$.

5   Numerical Results and Their Usage for Planning the Values of the Model's Structural Parameters

The load parameters of the model are λ_r and λ_d. The parameters that determine the use and allocation of the information transfer resource are the values C, c_r, c_1, c_2. The mathematical brute force method can be used to solve the problem of estimating any of these parameters while other values are fixed ones. It is natural to take the portion of denied requests for real-time traffic π_r and the portion of denied requests for elastic traffic π_d as a functional that determines the quality of service of incoming requests. The values of these characteristics are calculated after solving the system of equilibrium equations by the Gauss-Seidel iterative method. When solving the problem of planning the information transfer resource, both indicators can be normalized by different values. For simplicity, we assume that the standard value of the portion of lost requests is π, which is determined from the ratio $\pi = \max(\pi_r, \pi_d)$.

Denote by ρ the minimum potential loading coefficient for a cell resource unit. The value of ρ is calculated from the ratio

$$\rho = \frac{\lambda_r \cdot \frac{1}{\mu_r} \cdot c_r + \lambda_d \cdot F}{C}.$$

For evaluating the cell transfer resource sufficient to service the incoming traffic flows with the specified quality we assume $\pi = 0.05$. Parameters for the incoming traffic and its servicing are calculated by means of the correlation of $C = 100$ Mbps, $c_r = 3$ Mbps, $c_1 = 1$ Mbps, $c_2 = 5$ Mbps, $F = 16$ Mbps, $\mu_r = 1/300$, $\lambda_r = 0.05$, $\lambda_d = 5$ [5–7,15]. At the given parameter values, losses are determined from the equations $\pi_r = 0.3274$, $\pi_d = 0.1299$. It is seen that each of the values exceeds the standard value $\pi = 0.05$.

Calculate the dependence of the characteristics on increasing ρ in the range from 0.75 to 1.25. Calculations of losses of λ_r and λ_d requests depending on the growth of ρ are shown in Fig. 4. As expected, request losses increase when loading of the cell resource unit rises. The rate of loss growth becomes significant when the cell resource loading is close to one, the portion of lost requests for real time traffic transmission increases more quickly than that of lost requests for elastic traffic transfer.

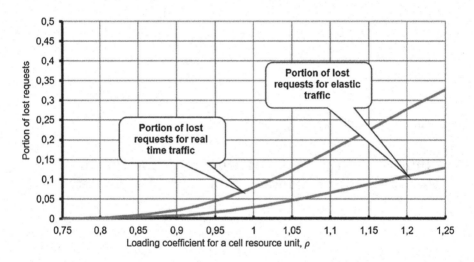

Fig. 4. Calculation of request losses while the minimum potential loading coefficient for a cell resource unit is increasing due to the arriving traffic

The conditions of building the model imply that data is conveyed at the maximum allowable rate when the cell load is low. In the considered example this parameter c_d will tend to 5 Mbps. Figure 5 shows the results of the calculated c_d. It can be seen that the cell load increases with the growing ρ, therefore, elastic traffic is gradually coming to be serviced at the minimum allowable rate $c_1 = 1$ Mbps [8–10].

We will consistently increase the information transfer rate C in the cell until the maximum loss value is less than 0.05. The results of the calculations are shown in Fig. 6, which shows the calculated data of request losses π_r and π_d while C is growing [11].

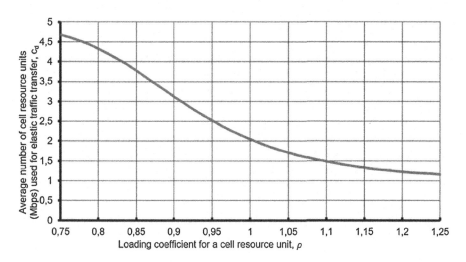

Fig. 5. Average use of cell resource for elastic traffic while the minimum potential loading coefficient for a cell resource unit is increasing due to the arriving traffic

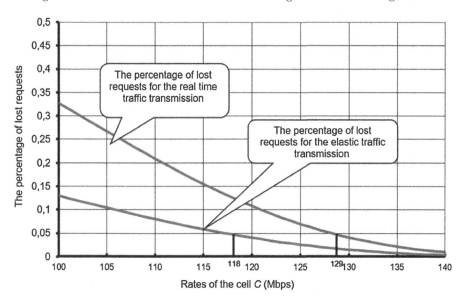

Fig. 6. The evaluation results of the rate in the cell providing the service of specified request flows of real time and non-real time traffic with the predetermined amount of maximum losses

As expected, the number of lost requests decreases with the growth of C. When the value C reaches 118 Mbps, losses of elastic traffic requests are less than 0.05, and when the value C arrives at 129 Mbps, losses of real-time traffic services are also becoming less than 0.05. Thus, the solution of the problem is obtained at $C = 129$ Mbps [12–14].

When determining the cell transfer resource sufficient for processing incoming traffic flows with a given quality it is often necessary to take into account the value of the average time for elastic traffic delivery. Let's assume that for the model, the calculation results of which are shown in Fig. 7, it is also necessary to ensure that the average transfer time T_d of a non-real time request should not be more than 5 seconds. Then the selection of the transfer cell resource will occur repeatedly until this restriction is fulfilled simultaneously with the condition $\pi = \max(\pi_r, \pi_d)$. The results of solving the problem are shown in Fig. 7, which shows the average transmission time T_d of the elastic request processing with C increasing. The graph shows that the value T_d falls while C grows. When the value C reaches 137 Mbps, the losses of elastic traffic requests and real-time service traffic are less than 0.05, and the average elastic traffic transfer time is less than 5 seconds. Thus, the solution of the problem is obtained at $C = 137$ Mbps [16].

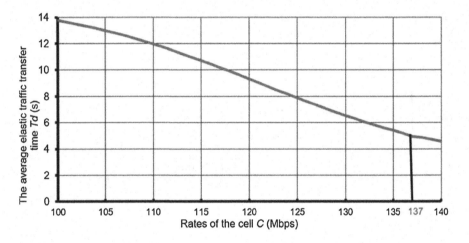

Fig. 7. The results of the rate estimation in the cell, which provides the service of the specified requests flows of elastic traffic and real time services traffic, with the specified values of the maximum losses and the average elastic traffic transfer time

Similarly, you can select the maximum allowable amount of traffic that can be serviced in a cell at a fixed bandwidth C and with a given maximum loss (you can also use the average elastic traffic transfer time limit as a normative indicator). For this purpose, the loading of the cell resource unit decreases until the restrictions on the service quality of received requests are reached while the remaining parameters in the model are fixed. The results of the estimation are shown in Fig. 8, which illustrates the calculation of losses of real time requests π_r and elastic traffic transmission requests π_d with a decrease in the load value ρ. As one would expect, the reduction of ρ results in decreasing of request losses. For ρ values less than 0.95, losses of non-real time requests and losses of real-time traffic requests become less than 0.05. Thus, the solution of the problem

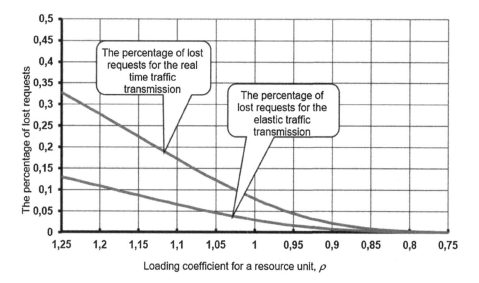

Fig. 8. The estimation results of the maximum allowable amount of traffic that can be processed in a cell at a fixed capacity of cell C with a given maximum loss

is obtained at $\rho = 0.95$. The specific values of λ_r and λ_d follow from the given above relation defining ρ. To solve the formulated problem, it is also necessary to know the ratio between the intensities of each of the analyzed traffic types.

The values of parameters have been derived from the system of equilibrium equations solved by the Gauss-Seidel iteration. The run time on a standard PC was up to several seconds. The number of unknowns in the system of equilibrium equations varied from 1 to 3000.

After consideration of the properties of simultaneous processing real time and elastic types of traffic we may start to plan allowable amounts of traffic and data transmission resource.

This model makes it possible to numerically evaluate the simultaneous transmission of two types of traffic, which allows to optimize cellular network operation and to control the rates of elastic traffic at the level of the radio interface without constant calls to the home subscriber server (to determine the priority of elastic traffic).

6   Conclusion

The paper proposes one of the possible schemes of resource allocation in a cellular data network with a limit on the maximum possible rate of elastic traffic transfer. A cell model for real time and elastic traffic transmission is constructed. A detailed numerical analysis of the model operation is carried out. The estimation results of the cell rates for processing specified flows of elastic traffic requests and real time traffic services with the specified values of the maximum loss and the average elastic traffic transfer time are obtained.

References

1. Skrynnikov, V.G.: UMTS/LTE radio subsystems. Theory and Practice. Moscow: Sport and Culture - 2000 Publishing House (2012)
2. Stepanov, S.N.: The model for servicing the real time traffic and data with a dynamically changeable speed of transmission. Autom. Remote Control **71**(1), 18–33 (2010)
3. Chowdhury, M.Z., Jang, Y.M., Haas, Z.J.: Call admission control based on adaptive bandwidth allocation for wireless networks. J. Commun. Netw. **15**(1), 15–24 (2013)
4. Tung, T., Walrand, J.: Providing QoS for real-time applications. In: Proceedings of Communications, Internet, and Information Technology (CIIT 2003), USA, pp. 121–130, November 2003
5. Stepanov, S.N.: Model of joint servicing of real-time service traffic and data traffic. I. Autom. Remote Control **72**(4), 121–132 (2011)
6. Stepanov, S.N.: Model of joint servicing of real-time service traffic and data traffic. II. Autom. Remote Control **72**(5), 139–147 (2011)
7. Antonova, V.M.: Channel resource assessment for multi-mode links on an LTE network segment. Nat. Tech. Sci. **10**, 356–358 (2014)
8. Antonova, V.M., Malikova, E.E.: The research of the influence of the service and the information traffic on each other in LTE networks. T-Comm - Telecommun. Transp. **9**, 17–19 (2014)
9. Antonova, V.M., Tsirik, I.A.: Access control of new requests in an LTE network segment. Fundam. Prob. Radioeng. Device Constr. **15**(5), 226–228 (2015)
10. Antonova, V.M., Gudkova, I.A., Markova, E.V., Abaev, P.O.: Analytical modeling and simulation of admission control scheme for non-real time services in LTE networks. In: Proceedings of the 29th European Conference on Modelling and Simulation, ECMS, Varna, pp. 1156–1163 (2015). ISBN: 978-0-9932440-0-1/ISBN: 978-0-9932440-1-8 (CD)
11. Kuznetsov, N.A., Myasnikov, D.V., Semenikhin, K.V.: Optimization of two-phase queuing system and its application to the control of data transmission between two robotic agents. J. Commun. Technol. Electron. **62**(12), 1484–1498 (2017)
12. Kuznetsov, N.A., Myasnikov, D.V., Semenikhin, K.V.: Two-phase queuing system optimization in applications to data transmission control. Procedia Eng. **201**, 567–577 (2017)
13. Kuznetsov, N.A., Minashina, I.K., Ryabykh, N.G., Zakharova, E.M., Pashchenko, F.F.: Design and comparison of freight scheduling algorithms for intelligent control system. Procedia Comput. Sci. **98**, 56–63 (2016). ISSN 1877-0509
14. Kuznetsov, N.A., Myasnikov, D.V., Semenikhin, K.V.: Optimal control of data transmission in a mobile two-agent robotic system. J. Commun. Technol. Electron. **61**(12), 1456–1465 (2016)
15. Kuznetsov, N.A., Semenikhin, K.V.: Parametric optimization of packet transmission with resending packets mechanism. In: SPIIRAS Proceedings, vol. 18, no. 4, pp. 809–830 (2019). http://proceedings.spiiras.nw.ru/index.php/sp/article/view/4141
16. Antonova, V.M.: The study of wireless network resources while transmitting heterogeneous traffic. In: Proceedings of the 24th Conference of FRUCT Association, 8–12 April 2019, Moscow, p. 78 (2019). ISSN 2305-7254. 565–571, ISBN 978-952-68653-8-6

On Failure Rate Comparison of Finite Multiserver Systems

Evsey Morozov[1,2](✉) [ID], Irina Peshkova[2] [ID], and Alexander Rumyantsev[1,2] [ID]

[1] Institute of Applied Mathematical Research of the Karelian Research Centre
of RAS, Pushkinskaya str. 11, Petrozavodsk 185910, Russia
emorozov@karelia.ru
[2] Petrozavodsk State University, Lenin str. 33, Petrozavodsk 185910, Russia

Abstract. In this work, we discuss various (stochastic) orderings
between random variables, with focus on the failure rate comparison.
Then these results are applied to construct a coupling of random variables. In particular, it is shown how to compare the failure rates of various
distributions, including Exponential, Weibull, Pareto and Burr distributions. These results are then applied to establish some monotonicity
properties of the multiserver queueing systems with finite waiting room.
These theoretical results are illustrated by a numerical simulation of
$M/G/c/n$ and $M/M/c/n$ queueing systems.

Keywords: Failure rate comparison · Coupling · Multiserver system ·
Queue size estimation

1 Introduction

It is a widely recognized problem that performance analysis of multiserver queueing system with general service time is a challenging problem. Explicit expressions for the stationary performance measures of such systems are available in a
few particular cases only, and the estimation based on the stochastic simulation
remains the only tool to evaluate the quality of service (QoS) of such systems
[1,2]. Even more problems arise when we want to evaluate the performance of
a finite system, that is the system with finite buffer for waiting customers, see
[13]. It follows from the celebrated *Kiefer-Wolfowitz recursion* for the workload
process in classic infinite buffer system $GI/GI/c$, that such systems obey some
monotonicity properties with respect to the input intervals and service times [1].
However, as a rule, some of these properties are false for finite buffer systems
[4,10,11,13]. Nevertheless, such systems obey a useful monotonicity property
provided the so-called *failure rate* function $r(t)$ of the service time distributions

The research is supported by the Ministry of Science and Higher Education of the Russian Federation (project No. 05.577.21.0293, unique identifier RFMEFI57718X0293).
The research is supported by Russian Foundation for Basic Research, projects No.
19-57-45022, 19-07-00303, 18-07-00156, 18-07-00147.

V. M. Vishnevskiy et al. (Eds.): DCCN 2019, LNCS 11965, pp. 419–431, 2019.
https://doi.org/10.1007/978-3-030-36614-8_32

satisfy an ordering. This is a strong ordering which is especially useful to compare the queueing systems with different inputs, see [13]. It is usually not easy to construct non-trivial comparison of the failure rates because this ordering involves operation taking "infimum" of $r(t)$, which equals to zero in the most important cases. In this note, we develop the monotonicity property of the multiserver system with finite buffer using the failure rates ordering of the exponential, Weibull, Pareto, Burr and Hyperexponential service time distributions. The main contribution of this research is that, using failure rates comparison, we compare the queueing processes in the system with different interarrival/service time distributions, while usually such a comparison is applied to establish monotonicity of the queueing processes in the system with the *same corresponding distributions* but different values of parameters. In turn, monotonicity properties of various queueing processes is a key element of the method of *regenerative envelops* recently developed by the authors in the paper [8]. This method opens new possibilities in the regenerative estimation of the QoS of complex systems. In this method, the monotonicity properties are used to derive the lower and upper bounds of the performance measures of the original system which in general does not possess regenerative property. These bounds relate respectively, to minorant and majorant queueing systems which have classic regenerations.

In this research, we present a preliminary study of the monotonicity properties of stochastic queueing processes in multiserver queueing systems based on comparison of the failure rate functions. It is worth mentioning that this approach can be useful both for an analytic study and for the QoS estimation based on regenerative simulation.

The structure of the paper is as follows. Section 2 describes some popular orderings of the random variables. In Sect. 3, we discuss the comparison and specific monotonicity properties of the failure rates, with focus on the distributions with long tails. In Sect. 4, we compare the queue sizes in the multiserver systems with finite buffers based on the ordering of the failure rate functions. Finally, in Sect. 5, we illustrate the results by numerical simulation.

2   Stochastic Comparison of Random Variables

We consider two non-negative random variables (r.v.) X, Y with distribution functions (d.f.) F_X, F_Y, and densities f_X, f_Y, respectively. Denote $\mathsf{E}X$ the expectation of X, and $\overline{F}_X(x) = 1 - F_X(x)$ the tail distribution. For each x such that the tail distribution $\overline{F}_X(x) = \mathsf{P}(X > x) > 0$, we can define the *failure rate* function as

$$r_X(x) = \frac{f_X(x)}{\overline{F}_X(x)}. \tag{1}$$

In the *survival theory*, $r_X(x)dx$ is the probability that, given survival until time x, the event occurs in an interval of length $(x, x + dx)$ [3]. The failure rate function is also known as the *force of mortality* in actuarial science [5] or the intensity function in the *extreme value theory* [6]. In the *reliability theory or queueing theory*, the failure rate can be thought of as the likelihood of an instant completion of the customer waiting time x, given that he has already been waiting for time x [3]. In what follows, to simplify notation, we assume that the r.v. X is unbounded and thus the failure rate $r(x)$ is defined over $[0, \infty)$. This assumption holds for all examples considered below, whereas in general $r(x)$ is defined for such x that $\overline{F}(x) > 0$.

We say that a r.v. X *is less than* a r.v. Y:

stochastically (in distribution), and denote it as $X \underset{st}{\leq} Y$, if

$$\overline{F}_X(x) \leq \overline{F}_Y(x), \ x \geq 0; \tag{2}$$

in *probability (almost surely)*, denoted $X \underset{a.s.}{\leq} Y$, if

$$\mathsf{P}(X \leq Y) = 1; \tag{3}$$

in *failure rate*, denoted $X \underset{r}{\leq} Y$, if

$$r_X(x) \geq r_Y(x), \ x \geq 0. \tag{4}$$

It is well known that [9]:

$$X \underset{r}{\leq} Y \text{ implies } X \underset{st}{\leq} Y, \text{ and } X \underset{st}{\leq} Y \text{ implies } \mathsf{E}X \leq \mathsf{E}Y. \tag{5}$$

In the performance and stability analysis the stochastic ordering plays often a key role. The main reason is that, by coupling technique, stochastic ordering is nicely related to a.s. ordering. We recall this relation (for more detail see [12]). Let $\tilde{X}$ be a (stochastic) copy of a r.v. X, i.e., $X \overset{st}{=} \tilde{X}$. A *coupling* of collection of the r.v.'s $X_i \ i \in M$ is a family of the r.v.'s $\{\tilde{X}_i, \ i \in M\}$ such that $X_i \overset{st}{=} \tilde{X}_i$ for all $i \in M$, and $\{X_i\}$ are defined on the *same probability space*, where M is an index set. Now denote F^{-1} *quantile function* defined as

$$F^{-1}(u) = \inf\{x \in \mathbf{R} : F(x) \geq u\}, \ u \in [0; 1].$$

Then the r.v. $\tilde{X} = F^{-1}(U)$, where r.v. U is uniformly distributed on $[0, 1]$, is a stochastic copy of X. There exists the following relation between stochastic ordering and a.s. ordering [9,12]:

$$X \underset{st}{\leq} Y \text{ if and only if } F_X^{-1}(u) \leq F_Y^{-1}(u), \ u \in [0; 1], \tag{6}$$

$$\text{if and only if there exists coupling } (\tilde{X}; \tilde{Y}) : \tilde{X} \underset{a.s.}{\leq} \tilde{Y}. \tag{7}$$

This result, in particular, allows to establish monotonicity properties of the stochastic processes in the system by performing sample paths comparison. We present such a result in Sect. 4. However, a direct stochastic comparison might be complicated. At the same time, failure rate ordering, implying stochastic ordering, can provide a more straightforward comparison of the two r.v. Thus, in what follows, we use the failure rate ordering to establish stochastic ordering of queueing systems as a consequence.

3 Failure Rate Comparison of Some Distributions

In this Section we demonstrate the stochastic ordering technique based on the failure rate comparison. To do so, we consider some useful classes of distributions based on the monotonicity properties of failure rate function and establish some auxiliary results used further in Sect. 4. Examples of the failure rate comparison of some well-known distributions conclude the section.

We call X a r.v. with *increasing (decreasing) failure rate distribution* (IFR, DFR, respectively) if $r_X(x)$ is an increasing (decreasing) function. However, not all the distributions possess monotonic failure rates (such as DFR or IFR), for example, the distributions which exhibit bathtub shaped model. A r.v. X has the *bathtub-shaped failure rate distribution* BFR (*upside-down bathtub-shaped failure rate distribution*, UBFR) if it is initially decreasing (increasing), then becomes constant and finally is increasing (decreasing):

$$
r_X(x) = \begin{cases} r_1(x), & r_1'(x) < (>) \ 0, \ x \le a_1; \\[2mm] r_2(x), & r_2'(x) = 0, \ a_1 < x \le a_2; \\[2mm] r_3(x), & r_3'(x) > (<) \ 0, \ x > a_2, \end{cases} \tag{8}
$$

where $a_1 \le a_2$ are *change points*. If $\alpha_1 = \alpha_2$, then BFR becomes *U-shaped* failure rate (UFR), while UUFR becomes *upside-down U-shaped* failure rate distribution.

Note that the DFR property is naturally related to the so-called *long-tailed distributions* defined as follows: X has long tail, if for any $y > 0$,

$$
\lim_{x \to \infty} P(X > x + y \mid X > x) = 1.
$$

This asymptotic property of the conditional overshoot of r.v. X means that, given $X > x$ for x large, it is quite likely that X will exceeds the threshold $x + y$ as well, for any $y > 0$. The long-tailed distributions have failure rate eventually decreasing to zero [7],

$$
\lim_{x \to \infty} r(x) = 0. \tag{9}
$$

An important example of the long-tailed distribution possessing DFR property is the two-parameter Pareto distribution, $Pareto(\alpha, x_0)$, with d.f. [7]

$$
F_X(x) = 1 - \left(\frac{x_0}{x_0 + x} \right)^\alpha, \ x \ge 0, \ x_0 > 0, \ \alpha > 0.
$$

It is easy to see that the failure rate of Pareto distribution,

$$
r(x) = \frac{\alpha}{x_0 + x}, \ x \ge 0, \tag{10}
$$

vanishes as $x \to \infty$. However, the asymptotic property (9) not necessary holds for DFR distributions. For example, a generalization of the Pareto distribution is the *Burr distribution*, $Burr(x_0, \alpha, k)$, with long-tailed d.f. [14]

$$F(x) = 1 - \left(\frac{x_0}{x_0 + x^k}\right)^\alpha, \ x \geq 0, \ x_0 > 0, \ \alpha > 0, \ k > 0, \tag{11}$$

which has failure rate function

$$r(x) = \frac{\alpha k x^{k-1}}{x_0 + x^k}. \tag{12}$$

Note that if $k \geq 2$, then the failure rate increases, reaches maximum value

$$r(x^*) = \frac{\alpha k (x_0(k-1))^{1-1/k}}{x_0 k} \tag{13}$$

at the point

$$x^* = (x_0(k-1))^{1/k},$$

and then decreases, and thus is a UFR distribution. For $k < 1$, the failure rate is monotonically decreasing in time. However, the property (9) is satisfied asymptotically for each $k > 0$.

On the other hand, there are DFR distributions with rate function that converges to positive value. A well-known practical example is *hyperexponential* r.v. (mixture of exponential distributions) with the d.f.

$$F(x) = 1 - \sum_{i=1}^{n} p_i e^{-\mu_i x}, \ x \geq 0, \tag{14}$$

where

$$p_i \geq 0, \ \mu_i > 0, \ i = 1, \ldots, n; \ p_1 + \cdots + p_n = 1.$$

The failure rate of hyperexponential distribution is

$$r(x) = \frac{\sum_{i=1}^{n} p_i \mu_i e^{-\mu_i x}}{\sum_{i=1}^{n} p_i e^{-\mu_i x}}. \tag{15}$$

Note that hyperexponential distribution belongs to class of the so-called *light-tailed distribution*. In particular then the failure rate does not converge to zero at infinity, that is, the property (9) is violated. However, $r(x)$ monotonically decreases to a positive constant,

$$\lim_{x \to \infty} r(x) = \min(\mu_1, \ldots, \mu_n) =: \mu^*. \tag{16}$$

Finally, there are distributions that exhibit both IFR and DFR properties, depending on the parameters. A natural example gives a r.v. Y with two-parameter *Weibull distribution*, $Weibull(\beta, k)$:

$$F_Y(x) = 1 - e^{-\beta x^k}, \ x \geq 0, \ k > 0, \ \beta > 0,$$

and monotone failure rate function

$$r_Y(x) = k\beta x^{k-1}, \ x \geq 0. \tag{17}$$

It is easy to verify that $Weibull(\beta, k)$ is IFR for $k > 1$, and DFR for $k < 1$.

The failure rate comparison is more restrictive than the stochastic ordering. However, in some cases (as we will demonstrate in Sect. 4) the following, even more restrictive, condition has to be applied. Namely, a *sufficient condition* for failure rate ordering $X \underset{r}{\geq} Y$ is the following:

$$\sup_{x \geq 0} r_X(x) \leq \inf_{x \geq 0} r_Y(x). \tag{18}$$

In particular, the following useful consequences follow from inequality (18).

1. Comparison (18) is useful in practice only if

$$\sup_{x \geq 0} r_X(x) < \infty \text{ and } \inf_{x \geq 0} r_Y(x) > 0.$$

2. Supremum/infimum of monotone failure rate (IFR/DFR) function is achieved at the border of domain of the tail $\overline{F}_X(x)$, e.g., for IFR,

$$\sup_{x \geq 0} r_X(x) = \lim_{u \to \infty} r_X(u).$$

At the same time, it follows from (8) that BFR (UBFR) achieves an extremum at the intermediate points $x \in (a_1, a_2]$.

3. For a long-tailed r.v. Y,

$$\inf_{x \geq 0} r_Y(x) = \lim_{x \to \infty} r_Y(x) = 0,$$

and thus, a r.v. Y can not be used at the r.h.s. of the inequality (18).

To illustrate the stochastic comparison technique, we first compare a r.v X with distribution $Pareto(\alpha, x_0)$ and an exponentially distributed r.v. Y with d.f. $F_Y(x) = 1 - e^{-\lambda x}$, denoted $Exp(\lambda)$, having *constant* failure rate $r_Y(x) = \lambda > 0$. If $\lambda \geq \alpha/x_0$, then it is easy to see from (10), that

$$r_X(x) \leq r_Y(x), \ x \geq 0, \tag{19}$$

and thus $X \underset{r}{\geq} Y$, which implies $X \underset{st}{\geq} Y$. Note also that then condition (18) is automatically satisfied, since it follows from (10) that

$$\sup_x r_X(x) = r_X(0) = \frac{\alpha}{x_0} \leq \lambda = \inf_x r_Y(x).$$

Now consider a r.v. W having distribution $Burr(x_0, \alpha, k)$, with $k > 1$ in d.f. (11), and a hyperexponential r.v. V with d.f. (14). Now we fix parameters

$$x_0, k, \ \mu_i, \ p_i, i = 1, \ldots, n$$

and try to find the critical value α^* that guarantees the following ordering of the failure rates,

$$V \leq_r W \text{ for } \alpha \leq \alpha^*.$$

However, to find α^* explicitly, we must find a unique solution of the equation $r_V(x) = r_W(x)$, for $x > 0$. Although the latter problem can be resolved numerically, the explicit solution is hardly available or impossible in a general case. Instead, it is easy to check that $r_V(x)$ approaches μ^* (defined in (16)) from above. Thus, it follows from (15), (12) and (13) that the value

$$\alpha^* = \frac{x_0 k}{k(x_0(k-1))^{1-1/k}} \mu^* \tag{20}$$

guarantees that the inequality, $r_W(x) \leq r_V(x)$ holds, implying finally the stochastic ordering

$$V \leq_{st} W.$$

For easy understanding, we graphically compare the failure rate functions for given parameter values on Fig. 1. Note that, in fact, we use (more restrictive) ordering (18) to derive (20), since $\inf_x r_V(x) = \mu^*$, while

$$\sup_x r_W(x) = r_W(x^*),$$

where x^* satisfies (13).

Concluding this Section, we note that condition (18) is useless for a r.v. X having $Weibull(\beta, k)$ distribution (with $k \neq 1$), since (17) implies $\inf_x r_X(x) = 0$ while $\sup_x r_X(x) = \infty$ for any $k > 0, k \neq 1$.

4   Comparing Multiserver Systems with Finite Waiting Room

In this section, we demonstrate how the failure rate comparison allows to stochastically compare the steady-state performance of the multiserver queueing systems with finite buffers. We consider two queueing systems, denoted $\Sigma^{(1)}$ and $\Sigma^{(2)}$, with $N^{(i)}$ servers working in parallel and finite waiting room of the size $C^{(i)}$, $i = 1, 2$. In what follows, the superscript (i) denotes the number of the system. The service discipline is assumed to be First-Come-First-Served, and if the waiting space is full upon arrival of a customer, then this customer leaves the system with no service. We denote $S_n^{(i)}$ the service time of customer n, $t_n^{(i)}$ the arrival instant of customer n, and then the interarrival times are defined as $\tau_n^{(i)} = t_{n+1}^{(i)} - t_n^{(i)}$, $n \geq 1$, $i = 1, 2$. We assume that $\{S_n^{(i)}, n \geq 1\}$ and $\{\tau_n^{(i)}, n \geq 1\}$

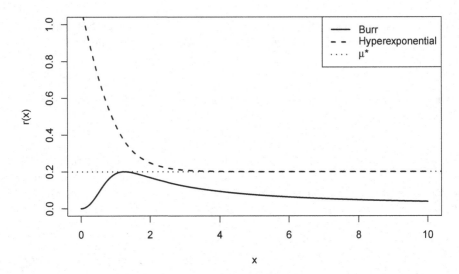

Fig. 1. Comparison of $Burr(x_0, \alpha^*, k)$ and hyperexponential failure rate functions for $x_0 = 1, k = 3, n = 2, p_1 = p_2 = 0.5, \mu_1 = 0.2, \mu_2 = 2$, and value $\alpha^* = 0.12$ satisfying (20).

are independent iid sequences. Moreover it is assumed that the service time distribution does not depend on the arrival process in both systems. Denote $S^{(i)}$ the generic service time, and $\tau^{(i)}$ the generic interarrival time, $i = 1, 2$.

Now we compare the steady-state queue-length processes in the systems $\Sigma^{(1)}$ and $\Sigma^{(2)}$. Let $Q_n^{(i)}$ be the number of customers in the ith system at the instant $t_n^{(i)}-$, and denote the limit (in distribution) $Q_n^{(i)} \Rightarrow Q^{(i)}, i = 1, 2$. The following statement is a small modification of Theorem 6 in [13], which establishes stochastic ordering of the queue sizes in the finite buffer systems.

Theorem 1. *Assume the following ordering between the buffer capacities and the number of servers in both systems*

$$C^{(1)} \geq C^{(2)}, \;\; C^{(1)} + N^{(1)} \leq C^{(2)} + N^{(2)}, \tag{21}$$

holds, and moreover, the following orderings hold true:

$$\tau^{(1)} \underset{st}{\geq} \tau^{(2)}, \tag{22}$$

$$\inf_{x \geq 0} r_{S^{(1)}}(x) \geq \sup_{x \geq 0} r_{S^{(2)}}(x). \tag{23}$$

Then the steady-state queue sizes at the arrival instants in both systems are ordered as

$$Q^{(1)} \underset{st}{\leq} Q^{(2)}. \tag{24}$$

The conditions (22) and (23) are rather restrictive but they are motivated by a non-monotone dependence of the performance measures of finite buffer systems on the system parameters. For instance, a small increase of the arrival rate may cause a considerable decrease of the departure rate due to loss of customers (see [13]). However, a disadvantage of these conditions is compensated by a possibility to compare queue sizes in systems with *different interarrival/service time distributions*.

The following statements directly follow from Theorem 1 and consequences of the inequality (18) discussed in Sect. 3.

Corollary 1. *The ordering* (24) *is valid if*

1. *Stochastic ordering* (22) *is replaced by* $\tau^{(1)} \underset{r}{\geq} \tau^{(2)}$.
2. *The infimum* $\inf_{x \geq 0} r_{S^{(1)}}(x)$ *in l.h.s. of* (23) *is replaced by*
   (a) $r_{S^{(1)}}(0)$, *if* $S^{(1)}$ *has IFR;*
   (b) $\lim_{x \to \infty} r_{S^{(1)}}(x)$, *if* $S^{(1)}$ *has DFR;*
   (c) $r_{S^{(1)}}(a_2)$, *if* $S^{(1)}$ *has BFR (UFR).*
3. *The supremum* $\sup_{x \geq 0} r_{S^{(2)}}(x)$ *in r.h.s. of* (23) *is replaced by*
   (a) $r_{S^{(2)}}(0)$, *if* $S^{(2)}$ *has DFR;*
   (b) $\lim_{x \to \infty} r_{S^{(2)}}(x)$, *if* $S^{(2)}$ *has IFR;*
   (c) $r_{S^{(2)}}(a_2)$, *if* $S^{(2)}$ *has UBFR (UUFR).*

The limits are assumed to exist, be finite and positive.

In the next section we illustrate by numerical examples the practical applicability of the obtained results.

5   Numerical Examples

In this section, by numerical simulation, we illustrate for the multiserver systems the comparison results obtained in Sect. 4.

First, we apply Theorem 1 (and Corollary 1) to compare stationary queue size distribution of $M/G/c/n$, c-server system with buffer size n, denoted $\Sigma^{(2)}$, with the stationary queue size distribution in the system $M/M/c/n$, denoted $\Sigma^{(1)}$. While there is no explicit expression for the steady-state queue size distribution in general case of the former system, the latter system is well-studied, with the following explicit expression for queue size distribution:

$$P(Q^{(1)} = i) = \begin{cases} \pi_0 \frac{\rho^i}{i!}, & i = 0, \ldots, c, \\ \pi_0 \frac{\rho^i}{c^{i-c}c!}, & i = c+1, \ldots, c+n, \end{cases} \tag{25}$$

where $\rho = \mathsf{E}S^{(1)}/\mathsf{E}\tau^{(1)}$ and

$$\pi_0 = \left[\sum_{k=0}^{c} \frac{\rho^k}{k!} + \frac{\rho^c}{c!} \sum_{k=c+1}^{c+n} \left(\frac{\rho}{c}\right)^{k-c} \right]^{-1}.$$

In our experiment, we compare the empirical distribution of steady-state queue size in the system $M/G/4/5$ and $Pareto(2.5, 1)$ service time to the explicit distribution (25) for the system $M/M/4/5$ with service time $Exp(2.5)$. We assume Poisson input with rate $\lambda = 2$ in both systems. The results of the comparison are depicted on Fig. 2. At that, queue size in system $\Sigma^{(1)}$ is stochastically dominated by queue size in $\Sigma^{(2)}$.

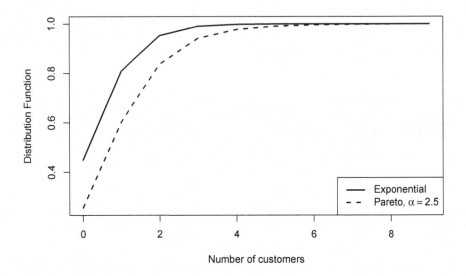

Fig. 2. Comparing queue size distributions in system $M/G/4/5$ with service time $Pareto(2.5, 1)$ and in the system $M/M/4/5$ with exponential $Exp(2.5)$ service time; input rate $\lambda = 2$ in both systems.

The purpose of the following simulation experiment is to demonstrate the dependence of the distance between the empirical d.f. of two stochastically ordered systems. We compare two $GI/GI/$ systems. The system $\Sigma^{(1)}$ has generic interarrival time $\tau^{(1)}$, distributed as $Burr(x_1, \alpha_1, k)$, for $k > 2$, and exponential service time $S^{(1)}$, $Exp(\lambda)$. The system $\Sigma^{(2)}$ has two-phase hyperexponential distribution of interarrival times with parameters μ_1, μ_2 and p_1, p_2. Finally, let the service time $S^{(2)}$ have long-tailed Pareto distribution, $Pareto(x_2, \alpha_2)$. To obtain stochastic ordering $Q^{(1)} \underset{st}{\leq} Q^{(2)}$, we use inequality (18) together with Corollary 1. Thus, we take the initial conditions as follows:

$$x_1 = 1, k = 3, p_1 = p_2 = 0.5, \mu_1 = 3, \mu_2 = 30,$$

and obtain $\alpha_1^* \approx 1.89$ from expression (20), where $\mu^* = 3$. Moreover, we take $\lambda = 2.5$ and $x_2 = 1$. Finally, we vary $\alpha_2 \in \{1.5, 2, 2.5\}$ such that inequality (19) is satisfied. As expected, the distance is larger for smaller values of α_2, as it is seen in Fig. 3.

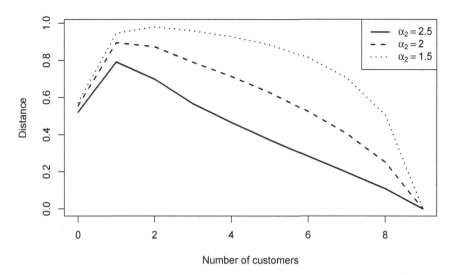

Fig. 3. Two $G/G/4/5$-type systems: estimated by simulation distance between the empirical d.f. of steady-state queue size in $Burr(x_1, \alpha_1, k)/Exp(\lambda)/4/5$ and $Hyperexponential/Pareto(x_2, \alpha_2)/4/5$ systems (for various α_2).

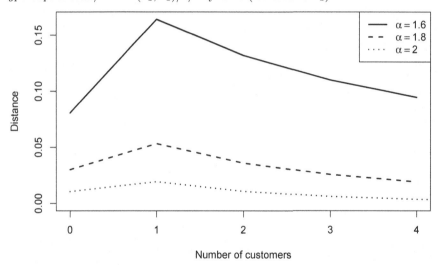

Fig. 4. Two $M/G/3$-type systems: the distance between empirical d.f.'s of steady-state queue size in systems with $Exp(\alpha)$ and $Pareto(\alpha, 1)$ service time distributions (for various α); Poisson input with rate 1.

Concluding this section, we note that the statement of Theorem 1 is valid for the infinite buffer systems, $C^{(i)} = \infty$. We set $N^{(i)} = 3$ and configure two systems with the same Poisson input process of rate $\lambda = 1$. Assume the system $\Sigma^{(1)}$ has service time distribution $Pareto(\alpha, x_0)$ with $x_0 = 1$. Then it follows that the ordering (19) holds for $\lambda = \alpha$, and thus we take service time distribution $Exp(\alpha)$ in the second system $\Sigma^{(2)}$ so as to guarantee the required ordering. Then we run simulation with $N = 10^5$ arrivals and construct the empirical d.f. for both systems. Finally, we plot on Fig. 4 the distance between the empirical d.f. of the system $\Sigma^{(1)}$ and the empirical d.f. of $\Sigma^{(2)}$.

6   Conclusion

In this paper, we discuss the applicability of the stochastic comparison of steady-state performance measures in the multiserver queueing systems with finite waiting space. We demonstrate a few refinements for the systems in which interarrival and/or service time distributions have specific monotonicity properties of the failure rate functions. The failure rate comparison allows to perform estimation of the unknown quantities for the complicated distributions of the governing sequences (input and service times). Some numerical results are also presented. The obtained results may be quite useful for the estimation of the performance measures using regenerative simulation, and we leave this discussion for a future research.

References

1. Asmussen, S.: Applied Probability and Queues. Stochastic Modelling and Applied Probability, 2nd edn. Springer-Verlag, New York (2003). https://doi.org/10.1007/b97236
2. Asmussen, S., Glynn, P.: Stochastic Simulation: Algorithms and Analysis. Springer, New York (2007). https://doi.org/10.1007/978-0-387-69033-9
3. Aven, T., Jensen, U.: Stochastic Models in Reliability. Stochastic Modelling and Applied Probability, p. 270. Springer-Verlag, New York (1999). https://doi.org/10.1007/b97596
4. Berger, A., Whitt, W.: Comparisons of multi-server queues with finite waiting rooms. Stoch. Models **8**(4), 719–732 (1992)
5. Borowiak, D.S., Shapiro, A.F.: Financial and Actuarial Statistics: An Introduction, 2nd edn, p. 392. CRC Press, Boca Raton (2013)
6. Embrechts, P., Klüppelberg, C., Mikosch, T.: Modelling Extremal Events for Insurance and Finance. Applications of Mathematics, p. 660. Springer, Heidelberg (1997). https://doi.org/10.1007/978-3-642-33483-2
7. Goldie, C., Klüppelberg, C.: Subexponential distributions. In: A Practical Guide to Heavy Tails: Statistical Techniques for Analysing Heavy Tails. Birkhauser, Basel (1997)
8. Morozov, E., Peshkova, I., Rumyantsev, A.: On regenerative envelopes for cluster model simulation. In: Vishnevskiy, V.M., Samouylov, K.E., Kozyrev, D.V. (eds.) DCCN 2016. CCIS, vol. 678, pp. 222–230. Springer, Cham (2016). https://doi.org/10.1007/978-3-319-51917-3_20
9. Ross, S., Shanthikumar, J.G., Zhu, Z.: On increasing-failure-rate random variables. J. Appl. Probab. **42**, 797–809 (2005)
10. Sonderman, D.: Comparing multi-server queues with finite waiting rooms, I: same number of servers. Adv. Appl. Probab. **11**, 439–447 (1979)

11. Sonderman, D.: Comparing multi-server queues with finite waiting rooms, II: different number of servers. Adv. Appl. Probab. **11**, 448–455 (1979)
12. Thorisson, H.: Coupling, Stationarity, and Regeneration. Springer, New York (2000)
13. Whitt, W.: Comparing counting processes and queues. Adv. Appl. Probab. **13**, 207–220 (1981)
14. Zimmer, W., Keats, J., Wang, F.: The Burr XII distribution in reliability analysis. J. Qual. Technol. **30**(4), 386–394 (1998)

Queue with Retrial Group for Modeling Best Effort Traffic with Minimum Bit Rate Guarantee Transmission Under Network Slicing

Ekaterina Markova[1]([✉])[iD], Yves Adou[1][iD], Daria Ivanova[1][iD], Anastasia Golskaia[1][iD], and Konstantin Samouylov[1,2][iD]

[1] Department of Applied Probability and Informatics, Peoples' Friendship University of Russia (RUDN University), 6 Miklukho-Maklaya St., Moscow 117198, Russian Federation
{markova-ev,1032135491,1032181905,golskaia-aa,samouylov-ke}@rudn.ru
[2] Institute of Informatics Problems, Federal Research Center "Computer Science and Control" of the Russian Academy of Sciences, 44-2 Vavilov St., Moscow 119333, Russian Federation

Abstract. Nowadays, information and communication technologies (ICT) are used in many areas of modern life. User equipment (UE), which uses the capabilities of wireless cellular networks, is becoming an integral part of modern social life. The increase in the number of users and services leads to an exponential growth of generated mobile traffic in the broadband wireless network which conducts radio resources lack problem. This problem is becoming major in the development of modern wireless technologies, mainly because users cannot be serviced with the required quality of service (QoS) level. Leading standardization organizations offer different approaches to solve this problem, for example, the use of network slicing (NS) technology, which allows a mobile network operator (MNO) to effectively share its resources with a mobile virtual network operator (MVNO). This paper proposes one of the possible radio resources allocation scheme models for multi-service wireless networks implementing NS technology, described as a queuing system with a buffer and its retrial group.

Keywords: 5G · Network slicing · Virtual network operator · BG service · Retrial group · Performance measures

The publication has been prepared with the support of the "RUDN University Program 5–100" (recipient Markova E.V., mathematical model development). The reported study was funded by RFBR, project number 18-37-00231 (recipients Golskaia A.A., Ivanova D.V., numerical analysis). The reported study was funded by RFBR, project number 19-07-00933 (recipient Samouylov K.E., problem formulation and analysis).

© Springer Nature Switzerland AG 2019
V. M. Vishnevskiy et al. (Eds.): DCCN 2019, LNCS 11965, pp. 432–442, 2019.
https://doi.org/10.1007/978-3-030-36614-8_33

1   Introduction

Modern information and communication technologies (ICT) provide users a wide range of broadband services which require different quality of service (QoS) levels. All services provided in wireless networks can be divided into nineteen different groups depending on the priority level, the bit rate guarantees and the corresponding traffic type [1,2]. The priority level can vary from 1 to 127, where 1 corresponds to services with the highest priority and 127 - to services with the lowest priority [3]. Bit rate guarantees are determined by the following categories:

- Guaranteed Bitrate (GB): minimum and maximum bit rates are assigned for services;
- Best effort with minimum Guaranteed (BG): only a minimum bit rate is assigned for services;
- Best Effort (BE): minimum and maximum bit rates are not defined for services.

Bit rate changes, for GB services, do not alter the service duration. Whereas bit rate changes, for BG or BE services, result in a variation of the service duration. These principles allow dividing traffics into two types: streaming and elastic. GB services correspond to streaming or real-time traffic and encompass voice, video calling, video streaming and music streaming. BG and BE services correspond to elastic or non-real time traffic. BG services are file sharing, web browsing, and social networking. BE services are smart metering and email.

Due to frequency range limitations, the currently used fourth-generation (4G) communication networks will very soon be unable to handle the exponentially increasing number of users [4]. In that regard, moving to the next stage of mobile wireless technologies appear to be the solution, i.e. the fifth-generation (5G) communication networks. Compared to 4G networks, 5G networks major improvements are the high bandwidth capacity (over 1 Gbps), the wide coverage and the ultra-low latency (1 ms) at a very high volume of generated traffic.

In this context, mobile operators need flexible solutions to efficiently manage radio resources. To this purpose, the company DOCOMO developed the network slicing (NS) technology which allows mobile network operators to share their radio resources with several mobile virtual network operators (MVNOs) or short virtual network operators (VNOs) which do not own any physical infrastructure [5–8]. In the future, VNOs are expected to provide numerous and diverse 5G services.

The NS is a virtual network architecture that uses the same principles as the software-defined networks (SDN) and the network function virtualization (NFV) on fixed networks. Currently, these two are commercially deployed to provide greater network flexibility allowing traditional network architectures to be divided into virtual elements that can be connected (including through software). The NS also allows the creation of multiple virtual networks on top of a common physical infrastructure. Therefore, virtual networks are configured to meet the specific requirements of the applications, the services, the devices, the customers or the operators. This approach will allow us to put into practice the flexibility to configure and manage infrastructure which is included in the requirements of the next 5G networks.

In this paper, we consider the new network slicing concept. Therefore, let us divide the network into slices, each of which will provide users multiple services (GB, BG, and BE) and have tenant (Fig. 1). By modeling this slicing concept, we want to describe the benefits of virtualized radio resource allocation following the various QoS level requirements.

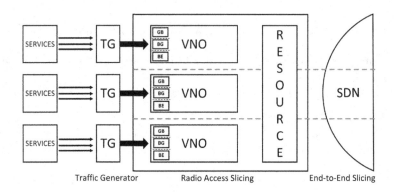

Fig. 1. Traffic service system architecture.

The aim of this paper is the development and the analysis of a user service model in a wireless network by one VNO providing BG services. The rest of the paper is organized as follows. In Sect. 2, we describe the model as a queuing system with a buffer and its retrial group with infinite capacities [9–12]. In Sect. 3, we consider the finite case of the buffer and its retrial group. In Sect. 4 is performed a numerical analysis. Conclusions are drawn in Sect. 5.

2   Description of BG Traffic Service Model for One VNO

Let us consider the work of a single VNO who rents a slice of a multi-service wireless network with total capacity R_{nc} Mbps from one MNO. We assume that the VNO provides BG services to its users, i.e services generating elastic traffic. The requests for starting elastic sessions arrive in the system according to the Poisson process with rate λ. The elastic sessions are parameterized by the exponentially distributed average file size θ MB and the minimum assignable bit rate b Mbps.

The radio admission control (RAC) is organized in such a way that when a user request arrives in the system, two outcomes are possible:

– session starts when the number of started sessions is less than $\lfloor R_{nc}/b \rfloor$;

– request waits in the buffer when the number of started sessions equals $\lfloor R_{nc}/b \rfloor$.

The requests waiting in the buffer can enter its retrial group after an exponentially distributed time β^{-1} and re-enter the buffer after an exponentially distributed time α^{-1}. We assume that the capacities of the buffer and its retrial group are infinite. Also, each started session is ended and the number of retries is unlimited. The corresponding scheme model is shown in Fig. 2. All the necessary further notations are listed in Table 1.

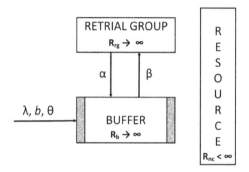

Fig. 2. Scheme model for infinite case.

Table 1. Description of parameters.

Parameter	Description
R_{nc}	Total network capacity, [Mbps]
λ	Arrival rate of requests, [sessions/s]
θ	Average file size, [MB]
b	Minimum assignable bit rate to session, [Mbps]
$\mu^{-1} = \theta_s/R_{nc}$	Average duration of ongoing session, [s]
β^{-1}	Average waiting time in the buffer before leaving for its retrial group, [s]
α^{-1}	Average waiting time in the buffer retrial group before re-entering the buffer, [s]
$N_{nc} = \lfloor R_{nc}/b \rfloor$	Maximum number of starting sessions, [sessions]
N_b	Maximum number of requests in the buffer, [sessions]
N_{rg}	Maximum number of requests in the buffer retrial group, [sessions]

According to above-described RAC rules and taking into account that the arrival process is Poisson in nature while service times are exponentially distributed, the behavior of the system could be described by the Markov chain $X(t) = \{(N(t), M(t)), t > 0\}$, where $N(t)$ is the total number of the started sessions and the requests waiting in the buffer, at instant time t; $M(t)$ - the number of requests in the buffer retrial group at instant time t; on the state space $\mathbf{X} = \{(n, m): n \geq 0, m \geq 0\}$, where n is the total number of the started sessions and the requests waiting in the buffer; m - the number of requests in the buffer retrial group. The corresponding state transition and central state transition diagrams are shown respectively in Figs. 3 and 4. Note that due to their elastic nature, the sessions are characterized by an uniformly-distributed channel rate R_{nc} between all the simultaneously served elastic flows. Wherein the elastic session bit rate is R_{nc}/n, provided that the total number of the started sessions is less or equal to N_{nc}. The service rate of the incoming sessions are defined as $C/n\theta$, where $n \leq N_{nc}$.

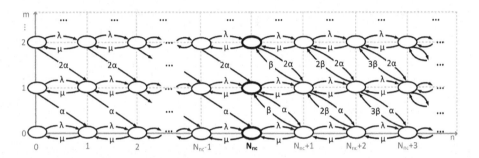

Fig. 3. State transition diagram for infinite case.

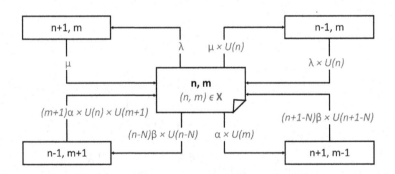

Fig. 4. Central state transition diagram for infinite case.

According to the central state transition diagram (Fig. 4), the corresponding system of equilibrium equations has the following form:

$$
\begin{aligned}
\big[\lambda + \mu \cdot U(n) + m\alpha \cdot U(m) + (n - N_{nc})\beta \cdot U(n - N_{nc})\big] \cdot p(n, m) &= \lambda \cdot U(n) \times \\
\times\, p(n - 1, m) + \mu \cdot p(n + 1, m) + (m + 1)\alpha \cdot U(n) \cdot U(m + 1) &\times \quad\quad (1) \\
\times\, p(n - 1, m + 1) + (n + 1 - N_{nc})\beta \cdot U(n + 1 - N_{nc}) \cdot p(n + 1, m - 1),
\end{aligned}
$$

where $p(n, m), (n, m) \in \mathbf{X}$ is the stationary probability distribution and U - the unit step function that equals 1 when the argument is positive and 0 otherwise.

Since the buffer and its retrial group have infinite capacities, the only way for finding the stationary probability distribution $p(n, m), (n, m) \in \mathbf{X}$ is to use a generating function-based approach [13]. Since this work presents the initial stage of the studied model, we tried to simplify its analysis as much as possible limiting the system capacity and avoiding the introduction of the generating function (Table 2).

3   BG Traffic Service Model with Finite Capacity

Let us turn to the model with finite capacity (Fig. 5), i.e. $N_b < \infty$ and $N_{rg} < \infty$.

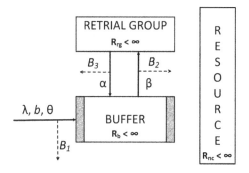

Fig. 5. Scheme model for finite case.

The RAC is defined in such a way that, when the user request arrives in system, three outcomes are possible:

- session starts when the number of started sessions is less than N_{nc};
- request waits in the buffer when the number of started sessions equals N_{nc} and the number of requests in the buffer is less than N_b;
- request is blocked otherwise.

Note that the requests leaving the buffer for its retrial group can be blocked when the number of requests in the buffer retrial group equals N_{rg}, as also the requests leaving the buffer retrial group for the buffer when the number of requests in the buffer equals N_b.

According to these RAC rules, the system behavior is described by a two-dimensional vector (n, m) on the state space $\mathbf{Y} = \{(n, m): 0 \leq n \leq N_{nc} + N_b, 0 \leq m \leq \leq N_{rg}\}$, where n is the total number of the started sessions and the requests in the buffer; m - the number of requests in the buffer retrial group.

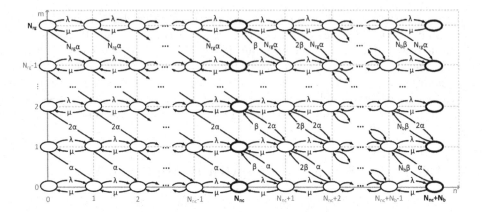

Fig. 6. State transition diagram for finite case.

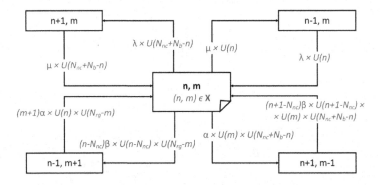

Fig. 7. Central state transition diagram for finite case.

The corresponding state transition and central state transition diagrams are shown respectively in Figs. 6 and 7.

The considered Markov process is described by the following system of equilibrium equations:

$$
\begin{aligned}
&[\lambda \cdot U(N_{nc} + N_b - n) + \mu \cdot U(n) + m\alpha \cdot U(m) \cdot U(N_{nc} + N_b - n) \\
&+ (n - N_{nc})\beta \cdot U(m - N_{nc}) \cdot U(N_{rg} - m)] \cdot q(n, m) \\
&= \mu \cdot U(N_{nc} + N_b - n) \cdot q(n + 1, m) + \lambda \cdot U(n) \cdot q(n - 1, m) \quad (2)\\
&+ (m + 1)\alpha \cdot U(n) \cdot U(N_{rg} - m) \cdot q(n - 1, m + 1) + (n + 1 - N_{nc})\beta \times \\
&\times U(n + 1 - N_{nc}) \cdot U(N_{nc} + N_b - n) \cdot U(m) \cdot q(n + 1, m - 1),
\end{aligned}
$$

where $q(n, m), (n, m) \in \mathbf{Y}$ is the stationary probability distribution.

Similarly to infinite case described in previous section, the process describing the system behavior is not reversible Markov process. Therefore, the stationary probability distribution $q(n, m)_{(n,m) \in \mathbf{Y}} = \mathbf{q}$ can be calculated by using a numerical solution of the system of equilibrium equations $\mathbf{q} \cdot \mathbf{A} = \mathbf{0}, \mathbf{q} \cdot \mathbf{1}^T = \mathbf{1}$, where $\mathbf{A}$

is the infinitesimal generator of the Markov process, elements $a((n, m)(n', m'))$ of which are defined as follows (3):

$$
\begin{cases}
\lambda, & \text{if } n' = n + 1, m' = m, n < N_{nc} + N_b; \\
\mu, & \text{if } n' = n - 1, m' = m, n > 0; \\
(n - N_{nc})\beta, & \text{if } n' = n - 1, m' = m + 1, n > N_{nc}, m < N_{rg}; \\
m\alpha, & \text{if } n' = n + 1, m' = m - 1, n < N_{nc} + N_b, m > 0; \\
*, & \text{if } n' = n, m' = m; \\
0, & \text{otherwise,}
\end{cases}
\tag{3}
$$

where $* = -(\lambda \cdot U(N_{nc} + N_b - n) + \mu \cdot U(n) + m\alpha \cdot U(N_{nc} + N_b - n) \cdot U(m) + (n - N_{nc})\beta \cdot U(n - N_{nc}) \cdot U(N_{rg} - m))$.

The main performance measure of the described model is the blocking probability. Due to the system limited capacity and according to the above-described RAC scheme, we consider the following types of blocking probabilities:

– Blocking probability B_1 of a new arriving request

$$
B_1 = \sum_{m=0}^{N_{rg}} q(N_{nc} + N_b, m);
\tag{4}
$$

– Blocking probability B_2 of a request arriving from the buffer to its retrial group

$$
B_2 = \sum_{n=0}^{N_{nc}+N_b} q(n, N_{rg});
\tag{5}
$$

– Blocking probability B_3 of a request arriving from the buffer retrial group to the buffer

$$
B_3 = \sum_{m=1}^{N_{rg}} q(N_{nc} + N_b, m);
\tag{6}
$$

– Total blocking probability B

$$
B = B_1 + B_2 - q(N_{nc} + N_b, N_{rg}).
\tag{7}
$$

4 Numerical Analysis

For the numerical analysis, let us consider a slice of one multi-service wireless network with total offered capacity of 40 Mbps. The average file size is 3 MB. We assume that the requests can leave the buffer for its retrial group each 10 s and the buffer retrial group for the buffer each 5 s. As an example of the physical interpretation of this model, we can consider the process of loading a web page by a user. Suppose that the loading happens with some delay. This delay in our model corresponds to finding the request in the buffer. The user does not like to be in standby mode and he reloads the page, thereby switching to the buffer retrial group. Then he returns to the buffer and starts again waiting for the page to load.

Since the aim of the paper is the model with infinite capacity analysis using the model with finite capacity, we need to set carefully the capacities of the buffer and its retrial group so that resulting blocking probability does not exceed a certain threshold value (i.e. 10^{-4}). For that, we set the maximum number of requests in the buffer equal to 13 and the maximum number of requests in the buffer retrial group equal to 10.

Table 2. Parameters of the numerical analysis.

–	R_{nc}	λ	θ	b
Scenario 1	40 Mbps	3 − 10	3 MB	5 Mbps
Scenario 2		3		5 − 15 Mbps

Scenario 1. We illustrate the behavior of all the blocking probabilities depending on the arrival rate λ (i.e. varies from 3 to 10) for a minimum assignable bit rate b equal to 5 Mbps (Fig. 8). According to the formulas (4) and (6), the blocking probabilities B_1 and B_3 are substantially equal (i.e. $B_1 \gtrless B_3$), which is denoted in the resulting plot.

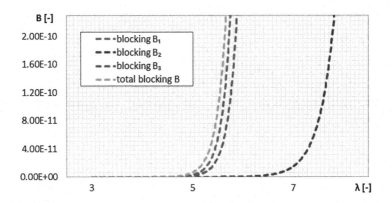

Fig. 8. Blocking probabilities depending on the arrival rate λ.

Scenario 2. We illustrate the behavior of all the blocking probabilities depending on the minimum assignable bit rate b (i.e. varies from 5 to 15 Mbps) for an arrival rate λ equal to 3 (Fig. 9). We remark that, the difference between the curves of the blocking probabilities increases as b gets higher.

From Both Resulting Plots. One can see that consequently to initial data sets, the blocking probability B_2 is significantly lower than the probabilities B_1 and B_3 (i.e $B_2 \ll B_3 < B_1$), because the average waiting time in the buffer retrial group before re-entering the buffer is less than the average waiting time in the buffer before leaving for its retrial group (i.e. $\beta < \alpha$).

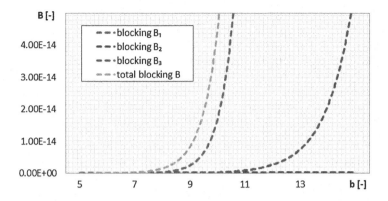

Fig. 9. Blocking probabilities depending on the minimum assignable bit rate b.

5   Conclusion

In this paper, a mathematical model for radio resources allocation in a wireless communication network implementing NS technology is proposed. The model was described by a buffer and its retrial group. For the model performance analysis, another model with finite capacities of the buffer and its retrial group was considered. In both models, systems of equilibrium equations were obtained. For the finite model, formulas were proposed for calculating its main performance measures - blocking probabilities and a numerical analysis of them was provided. In the future, it is expected to conduct an investigation of the infinite case using the generating function-based approach. And also consider the classic case of the retrial group model, in which the introduction of the buffer will determine the priorities for servicing applications.

References

1. 3GPP TS 23.501 V15.4.0 - System architecture for the 5G System (5GS)
2. Khatibi, S., Correia, L.M.: Modelling virtual radio resource management in full heterogeneous network. EURASIP Journal on Wireless Communications and Networking. vol. 2017. Art. no. 73
3. TS 129 512–V15.3.0 - 5G; 5G System; Session Management Policy Control Service; Stage 3 (3GPP TS 29.512 version 15.3.0 Release 15) - ETSI
4. Cisco Visual Networking Index: Global Mobile Data Traffic Forecast Update, 2017–2022 White Paper
5. Ordóez, J., et al.: The Creation Phase in Network Slicing: From a Service Order to an Operative Network Slice. 1–36 (2018). https://doi.org/10.1109/EuCNC.2018.8443255
6. Kułacz, Ł., Kryszkiewicz, P., Kliks, A.: Waveform flexibility for network slicing. Wirel. Commun. Mob. Comput. **2019**, 1–15 (2019). https://doi.org/10.1155/2019/6250804

7. Khan, H., Luoto, P., Samarakoon, S., Bennis, M., Latva-aho, M.: Network slicing for vehicular communication. Trans. Emerg. Telecommun. Technol. (2019). https://doi.org/10.1002/ETT.3652
8. Campolo, C., Fontes, R.R., Molinaro, A., Rothenberg, C.E., Iera, A.: Slicing on the road: Enabling the automotive vertical through 5G network softwarization. Sensors (Switzerland), vol. 18 (12). Art. no. 4435 (2018)
9. Korenevskaya, M., Zayats, O., Ilyashenko, A., Muliukha, V.: Retrial queuing system with randomized push-out mechanism and non-preemptive priority. Procedia Comput. Sci. **150**, 716–725 (2019). https://doi.org/10.1016/j.procs.2019.02.016
10. Devos, A., Walraevens, J., Bruneel, H.: A priority retrial queue with constant retrial policy (2018). https://doi.org/10.1007/978-3-319-93736-6_1
11. Springer, New York. https://doi.org/10.1007/s11134-019-09608-z
12. Ammar, S., Rajadurai, P.: Performance analysis of preemptive priority retrial queueing system with disaster under working breakdown services. Symmetry **11**, 419 (2019). https://doi.org/10.3390/sym11030419
13. Bocharov, P.P., D'Apice, C., Pechinkin, A.V., Salerno, S.: Queueing Theory (2003). https://doi.org/10.1515/9783110936025

Reliability Model of a Homogeneous Warm-Standby Data Transmission System with General Repair Time Distribution

H. G. K. Houankpo[1] and Dmitry Kozyrev[1,2(⊠)]

[1] Department of Applied Probability and Informatics,
Peoples' Friendship University of Russia (RUDN University),
6 Miklukho-Maklaya Street, Moscow 117198, Russian Federation
gibsonhouankpo@yahoo.fr, kozyrev-dv@rudn.ru
[2] V. A. Trapeznikov Institute of Control Sciences of Russian Academy of Sciences,
65, Profsoyuznaya Street, Moscow 117997, Russian Federation

Abstract. We consider a mathematical model of a repairable data transmission system as a model of a closed homogeneous system $\langle M_2/GI/1 \rangle$ of two elements in a warm-standby with an exponential distribution function of the time to failure, and with an arbitrary distribution function of the repair time of its elements with one restoring element. Explicit analytical expressions for the stationary probability distribution of the system states and for the stationary failure probability of the system are obtained. Comparison of the reliability of a cold-standby system with hot-standby system through a warm-standby model is performed. A simulation model for the most reliable cold-standby case is developed.

Keywords: System-level reliability · Steady state probabilities · Warm standby · Mathematical modeling · Simulation · Sensitivity

1 Introduction

Recently the functioning of various aspects of modern society has become critically reliant on communication networks [1, 2]. With the migration of critical facilities to the communication networks, it has become vitally important to ensure the reliability and availability of networks and data transmission systems. A series of previous studies [3–8] were focused on reliability-centric analysis of various complex telecommunications systems. Particularly, the reliability study of cold-standby data transmission systems was conducted. In [9] the creation of simulation models and means to support computer-aided design of highly reliable distributed computer systems was considered. In [10] there was introduced a transmission line failure model that is enhanced with the dynamic thermal rating (DTR) system, and there were investigated the uncertainty effects of the line failure model parameters, effects of the DTR system reliability, and the

© Springer Nature Switzerland AG 2019
V. M. Vishnevskiy et al. (Eds.): DCCN 2019, LNCS 11965, pp. 443–454, 2019.
https://doi.org/10.1007/978-3-030-36614-8_34

effects of the weather data correlation on the reliability performance of the power system. In [12] a reliability analysis was presented of a combined cycle gas turbine CCGT power plant. In [13] a threat modeling framework was introduced using the Icelandic transmission system as an example, highlighting the need for improved data collection and failure rate modeling. In [14] the result of the practice of data storage was discussed and a solution was proposed for finding transition rate matrix for a component model when data is available to calculate the state probabilities when the data for the estimation of transition rates is either incomplete or unavailable. In [15] there was developed a vendor independent model for system reliability studies and sensitivity analysis of system availability. Paper [16] was focused on the above problem and proposed the refining modeling method based on the probabilistic risk assessment (PRA) method, which can synthetically adopt the event tree (ET), fault tree (FT), dynamic fault tree (DFT) and Bayesian networks (BN) to model and analyze the above features, the example in this article shows the effectiveness of the proposed method, and this method can be as a reference to model and assess the satellite life and reliability. In [17] an improved method was proposed for tracking performance degradation during a long time operation. Paper [18] considered a simulation modeling method for complex system mission reliability simulation via mission profile modeling, environment profile modeling, and dynamic reliability modeling techniques. In [19] modeling and estimation techniques were presented permitting the temperature-aware optimization of application-specific multiprocessor system-on-chip (MPSoC) reliability. In [20] a new reliability model was introduced for a warm standby redundant configuration with units that are originally operated in active mode, and then, upon turn-on of originally standby units, are put into warm standby mode.

In the current paper, the results of previous studies are summarized for the case of the so-called warm redundancy of the system $\langle M_2/GI/1 \rangle$ and the calculation and comparison of the stationary reliability characteristics for different types of redundancy is performed. The aim of the work is to perform analytical modeling and simulation of a warm-standby redundant system $\langle M_2/GI/1 \rangle$ with a general distribution function $B(x)$ and the corresponding distribution density $b(x)$ of the repair time of its elements, and the exponential distribution of the failure-free time of its elements with a parameter $\lambda_i = \alpha + (1-i)\gamma;\ i = 0,1$.

2   Mathematical Model and Analytical Results

Consider a stochastic process $v(t)$—the number of failed elements at time t, with the set of states $E = \{0,1,2\}$.

To solve the stated problem, we consider an approach which is based on the Markovization principle [9]. To describe the behavior of the system using a Markov process, we introduce a supplementary variable $x(t) \in \mathbb{R}_+$—the overall duration spent at time t for recovery of the failed element. We obtain a two dimensional [10] process $(v(t), x(t))$, with an extended phase space $\mathcal{E} = \{(0), (1,x), (2,x)\}$.

We denote by $p_0(t)$ the probability that at time t the system is in the state $i = 0$, and by $p_i(t, x)$ the probability density function (in continuous component) that at time t the system is in state i $(i = 1, 2)$, and the time taken to repair the failed element is in the range $(x, x + dx)$.

$$p_0(t) = P\{v(t) = 0\}, \tag{1}$$

$$p_1(t, x)dx = P\{v(t) = 1, x < x(t) < x + dx\}, \tag{2}$$

$$p_2(t, x)dx = P\{v(t) = 2, x < x(t) < x + dx\}. \tag{3}$$

With the help of the total probability rule we move to the Kolmogorov's forward system of differential equations, which makes it possible to find the stationary state probabilities [11] of the considered system. The state-related equations are as follows:

$$\begin{cases} p_0(t + \Delta) = (1 - \lambda_0\Delta)p_0(t) + \int_0^t p_1(t, x) \cdot \delta(x)\Delta dx \\ p_1(t + \Delta, x + \Delta) = p_1(t, x) \cdot (1 - \lambda_1\Delta)(1 - \delta(x)\Delta) \\ p_2(t + \Delta, x + \Delta) = p_2(t, x) \cdot (1 - \delta(x)\Delta) + p_1(t, x)\lambda_1\Delta, \end{cases} \tag{4}$$

with the boundary condition:

$$p_1(t + \Delta, \theta\Delta) = P\{v(t + \Delta) = 1, 0 < x(t + \Delta) < \Delta\}, \tag{5}$$

$$p_1(t + \Delta, \theta\Delta)\Delta = p_0(t)\lambda_0\Delta + \int_0^t p_2(t, x) \cdot \delta(x)\Delta dx, \tag{6}$$

where $\delta(x) = \frac{b(x)}{1-B(x)}$ is the conditional probability density function of the residual repair duration of the element being repaired at time t [12]. Passing to the limit $\Delta \to 0$, and under the assumption that the described process has the stationary probability distribution, as $t \to \infty$ the transformed equations take the following form:

$$\begin{cases} \lambda_0 \cdot p_0 = \int_0^\infty p_1(x) \cdot \delta(x)dx \\ \frac{\partial p_1(x)}{\partial x} = -(\lambda_1 + \delta(x)) \cdot p_1(x) \\ \frac{\partial p_2(x)}{\partial x} = -\delta(x) \cdot p_2(x) + \lambda_1 \cdot p_1(x), \end{cases} \tag{7}$$

with the corresponding boundary condition:

$$p_1(0) = \lambda_0 \cdot p_0 + \int_0^\infty p_2(x) \cdot \delta(x)dx. \tag{8}$$

From this, we turn to the solution of the resulting system of differential equations applying the parameters variation method [21], and we obtain the following stationary state probabilities of the considered warm-standby system:

$$p_0 = C_1 \cdot \frac{\tilde{b}(\lambda_1)}{\lambda_0}, \tag{9}$$

$$p_1 = C_1 \cdot \left(\frac{1 - \tilde{b}(\lambda_1)}{\lambda_1} \right), \tag{10}$$

$$p_2 = C_1 \cdot \left(\frac{\rho^{-1} - (1 - \tilde{b}(\lambda_1))}{\lambda_1} \right). \tag{11}$$

where $\rho^{-1} = \frac{EB}{EA} = b \cdot \lambda_1$, b – mean repair time of a failed element and $\tilde{b}(\lambda_1)$ – Laplace transform of the probability density function $b(x)$ of the repair time distribution.

Using the normalization condition $p_0 + p_1 + p_2 = 1$, we obtain the constant C_1:

$$C_1 = \left(\frac{\tilde{b}(\lambda_1)}{\lambda_0} + \frac{\rho^{-1}}{\lambda_1}\right)^{-1}. \tag{12}$$

From this, we obtain the stationary state probabilities for the cases of the cold-standby ($\gamma = 0$) and hot-standby ($\gamma = \alpha$) redundancy.

Obviously, there is a dependency of stationary state probabilities of the system on the type of the distribution of repair time of its elements.

Table 1 shows the values of the system failure probability p_2 for different distribution functions of repair time (we considered for example: Exponential (M), Weibull (WB) and Pareto (PAR)) for different values of the model parameter ρ, calculated with the use of the elaborated analytical model.

Table 1. Analytical results for the steady-state failure probability p_2 of the system $\langle M_2/GI/1 \rangle$ calculated for different values of the model parameter $\rho = 1, 10, 100$.

ρ	M		WB		PAR	
	$\gamma = 0$	$\gamma = \alpha$	$\gamma = 0$	$\gamma = \alpha$	$\gamma = 0$	$\gamma = \alpha$
1	0.3333	0.4	0.3960	0.4938	0.2843	0.3315
10	0.0090	0.0164	0.0204	0.0371	0.006	0.0109
100	$9.89 \cdot 10^{-5}$	0.0002	0.0003	0.0006	$6.6 \cdot 10^{-5}$	0.0001

One can see that for all distributions, the steady-state failure probability p_2 under hot redundancy is greater than that under the cold one, that is, the most reliable model is the cold-standby one.

Figures 1 and 2 show the corresponding plots of the steady-state probability of failure of the cold and hot-standby systems $\langle M_2/GI/1 \rangle$ versus the relative recovery rate for various "repair" time distributions (for example: Exponential (M), Weibull (WB) and Pareto (PAR)).

Evidently, this dependency becomes vanishingly small under a "fast" recovery [13–18] of the system elements.

In these figures, it is very difficult to see the difference between the cold and hot standby cases, so we decided to find out the first value of the system uptime probability $1 - p_2 GI(1)$ and the last value of the system uptime probability through probability of system failure $1 - p_2 GI(50)$ to confirm the most reliable reserve (Table 2).

Predictably, the system uptime probability for the cold standby case is greater than the one for the hot standby case.

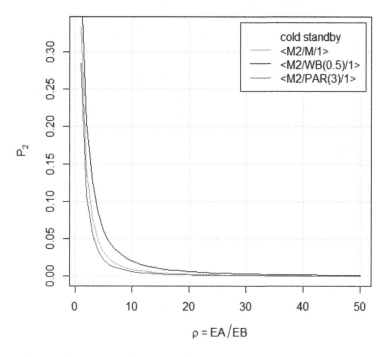

Fig. 1. Steady-state failure probability of a cold standby system.

Table 2. Values of the system's uptime probability $1 - p_2$.

$1\text{-}p_2 GI$	Cold-standby	Hot-standby
$1\text{-}p_2\mathrm{M}[1]$	0.6666667	0.6644737
$1\text{-}p_2\mathrm{WB}[1]$	0.6039816	0.6008824
$1\text{-}p_2\mathrm{PAR}[1]$	0.7156417	0.7140474
$1\text{-}p_2\mathrm{M}[50]$	0.9996080	0.9992314
$1\text{-}p_2\mathrm{WB}[50]$	0.9989058	0.9978546
$1\text{-}p_2\mathrm{PAR}[50]$	0.9997400	0.9994902

It is not always possible to obtain the explicit analytical expressions for stationary distribution of the system under consideration. Therefore, the task of constructing the simulation model approximating the analytical model of the system.

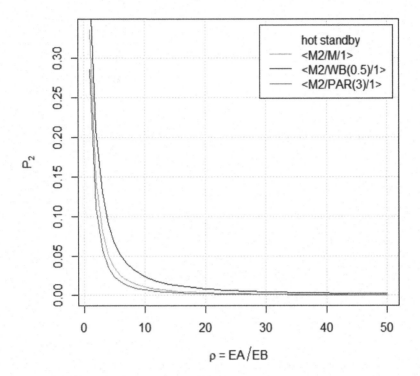

Fig. 2. Steady-state failure probability of a hot standby system.

3   Simulation Model

We've developed a simulation model for the most reliable "cold-standby" case. We define the following states of the simulated system:

- state 0: one (main) element is working, the second one is in cold reserve;
- state 1: one element has failed and is being repaired, the second one is working;
- state 2: both elements have failed, one is under repair, the other one is waiting for its turn to be repaired.

The simulation of the system $\langle M_2/GI/1 \rangle$ is carried out on the basis of a discrete-event approach, according to which a change in the state of the system occurs instantaneously (in model time) upon the occurrence of a particular event.

To describe the algorithm for modeling the system reliability assessment $\langle M_2/GI/1 \rangle$ the following variables are introduced:

- double t—hours of model time, change in case of failure or restoration of system elements;
- int i, j—system state variables; when an event occurs, the transition from i to j;
- double $t_{nextfail}$—service variable, which stores the time until the next element failure;

- double $t_{nextrepair}$—service variable, which stores the time until the next repair of the failed item;
- int k—count of iterations of the main loop.

For clarity, the simulation model is presented graphically in Fig. 3 in the form of a flowchart.

The stop condition for the main cycle of the model is reaching the maximum model execution time T.

For a better understanding and reproducibility of the simulation model, in addition to the flowchart, the algorithm of the discrete-event process of simulation modeling is also provided in the form of a pseudo-code with comments (Algorithm 1).

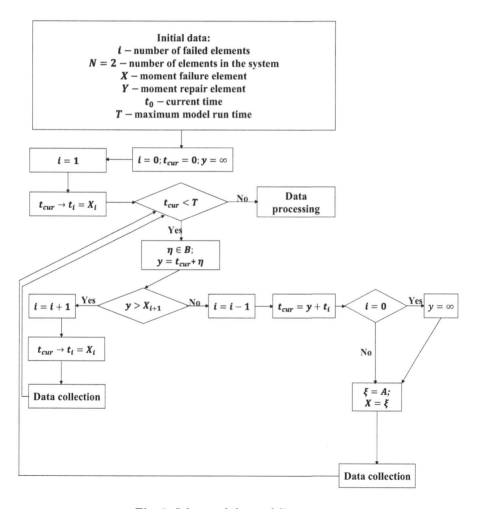

Fig. 3. Scheme of the modeling system

Algorithm 1. Pseudo-code of the system simulation process $\langle M_2 GI/1 \rangle$
Input: a1, b1, N,T, NG,"GI".

$a1$ - Mean time between element failures
$b1$ - Average repair time
$N = 2$ - The number of elements in the system
$T = 1000$ - Maximum model run time
$NG = 10000$ - Number of Path Graphs
"GI" - Distribution function.

Output: stationary state probabilities P_0, P_1, P_2.

Begin
array $r[] := [0,0,0]$; // multidimensional array containing results, k-steps of the main loop
double $t := 0.0$; // time clock initialization
int $i := 0$; $j := 0$; // system state variables
double $t_{nextfail} := 0.0$; // variable in which time until the next element failure
double $t_{nextrepair} := 0.0$; // variable in which time is stored until the next repair is completed
int $k := 1$; // count of iterations of the main loop
$s := rf\_exp(1/a1)$; // generation of an arbitrary random variable s – time to the first event (failure)
$ss := rf\_GI(\delta(x))$; // generation of an arbitrary random variable ss – time of repair of the failed element)
$t_{nextfail} := t + s$;
$t_{nextrepair} := t + ss$;
while $t < \infty$ **do**
    if $i = 0$ **then**
        $t_{nextrepair} := \infty$; $j := j + 1$; $t := t_{nextfail}$;
    else if
        if $i = 1$ **then**
            $S_1 := rf\_exp(1/a1); S_2 := rf\_GI("\delta(x)")$;
            $t_{nextfail} := t + S_1$; $t_{nextrepair} := t + S_2$;
            if $t_{nextfail} < t_{nextrepair}$ **then**
                $j := j + 1$; $t := t_{nextfail}$;
            else
                $j := j - 1$; $t := t_{nextrepair}$;
            end
        else
            $i = 2$; $t_{nextfail} := \infty$; $j := j - 1$; $t := t_{nextrepair}$;
        end
        if $t > T$ **then**
            $t = T$
        end
        $r[,,k] := [t,i,j]$; $i := j$; $k := k + 1$;

end do

Estimated duration of stay in each state $i, i = 0, 1, 2.$; calculation of stationary probabilities

$$\widehat{P}_i = \frac{1}{NG} \sum_{j=1}^{NG} (\text{ length of stay } i/T)_j$$

end

Software implementation of Algorithm 1 was performed in the programming language R.

Table 3 shows the values of stationary state probabilities, calculated both via the simulation approach and by the analytical formulas given earlier in Sect. 2. For the analysis and comparison of the results, the following distribution of repair times of the system's elements were chosen: exponential, Weibull-Gnedenko and Pareto. However, the developed simulation model is not limited to the choice of the repair time distribution. As a model parameter the value $\rho = \frac{a_1}{b_1}$ is considered.

For analyzing the sensitivity [22] of the model to the form of the distribution function, the magnitude values increase $\rho = 1, 10, 100$, where $b_1 = 1$ in all the considered cases, the distribution parameters are chosen in such a way as to correspond to the model's parameter value.

Table 3. Analytical and simulation results for the steady-state failure probability p_2 of the system $\langle M_2/GI/1\rangle$ calculated for different values of the model parameter $\rho = 1, 10, 100$.

ρ	M		WB		PAR	
	Analytic	Simulation	Analytic	Simulation	Analytic	Simulation
1	0.3333	0.3330	0.3960	0.3945	0.2844	0.284
10	0.009	0.009	0.0204	0.0202	0.006	0.0061
100	0.0001	0.0001	0.0003	0.0003	0.000066	0.000067

As Table 3 shows, the simulation results are in good agreement with the results obtained using the explicit analytical formulas. It is also seen that with increasing ρ the differences in the values of p_2 are leveled. For greater clarity, we compare the analytical and simulation results graphically.

Figure 4 shows the plots of the probability of system failure versus the relative recovery rate ρ.

It is evident that the differences between the analytical and simulation curves become vanishingly small as the relative recovery rate increases.

It can be concluded that the constructed simulation model well approximates the analytical model of the $\langle M_2/GI/1\rangle$ system, which means it can be used in such cases (for such repair time distribution functions) when it's hard or

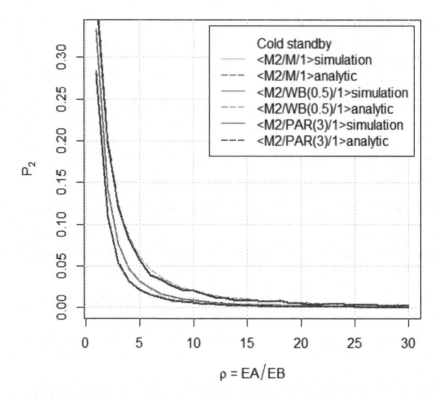

Fig. 4. Steady-state failure probability of a cold-standby system versus the relative recovery rate ρ. Dashed curves are analytical results and solid curves show the corresponding simulation results.

impossible to derive the expressions for stationary probabilities of system states in explicit analytical form.

4   Conclusion

Explicit analytical expressions were obtained for calculation of the steady-state probability distribution of the states of the warm-standby system and for calculation of the stationary failure probability of the considered system. We calculated the stationary failure probability of the considered system for the special cases of cold-standby and hot-standby redundancy. The obtained formulas show the presence of an explicit dependency of these characteristics to the types of the cumulative distribution function of the repair time of the system's elements. However, the numerical studies and the analysis of the constructed plots showed that this dependency becomes vanishingly small under the "fast" recovery, that is, with the increase in the relative recovery rate ρ, and also showed that the most reliable model for all the distributions is the model with a cold type of redundancy. Discrete-event simulation model was constructed for the most reliable "cold-standby" case that approximates the analytical results. The results obtained with both analytical and simulation approaches are in close agreement and therefore can be used in cases when it is not possible to get expressions for stationary probabilities of system states in an explicit analytical form.

Acknowledgments. The publication has been prepared with the support of the "RUDN University Program 5–100" (D.V.Kozyrev, mathematical model development, H.G.K. Houankpo, simulation model development). The reported study was funded by RFBR, project number 17-07-00142 (recipient D.V.Kozyrev, numerical analysis).

References

1. Ahmed, W., Hasan, O., Pervez, U., Qadir, J.: Reliability modeling and analysis of communication networks. J. Netw. Comput. Appl. **78**, 191–215 (2017). https://doi.org/10.1016/j.jnca.2016.11.008
2. Ometov, A., Kozyrev, D., Rykov, V., Andreev, S., Gaidamaka, Y., Koucheryavy, Y.: Reliability-centric analysis of offloaded computation in cooperative wearable applications, wireless communications and mobile computing, 2017, Article ID 9625687, 15 p. (2017). https://doi.org/10.1155/2017/9625687
3. Rykov, V., Kozyrev, D., Zaripova, E.: Modeling and simulation of reliability function of a homogeneous hot double redundant repairable system. In: Proceedings of the 31st European Conference on Modelling and Simulation, ECMS 2017, Germany, Digitaldruck Pirrot GmbH, pp. 701–705 (2017). https://doi.org/10.7148/2017-0701
4. Houankpo, H.G.K., Kozyrev, D.V.: Sensitivity analysis of steady state reliability characteristics of a repairable cold standby data transmission system to the shapes of lifetime and repair time distributions of its elements. In: CEUR Workshop Proceedings, vol. 1995, pp. 107–113 (2017). http://ceur-ws.org/Vol-1995/
5. Rykov, V.V., Ngia, T.A.: On sensitivity of systems reliability characteristics to the shape of their elements life and repair time distributions. Bull. Peoples' Friendship Univ. Russia. Ser. Math. Inf. Sci. Phys. **3**, 65–77 (2014). (In Russian)
6. Efrosinin, D.V., Rykov, V.V.: Sensitivity analysis of reliability characteristics to the shape of the life and repair time distributions. Commun. Comput. Inf. Sci. **487**, 101–112 (2014)
7. Efrosinin, D.V., Rykov, V.V., Vishnevskiy, V.V.: Sensitivity of reliability models to the shape of life and repair time distributions. In: 9-th International Conference on Availability, Reliability and Security (ARES 2014), pp. 430–437. IEEE (2014). Published in CD: 978-I-4799-4223-7/14, https://doi.org/10.1109/ARES.2014.65
8. Rykov, V.V., Kozyrev, D.V.: Analysis of renewable reliability systems by markovization method. In: Rykov, V.V., Singpurwalla, N.D., Zubkov, A.M. (eds.) ACMPT 2017. LNCS, vol. 10684, pp. 210–220. Springer, Cham (2017). https://doi.org/10.1007/978-3-319-71504-9_19
9. Parshutina, S., Bogatyrev, V.: Models to support design of highly reliable distributed computer systems with redundant processes of data transmission and handling. In: 2017 International Conference "Quality Management, Transport and Information Security, Information Technologies" (IT&QM&IS) (2017). https://doi.org/10.1109/ITMQIS.2017.8085772

10. Teh, J., Lai, C.-M., Cheng, Y.-H.: Impact of the real-time thermal loading on the bulk electric system reliability. IEEE Trans. Reliab. **66**(4), 1110–1119 (2017). https://doi.org/10.1109/TR.2017.2740158

11. Sevastyanov, B.A.: An Ergodic theorem for Markov processes and its application to telephone systems with refusals. Theor. Probab. Appl. **2**(1), 104–112 (1957)

12. Lisnianski, A., Laredo, D., Haim, H.B.: Multi-state Markov model for reliability analysis of a combined cycle gas turbine power plant. In: 2016 Second International Symposium on Stochastic Models in Reliability Engineering, Life Science and Operations Management (SMRLO) (2016). https://doi.org/10.1109/SMRLO.2016.31

13. Perkin, S., et al.: Framework for threat based failure rates in transmission system operation. In: 2016 Second International Symposium on Stochastic Models in Reliability Engineering, Life Science and Operations Management (SMRLO) (2016). https://doi.org/10.1109/SMRLO.2016.34

14. Singh, C.: Assigning transition rates to unit models with incomplete data for power system reliability analysis. In: 2015 Annual IEEE India Conference (INDICON) (2015). https://doi.org/10.1109/INDICON.2015.7443163)

15. Tourgoutian, B., Yanushkevich, A., Marshall, R.: Reliability and availability model of offshore and onshore VSC-HVDC transmission systems. In: 11th IET International Conference on AC and DC Power Transmission, 13 July 2015. https://doi.org/10.1049/cp.2015.0101

16. Li, X., Ao, N., Wu, L.: The refining reliability modeling method for the satellite system. In: 2014 10th International Conference on Reliability, Maintainability and Safety (ICRMS) (2014). https://doi.org/10.1109/ICRMS.2014.7107244

17. Xu, M., Zeng, S., Guo, J.: Reliability modeling of a jet pipe electrohydraulic servo valve. In: 2014 Reliability and Maintainability Symposium (2014). https://doi.org/10.1109/RAMS.2014.6798480

18. Cao, J., Wang, Q., Shen, Y.: Research on modeling method of complex system mission reliability simulation. In: 2012 International Conference on Quality, Reliability, Risk, Maintenance, and Safety Engineering (2012). https://doi.org/10.1109/ICQR2MSE.2012.6246242

19. Gu, Z., Zhu, C., Shang, L., Dick, R.: Application-specific multiprocessor system-on-chip reliability optimization. IEEE Trans. Very Large Scale Integr. (VLSI) Syst. **16**(5) (2008), https://doi.org/10.1109/TVLSI.2008.917574

20. Huang, W., Loman, J., Song, T.: Reliability modeling of a warm standby redundancy configuration with active → standby → active units. In: 2014 Reliability and Maintainability Symposium, Colorado Springs, CO, pp. 1–5 (2014). https://doi.org/10.1109/RAMS.2014.6798473

21. Petrovsky, I.G.: Lectures on the theory of ordinary differential equations, Lectures on the theory of ordinary differential equations, Moscow, GITTL, 232 p. (1952). (In Russian)

22. Rykov, V., Kozyrev, D.: On sensitivity of steady-state probabilities of a cold redundant system to the shapes of life and repair time distributions of its elements. In: Pilz, J., Rasch, D., Melas, V., Moder, K. (eds.) Statistics and Simulation, IWS 2015. Springer Proceedings in Mathematics & Statistics, vol. 231, pp. 391–402. Springer, Cham (2018). https://doi.org/10.1007/978-3-319-76035-3_28

Distributed Systems Applications

Methodology for Data Processing in Modular IoT System

Kristina Dineva and Tatiana Atanasova$^{(\boxtimes)}$ (iD)

Institute of Information and Communication Technologies, BAS, Sofia, Bulgaria
{k.dineva,atanasova}@iit.bas.bg

Abstract. The Internet of Things (IoT) is a technological paradigm in the sphere of networks that has the potential to influence how we live and how we work. This technology allows communication between all types of physical objects over the Internet that includes data sharing and also allows data to be collected and actions taken based on the information received. "Things" in the "Internet of Things" consist of a variety of hardware specifications, communication capabilities and service qualities, making IoT heterogeneous in its nature. The lack of Reference Modelling IoT Architecture prevents a common approach to processing the generated data. The need to retrieve and analyze this data from the Internet based complex systems in real-time requires applying of statistical data analysis and machine learning (ML) techniques as well as a sufficient amount of computational resources. In the paper a methodology to deal with a variety of data is proposed. A modular IoT system is considered as an instance for implementation of several methods for processing of heterogeneous data. The approaches for resolving problems that can affect the creation of predictive models are outlined.

Keywords: IoT · Heterogeneous data · Data processing · Imbalanced datasets

1 Introduction

IoT is the layer of digital intelligence that makes the device smarter than what it would be by itself [1]. When the communication option is added to a device, the elements become known as "smart". However, not just the devices (*things*) that make the Internet of Things (IoT) - but the data collected and the actions taken based on these data.

On the base of IoT can be seen such technologies as Sensor Networks, Machine to Machine, Semantic Data Integration and Search Engines, Data Transmission Protocols and many others [2–4]. They could be grouped into three categories:

1. Technologies that allow objects *to acquire* information.
2. Technologies that allow objects to *process* information.
3. Technologies to improve *security and privacy*.

© Springer Nature Switzerland AG 2019
V. M. Vishnevskiy et al. (Eds.): DCCN 2019, LNCS 11965, pp. 457–468, 2019.
https://doi.org/10.1007/978-3-030-36614-8_35

The first two categories are building blocks that are needed to create object "intelligence". The third category is a requirement for the distribution and management of IoT systems.

Most Internet applications of Things produce or rely on large streams of data that need to be analyzed in real time to gain knowledge and make decisions. Data from various sensors on the Internet of Things is generated in the form of real-time streams, which often form complex models that have to be interpreted with minimal latency to be applied for decision-making in the context of the current situation. The need to process, analyze, and retrieve from these real-time Internet based complex systems requires statistical data analysis and machine learning (ML) as well as a sufficient amount of computational resources.

In the paper a modular IoT system is considered as an example and instance for implementation of several methods and approaches for processing of heterogeneous data [5]. Some problems which arise in data preparation and processing together with ways for their resolving are outlined.

2   Collection, Organization and Analysis of Data

Mainly the communication aspects of the Internet of Things attract the researcher's focus. Little attention is devoted to the integration of IoT devices with software components as resources in various information processes and security. There is currently no standard methodology for modelling and dealing with real complex scenarios using the Internet of Things.

The IoT software system can be installed in different types of devices, which can be physically located in different places - terrain, cloud environment and user interfaces.

There are various hardware devices to collect heterogeneous data from sensors with the aim of store information and perform operations where necessary. Some of the devices that store the information have also properties for acquiring it according to the needs of the particular system. In the cloud environment it is a part of software system that is responsible for the specific intelligent processing of collected data and its preparation for the user-friendly visualization. Moreover, in the cloud environment it is the back-end of the software system that is responsible for data presenting on the user interfaces. The user interfaces themselves can be installed on mobile and desktop devices as standalone applications, or the data can be accessed by web-based interfaces (browsers).

The nodes in the IoT system are connected to microcomputer modules, thus forming logical blocks where the data is cleared and normalized. This is an important part of the overall sequence of data handling. So the processed datasets have a proper structure and future work with them is much easier.

This structure is prepared according to the following four rules: the values of each measured variable are placed in a single column, the values of each subsequent observation on that variable are in different rows, for each node there is only one table with its values, and there is a mandatory column to link them to. The logic blocks can function independently, but can also be part of a larger system depending on the specific needs of the user.

The collected, cleaned and ordered data is then transferred to a separate logical block where data summation and aggregation, combining datasets and data transformation, is performed. In this logical block, data is stored until it is transferred to a cloud environment for analysis and visualization. In this logical block of the modular system, the configuration files are sent through which the various logical modules are managed - what information to collect, how long to preserve it, when to transfer the data, which are the access points for the formation, and much more.

The collection, transmission, intelligent processing and visualization of gathered data from the intelligent monitoring system is divided into four parts (see Fig. 1):

- collecting data from physical devices;
- organizing and grouping data according to predefined rules and user needs;
- data processing and analysis in cloud environment;
- the results obtained are rendered as visualization logical blocks of data in the different user interfaces.

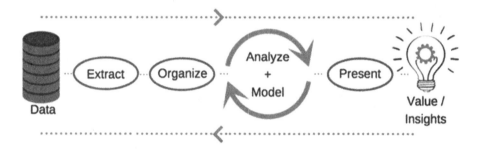

Fig. 1. Collection, processing and visualization of gathered data.

A standardized data processing process - OSEMN - has been selected for the implementation of the entire work cycle [6]. Using the OSEMN process gives a clear order of activities - Obtain, Scrub, Explore, Model data and iNterpret the data [7]. By following these steps, the process of data mining can be well planned and organized to their visualization.

2.1 Extract Data

Data is collected by sensors. Especially for the needs of the system, cloud-based software has been developed to analyze and intelligently process the data collected using the intelligent IoT monitoring system. The application connects to the logical block that stores the aggregated data over a predetermined time interval and through the SFTP protocol takes the necessary data. If the operation is unsuccessful, a mechanism is built for retrying the data from the logic block and notifying the user of the problems encountered when connecting to the system.

2.2   Organize Data

Before processing and analyzing the data obtained, several actions need to be carried out: merging the individual data columns into a single table, clearing the data from invalid values, normalizing the data and processing the extreme values:

Merging the Individual Data Columns into a Single Table - Firstly the data received from IoT sensor devices need to be merged into a single table. Each column of this table represents a variable.

Clearing Data from Invalid Values - Real-world data obtained from real observations often contains lots of invalid values - null, NaN or NA. They may cause the continuation of data processing to stop and to threat the proper flow of the predicting algorithms. In this case, the value that the analysis needs is unknown for some reasons, and as a result the analysis may be incorrect and the data models compiled incorrectly because not every predictive algorithm can work with missing data. If the dataset is stacked then the missing data can be replaced with the next closest value in ascending or descending order. However, often it is important to have all the data sequences and to not perform the array sorting.

Methods for cleaning the missing data from the dataset:

- Clean Missing Data with MICE (Multivariate Imputation using Chained Equation);
- Custom substitution value;
- Replace with mean;
- Replace with median;
- Replace with mode;
- Remove entire row;
- Remove entire column;
- Replace using Probabilistic PCA (Principal Component Analysis). PCA removes the missing values using a linear model that analyzes the correlations between the columns.

Data Normalization - Because of the different types and ranges of data obtained from the IoT system, they need to be normalized. The most commonly used normalization methods are Zscore, MinMax and LogNormal.

Zscore is the number of standard deviations from the mean data point. Zscore ranges from (-3) standard deviations (which would fall to the leftmost of the normal distribution curve) to $(+3)$ standard deviations (which would fall to the far right of the normal distribution curve).

Min-max normalization performs a linear transformation to the initial data and MinMax linearly resizes each function in $[0, 1]$ interval.

LogNormal converts all values to a logarithmic scale.

Extreme Value Processing - Extreme values and outliers are another factor influencing the correctness of data processing results. A RANSAC (RANdom

SAmple Consensus) algorithm is used to determine the extreme values obtained from the collected data. It takes a random number of all records in the database and performs linear regression by:

- filtering of noise samples (i.e., deviations);
- selecting of the best features (i.e. descriptors);
- extracting a QSAR model from a set of training examples;
- predicting the samples from test sets by resorting to the concept of applicability.

Outliers (or large differences in values) are extreme values that are far from the other observations. They can have an impact on forecasts or estimates. Causes of outliers may be:

- human mistake;
- malfunction of the measuring equipment;
- error in data transmission or transcription;
- specific behaviour of the system;
- natural outliers;
- sampling error, etc.

High limit values errors in the initial data processing, incorrect operation of the data-generated sensors, etc. may cause violation on the quality of data processing. The presence of such a problem makes the statistical analyzes inaccurate, which directly affects the creation of predictive models. Several methods for resolving these issues are as follows:

- Distribution histogram, boxplot and scatterplot graphs showing the distribution of values assist in the selection of appropriate statistical tools and techniques;
- Such values can be removed; the values can be transformed by performing some mathematical operation (log, sqrt) on them and thus reducing their importance for the common dataset.
- Another solution is to group these values separately and carry out a separate analysis on them.

3   Analysis of Unbalanced Data

Frequently, real-world data sets are mostly composed of "normal" examples with a small percentage of "unusual" or "interesting" examples. The cost of incorrectly classifying an unusual (interesting) example as a normal example is often much higher than the cost of making a reverse mistake.

3.1 Imbalance in the Classes

The data are unbalanced if the classification categories are not represented approximately equally. Data balancing can be done by reducing the majority class or by increasing the minority class examples.

Regulation of the Imbalance in the Classes - there are 4 ways to deal with the imbalance problems in the classes:

- Synthesis of a new instance of the minority class;
- Taking samples of the minority class;
- A smaller sample of the majority;
- Set up an evaluation function to make the wrong classification of minority instances more important than the wrong classification of the majority.

The ways of dealing with these kinds of problems are considered in the next section in more details.

3.2 Applying the SMOTE Model

Let's see a set of data received from the IoT system. The green points provide an observation of the parameters of good features from IoT sensors. Black points represent the observation of the parameters of poor features values. On the histogram (see Fig. 2) it can be clearly seen the imbalance in the IoT data received and the experimentally observed.

Often because of the specificity and difficulty of obtaining data, the reduction of majority class data is not an acceptable option. This kind of balancing can lead to overfit of the model due to the small number of data the study is using. For these reasons, a choice has been made to balance the data by increasing the number of minority class data. In this particular case, the minority class data are of significant importance.

The Boosted Decision Tree model, which is sensitive to the proportions of the different classes, was chosen for the study. As a consequence, this algorithm tends to prefer the class with the largest share of observations (known as the majority), which can lead to misleading accuracy of the model. This is especially problematic in cases like ours - when we are interested in the proper classification of the rare class (aka minority).

Given that this type of algorithm is designed to minimize the overall error rate instead of paying special attention to the minority class, it may fail to accurately predict this class if it does not get enough information about it.

For these reasons, it is necessary to balance the data by using SMOTE procedure.

SMOTE (Synthetic Minority Oversampling Technique) [10] synthesizes new minority cases between existing cases. Relations are being created between existing minorities and new ones are introduced. In this way the imbalance shrinks. This is a statistical technique for increasing the number of cases in the dataset in a balanced way. The algorithm takes samples from the functional space for each target class and its closest neighbours and generates new examples that combine the characteristics of the target class with the characteristics of its neighbours. This approach increases the capabilities of each class and makes the samples more general.

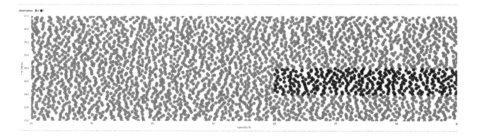

Fig. 2. Data histogram before SMOTE.

Steps in SMOTE process:

1. Identify the feature vector and its nearest neighbour;
2. Take the difference between the two vectors;
3. Multiply the difference with a random number between 0 and 1;
4. Identify a new point on the line segment by adding the random number to feature vector;
5. Repeat the process for identified feature vectors.

Figure 3 presents the data balancing process by applying the SMOTE model, realized with Azure Machine Learning Studio. By using the SMOTE model, copies of the minority class were created. The objective of increasing the proportion of minority classes to majority examples has been successfully achieved (Fig. 4). Copies were created to preserve the distribution of the minority class. The number of data has been increased without any new data being received that could influence the retrieval of minority class information.

Figure 5 shows the difference in the percentage of the information before applying the SMOTE model and after its use. After the application of the model for the increase of the minority class, there are increases in this class from 10% to 32%, which is equivalent to 22% generated copies of existing data.

Two cases have to be considered:

1. Over-sampling of minority class treatment:
   - Random addition of more minority observations by replication;
   - No loss of information;
   - Tend to excessive adjustment due to copying the same information.
2. Under-sampling of the majority class
   - The majority class is not sampled by random sampling of the majority, while the minority class does not become a certain percentage of the majority.

- Observations by the majority are removed at random;
- Helps balance the data set;
- Rejected observations may have important information;
- It may lead to deviation.

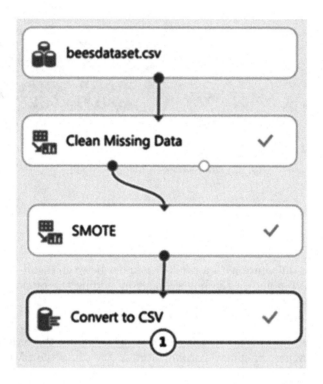

Fig. 3. SMOTE workflow.

Fig. 4. Data histogram after applying SMOTE.

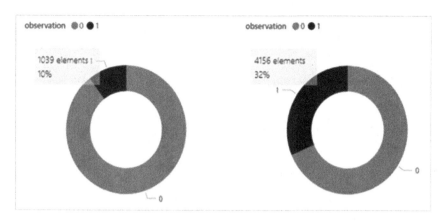

Fig. 5. Percentage ratio of data before application of the SMOTE model and after its use.

4   Data Modelling and Presentation

Modelling in IoT system is used for classification, forecasting and abnormally detection. Constructed models allow to create hypotheses about the causes of observed phenomena and to build a basis for future data collection through surveys and experiments.

The need to retrieve, analyze data from the Internet based systems and create complex models requires machine learning (ML) [8,9]. Machine learning algorithms in IoT aims at extracting previously unknown interesting information, such as dependencies, groups of data objects and categories of new observations.

In this investigation the Boosted Decision Tree is used to construct predictive models (Fig. 6). The results are shown in Table 1. Graphics of ROC, Precision/Recall and LIFT are presented on Fig. 7. The lines in red are results before balancing the data, and the lines marked in blue are results after a SMOTE model implements a controlled balancing of data by increasing the "minority" class.

ROC curve shows the effectiveness of the classification model before and after applying the SMOTE model. This curve represents two parameters - a true positive rate and a false positive value. An important part of this chart is the AUC.

AUC is the area under the ROC curve; it measures the entire 2-dimensional area under the ROC curve. It shows how well the forecasts are valued rather than their absolute values. AUC is the invariant classification threshold; it measures the quality of the model predictions regardless of the selected classification threshold. The AUC value before data balancing is 0.859 and after balancing it increases to 0.917.

The second graphic is Precision/Recall of the model before and after the data balance. Precision examines the right positive observations, and Recall considers the number of right positive observations divided by the total number of elements

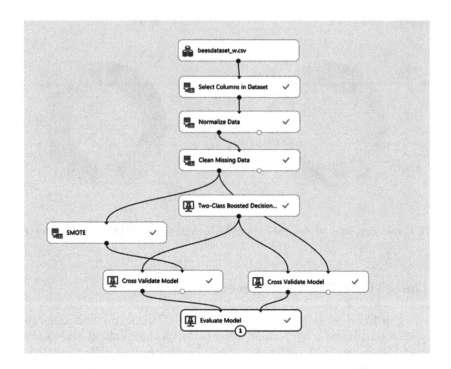

Fig. 6. Machine Learning workflow with data balancing.

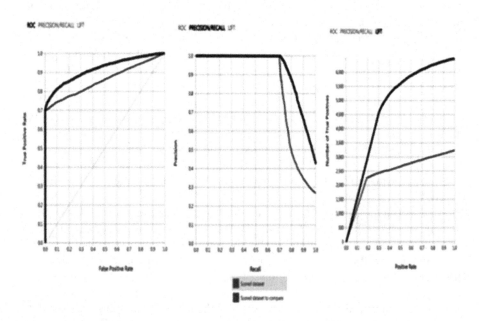

Fig. 7. Comparative results with and without SMOTE application.

belonging to the positive class. From Table 1 it can be seen that after balancing the data, the Precision increases from 0.807 to 0.866 and the Recall is improved (0.688 to 0.716), but the indicator best summarizing these two parameters is F1 Score. Its value before balancing the data is calculated at 0.743, and then is increased to 0.784.

The third graph (LIFT curve) shows the quality of the classification model. The curve is calculated on the basis of the ratio between the right positive observations, the correct negative observations, the false positive observations and the false negative observations. The efficiency of the classifier is higher when the curve approaches the upper left corner. This shows maximizing the true positive results by minimizing the value of the counterfeit ones.

All metrics are calculated and summarized in one parameter - accuracy. Accuracy shows the total number of correctly defined elements and is the main criteria for determining the success of the model used. Following the implementation of the SMOTE model, the prediction accuracy increased from 0.879 to 0.893, indicating that raising the "minority" class helped and improved the model's accuracy.

Table 1. Comparative results.

	With SMOTE	Without SMOTE
Accuracy	0.893	0.879
Precision	0.866	0.807
Recall	0.716	0.688
F1 score	0.784	0.743

5 Conclusion

The Internet of Things (IoT) has dramatically increased the number of Internet-connected devices ranging from sensors and smartphones to more and more diverse data or sources of information such as observing crowds or sensors. The availability of data generated by these diverse data sources has opened up new opportunities for innovative applications in various fields and overall decision-making systems.

The solutions to manage and utilize the massive volume of data produced by these objects are yet to mature.

In the paper a methodology to deal with a variety of data from Internet-connected devices is proposed. A modular IoT system is considered as an instance for implementation of several methods for processing of heterogeneous data. The approaches for problems resolving that can affect the creation of predictive models are outlined. In order to extract more valuable information from this data, machine learning needs to be applied to them.

References

1. Internet of Things (IoT): Cheat sheet. https://www.techrepublic.com/article/internet-of-things-iot-cheat-sheet. Accessed 8 Aug 2019
2. Mezghani, E., Expósito, E., Drira, K.: A model-driven methodology for the design of autonomous and cognitive IoT-based systems: application to healthcare. IEEE Trans. Emerg. Top. Comput. Intell. **1**(3), 224–234 (2017)
3. Bali, A., Al-Osta, M., Abdelouahed, G.: An ontology-based approach for iot data processing using semantic rules. In: Csöndes, T., Kovács, G., Réthy, G. (eds.) SDL 2017. LNCS, vol. 10567, pp. 61–79. Springer, Cham (2017). https://doi.org/10.1007/978-3-319-68015-6_5
4. Abu-Elkheir, M., Hayajneh, M., Abu Ali, N.: Data management for the internet of things: design primitives and solution. Sensors (Basel) **13**(11), 15582–15612 (2013)
5. Dineva, K., Atanasova, T.: ICT-based beekeeping using IoT and machine learning. In: Vishnevskiy, V.M., Kozyrev, D.V. (eds.) DCCN 2018. CCIS, vol. 919, pp. 132–143. Springer, Cham (2018). https://doi.org/10.1007/978-3-319-99447-5_12
6. Janssens, J.: Data Science at the Command Line. https://www.oreilly.com/library/view/data-science-at/9781491947845/ch01.html. Accessed 8 July 2019
7. Dineva, K., Atanasova, T.: OSEMN process for working over data acquired by IoT devices mounted in beehives. Curr. Trends Nat. Sci. **7**(13), 47–53 (2018). University of Pitesti, Romania
8. Tashev, T., Monov, V.: Large-scale simulation of uniform load traffic for modeling of throughput on a crossbar switch node. In: Lirkov, I., Margenov, S., Waśniewski, J. (eds.) LSSC 2011. LNCS, vol. 7116, pp. 638–645. Springer, Heidelberg (2012). https://doi.org/10.1007/978-3-642-29843-1_73
9. Balabanov, T., Zankinski, I., Shumanov, B.: Slot machines RTP optimization with genetic algorithms. In: Dimov, I., Fidanova, S., Lirkov, I. (eds.) NMA 2014. LNCS, vol. 8962, pp. 55–61. Springer, Cham (2015). https://doi.org/10.1007/978-3-319-15585-2_6
10. Chawla, N.V., Bowyer, K.W., Hall, L.O., Kegelmeyer, W.P.: SMOTE: synthetic minority over-sampling technique. J. Artif. Intell. Res. **16**, 321–357 (2002)

Effect of Heterogeneous Traffic on Quality of Service in 5G Network

Omar Abdulkareem Mahmood[1,2], Abdukodir Khakimov[1,3],
Ammar Muthanna[1,3(✉)], and Alexander Paramonov[1]

[1] The Bonch-Bruevich State University of Telecommunications,
St. Petersburg, Russia
alex-in-spb@yandex.ru, mahmood-omar@list.ru
[2] Department of Communications Engineering, College of Engineering,
University of Diyala, Diyala, Iraq
[3] Peoples' Friendship University of Russia (RUDN University),
6 Miklukho-Maklaya St., Moscow 117198, Russia
Khakimov-aa@rudn.ru, ammarexpress@gmail.com

Abstract. In this work we provide an analysis of Internet of Things concept and the Tactile Internet, as well as trends in traffic served by communication networks, build a real time interactive system between the human and the machine and introduce a new evolution in human-machine (H2M) communication. The most important challenge in realizing Tactile Internet is the latency requirement. In this paper we consider the impact of Internet of Things and tactile internet traffic, formed by monitoring and dispatching control systems or other systems, when the properties of this traffic are described by regular data stream properties. We estimate the impact of this traffic on such key QoS indicators as data delivery delay and probability of loss. As a model of a communication network, we consider a queuing system (QS) with a combined service discipline.

Keywords: Internet of Things · Traffic · Data transmitting · Queuing system · Quality of service

1 Introduction

The organization of the Internet of Things (IoT) is one of the priorities of the development of infocommunication system [5–11], which construction concept is reflected in the [1]. The development of IoT is an extremely important step, as it affects almost all areas of human activity. The penetration of the Internet of Things will contribute to the availability of more and more information, the growth of its analysis capabilities, the formation of decisions and actions based on its results. The second important direction of development of infocommunication system is the concept of the Tactile Internet (TI) [2], which implies a substantial increase in the requirements for quality of service (QoS) of traffic, which are presented by new interactive services. As an example, we can cite the

© Springer Nature Switzerland AG 2019
V. M. Vishnevskiy et al. (Eds.): DCCN 2019, LNCS 11965, pp. 469–478, 2019.
https://doi.org/10.1007/978-3-030-36614-8_36

construction of monitoring and dispatching control systems [3], in terms of development of IoT, telemedicine applications and unmanned vehicles. Comparing the construction of IoT with the construction of telemetry and telemechanical systems [4], we can notice a lot in common. The principal novelty of these directions consists, first of all, in the potentially possible number of monitoring and control devices and their penetration the possibilities to the most diverse levels of technological and other processes, as well as the QoS requirements, in particular, to the probabilistic and time parameters. The data delivery requirement between a large number of devices, which potentially can significantly exceed the number of subscribers of existing communication networks, sets the task of ensuring the availability, QoS, reliability and stability of the operation of communication networks in such conditions.

2 Problem Statement

According to [3,5,8], the traffic generated by Internet of Thing devices can be divided into three characteristic types: deterministic - produced by devices operating on a fixed schedule; deterministic technological - necessary to maintain the functioning of the system; mediated i.e. generated as a reaction to some external events. The traffic generated by the Internet of Things devices can be served together with the traffic of other communication services, for example, base stations, wireless access points and other network nodes. Due to the fact that the nature of Internet of Thing traffic in general is different from the traffic of other services, it makes sense to assess its characteristics and impact on QoS. For the analysis, the model shown in Fig. 1 was chosen. The model consists of an IoT traffic generator that simulates the operation of one or a group of IoT devices, Traffic generator of traditional communication services and TI traffic, designated as H2H + TI (H2H - Human to Human, TI - tactile Internet). The produced incoming traffic streams to communication node, the model of which is presented as queuing system with Combined Service Discipline (with delay-basis and failure-basis system). The average service time of a packet (message) is equal to $\bar{t}$.

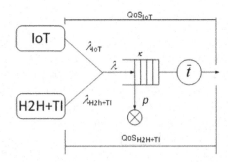

Fig. 1. Service model of aggregated traffic

The intensity of the Internet of things traffic is denoted by λ_{IoT}, H2H traffic – λ_{h2h}, aggregated stream $\lambda = \lambda_{h2h} + \lambda_{IoT}$. The packet arrives with some probability p, at the input of the system in which all the positions in the queue are occupied, and receives a failure (losses occur). The aggregated stream at the system output has a total intensity λ. The properties of the mixed stream at the system input are determined by the properties of both streams, therefore, in general, it differ from the properties of both traditional traffic and Internet of Things traffic.

The system will be characterized by indicators of QoS: the probability of loss (failure) of packets (messages) and the delay in delivery of the packet (waiting time in the queue and service time). Different services that generate traffic in a communication network have specific QoS performance requirements. The process of serving packets (messages) affects the properties of the served traffic, which then goes to other network elements, so the properties of the served traffic at the output of the system are also of considerable interest.

In the study of the mutual influence of considered traffic stream we will estimate the QoS indicators separately for the IoT and H2H + TI traffic streams.

3   Service System Model

The QS model described above can be represented as a G/G/1/k system. For this system, there are no accurate analytical models to estimate the probability of packet loss and delivery delay (waiting time in the queue). In [13], the diffusion approximation method is used to estimate the probability of losses with known distribution parameters describing the traffic at the entrance and the packet servicing process, and the following expression is obtained for an approximate estimate:

$$p = \frac{1-\rho}{1-\rho^{\frac{2}{C_a^2+C_s^2}n_b+1}}\rho^{\frac{2}{C_a^2+C_s^2}n_b} \tag{1}$$

where C_a^2 and C_s^2 – quadratic coefficients of variation of the distributions of the incoming stream and service time, respectively; n_b – buffer size, ρ – system load.

An approximate estimate of the average package delivery time can be obtained using the expression [?]:

$$T = \frac{\rho\bar{t}}{2(1-\rho)}\left(\frac{\sigma_a^2+\sigma_s^2}{\bar{t}^2}\right)\left(\frac{\bar{t}^2+\sigma_s^2}{\bar{a}^2+\sigma_s^2}\right) + \bar{t} \tag{2}$$

where σ_a^2, σ_s^2 – the values of the variance of the time interval between packets and the service time, $\bar{a}$ – the average value of the interval between packets, $\bar{t}$ – average service time.

Since it is of interest to have a separate QoS estimate for H2H + TI and IoT traffic, it makes sense to investigate the applicability of the above approximate solutions for estimating the QoS of the aggregated traffic stream. We will assume that the stream of background traffic H2H has the properties of a self-similar flow (Hurst coefficient H = 0.7 ... 0.9). The assumption is based on the fact that a large proportion of traffic in modern communication networks is video transmission. As a rule, its reproduction by modern players generates self-similar (pack) traffic. Thus, the assumption about the properties of subscriber traffic is quite acceptable.

We also make the assumption that M2M traffic is a deterministic flow, defined as a periodic process of sending monitoring system data. This assumption is based on the fact that in many cases M2M traffic is generated by monitoring and dispatch control systems (SCADA).

4   Simulation Model

For the construction of the model, the AnyLogic simulation system was chosen, which allows creating discrete event simulation models. To simulate a self-similar stream, a generator of a sequence of independent events was used, the time intervals between which are random and have a Pareto distribution.

$$f(x) = \begin{cases} \frac{kx_m^k}{x^{k+1}} & x \geq x_m \\ 0 & x < x_m \end{cases} \tag{3}$$

where x_m and k are distribution parameters.

Expectation and variance are determined according to

$$E(x) = \frac{kx^m}{k-1}, \quad D(x) = \left(\frac{x_m}{k-1}\right)^2 \frac{k}{k-2} \tag{4}$$

Figure 2 shows examples of the implementation of the simplest (H = 0.50) and self-similar (H = 0.75) streams.

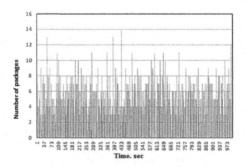

Fig. 2. The simplest and self-similar stream

Deterministic stream is a regular stream with a given packet intensity. Figure 3 shows the implementation of aggregated traffic (H2H + M2M) with the value of the Hurst coefficient H = 0.8.

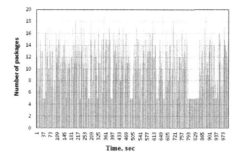

Fig. 3. Implementation of aggregated traffic (H = 0.7)

The Hurst coefficient is estimated by analyzing the variance change. Graphs of the dependences of the variance of the incoming and outgoing streams on the interval of stream aggregation are shown in Fig. 4.

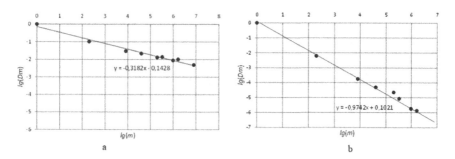

Fig. 4. Estimation of the Hearst coefficient for incoming and servicing streams (H = 0.80 and H = 0.51, respectively)

The following example is a relatively high traffic load (0.9 Erlang). As shown below, the Hurst coefficient of the stream served depends on the intensity of the load, for which a high value properties of the output stream are determined by the properties of the service process.

An example is given for relatively high traffic intensity (0.9 Erl). As will be shown below, the coefficient of the Hurst served flow depends on the intensity of the load, for which a high value the properties of the output stream are determined by the properties of the service process.

For model, it was assumed that the service time should reflect the time of transmission of a packet over the communication line, which is determined by the size of the packet and the speed of data transmission on the line. If the latter is constant (such an assumption can be taken for wired communication lines), then the transmission time is determined only by the packet size, and the distribution function is determined by the packet length distribution function.

We will assume that the minimum and maximum packet sizes are limited. Based on the analysis of the results of a sufficiently large number of measurements, it can be concluded: a large proportion of packets in wired communication networks have either a maximum or a relatively small length. Packages that have an intermediate length value are a significantly smaller part. Therefore, when approximating the distribution of the packet length, the beta distribution was chosen for the simulation.

$$f(x) = \frac{1}{B(u, v)} x^{u-1}(1 - x)^{v-1} \tag{5}$$

where u, v are form parameters, $B(u, v)$ is a beta function.

The empirical histogram of the service time obtained in the simulation is shown in Fig. 5.

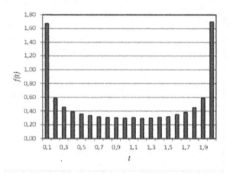

Fig. 5. Histogram of service time

We assume that such a distribution of the packet length is characteristic of Internet of things traffic. Different Internet of Things applications can create packets of different lengths; however, in this model, we focus on monitoring and dispatching services, the implementation of which currently uses packets of equal length, which is necessary to present telemetry data. Here we assume that the length of Internet of things packages is constant ("short" packages).

5   Analysis Results

In the simulation, the empirical dependences of the packet loss probability on the traffic intensity were obtained differentially for the Internet of things traffic and background traffic. These dependences are shown in Fig. 6 for different buffer sizes.

Also the figures show dependences obtained according to the approximate formula (1). The simulation results represented that estimated using (1) gives a slightly overestimated loss factor, with the largest error (about 2 times) occurring

Fig. 6. The dependence of the probability of loss on intensity of load at different lengths of buffer ($n = 2$; 10, $\rho \neq 1$)

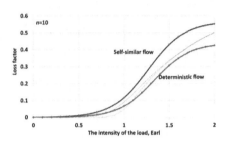

Fig. 7. The dependence of the probability of loss on intensity of load at different lengths of buffer ($n = 2$; 10, $\rho \neq 1$)

for a self-similar stream with average values of the load intensity. It is also seen from the graphs that the loss ratio for regular stream requests is significantly less than the loss ratio of self-similar stream in aggregated traffic.

Figure 7 shows the empirical dependences of packet delivery delays for self-similar and regular stream in aggregated traffic on the load intensity with different buffer sizes (2 and 10, respectively). Dependencies are obtained from simulation results. For comparison, the figures show the estimates (dashed line) obtained using the approximate model (2).

As can be seen from the above results, the average delay in packet delivery for a self-similar stream exceeds the delay in delivery of a packet in a regular stream, the difference of values does not exceed 20%. The analytical model (2) for the aggregated stream rather accurately describes the packet delivery delay in the interval of load intensity values, in which the packet loss is close to zero (up to 0.5 Earl – for buffer length n = 2 and 0.8 Earl – for n = 10). Moreover, its values are closest to the delays for the self-similar stream (Fig. 9).

Thus, requests for regular stream (IoT) are serviced with higher quality, and this is most pronounced in the growth of the loss coefficient for served traffic.

When analyzing the properties of traffic at the output of the QS, the dependence of the Hurst coefficient [15] on the intensity of the load was studied. Figure 8 shows the results of simulation modeling of a QS of the G/M/1/k type,

Fig. 8. Dependence of package delivery delay on the load intensity with different buffer lengths ($n = 2;\ 10,\ \rho \neq 1$)

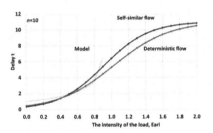

Fig. 9. Dependence of package delivery delay on the load intensity with different buffer lengths ($n = 2;\ 10,\ \rho \neq 1$)

at the input of which an aggregated stream obtained by combining self-similar and regular streams. Hurst coefficient of the input stream $H_{in} = 0,77$.

With an increase in the intensity of the load at the entrance of the QS, a decrease in the Hurst coefficient of the flow served at the output of the QS is observed. At small and medium values of the intensity of the load at the input (from 0 to 0.5 Erl), the Hurst coefficient of the output stream is almost equal to the analogous parameter of the input stream. The obtained dependence can be explained by the fact that for high values of load intensity, the properties of the served stream are determined to a greater degree by the law of service time distribution than the properties of the input stream, which coincides with the results of the study.

With a high intensity of the load, the distribution of the time interval between the arrivals of packets tends to the distribution of the service time, i.e. to the beta distribution, which of course.

Consequently, the time intervals between packets take a limited range of values, while a self-similar stream is characterized by a distribution that has a "long tail" (for example, a Pareto distribution). Thus, with the selected service time model, an increase in the intensity of the load leads to a decrease in the self-similarity properties of the served traffic (Fig. 10).

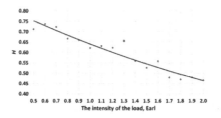

Fig. 10. The dependence of the coefficient of Hurst served stream to the system outputs of the load intensity

6   Conclusion

1. Analysis of the infocommunication system development trends shows that in the promising communication networks the proportion of IoT traffic will significantly increase, which will lead to its impact on QoS. Given that traffic on the network will also contain TI traffic, it can have a significant impact on the quality of service.
2. The proposed Internet of Things traffic model considers the component obtained by monitoring and dispatching services. As its model, a regular stream of requests was chosen, and models for background traffic (H2H and TI) - a self-similar stream.
3. The simulation results showed that when servicing an aggregated stream, the QoS parameters of the Internet of Things traffic and background traffic are significantly different.
4. An analysis of the simulation results of the aggregated stream service process showed that the probability of loss for regular stream packets is less than the random stream (H2H + TI). Moreover, this difference increases with increasing intensity of the incoming load.
5. The results obtained revealed the range of applicability of the known approximate models of the $G/G/1/k$ and $G/G/1$ systems for describing the loss coefficient and packet delivery delay.
6. Analysis of the dependency of the self-similarity properties of the served stream from the load intensity showed that the properties of the output and input stream are close at small and medium values of the intensity of the input load. For large values, the properties of the served streams are determined by the distribution of the service time.
7. When choosing network solutions for joint servicing of regular Internet of Things traffic and traffic of other services, greater "resilience" of Internet of Things traffic to losses should be considered, while random flows, in particular TI traffic, are more susceptible to losses during joint servicing.

Acknowledgment. The publication was prepared with the support of the "RUDN University Program 5-100".

References

1. ITU-T—E.800: Terms and definitions related to quality of service and network performance including dependability (1994)
2. ITU-T The Tactile Internet, ITU-T Technology Watch Report, August 2014
3. Muthanna, A., Khakimov, A., Ateya, A.A., Paramonov, A., Koucheryavy, A.: Enabling M2M communication through MEC and SDN. In: Vishnevskiy, V.M., Kozyrev, D.V. (eds.) DCCN 2018. CCIS, vol. 919, pp. 95–105. Springer, Cham (2018). https://doi.org/10.1007/978-3-319-99447-5_9
4. Marro, G.: Controlled and Conditioned Invariants in Linear System Theory
5. Paramonov, A., Koucheryavy, A.: M2M traffic models and flow types in case of mass event detection. In: Balandin, S., Andreev, S., Koucheryavy, Y. (eds.) NEW2AN 2014. LNCS, vol. 8638, pp. 294–300. Springer, Cham (2014). https://doi.org/10.1007/978-3-319-10353-2_25
6. Chornaya, D., Paramonov, A., Koucheryavy, A.: Investigation of machine-to-machine traffic generated by mobile terminals. In: 2014 6th International Congress on Ultra Modern Telecommunications and Control Systems and Workshops, ICUMT 2014, pp. 210–213 (2015)
7. OECD: OECD Science, Technology and Industry Scoreboard 2015: Innovation for growth and society. OECD Publishing, Paris (2015). https://doi.org/10.1787/sti-scoreboard-2015-en
8. Muthanna, A.: Secure and reliable IoT networks using Fog computing with software-defined networking and blockchain. J. Sens. Actuator Netw. 8, 15 (2019)
9. Cisco Visual Networking Index: Global Mobile Data Traffic Forecast Update, 2015–2020 White Paper
10. Masek, P., Fujdiak, R., Zeman, K., Hosek, J., Muthanna, A.: Remote networking technology for IoT: cloud-based access for AllJoyn-enabled devices. In: 2016 18th Conference of Open Innovations Association and Seminar on Information Security and Protection of Information Technology (FRUCT-ISPIT), pp. 200–205 (2016)
11. Muthanna, A., Prokopiev, A., Koucheryavy, A.: Ubiquitous sensor networks traffic models for image applications (2013)
12. Volkov, A., Khakimov, A., Muthanna, A., Kirichek, R., Vladyko, A., Koucheryavy, A.: Interaction of the IoT traffic generated by a smart city segment with SDN core network. In: Koucheryavy, Y., Mamatas, L., Matta, I., Ometov, A., Papadimitriou, P. (eds.) WWIC 2017. LNCS, vol. 10372, pp. 115–126. Springer, Cham (2017). https://doi.org/10.1007/978-3-319-61382-6_10
13. Iversen, V.B.: Teletraffic Engineering Handbook. COM Center Technical University of Denmark Building 343, DK-2800 Lyngby Tlf.: 4525 3648. www.tele.dtu.dk/teletra

Flying Ad-Hoc Network for Emergency Based on IEEE 802.11p Multichannel MAC Protocol

Truong Duy Dinh[1], Duc Tran Le[2(✉)], Thi Thu Thao Tran[3],
and Ruslan Kirichek[1,4]

[1] The Bonch-Bruevich Saint - Petersburg State University of Telecommunications,
22 Prospekt Bolshevikov, St. Petersburg 193232, Russia
din.cz@spbgut.ru, kirichek@sut.ru
[2] The University of Danang - University of Science and Technology,
Nguyen Luong Bang 54, Danang, Vietnam
letranduc@dut.udn.vn
[3] The University of Danang - University of Economics,
Ngu Hanh Son 71, Danang, Vietnam
thaotran@due.udn.vn
[4] Peoples' Friendship University of Russia (RUDN University),
Miklukho-Maklaya St. 6, Moscow 117198, Russian Federation

Abstract. Flying ad-hoc networks are widely used in various fields, especially in searching and rescuing of people by using unmanned aerial systems, which includes one or more mobile base stations and mission-oriented UAVs. Thanks to the mobility of UAVs, we can create a communication of Flying Network for Emergencies to support quickly and ensure strict conditions of the time in searching and rescuing. In this paper, we propose an architecture that supports the communication among rescuers or between rescuers with victims or between victims with their relatives by using the flying network for emergency over satellite systems. We particularly propose a MAC protocol based on IEEE 802.11p and IEEE 1609.4 protocols called Cluster-based Multichannel MAC IEEE 802.11p protocol to support communication in flying ad-hoc network for emergency.

Keywords: FANET · Flying network for emergencies · UAV · IEEE 802.11p

1 Introduction

In recent years, flying ad-hoc networks (FANET) based on unmanned aerial vehicles (UAVs) are widely used in various fields: military, commercial, agricultural, etc. A feature of such networks is searching and rescuing of people affected

The publication has been prepared with the support of the RUDN University Program 5-100 and funded by RFBR according to the research project No. 20-37-70059.

by natural disasters [1]. FANET is a special kind of self-organizing network based on UAVs, which connects and communicates with each other in certain airspace through wireless channels, and has characteristics of distributed, non-central, rapid deployment, self-organization and strong self-healing. As practice has shown, the rapid deployment of such networks is very importance in eliminating the consequences and saving lives. This network is called the flying network for emergencies [2].

Flying ad-hoc networks consist of two segments: the flying segment and the terrestrial segment [12]. It is necessary for FANET to be easily deployable and self-configured ad-hoc UAVs network to connect with the emergency services on the terrestrial segment. All the UAVs must be communicated with each other and to the emergency services simultaneously without having any pre-defined fixed infrastructure [13] with the help of a multi-hop ad-hoc networks scheme. In this way, it not only delivers the aggregated data to the base station instantly but also can share data among the connected UAVs. Moreover, during the operation, if some of the UAVs are disconnected due to the objective conditions, it can still make their connectivity to the network through the other UAVs.

Also, due to the ad-hoc networking among the UAVs, it can solve the complications like short range, network failure and limited guidance which arise in a single UAV system [14]. Even though such distinctive attributes make FANETs an appropriate solution for different types of scenarios, but they also bring some challenging issues such as communications and networking of the multiple UAVs [15]. Also, it is necessary to consider the quality of services (QoS) for several priority services in a flying network. The Medium Access Control (MAC) layer, which serves as the main link layer mechanism, is an important component in a flying network. It has a direct impact on throughput system, data rate, transmission delay and must meet many strict requirements [4].

IEEE 802.11p protocol [5] is a modified version of the familiar IEEE 802.11 (Wi-Fi) standard. The use of that protocol in a flying network has been reviewed in many articles and has proven to be the most suitable for a flying network [6,7]. It was originally developed for VANET - Vehicular ad hoc networks. IEEE 802.11p standard was adopted as Medium Access Control (MAC) and Physical Layer (PHY) specifications for the lower-layer Dedicated Short-Range Communication standard (DSRC), which has characteristics such as frequency range - 5.9 GHz (5.85–5.925 GHz), wide coverage (up to 1000 m), fast transmission rate (up to 27 Mbps), self-organization and fast convergence. It is expected that these characteristics will demonstrate excellent performance not only in vehicles on the ground but also in the interaction of unmanned aerial vehicles in the air, respectively, VANET is supposed to be moved into the air and used for UAVs.

Also, IEEE 1609.4 protocol [8] allowed adding a seven-channel MAC scheme to the MAC layer of IEEE 802.11p (Fig. 1). Each channel has a total bandwidth of 10 MHz. They include one control channel (CCH - Control Channel: 178) for managing the network and transmitting safety messages and six service channels (SCH - Service Channels: 172, 174, 176, 180, 182, 184) for other traffics. All vehicles must monitor the control channel with safety/control messages during

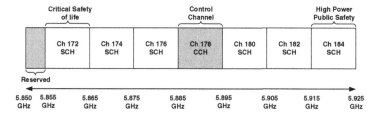

Fig. 1. Multiple channel IEEE 1609.4 protocol.

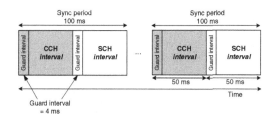

Fig. 2. Division of time into CCH intervals and SCH intervals.

the CCH period and will be able to switch to the service channel to exchange only non-safety applications. The channel access time is equally divided into repeating synchronization intervals of 100 ms, and each synchronization interval is divided into CCH Intervals (CCHI) of 50 ms and SCH Intervals (SCHI) of 50 ms (Fig. 2). Synchronization between vehicles is achieved by receiving the coordinated universal time (UTC) provided by the GPS equipped in each vehicle. It should be noted that although 6 SCH channels can be used, in practice channel 172 is used for accident avoidance safety of life, and channel 184 is used for high power, long-range services [11], so we only consider 4 SCH channels: 174, 176, 180, 182. The combination of IEEE 802.11p and IEEE1609.x is called wireless access in vehicular environments (WAVE) standard [9].

IEEE 802.11p protocol uses the Enhanced Distributed Channel Access (EDCA) mechanism [10], which uses Carrier Sense Multiple Access with Carrier Sense and Conflict Prevention (CSMA/CA) to support various types of applications with quality of services. The EDCA mechanism allows messages, which have a higher priority to have a better chance of being transmitted than messages with a lower priority. There are four types of access categories: background traffic (AC0), traffic from the best attempt (AC1), video traffic (AC2) and voice traffic (AC3). Prioritization is achieved by varying the EDCA parameter set, including Contention Windows (CWs) and the Arbitration Inter-Frame Spaces (AIFS), which increase the probability of successful medium access for real-time messages.

In this paper, we propose a new MAC protocol based on IEEE 802.11p and IEEE 1609.4 protocols to perform communications between UAVs as well as between groups of UAVs. UAVs will be divided into different groups called clusters to move in rescue areas and perform missions. The proposed protocol includes three different protocols: cluster management protocol (CMP), intra-cluster communication protocol (IntraCP) and inter-cluster (InterCP) communication protocol. These protocols use different modified WAVE Service Advertisement (WSA) and WAVE Short Message (WSM) packets [16] to update information, transmit data within and between clusters.

The rest of this paper is organized as follows: Sect. 2 briefly presents the related works. In Sect. 3, we present the assumptions and problem statement. Section 4 describes the proposed protocol CMMpP. Section 5 finally concludes this paper and future works.

2   Related Works

The research of methods of transmission data in flying networks has been devoted to many research works. In the paper [17] the authors discussed using IEEE 802.11n and IEEE 802.15.4 for transmitting data in the flying network with a different number of UAVs in star and mesh topologies. The result showed that star network topology is affected by high UAV density and speed, which impact negatively in the packet delivery rate and the end-to-end delay and using mesh topology with IEEE 802.11n are safer than using star topology with the same protocol. Although the performance of IEEE 802.15.4 was not as better as IEEE 802.11n in a mesh topology.

The authors in [18] address the experimental evaluation of such high throughput enabling technologies, such as IEEE 802.11n and IEEE 802.11ac in real-world scenarios, using outdoor experiments in a UAV setting with only two UAVs.

In the paper [19] multi-UAV-aided vehicular network is presented, two UAVs are deployed to cooperate with three ground vehicles, UAV acts as a guide for vehicle. These UAVs can communicate with each other through WiFi (IEEE 802.11a) and ZigBee modules, where the WiFi module is for the image transmission and the ZigBee module is for command message delivery. The results showed the performance of delay and throughput of the air-to-air link is better than the air to ground link, but it is also evaluated in terms of average latency and throughput; a relatively large latency can be observed when video data are transmitted. IEEE 802.11s also used in [20] but the results are limited to single-hop and two-hop mesh networks.

In addition, there are many studies focused on creating MAC protocols for VANET [21–23]. All these ideas can be considered to apply for FANET communication.

3   Assumptions and Problem Statement

As mentioned above, UAVs are more widely used to serve special civilian purposes in searching and rescuing when people are lost in the forest, victims in natural disasters such as earthquakes, forest fires, tsunamis et al. The common point of these problems is the lack of telecommunication infrastructure or the complete or partial destruction of telecommunication infrastructure. Therefore, communication between rescue workers or with the victims is almost impossible. To overcome this problem, we propose a solution to use UASs (unmanned aerial systems), which includes one or more mobile base stations (MBS) and mission-oriented UAVs, which can be considered as a mobile heterogeneous gateway [2]. Thanks to the mobility of UAVs, we can create a communication network – Flying Network for Emergencies to support quickly and ensure strict conditions of the time in searching and rescuing [3].

One of the reasons for creating a network of UAVs is the coverage area is usually not large when using a high data rate. Look at Fig. 3 we can calculate the maximum distance between two UAVs related to the data rate as follow:

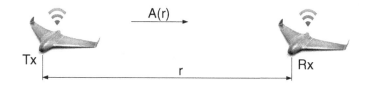

Fig. 3. The distance between two UAVs related to the data rates.

It is assumed that the antennas are aligned along with the polarization, and both antennas have unity gain [25].

$$a(r) = \frac{P_r}{P_t} = G_t \cdot G_r \cdot \left(\frac{\lambda}{4 \cdot \pi \cdot r}\right) \tag{1}$$

Where,

$A(r)$ - Damping according to the Friis formula in distance r,
G_t - Transmit antenna gain, when the omnidirectional transmission $G_t = 1$,
G_r - Receiver antenna gain, when the omnidirectional transmission $G_r = 1$,
P_t - Transmitting antenna power (without losses) (W),
P_t - Receiving antenna power (without losses) (W),
r - Distance between antennas in meters.
$\lambda = \frac{c}{f}$ (m) - Wavelength in meters corresponding to the transmission frequency.
$A(r) = 10 \cdot \lg(a(r))$ - Damping according to the Friis formula in distance r in logarithm form.
$P_{tx} = 10 \cdot \lg \frac{P_t}{10^{-3}} = 28.8$ (dBm) - Transmitting antenna power level [26].
$P_{rx} = 10 \cdot \lg \frac{P_r}{10^{-3}}$ (dBm) - Receiving antenna power level.
$P_n = -90$ (dBm) - Noise power over 10 MHz.
$P_{rx} = P_{tx} - P_n$

From these formulas, we have Table 1

Table 1. Relation between maximum distance and data rate.

Data rate (*Mbps*)	Maximum distance (*m*)
3	1115
6	790
12	445

We can see that if the data rate is 12 Mbps, the distance between the two UAVs is only 445 m. Thus, the coverage area is not large. That is the reason why we should create a network of UAVs (Fig. 4).

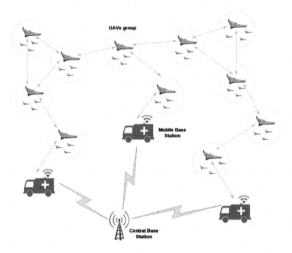

Fig. 4. Network of UAVs for emergencies.

It should be recalled that, in the proposed mission-oriented flying network, the UAVs move with specific purposes and are subject to geographical restriction. In a mission-oriented FANET environment, UAVs cooperatively communicate with each other to receive information on mission assignment from the base station or forward the collected data to the base station [24].

In particular, MBSs will deploy UAVs groups to areas around MBS to gather information, which will be disseminated between groups and transmitted to the base station. Not only that, but the UAVs groups also act as the intermediaries to make connections and transfer data between the base stations together to create a timely, efficient and effective rescue system and increase the coverage area of the flying networks.

In addition, in searching and rescuing, it is very necessary to have communicate among rescuers or between rescuers with victims or between victims with their relatives. In order to support these requirements when the telecommunications infrastructure is destroyed, we propose an architecture that uses the flying network for emergency over satellite systems. Satellite systems covering

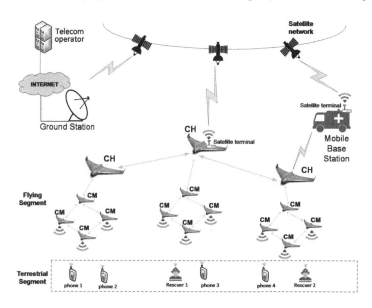

Fig. 5. An architecture for flying ad-hoc network for emergency.

the entire earth can connect to almost every point on earth. Satellite user terminals are mobile devices that help quickly connect to satellites. In order to connect to the victim's devices as well as transfer data to MBS or to satellites, on UAVs equipped: Satellite user terminal, IEEE 802.11p module, IEEE 802.11n/ac module. In this architecture, voice traffic can be transmitted from subscriber over UAVs via VoWiFi to the MBS (or to UAV, which can connect to satellite) and over satellite systems to telecom operator (Fig. 5).

In this paper, we focus on creating relevant communication protocols, which are cluster management protocol (CMP), intra-cluster communication protocol (IntraCP) and inter-cluster communication protocol (InterCP) for only one MBS in the flying network (Fig. 6). We call the combination of those protocols as Cluster-based Multichannel MAC IEEE 802.11p protocol (CMMpP). It should be noted that, in this paper, we call a group of UAVs as cluster and UAV as node alternatively.

In order not to complicate the problem, we have some assumptions as follows:

- The nodes will be divided into different clusters at the beginning of the mission, and each cluster will have different flight directions to gather information as well as increase the coverage of the network. Each node in the cluster also has its flight speed and flight direction, but this difference is within the allowable limit to ensure cluster stability. Also, during the task implementation, there will be no node left the cluster or join the cluster.
- All UAVs is equipped with a GPS device to define the location, and time popularity coordinates of UAV is small enough so that calculation errors can be ignored.

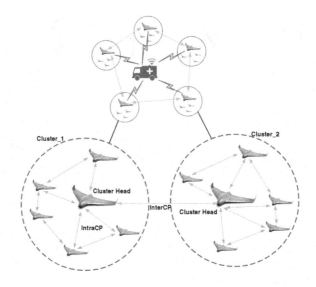

Fig. 6. Groups of mission-oriented UAVs around a MBS.

– We use two transceivers on each UAV to support multichannel transmission at the same time. Because negative affects of channel switching on spectrum efficiency and performance of safety applications do not make the single-radio approach a good candidate for multi-channel MAC architectures [23]. This is acceptable and has been proposed in several other studies [27,28].

The problem statement is as follows:

First, the MBSs will move to the rescue area. There they will deploy groups of mission-oriented UAVs (clusters) to collect information, make connections and exchange data. These clusters will use Cluster Management Protocol (CMP) to elect their cluster head (CH) node. This CH node may change during the execution of the task depending on the location, speed of the nodes in the cluster. The CH node is responsible for collecting information from the cluster members (CM) and then transferring it to the base station or passing it to CHs of other clusters. The CH node also acts as intermediate for data transmission, scheduling, channel assignment through message controls.

Nodes will use IntraCP and InterCP to transmit data. These protocols employ multichannel MAC protocol IEEE 1609.4 standard and IEEE 802.11p protocol in combination with TDMA to conduct every communication. In particular, IntraCP protocol performs the following main tasks:

– Collecting/Delivering control messages from/to CM on CCH channel.
– Allocating the available SCH channels to CM for data traffic.
– Contending to transmit data on the SCH in the cluster.

Meanwhile, the InterCP protocol is responsible for making communications between different clusters (transmitting both control messages on the CCH channel and data traffic on the SCH channel).

4   Proposed Work

4.1   Cluster Management Protocol (CMP)

The CMP protocol is responsible for maintaining and electing the cluster head for the current cluster. In the beginning, a node in the center of the cluster will be selected as the CH. This ensures that nodes from the beginning can exchange information with each other.

The decision which one stays CH is based on a total weighted factor F, which takes into account different metrics. These metrics involve the mobility information of each node, including the speed of a node (V) and the distance to the neighbors (D). A node identifies its neighbors by sharing this mobility information through control messages by using IntraCP protocol (in the next subsection).

The speed weight of the ith node can be expressed as:

$$f_v(i) = \frac{1}{|V_i - V_{avg}|} \tag{2}$$

where:

$$V_{avg} = \frac{V_1 + V_2 + V_3 + ... + V_n}{n} \tag{3}$$

n is the number of neighbours, V_i is the speed of ith node.

The denominator of Eq. (2) represents the average difference of speed between one node and all its cluster members. The idea here is that a node with similar velocity to most nodes in its neighborhood will cause less change in the cluster membership than a node, which is much faster or slower than the rest.

Next, we calculate the central position by:

$$D_{avg} = \frac{1}{n} \cdot \left(\sum_{i=1}^{n} x_i, \sum_{i=1}^{n} y_i, \sum_{i=1}^{n} z_i \right) \tag{4}$$

where:

(x_i, y_i, z_i) is the coordinate of the ith node and we can express the distance weight of the ith node as:

$$f_d(i) = \frac{1}{\sqrt{(x_i - D_{avg})^2 + (y_i - D_{avg})^2 + (z_i - D_{avg})^2}} \tag{5}$$

Equation (5) represents the average distance between all neighbors and itself. It means how close the neighbors to one node are. Here also a great value is preferable, as it can be expected that nodes, which are placed closer together, will stay longer in each other transmission range.

Combining these measures, the total weight factor F is obtained, which shows the suitability of a node to become CH. The greater F is, the better it is qualified to be CH. The factors w_v, w_d can be chosen between 0 and 1 according to the different scenarios. Their sum has to result in 1.

$$F = w_v \cdot f_v(i) + w_d \cdot f_d(i) \tag{6}$$

4.2　Sharing Information

For the protocols to take place, each node will periodically update the information about the location of the node, speed of node in the cluster, ClusterID, SCH channel requests (if there is data to send) and Priority of service (with EDCA parameter sets) to all members in its cluster through control messages. Where,

- Location is defined by the GPS system.
- ClusterID is the identification of the cluster.
- SCH channel request is information, which informs that this node needs to be assigned a SCH channel to transmit data. It also specifies whether it needs SCH channel for intra-cluster or inter-cluster communication.
- Priority indicates the type of traffic transmitted in the SCH channel in the next interval. In this paper, we use four levels of priority, as shown in Table 2 [29].

Table 2. Delay parameter requirements for each type of traffic.

Priority	Traffic type	Acceptable delay value
1	Voice, Highly interactive video	100 ms
2	Interactive video	100–400 ms
	Video streaming	<1 s
3	Best effort text follows	Not normalized
4	Background text follows	Not normalized

4.3　Cluster-Based Multichannel MAC IEEE 802.11p Protocol (CMMpP)

Since both the IntraCP and InterCP protocols support each other as well as can happen at the same time, we will not present each protocol separately but combine it as CMMpP protocol.

In our proposed CMMpP protocol, each node is equipped with two transceivers, which are denoted by Trans1 and Trans2, respectively, which can operate simultaneously on different channels. Trans1 is always tuned on the CCH to monitor and perform transmission of control messages on the CCH channel while Trans2 is tuned to any SCH (channels 174, 176, 180, 182) to perform data transmission. Therefore, instead of using CCH intervals (CCHI) and SCH intervals (SCHI) with 50 ms duration for each one, we can use the whole synchronization interval 100 ms (Fig. 2). In addition to using CCH Channel 178 for

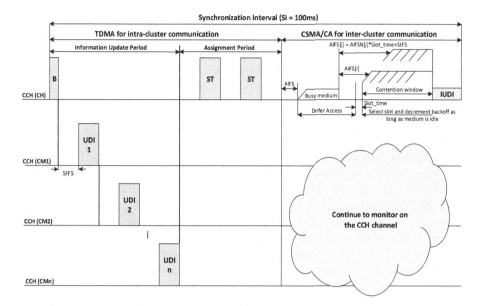

Fig. 7. Operation of CMMpP protocol including IntraCP and InterCP protocols on CCH channel.

control messages, we propose to use SCH channel 182 exclusively for the inter-cluster transmission of data. Thus, the SCH channels 174, 176, 180 will be used to assign to CMs for the intra-cluster transmission of data.

Operation of Trans1. First, CH will send a beacon message (B) packet to notify about starting Synchronization Interval (SI) on CCH channel to all members CMs of the cluster. After receiving the notification, the CMs will reply to update all information of nodes using UDI packets in Information Update Period. The information in this UDI packet will be used to redefine CH for the next synchronization interval if necessary and update the requests for using the SCH channels of CMs. Based on CH channel requests, CH will create schedules for CM intra- and inter- transmissions in the next SI interval. Next, in the Assignment Period, CH broadcasts a packet to announce schedule of transmissions (ST packets) to the CMs. The ST packet indicates which node will use which SCH channel to transmit data in the cluster or between different clusters. In addition, ST also shows information of the CH node for the next interval. To increase reliability and ensure CMs receive ST packets, CH will send ST packets two times. All these processes occur with TDMA mechanism on CCH channel under IntraCP protocol. Figure 7 shows the operation of CMMpP protocol.

The principle of assigning SCH channels for CMs is as follows: CH receives the SCH channel request information from CM. CH then searches for a rarely used (to reduce interference) SCH channel in the list of available SCH channels in accordance with the purpose of intra-cluster or inter-cluster communication.

Then SCH channel will be assigned to CM through ST packet. After ending the intra-cluster communication period, the CH performs inter-cluster communication using the contention-based CSMA/CA mechanism in IEEE 802.11p to compete with the CHs of other clusters. When the environment is idle, it will broadcast the inter-cluster communication packet IUDI to disseminate all information of the cluster, list of active nodes and assigned SCH channels to other CHs (Fig. 7).

Operation of Trans2. At the same time, with the processes on the CCH channel, nodes also transmit data on SCH channels. On SCH channels, nodes including CH compete for data transmission via CSMA/CA mechanism in EDCA of IEEE 802.11p all the time. It means those nodes that participate in data transmission but have the same assigned SCH channel will compete based on updated priority information and respective EDCA parameter sets. We use priority by assigning different EDCA parameters to make sure that traffic tolerated to delay is transmitted before other traffics (see Table 2). It should be noted that transmissions taking place in the cluster (IntraCP) can use only SCH channels 174, 176, 180. Meanwhile, inter-cluster communication (InterCP) will take place on SCH channel 182. These operations occur on non-overlapped channels, so they do not affect each other.

We modify the WAVE Service Advertisement (WSA) packet format [16] by using optional fields to create a UDI packet. Figure 8 shows the format of the UDI packet.

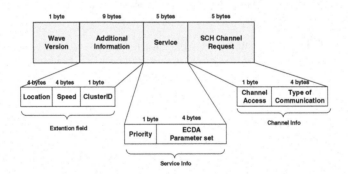

Fig. 8. UDI packet format.

Wave Version field defines the format of this WSA. The version number associated with the current standard is 1.

Additional Information field includes:

- *Location and Speed* subfields indicate the location and speed of node in the cluster.
- *ClusterID subfield* identifies the cluster.

Service field includes *Priority* subfield and *EDCA Parameter* set subfield. Priority indicates the priority of traffic that the current node needs to send, as shown in Table 2. EDCA Parameter set indicates the predefined EDCA parameters according to the priority. There are some studies that focus on this parameter to support the quality of service in a contention-based mechanism [30–32]. *SCH Channel Request* field includes two subfields to notify the CH about transmission demand on SCH channel in next interval.

- *Channel Access* subfield is used to indicate that this node needs a SCH channel assignment or not. In particular, if the last bit is 1, it demands CH to assign the SCH channel; if the last bit is 0, it does not need a SCH channel in the next interval.
- *Type of Communication* subfield indicates intra-cluster communication (the last bit is 1) or the inter-cluster communication (the last bit is 0). CH node uses this information to assign accordingly.

For ST packet format, we modify the WAVE Short Message (WSM) packet format [16] as shown in Fig. 9.

1 byte	1-4 bytes	1 byte	1 byte	1 byte	2 bytes	variable
Version	PSID	SCH Channel assignment	Next CH	WSM WAVE Element ID	Length	WSM Data (Payload)

Fig. 9. ST packet format.

Version, PSID (Provider Service IDentifier), WSM WAVE Element ID, Length and WSM Data (Payload) are required fields of WSM packet. We add the *SCH channel assignment* field to indicate the list of assigned SCH channels to CMs.

Next CH field is used to inform about new CH node in the next SI interval.

In this paper, we propose to use broadcast packets IUDI to inform the CHs of other clusters about the information of its cluster (Fig. 10). This packet is similar to the ST packet, and we add some of the following fields.

1 byte	1-4 byte	1 byte	1 byte	1 byte	5 bytes	1 byte	1 byte	2 bytes	variable
Version	PSID	ClusterID	SCH Channel assignment	Active nodes	Service	Next CH	WSM WAVE Element ID	Length	WSM Data (Payload)

Fig. 10. IUDI packet format.

Version, PSID, WSM Wave Element ID, Length, WSM Data (payload) fields are required fields by the standard.

Active nodes field indicates the list of active nodes, which have data to transmit on the SCH channels.

Service field is similar to the Service field in UDI packet format.

5 Conclusion

In this paper, we proposed an architecture that supports the communication among rescuers or between rescuers with victims or between victims with their relatives by using the flying network for an emergency over satellite systems. In particular, a MAC protocol based on IEEE 802.11p and IEEE 1609.4 protocols called CMMpP to support communication in flying ad-hoc network for emergency was developed. CMMpP protocol includes three components: Cluster Management Protocol, Intra-Cluster Communication Protocol and Inter-Cluster Communication Protocol. They use TDMA and CSMA/CA mechanisms to transmit control messages on the CCH channel and different kinds of traffics on SCH channels. We modify the WVA and WSM packets format and add some fields/subfields to those packets format to support the new protocol.

In the future, we will perform evaluations of the proposed protocol using simulation to consider its advantages and disadvantages. At the same time, we will focus on solving the problem of ensuring QoS for FANET for emergency network.

References

1. Ahn, T., Seok, J., Lee, I., Han, J.: Reliable flying IoT networks for UAV disaster rescue operations. Mob. Inf. Syst. **2018**, 1–12 (2018)
2. Dinh, T.D., Pham, V.D., Kirichek, R., Koucheryavy, A.: Flying network for emergencies. In: Vishnevskiy, V.M., Kozyrev, D.V. (eds.) DCCN 2018. CCIS, vol. 919, pp. 58–70. Springer, Cham (2018). https://doi.org/10.1007/978-3-319-99447-5_6
3. Dinh, T.D., et al.: Unmanned aerial system-assisted wilderness search and rescue mission. Int. J. Distrib. Sens. Netw. **15**(6) (2019). https://doi.org/10.1177/1550147719850719
4. Xu, D., Zhang, H., Zheng, B., Xiao, L.: A priority differentiated and multi-channel MAC protocol for airborne networks. In: 2016 8th IEEE International Conference on Communication Software and Networks (ICCSN), pp. 64–70. IEEE (2016)
5. IEEE Standards Association. 802.11p-2010-IEEE standard for information technology-local and metropolitan area networks-specific requirements-part 11: wireless LAN medium access control (MAC) and physical layer (PHY) specifications amendment 6: wireless access in vehicular environments (2010). http://standards.ieee.org/findstds/standard/802.11p-2010.html
6. Park, J.H., Choi, S.C., Kim, J., Won, K.H.: Unmanned aerial system traffic management with WAVE protocol for collision avoidance. In: 2018 Tenth International Conference on Ubiquitous and Future Networks (ICUFN), pp. 8–10. IEEE (2018)
7. Pu, C.: Jamming-resilient multipath routing protocol for flying ad hoc networks. IEEE Access **6**, 68472–68486 (2018)
8. IEEE 1609 Working Group. IEEE Standard for Wireless Access in Vehicular Environments (WAVE)-Multi-Channel Operation. IEEE Std (2016): 1609-4

9. Uzcategui, R.A., De Sucre, A.J., Acosta-Marum, G.: Wave: a tutorial. IEEE Commun. Mag. **7**(5), 126–133 (2009)
10. Sun, W., Zhang, H., Pan, C., Yang, J.: Analytical study of the IEEE 802.11p EDCA mechanism. In: 2013 IEEE Intelligent Vehicles Symposium (IV), pp. 1428–1433. IEEE (2013)
11. Eichler, S.: Performance evaluation of the IEEE 802.11p WAVE communication standard. In: 2007 IEEE 66th Vehicular Technology Conference, pp. 2199–2203. IEEE (2007)
12. Koucheryavy, A., Vladyko, A., Kirichek, R.: State of the art and research challenges for public flying ubiquitous sensor networks. In: Balandin, S., Andreev, S., Koucheryavy, Y. (eds.) ruSMART 2015. LNCS, vol. 9247, pp. 299–308. Springer, Cham (2015). https://doi.org/10.1007/978-3-319-23126-6_27
13. Bekmezci, I., Ermis, M., Kaplan, S.: Connected multi UAV task planning for flying ad hoc networks. In: 2014 IEEE International Black Sea Conference on Communications and Networking (BlackSeaCom), pp. 28–32. IEEE (2014)
14. Sharma, V., Kumar, R.: A cooperative network framework for multi-UAV guided ground ad hoc networks. J. Intell. Robot. Syst. **77**(3–4), 629–652 (2015)
15. Bekmezci, I., Sahingoz, O.K., Temel, Ş.: Flying ad-hoc networks (FANETs): a survey. Ad Hoc Netw. **11**(3), 1254–1270 (2013)
16. Kenney, J.B.: Dedicated short-range communications (DSRC) standards in the United States. Proc. IEEE **99**(7), 1162–1182 (2011)
17. Marconato, E.A., Maxa, J.A., Pigatto, D.F., Pinto, A.S., Larrieu, N., Branco, K.R.C.: IEEE 802.11n vs. IEEE 802.15.4: a study on Communication QoS to provide Safe FANETs. In: 2016 46th Annual IEEE/IFIP International Conference on Dependable Systems and Networks Workshop (DSN-W), pp. 184–191. IEEE (2016)
18. Hayat, S., Yanmaz, E., Bettstetter, C.: Experimental analysis of multipoint-to-point UAV communications with IEEE 802.11n and 802.11ac. In: 2015 IEEE 26th Annual International Symposium on Personal, Indoor, and Mobile Radio Communications (PIMRC), pp. 1991–1996. IEEE (2015)
19. Zhou, Y., Cheng, N., Lu, N., Shen, X.S.: Multi-UAV-aided networks: aerial-ground cooperative vehicular networking architecture. IEEE Veh. Technol. Mag. **10**(4), 36–44 (2015)
20. Yanmaz, E., Hayat, S., Scherer, J., Bettstetter, C.: Experimental performance analysis of two-hop aerial 802.11 networks. In: 2014 IEEE Wireless Communications and Networking Conference (WCNC), pp. 3118–3123. IEEE (2014)
21. Rawashdeh, Z.Y., Mahmud, S.M.: Media access technique for cluster-based vehicular ad hoc networks. In: 2008 IEEE 68th Vehicular Technology Conference, pp. 1–5. IEEE (2008)
22. Almalag, M.S., Olariu, S., Weigle, M.C.: TDMA cluster-based MAC for VANETs (TC-MAC). In: 2012 IEEE International Symposium on a World of Wireless, Mobile and Multimedia Networks (WoWMoM), pp. 1–6. IEEE (2012)
23. Torabi, N., Ghahfarokhi, B.S.: Survey of medium access control schemes for inter-vehicle communications. Comput. Electr. Eng. **64**, 450–472 (2017)
24. Park, J.H., Choi, S.C., Hussen, H.R., Kim, J.: Analysis of dynamic cluster head selection for mission-oriented flying ad hoc network. In: 2017 Ninth International Conference on Ubiquitous and Future Networks (ICUFN), pp. 21–23. IEEE (2017)
25. Friis, H.T.: A note on a simple transmission formula. Proc. IRE **34**(5), 254–256 (1946)
26. Bazzi, A., Masini, B.M., Zanella, A., Thibault, I.: On the performance of IEEE 802.11p and LTE-V2V for the cooperative awareness of connected vehicles. IEEE Trans. Veh. Technol. **66**(11), 10419–10432 (2017)

27. Mammu, A.S.K., Hernandez-Jayo, U., Sainz, N.: Cluster-based MAC in VANETs for safety applications. In: 2013 International Conference on Advances in Computing, Communications and Informatics (ICACCI), pp. 1424–1429. IEEE (2013)

28. Hadded, M., Muhlethaler, P., Laouiti, A., Zagrouba, R., Saidane, L.A.: TDMA-based MAC protocols for vehicular ad hoc networks: a survey, qualitative analysis, and open research issues. IEEE Commun. Surv. Tutor. **17**(4), 2461–2492 (2015)

29. ITU-T Rec. Y.1541: Network performance objectives for IP-based services. International Telecommunication Union, ITU-T (2003)

30. Serrano, P., Banchs, A., Patras, P., Azcorra, A.: Optimal configuration of 802.11e EDCA for real-time and data traffic. IEEE Trans. Veh. Technol. **59**(5), 2511–2528 (2010)

31. Serrano, P., Banchs, A., Kukielka, J.F.: Optimal configuration of 802.11e EDCA under voice traffic. In: IEEE GLOBECOM 2007-IEEE Global Telecommunications Conference, pp. 5107–5111. IEEE (2007)

32. Banchs, A., Vollero, L.: Throughput analysis and optimal configuration of 802.11e EDCA. Comput. Netw. **50**(11), 1749–1768 (2006)

Minimizing the IoT System Delay
with the Edge Gateways

Van Dai Pham[1], Trung Hoang[2], Ruslan Kirichek[1,3], Maria Makolkina[1,3(✉)],
and Andrey Koucheryavy[1]

[1] The Bonch-Bruevich Saint-Petersburg State University of Telecommunications,
22 Prospekt Bolshevikov, St. Petersburg, Russian Federation
daipham93@gmail.com, kirichek@sut.ru, makolkina@list.ru, akouch@mail.ru
[2] The Authority of Radio Frequency Management,
115 Tran Duy Hung, Ha Noi, Vietnam
trunghl@rfd.gov.vn
[3] Peoples Friendship University of Russia (RUDN University),
6 Miklukho-Maklaya St., Moscow, Russian Federation

Abstract. Nowadays, the number of Internet of Things (IoT) devices
has grown much faster. Many different devices are connected to the Internet through various network interfaces. Thus, these such devices create
the heterogeneous networks and the edge heterogeneous gateways provide communication between them and cloud servers. In this article, we
consider the functional framework of the edge gateway and evaluate the
efficiency of using it to minimize the IoT system delay. The simulation
results show that the system delay is reduced when using edge gateways.

Keywords: IoT · Internet of Things · Delay minimization · Edge
gateway · Edge computing

1 Introduction

Currently, many systems are integrated with the Internet of Things technologies
[5,10,11]. Various sensors are used to collect the data of devices and send these
data to cloud servers for further processing [4,6,13]. In parallel, the communication technologies are also being developed for connecting devices among themselves and to the global network. IoT devices can interact in the local network
by using different technologies, such as WiFi, Zigbee, Bluetooth BLE, Modbus
or CAN bus. The principles of connecting these devices to the external network
are the usage of gateways, which support end-user communication interfaces.
The main goal of the IoT gateway is to provide connectivity between devices in
the local network, and connectivity to the cloud servers via global network.

In transportation networks VANET, the mobile gateways are introduced to
aggregate the traffic of other vehicles and to communicate with the LTE infrastructure network in their coverage are of IEEE 802.11p. In [1], a selection algorithm was proposed, which dynamically selects a small number of QoS-enabled

© Springer Nature Switzerland AG 2019
V. M. Vishnevskiy et al. (Eds.): DCCN 2019, LNCS 11965, pp. 495–507, 2019.
https://doi.org/10.1007/978-3-030-36614-8_38

gateways, depending on the load state of the LTE network and the QoS requirement of VANET applications. In order to reduce latency in VANET application, the distributed system was proposed to compute data at the edge [12], the data processing at the edge is more interesting and considered in the IoT system [3].

A particular IoT application is Industrial Internet of Things, which is receiving much attention in the process to Era of Industry 4.0. Authors in [7] considered the architecture of a heterogeneous gateway in Industrial IoT for converting packets that are formed according to various industrial protocols.

In the heterogeneous networks, authors in [9] proposed an emulated gateway for connecting sensor nodes to the external network. The emulation of the gateway brings the possibility of studying the IoT system from the sensor layer to the layer of cloud computing. In addition, at the edge/fog layer a method was proposed to ensure security IoT application [2,8].

With the addition of functionality to IoT gateways, in this article we analyze the functional framework of the edge gateway that is located between the end devices and the external network. In order to minimize the IoT system delay, using the edge gateways in IoT system is analyzed with minimizing delay on different fragments of network.

2 Functional Framework of Edge Gateway

2.1 Analyzing of Requirements for Edge Gateway

For IoT applications, there are various types of devices, such as electrical equipment, cars, lighting systems, automatic doors or windows, cameras and sensors. Such devices can be access to external networks through various types of network interface (WiFi, Zigbee, LoRa, ...). Hence the requirements are considered for the edge gateway.

Gateway: there are many ways to connect IoT devices to external networks. However these networks don't follow the same protocol. Different types of devices working with various communication protocols can communicate with each other through a gateway. Therefore, the gateway should support network protocols over which end devices work.

Data Transfer Protocol: for different types of use, different devices use different data transfer protocol. These protocols can be in IP networks, sensor networks or in low energy-efficient networks, etc. Effective data transfer protocols can ensure the efficient use of limited resources in these networks. According to the OSI model, at the physical layer there are technologies such as IEEE 802.11 (WiFi), IEEE 802.15.4 (Zigbee), 6LoWPAN, LPWAN (LoRaWAN), Modbus, CAN bus, Ethernet, LTE, etc, and at the application layer there are protocols such as HTTP/1.1/2.0, MQTT, CoAP, AMQP, XMPP, etc.

Security: security is a more serious problem in many networks. Especially in networks with many devices, high security is required.

Controller: sensor networks generate a lot of raw data that requires further processing. However, sensor devices are very limited in resources and cannot execute such operations. In addition, the transmission of this raw data is necessary for computing and storage in the cloud, but it takes more time. This can be expensive for sensor devices, because the cloud can work with different protocols. In time-sensitive applications, such as in industry, if computation is processed in the cloud in order to raise the fire alarm, this can lead to serious damage and even risk of life. In addition, growing amount of data generated by billions of hybrid IoT devices is already a concern in the big community, along with growing volume of data traffic. However, most of the data generated by these devices is redundant and can be deleted with some pre-processing or processed before sending to the cloud. There is a need for a dedicated control node that can perform or process complex operations. The controller host also performs an almost transparent traffic flow between end devices and the clouds.

Resource Optimization: the edge gateways receive different types of traffic. Traffic flows between end devices and clouds, or between gateways. It is necessary that the gateway can balance traffic bandwidth.

Predictive Maintenance: analytic and forecasting methods are applied at the gateway. Such methods are trained based on the data collected from the end devices in order to predict the service cycle time or operation errors of these devices.

Real-Time Monitoring: data on the gateway is always available in real time, i.e. users can access gateways anytime and anywhere.

In the more complex reality of IoT with larger-scale IoT projects, multiple data transfer protocols, industry standards and IoT standards and many types of sensors and data located everywhere (all depending on the size of the project), it is required:

- to connect all devices and enable these various things, sensors and data to "talk" to each other (despite the fact that they communicate in another language) and;
- to understand all this before sending all this data somewhere else, where it really - ideally - leads to a real meaning, effective understanding and actions in any form.

A high level of interoperability, connectivity, data pre-processing, data aggregation, remote monitoring and control leads to the need for edge gateways. IoT gateways should provide scalability, manageability, increased security, performance, and faster and more flexible deployment.

2.2   Functionality of Edge Gateway

The goal of edge gateway is to connect devices, expand their accessibility and control, integrating them with the Internet. The gateway provides the necessary intelligence and processing capabilities of collected data from the end devices. In addition, users can access and control the IoT devices from the local or global network.

Controller Functionality: the edge gateway works as a controller. It can periodically collect data from sensor devices and can response to requests from the Internet and other connected networks. The edge gateway can receive control commands and manage connected devices. As the traffic between the cloud and sensor network increases, the load on gateway also increases. Therefor, it is necessary to use several gateways to distribute the computation load and share storage location. As traffic exchanges between the clouds and sensor networks increase, the load on the edge gateway also increases. Therefore, multiple gateways are necessary to distribute the data traffic.

Transmission Protocol: each subnet of sensors and actuator is associated with one or more gateways that are responsible for seamless integration with the Internet. Endpoint traffic will pass through the gateway to the cloud servers. Thus, gateways support protocols used by both the Internet and sensor networks to exchange data between them.

Principles of Data Exchange: generality for data exchange there are two main principles: request/response and publisher/subscriber. The edge gateway can support both methods for exchanging data.

Secured Registration: To connect new devices, each device needs to go through the registration procedure. Each device must register with the gateway. A edge gateway provides an identifier to each newly added device. Since in many applications devices are usually in sleep mode, depending on the location of the device, the controller in the gateway sends information to the devices about standby modes for configurations to ensure a stable network at any time. This is important because in a multi-hop network, all intermediate devices must be active at the same time to ensure successful communication.

Data Processing: when receiving a large volume of data, it is necessary to process it before using or sending to cloud servers, in order to save traffic between gateways and clouds. The edge gateway can pre-retrieve and cache content that may be needed in the near future. In addition, the gateway can perform some data operations, such as filtering raw data, aggregation, autoregression, etc. The large volume of data generated by IoT devices is already a big data problem. However, not all data generated by sensor devices is important to transmit and store, for example, devices that periodically measure the temperature of equipment should only store deviations. Thus, the required storage can be significantly reduced by pre-processing and defining important data for storage. Data can be sent to the clouds without processing, or processed before sending. Developers can use their own scripts to process specific data on the gateway.

Security: the edge gateway requires the secure registration of new devices and also the secure communication between devices. Since security is an important requirement, the edge gateway need to satisfy this requirement, in order to protect the data and the communication channel. If devices cannot successfully pass the registration procedure, they will be refuse to service. Thus, it needs to be ensured for the access security to end devices, the protection of sensor data at the gateway, and the communication to the cloud servers.

3   Delays in IoT System

3.1   System Architecture with Edge Gateways

The edge gateways can be connected with each other and also with cloud server. Figure 1 shows the interaction architecture of IoT system, which is divided into 4 layers.

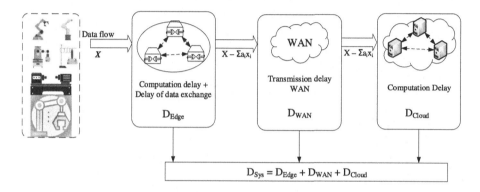

Fig. 1. IoT system model

The first layer is the layer of the endpoint IoT devices, which includes IoT devices and user terminals. Such devices connect to gateways according to user requirements for accessing the Internet and for interacting with other devices in a common network. At the end device layer, a data flow X is generated, and comes in the next layer.

The second layer is the edge layer, which include the edge gateways. The edge layer provides communication for the end devices with other devices and with cloud servers. The edge layer also plays the role of data pre-processing before being sent to the clouds. Assume that N edge gateways receive and process the data flow X. With a large mount of data, it is necessary to take into account data processing time (computation delay) and data exchange time between gateways.

The global network is the third layer, which takes a delay of data transmission between the edge layer and cloud servers.

The fourth layer is the cloud layer, which includes cloud servers. The cloud layer has powerful computational capabilities and a large database. The data flow, which isn't processed at the edge layer, is redirected to the cloud servers. At the cloud layer, there is a set of M cloud servers for processing residual data coming from the edge layer. At this layer, the computation delay for processing residual data is taken into account.

Thus, the system delay from the edge layer to the cloud layer includes the delay of data transmission between the edge gateways, between the edge layer and the cloud, the delay for data processing at the edge layer and at the cloud layer, as shown in Fig. 1. The formulation of system delay is defined as:

$$D_{Sys} = D_{Edge} + D_{WAN} + D_{Cloud} \tag{1}$$

3.2 Delay on the Edge Layer

The edge gateway includes a set of N edge gateways. These gateways are interconnected and connected to cloud servers.

The data flow X from the end device layer enters the edge layer. Thus, X is a computational task for the edge gateways. In order to reduce the amount of data transmitted from the end device layer to the cloud layer, and to reduce the delay in data transmission, it is necessary to expand the computing capabilities of the edge nodes and process the data at the edge layer as much as possible. Therefore, we propose a scheme of interaction between edge gateways. In the process of data processing and data exchange, the edge layer receives the amount of data X for calculation from the end devices and is divided into several subtasks x_i, i.e. the subtask will be processed by the edge node GW_i.

Thus, the delay at the edge layer includes the delay in the exchange of data between the edge nodes, and the delay in the calculation of the edge nodes.

The edge gateway has the computational capacity c_i, so we can calculate the required delay to process the amount of data x_i according to the formula:

$$D_i^{Edge} = \frac{a_i \cdot x_i}{c_i} \tag{2}$$

where:

a_i—percentage of data processed by the edge gateway GW_i.
x_i—amount of data to process at the edge node GW_i.
c_i—computational capacity of the edge node GW_i.

Thus, the delay in computing N edge gateways can be represented as follows:

$$D_{Edge}^{com} = min(\sum_{i=1}^{N} \frac{a_i \cdot x_i}{c_i}) \tag{3}$$

with condition:

$\sum_{i=1}^{N} x_i = X,$ and $0 < x_i < x_i^{max}$

where: X is the total amount of data entering the edge layer for processing.
The amount of data X can be represented in the form of a vector:

$$X = [x_1, x_2, ..., x_N]$$

x_i^{max} is the maximum amount of data that the edge gateway GW_i can handle.

Thus, the following objective function is set for the delay generated at the edge layer:

$$D_{Edge} = min(\sum_{i=1}^{N} \frac{a_i \cdot x_i}{c_i} + W(T))$$ (4)

with condition:
$$\sum_{i=1}^{N} x_i = X,$$
$$0 < x_i < x_i^{max}$$

3.3 Delay over WAN

The unprocessed data from the edge gateway layer is redirected to the cloud servers through the WAN. Assuming that there are M cloud servers serving the N edge gateways. Thus, the amount of data from the edge gateway GW_i transferred to the cloud server j is denoted by λ_{ij} and the delay in the exchange of data between them is denoted by d_{ij}.

If losses are not taken into account over the channels, the delay in data transmission through the WAN from the edge gateway GW_i to the cloud server j is calculated as follows:

$$D_{ij}^{WAN} = d_{ij} \cdot \lambda_{ij}$$ (5)

where:

$i = 1, ..., N$—number of edge gateways.
$j = 1, ..., M$—number of cloud servers.

Thus, the total delay in the data transmission from the edge to the cloud layer can be expressed by the following formula:

$$D_{WAN} = \sum_{i=1}^{N} \sum_{j=1}^{M} d_{ij} \cdot \lambda_{ij}$$ (6)

with condition:

$\sum_{j=1}^{M} \lambda_{ij} = Y$—the total amount of data enters to the cloud layer for processing.
$0 < \lambda_{ij} < \lambda_{ij}^{max}$—$\lambda_{ij}$ is in the limit of receiving cloud server.
λ_{ij}^{max}—The maximum received traffic of edge server.

3.4 Delay on Cloud Layer

Delay on the cloud layer includes time of data processing. Assuming that there are M cloud servers, the total amount of data Y will be processed. For the cloud server j, if the amount of data y_j needs to process, and the computational capacity is v_j, then the computation delay can be expressed as:

$$D_j^{com} = \frac{y_j}{v_j}, \quad j = 1, 2, ..., M$$ (7)

Thus, for M cloud servers, the total delay for processing the amount of data Y is calculated by the formula:

$$D_{Cloud} = \sum_{j=1}^{M} \frac{y_j}{v_j} \tag{8}$$

with condition:

$\sum_{j=1}^{M} y_j = Y,\; y_j > 0$—the amount of data from the edge layer.

4 Methods to Minimize Delay

4.1 Queueing Management of the Edge Gateway

As it was considered that the edge gateway supports several communication technologies, thus several flows of different traffic come in the gateway. Such flows are generated from end devices with network interfaces WiFi, Zigbee, BLE, Modbus, CAN bus, etc. At the beginning, incoming flows are in the queue to wait for serving, each flow has its own queue. Then, the processor takes the request from each queue for the next processing. The principle of processing data from buffers is an important task, since each type of traffic has different service requirements. For example, applications using IEEE 802.11 (WiFi) technology require faster action, or some industrial devices also require low latency and faster computation. This section discusses queueing management algorithms to minimize data loss and latency. The queue of data processing can be controlled by following methods.

Fair Queueing (FQ): this approach tries to read all traffic sources at the same time. Therefore, it must divide the computing capacity between all incoming sources (i.e. the amount of data that can be read during each cycle).

Weighted Fair Queuing (WFQ): the fair queueing works well when all incoming sources have the same characteristics (they have the same priority and the same bandwidth/frequency). When they don't have, method WFQ allocates computing bandwidth according to the characteristics of the incoming traffic.

Queueing management algorithms are simulated as follows:

- The first task is to create several traffic sources. We decide not to do this in real time, but instead generate timestamps (when packets arrive) and data packet sizes in bytes. Each traffic source will be stored in a separate buffer (queue).
- Then we run the simulation and count the cycles as unit. Simulation continues until all incoming packets have been read. During each cycle, we check whether we will clear any queues (and clear them) or whether new packets have arrived. Each cycle is checked by a timestamp for processing data from which queue. Statistics are updated—which packets were read and how full the queues are.

4.2   Minimizing System Delay

According to Fig. 1, the system delay includes the communication delay between the edge gateways, between the edge layer and the cloud layer, the delay of data processing at the edge layer and at the cloud layer. Thus, the task of system delay minimization is set as:

$$\min_{x_i, \lambda_i, y_i} \{D_{Sys}\} \tag{9}$$

where: x_i, λ_i, y_i are in the conditions from formulas (4, 6, 8).

The solution of minimization (9) can be found by converting the problem (9) to linear programming with constraint (4, 6, 8), which can be solved by existing algorithms.

5   Performance Evaluation

5.1   Comparison of Queueing Management Methods

For simulation, 5 different types of traffic are generated (Table 1).The weights of each traffic are $W1, W2, W3, W4, W5$, where $W1 + W2 + W3 + W4 + W5 = 1$. In each cycle, central processor can process 2000 data units. Each type of traffic generates 100 packages, and the maintenance process is carried out with resending the package when it has not been processed. The simulation results of FQ and WFQ methods are shown in Figs. 2 and 3.

Table 1. Simulation parameters

Traffic	Delay between packages	Package size	Buffer size	Traffic weight in WFQ
Traffic 1 (WiFi)	Random [1, 5]	Random [100, 1000]	20000	$W1 = 0.3$
Traffic 2 (Modbus)	Random [1, 10]	Random [50, 500]	20000	$W2 = 0.25$
Traffic 3 (CAN bus)	Random [1, 10]	Random [50, 500]	20000	$W3 = 0.25$
Traffic 4 (LoRa)	Random [8, 20]	Random [20, 200]	20000	$W4 = 0.1$
Traffic 5 (Zigbee)	Random [8, 20]	Random [20, 200]	20000	$W4 = 0.1$

According to the simulation results, the WFQ method works more efficiently for queueing management. In Fig. 2 we see that there are busy places in the buffers on channels 1, 2 and 3, i.e. traffic packages are waiting to be processed. And in Fig. 3 the busy places in the buffers on channels 2 and 3 are zero. Thus, the WFQ method allows reducing the waiting time for service and the incoming packages are serviced quickly. Therefore, according to the characteristics of application and incoming traffic, the weight of traffic can be controlled to manage the package maintenance queues.

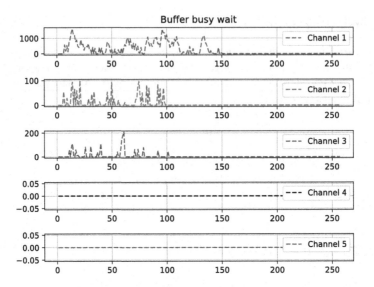

Fig. 2. Simulation of FQ method

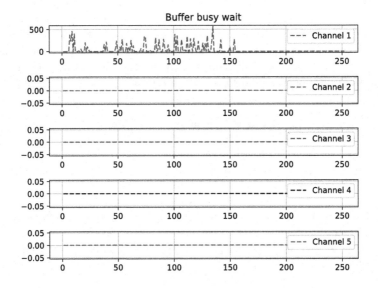

Fig. 3. Simulation of WFQ method

5.2   System Delay

The experiments are carried out in Matlab with parameters: $N = 6$ edge gateways with the computation capacity of $c = [1, 1.5, 1.2, 1.6, 1.5, 1.3]$ GHz, $M = 5$ cloud servers with the computation capacity of 10 GHz for each one, and 30% the amount of data is processed by each gateway.

The influence on system delay is considered and shown in Fig. 4 when the required processed data is changed. When using edge gateways the system delay is reduced, and also for the load of channels and cloud servers.

The experiment is carried out with changing the number of edge gateways for processing the amount of data 5. For a certain amount of data, more edge gateways can reduce the system delay. The influence of the number of edge

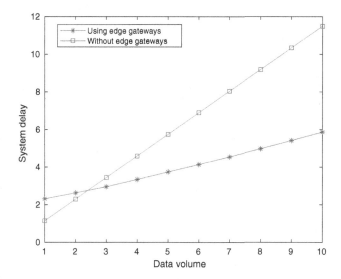

Fig. 4. The influence of data volume on the system delay

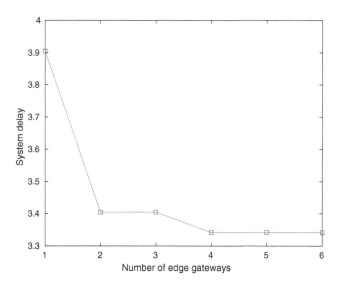

Fig. 5. The influence of the number of edge gateways on the system delay

gateways on the system delay is shown in Fig. 5. The system delay decrease when increasing the number of gateways to 4. After that, the system delay doesn't change. This means that for each specific amount of data there is a threshold number of edge gateways that can provide the minimum system delay.

6   Conclusion

In this article, the functionality requirements were considered for the edge gateway that is proposed for the Internet of Things system. In this case, the system delay was analyzed as a objective task to minimize this delay. To solve this problem, the methods of queueing management in the gateway and methods for minimizing delay on the different network fragments were proposed. The simulation results showed that using edge gateways with enhanced functionality helps reduce the delay and load of the IoT system.

Acknowledgment. The publication has been prepared with the support of the "RUDN University Program 5-100" and funded by RFBR according to the research projects No. 12-34-56789 and No. 12-34-56789.

References

1. Alawi, M., Sundararajan, E., Alsaqour, R., Ismail, M.: QoS-enable gateway selection algorithm in heterogeneous vehicular network. In: 2017 6th International Conference on Electrical Engineering and Informatics (ICEEI), pp. 1–6. IEEE (2017)
2. Bragadeesh, S., Arumugam, U.: A conceptual framework for security and privacy in edge computing. In: Al-Turjman, F. (ed.) Edge Computing. EAISICC, pp. 173–186. Springer, Cham (2019). https://doi.org/10.1007/978-3-319-99061-3_10
3. Grover, J., Garimella, R.M.: Optimization in edge computing and small-cell networks. In: Al-Turjman, F. (ed.) Edge Computing. EAISICC, pp. 17–31. Springer, Cham (2019). https://doi.org/10.1007/978-3-319-99061-3_2
4. Jagannath, U.R., Saravanan, S., Suguna, S.K.: Applications of the internet of things with the cloud computing technologies: a review. In: Al-Turjman, F. (ed.) Edge Computing. EAISICC, pp. 71–89. Springer, Cham (2019). https://doi.org/10.1007/978-3-319-99061-3_5
5. Jeschke, S., Brecher, C., Meisen, T., Özdemir, D., Eschert, T.: Industrial internet of things and cyber manufacturing systems. In: Jeschke, S., Brecher, C., Song, H., Rawat, D. (eds.) Industrial Internet of Things. SSWT, pp. 3–19. Springer, Cham (2017). https://doi.org/10.1007/978-3-319-42559-7_1
6. Kirichek, R., Koucheryavy, A.: Internet of things laboratory test bed. In: Zeng, Q.A. (ed.) Wireless Communications, Networking and Applications. LNEE, vol. 348, pp. 485–494. Springer, New Delhi (2016). https://doi.org/10.1007/978-81-322-2580-5_44
7. Kulik, V., Kirichek, R.: The heterogeneous gateways in the industrial internet of things. In: 2018 10th International Congress on Ultra Modern Telecommunications and Control Systems and Workshops (ICUMT), pp. 1–5. IEEE (2018)
8. Muthanna, A.: Secure and reliable IoT networks using Fog computing with software-defined networking and blockchain. J. Sens. Actuator Netw. **8**(1), 15 (2019)

9. Rajaram, K., Susanth, G.: Emulation of IoT gateway for connecting sensor nodes in heterogenous networks. In: 2017 International Conference on Computer, Communication and Signal Processing (ICCCSP), pp. 1–5. IEEE (2017)
10. Serpanos, D., Wolf, M.: Internet-of-Things (IoT) Systems: Architectures, Algorithms, Methodologies. Springer, Cham (2017). https://doi.org/10.1007/978-3-319-69715-4
11. Vladimirov, S., Kirichek, R.: The IoT identification procedure based on the degraded flash memory sector. In: Galinina, O., Andreev, S., Balandin, S., Koucheryavy, Y. (eds.) ruSMART 2017, NsCC 2017, NEW2AN 2017. LNCS, vol. 10531, pp. 66–74. Springer, Cham (2017). https://doi.org/10.1007/978-3-319-67380-6_6
12. Vladyko, A., Khakimov, A., Muthanna, A., Ateya, A.A., Koucheryavy, A.: Distributed edge computing to assist ultra-low-latency VANET applications. Future Internet 11(6), 128 (2019)
13. Volkov, A., Khakimov, A., Muthanna, A., Kirichek, R., Vladyko, A., Koucheryavy, A.: Interaction of the IoT traffic generated by a smart city segment with SDN core network. In: Koucheryavy, Y., Mamatas, L., Matta, I., Ometov, A., Papadimitriou, P. (eds.) WWIC 2017. LNCS, vol. 10372, pp. 115–126. Springer, Cham (2017). https://doi.org/10.1007/978-3-319-61382-6_10

States with Minimum Dispersion
of Observables in Kuryshkin-Wodkiewicz
Quantum Mechanics

A. V. Zorin[1] , L. A. Sevastianov[1,2](✉) , and N. P. Tretyakov[3,4]

[1] Peoples' Friendship University of Russia (RUDN University),
6 Miklukho-Maklaya St, Moscow 117198, Russia
{zorin-av,sevastianov-la}@rudn.ru
[2] Joint Institute for Nuclear Research (Dubna),
Joliot-Curie, 6, Dubna 141980, Moscow Region, Russia
[3] Department of Informatics and Applied Mathematics,
Russian Presidential Academy of National Economy, Moscow, Russia
trn11-2011@mail.ru
[4] Department of Applied Information Technologies,
Russian State Social University, Moscow, Russia

Abstract. We address the question about the fundamental limitations on the possibility of transmitting messages imposed by the nature of the physical carrier of information. For the presentation of rigorous results for quantum communication channels (including their throughputs), it is necessary to introduce a quantum theory into the statistical structure, which is the basis for the quantum information theory. The minimum requirements to the description of a quantum communication channel are formulated. We use the Kuryshkin-Wodkiewicz model of quantum measurements and construct the expressions for operators dispersions and formulate the problem of finding minimum-dispersion states. The case of an electron in a hydrogen-like atom is considered.

Keywords: Quantum communication channels · Quantum mechanics · Non-negative quantum distribution function · Hydrogen-like atom

1 Introduction to the Quantum Information Theory

The presentation of this section largely follows the presentation of similar aspects of the quantum theory of measurement [1–3]. The main result of the classical information theory is the establishment of the possibility of transmitting information at speeds that do not exceed a certain value, called throughput (carrier capacity). Almost simultaneously with the creation of mathematical foundations of information theory, the question arose about the fundamental limitations on

L. A. Sevastianov—The publication has been prepared with the support of the RUDN University Program 5-100.

V. M. Vishnevskiy et al. (Eds.): DCCN 2019, LNCS 11965, pp. 508–519, 2019.
https://doi.org/10.1007/978-3-030-36614-8_39

the possibility of transmitting messages imposed by the nature of the physical carrier of information. The problem of determining the capacity of a quantum communication channel was formed in the 1960s and was based on the works of Gabor and Brillouin who raised the question of the quantum-mechanical limits of the accuracy and speed of information transfer. The principal steps in this direction were made in the 1970s, when a non-commutative statistical decision theory was constructed. For the presentation of rigorous results for quantum communication channels (including their throughputs), it is necessary to introduce a quantum theory into the statistical structure, which is the basis for the quantum information theory.

Quantum information theory deals with quantum-mechanical systems, which are described by finite-dimensional Hilbert spaces H. Even for this case, the main differences of quantum statistics are seen. In the quantum theory of information transmission, "systems with continuous variables" are described by infinite-dimensional spaces $\tilde{H}$. The state of the quantum-mechanical system, representing a statistical ensemble of equally prepared system instances, is described by the density operator (the density matrix in a fixed basis) S in H.

Shannon's theorem provides the basis for the introduction of such a concept as the bandwidth of a classical channel with noise (the maximum speed of asymptotically error-free transmission of information through a channel). The simplest model of a quantum channel assumes that there is a classical parameter x, that takes values from the (finite) input alphabet, and a mapping $x \to S_x$ to quantum states at the channel output.

At the output, some observable M is measured in the space H (having obtained the outcome of the measurement of x, one can assume that x was sent). As a result, the receiver replies about obtained decision; so the resolution of identity in the space H describes the statistics of the entire decision procedure, which includes the physical measurement and the subsequent classical processing of its results The choice of the observed M is formally analogous to the choice of the decision procedure in the classical case, but how plays a more important role. After M is selected, one can get the classical channel $p_M(y|x) = \mathrm{Tr} S_x M_y$. In any physical experiment, there are two main stages: the preparation of the state and the measurement. Even if a pure quantum state is prepared, where there is no classical stochasticity, the measurement result can still be accidental.

The ideal measurement has the property of reproducibility: after repeating the measurement immediately after obtaining some outcome, we must with confidence obtain the same outcome. A number of paradoxes and dead ends in the standard formulation of quantum mechanics are due to the fact that only the ideal procedures that satisfy the reproducibility condition are considered as a mathematical model of measurements. These difficulties are largely removed in a generalized formulation that allows us to cover "approximate" measurements, the impact of which on the system can be arbitrarily weak.

Let us recall the minimum requirements to the theoretical description of a quantum communication channel. Every channel must convert quantum states from $S(H_1)$ into quantum states (generally speaking, in another space) $S(H_2)$.

A necessary requirement for consistency with the statistical interpretation of quantum mechanics is that the mixtures of input states must pass into the same mixture of output states, that is, the channel must be defined by an affine mapping $\Phi(\sum p_j S_j) = \sum p_j \Phi(S_j)$. The description of quantum evolution in terms of states corresponds to the Schrödinger picture.

In quantum information theory, a non-orthogonal resolution of identity is often considered, whose eigenvalues define the transition probabilities from X in Y. The set of all transition probabilities is convex. The extreme points of this set correspond exactly to the orthogonal resolution of identity.

A complete system of vectors is said to be overflowed if it consists of non-normalized and linearly dependent vectors. With each overflowed system, the resolution of identity is connected. This resolution is an extreme point of the convex set of all resolutions of identity if and only if the vectors are linearly independent. Overcrowded non-orthogonal systems have no analog in the classical statistics.

According to Naimark's theorem, any observable can be realized as a standard observable in a composite system by adding an auxiliary system in a fixed, pure state. This way of realization is called the quantum randomization. In the classical statistics the randomization does not increase information about the state of the observed system. The quantum randomization makes it possible to extract more information about the observed system than is contained in standard observables that do not use an auxiliary system.

So, a random variable x is measured, the distribution $\mu_S^M(x)$ of which depends on the preparation of the ensemble S and on the measuring instrument M. Naturally, mixing ensembles leads to the same mixing of distributions, that is if $S = \sum p_j S_j$, then $\mu_S^M(x) = \sum p_j \mu_{S_j}^M(x)$. So, the probabilities of measurement outcomes are affine state functions. This restriction is sufficient to derive the generalized Born statistical formula.

Wigner's Theorem. If Φ is the affine one-to-one mapping of $S(H)$ on $S(H)$, then $\Phi[S] = USU*$, where U – is the unitary (or antiunitary) operator in H.

The ideal measurement in quantum mechanics is described by the orthogonal resolution of identity $\sum E_j = I$. Then, according to the postulate of Lunders-von Neumann, after the measurement the j-th outcome is obtained with probability $p_j = \text{Tr} S E_j$ and density operator $S_j = \frac{E_j S E_j}{\text{Tr}(SE_j)}$. So

$$\Phi(S) = \sum p_j E_j. \tag{1}$$

Consequently, in quantum mechanics, unlike the classical one, an ideal measurement does not reduce to a simple reading of the value of the observed quantity, but always involves an impact on the observable system that changes its state.

This interaction with the environment is accounted for by the construction of a composite system $S \otimes S_0$ of the measured system in H in the state S and the trial (sample) system in H_0 in the state S_0. The effect of the measuring procedure on the measured object is described by the transformation $U(S \otimes S_0)U*$. And averaging the evolution of the effects of the measurement procedure, one can get $\mathrm{Tr}_{H_0}[U(S \otimes S_0)U*] = \Phi[S]$.

Let's consider the process of indirect measurement, when the system interacts with the trial system H_0, and then an ideal measurement of the observed $\{E_j^0\}$ is made over the trial system. Then the probability to obtain the outcome j is equal to $p_j = \mathrm{Tr}U(S \otimes S_0)U * (I \otimes E_j)$, and the state of that fraction of the statistical ensemble in which this outcome is obtained is described by a density operator S_j, such that

$$p_j S_j = \mathrm{Tr}_{H_0}(I \otimes E_j)U(S \otimes S_0)U * (I \otimes E_j). \tag{2}$$

So, the formula $\mathrm{Tr}_{H_0}[U(S \otimes S_0)U*]$ takes the form

$$\Phi[S] = \sum_j p_j S_j. \tag{3}$$

Suppose, for simplicity, the state $S_0 = |\psi_0\rangle\langle\psi_0|$ to be pure, and the measurement to be complete, that is $E_j = |e_j^0\rangle\langle e_j^0|$, where e_j – is the orthonormal basis in H_0. By introducing operators V_j, defined by $\langle\varphi|V_j\psi\rangle = \langle\varphi \otimes e_j^0|U\psi \otimes \psi_0\rangle$, $\varphi, \psi \in H$, one can obtain

$$p_j = \mathrm{Tr}V_j SV_j^*, \qquad S_j = \frac{V_j SV_j^*}{p_j}, \tag{4}$$

and, as a result,

$$\Phi[S] \sum_j V_j SV_j^*. \tag{5}$$

This formula generalizes the relation (1) for an ideal measurement, and describes the statistics (probability of outcomes and a posteriori states) of a more general process involving measurements, that do not satisfy the reproducibility requirement (in particular, approximate measurements). This and other arguments show that the property of complete positivity should be included in the channel definition.

Definition. The channel (in the state space) is a linear mapping Φ^* from $B(H_1)$ into $B(H_2)$, hat preserves the trace and such, that Φ^* – is a completely positive mapping. The channel in the space of observables is a linear completely positive mapping from $B(H_2)$ into $B(H_1)$, transferring identity into identity.

According to the Strinespring-Naimark theorem [3], every completely positive mapping of Φ^* algebra $B(H_1)$), preserving the identity, can be extended to an invertible transformation of the composite system in where the second system (the trial system of the measurement procedure) is in the pure state $|\psi_0\rangle\langle\psi_0|$.

2   Quantum Mechanics with Non-negative Distribution Function

The papers [4–11] present the Kuryshkin-Wodkiewicz constructive model of quantum measurements, in which for any mixed state (written in the basis of

Sturm states) $\rho_{app} = \sum_k a_k |\psi_k\rangle\langle\psi_k|$ of the quantum-mechanical part of the measurement device and any mixed state of the object $\rho_{obj} = \sum_k c_k |\psi_k\rangle\langle\psi_k|$ a "result of the measurement of the variable $A(q,p)$" is the action of the operator $O_{\rho_{app}}$ on the state ρ_{obj}

$$O(A)U(q,t) = (2\pi\hbar)^{-N} \int \varphi(q - \xi, p - \eta) e^{\frac{i}{\hbar}(q-q')p} A(\xi, \eta, t) U(q', t) d\xi d\eta dp dq', \tag{6}$$

with the QDF $F_\rho(q,p)$:

$$F_\rho(q,p,t) = (2\pi\hbar)^{-N} \sum \left| \int \varphi_k^*(q - \xi, t)\psi(\xi, t) e^{-\frac{i}{\hbar}d\xi} \right|^2 \tag{7}$$

According to the Kuryshkin-Wodkiewicz quantization rule, they are in one-to-one correspondence with each other and unambiguously depend on the auxiliary functions $\rho_{app} = \{\varphi_k\}$ as

$$\Phi(q, p, t) = (2\pi\hbar)^{N/2} e^{-\frac{i}{\hbar}(qp)} \sum_k \varphi_k(q,t)\tilde{\varphi}_k(p,t),$$

$$\tilde{\varphi}_k(p,t) = (2\pi\hbar)^{N/2} \int e^{-\frac{i}{\hbar}(qp)} \varphi_k(q,t) dq.$$

3   Results on States with Minimum Dispersion of Observables in QMK

The non-Neumannian quantization rule $O(f(A)) \neq f(O(A))$ leads to the relation

$$O(A^2) = O^2(A) + D(A), \tag{8}$$

which determines the operator $D(A)$ for any observable quantity A, which generally does not vanish [5]. The mean (expected) value of $\langle A\rangle_\Psi$ in the Ψ state of A is also characterized by the mean-square deviation (second central moment), i.e, the dispersion $\langle(\Delta A)^2\rangle_\Psi$ in the Ψ state

$$\langle(\Delta A)^2\rangle_\Psi \equiv \langle O((A - \langle A\rangle_\Psi)^2)\rangle_\Psi = \langle O(A) - \langle A\rangle_\Psi)^2\rangle_\Psi + \langle D(A)\rangle_\Psi \geq 0. \tag{9}$$

On the other hand [5],

$$\langle O(A^2)\rangle_\Psi - 2\langle O(A)\rangle_\Psi\langle A\rangle_\Psi + \langle A\rangle_\Psi = \langle A^2\rangle_\Psi - \langle A\rangle_\Psi^2 \geq 0. \tag{10}$$

Thus, in the states described by the eigenvectors of $O(A)$:

$$O(A)|\Psi_\alpha\rangle = \alpha|\Phi_\alpha\rangle \tag{11}$$

with an eigenvalue of α, coinciding with $\langle A\rangle_{\Psi_\alpha} = \alpha$. Unlike conventional quantum mechanics, the value of A is not set strictly, but distributed with dispersion

$$\langle(\Delta A)^2\rangle_{\Psi_\alpha} = \langle\Psi_\alpha|D(A)|\Psi_\alpha\rangle. \tag{12}$$

Thus, the operator $D(A)$ is positively defined on the eigenvectors Φ_α of the operator $O(A)$ and has the sense of a dispersion operator [5].

If we formulate the problem of finding minimum-dispersion states

$$\langle (\Delta A)^2 \rangle_{\Psi_\alpha} \xrightarrow[\Psi]{} \min, \tag{13}$$

or in more detail (see [12]),

$$\left\{ \frac{\langle |O(A^2)|\Psi \rangle}{\langle \Psi|\Psi \rangle} - \left(\frac{\langle \Psi|O(A)|\Psi \rangle}{\langle \Psi|\Psi \rangle} \right)^2 \right\} \xrightarrow[\Psi]{} \min, \tag{14}$$

then the necessary condition of extremum (14) is the Euler-Lagrange equation

$$\left\{ O(A^2) - 2\alpha O(A) + \alpha^2 \right\} |\Psi \rangle = d^2 |\Psi \rangle, \tag{15}$$

where

$$\alpha \equiv \langle \Psi|O(A)|\Psi \rangle, \quad d^2 = \langle \Psi|O((\Delta A)^2)|\Psi \rangle \equiv \langle A^2 \rangle_\Psi - \langle A \rangle_\Psi^2. \tag{16}$$

If the vector Ψ is a solution of the Eq. (15), then the vector $c\Psi$ with any numerical factor C is also a solution of the Eq. (15). Thus, all the (one-dimensional) suspace is a set of solutions of (15), so that one can seek the solutions normalized to 1.

If the operators $O(A^2)$ and $O(A)$ commute,

$$[O(A^2), O(A)] = \hat{O}, \tag{17}$$

then the sets of solutions of the Eqs. (11) and (15) coincide.

Note. If the eigenvectors of an individual operator $O(A)$ do not possess this property, i.e., the simplicity of spectrum is absent, then the classical observable $A(q, p)$ itself does not form a complete set of the first integrals of the system in involution. Adding such integrals $A_2, A_3, ..., A_n$ according to the number "n" of the degrees of freedom of the physical system, we get additionally $O(A_2), ..., O(A_n)$, a set of operators that commute with $O(A)$ and constitute with $O(A)$ a "complete set of observables".

The paper [13] contains the following Theorem. If the family of solutions of Eq. (15) forms a complete orthnormalized system of vectors and the spectrum of the operator $O(A)$ is multiplicity-free, then the set of solutions of the Eqs. (11) and (15) coincide.

In both cases of the coincidence of sets of solutions of Eqs. (11) and (15) The dispersion operator $D(A)$ is positively defined.

4 "Complete Set of Observables" of an Electron in the Hydrogen-Like Atom in QMK

Consider the case of an electron in a hydrogen-like atom. A classical analogue of this problem is the three-dimensional Kepler problem. This problem possesses a large set of first integrals:

– the total energy

$$H(\boldsymbol{r}, \boldsymbol{p}) = \frac{(\boldsymbol{p})^2}{2\mu} - \frac{Ze^2}{|\boldsymbol{r}|}; \tag{18}$$

– the angular momentum

$$\boldsymbol{L} = [\boldsymbol{r} \times \boldsymbol{p}]; \tag{19}$$

– the 3D integral, i.e. three integrals L_x, L_y, L_z;
– the square of angular momentum

$$\boldsymbol{L}^2 = ([\boldsymbol{r} \times \boldsymbol{p}], [\boldsymbol{r} \times \boldsymbol{p}]) = L_x^2 + L_y^2 + L_z^2 \tag{20}$$

The integrals L_x, L_y, L_z are not involutive with respect to Poisson brackets

$$\{L_x, L_y\} = L_z; \{L_y, L_z\} = L_x; \{L_z, L_x\} = L_y. \tag{21}$$

On the other hand, H, $\boldsymbol{L}^2$ and L_z are in the involition:

$$\{H, (\boldsymbol{L})^2\} \equiv \{H, L_z\} \equiv \{(\boldsymbol{L})^2, L_z\} \equiv 0 \tag{22}$$

In the orthodox quantum mechanics these three first integrals of motion correspond to the operators

$$\hat{H} = \frac{\hat{\boldsymbol{p}}}{2\mu} - \frac{Ze^2}{|\boldsymbol{r}|} \equiv -\frac{\hbar^2}{2\mu}\Delta - \frac{Ze^2}{|\boldsymbol{r}|}, \ \hat{H} = \hat{T}_r + \frac{\hat{\boldsymbol{L}}^2}{2\mu r^2} + U(r), \tag{23}$$

where

$$\hat{T}_r = -\frac{\hbar^2}{2\mu}\frac{1}{r^2}\frac{\partial}{\partial r}\left(r^2\frac{\partial}{\partial r}\right); \tag{24}$$

$$\hat{\boldsymbol{L}}^2 = -\hbar^2\nabla_{\theta\varphi}^2, \tag{25}$$

where

$$\nabla_{\theta\varphi}^2 = \frac{1}{\sin\theta}\frac{\partial}{\partial\theta}\left(\sin\theta\frac{\partial}{\partial\theta}\right) + \frac{1}{\sin^2\theta}\frac{\partial^2}{\partial\varphi^2}, \tag{26}$$

so that

$$\nabla^2 = \frac{1}{r^2}\frac{\partial}{\partial r}\left(r^2\frac{\partial}{\partial r}\right) + \frac{\nabla_{\theta\varphi}^2}{r^2};$$

$$\hat{L}_z = -i\hbar\frac{\partial}{\partial\varphi}; \tag{27}$$

commuting with each other

$$[\hat{H}, (\hat{\boldsymbol{L}})^2] = [\hat{H}, \hat{L}_z] = [(\hat{\boldsymbol{L}})^2, \hat{L}_z] = \hat{O}. \tag{28}$$

They form a complete set of observable quantities, possessing mutual multiplicity-free (simple) spectrum

$$\hat{H}\Psi_{nlm}(r, \theta, \varphi) = -\frac{E_1}{n^2}\Psi_{nlm}(r, \theta, \varphi); \quad E_1 = \frac{Z^2\mu e^4}{\hbar^2}; \tag{29}$$

$$(\hat{\boldsymbol{L}})^2 \Psi_{nlm}(r,\theta,\varphi) = \hbar^2(l+1)l\Psi_{nlm}(r,\theta,\varphi); \tag{30}$$

$$\hat{L}_z \Psi_{nlm}(r,\theta,\varphi) = \hbar m \Psi_{nlm}(r,\theta,\varphi). \tag{31}$$

The corresponding operators in Kuryshkin quantum mechanics have the form:

$$O(A) = \sum_{j=1}^{5} a_j^2 O_j(A). \tag{32}$$

The Hamiltonian operator looks like [14]:

$$O_1(H) = \hat{H} + \frac{\hbar^2}{2\mu b^2} + \frac{Z}{b}\left(1 + \frac{b}{r}\right)e^{-2r/b}; \tag{33}$$

$$O_2(H) = \hat{H} + \frac{\hbar^2}{2\mu b^2} + \frac{Z}{b}\left(\frac{b}{r} + \frac{3}{2} + \frac{r}{b} + \frac{r^2}{b^2}\right)e^{-2r/b}; \tag{34}$$

$$O_3(H) = \hat{H} + \frac{\hbar^2}{2\mu b^2} + \frac{Z}{b}\left(-\frac{2b}{r} - \frac{1}{2} - \frac{3b^2}{r^2} - \frac{3b^3}{2r^3} + \frac{3b^3}{2r^3}e^{2r/b}\right)e^{-2r/b} + \tag{35}$$
$$\frac{Z}{b}\left(\frac{9b}{r} + 6 + \frac{9b^2}{r^2} + \frac{9b^3}{2r^3} + \frac{r^2}{b^2} + \frac{3r}{b} - \frac{9b^3}{2r^3}e^{2r/b}\right)e^{-2r/b}\cos^2(\theta);$$

$$O_4(H) = \hat{H} + \frac{\hbar^2}{2\mu b^2} + \frac{Z}{b}\left(\frac{5b}{2r} + \frac{5}{2} + \frac{3b^2}{2r^2} + \frac{3b^3}{4r^3} + \frac{3r}{2b} + \frac{r^2}{2b^2} - \frac{3b^3}{4r^3}e^{2r/b}\right)e^{-2r/b} + \tag{36}$$
$$\frac{Z}{b}\left(-\frac{9b}{2r} - 3 - \frac{9b^2}{2r^2} - \frac{9b^3}{4r^3} - \frac{r^2}{2b^2} - \frac{3r}{2b} + \frac{9b^3}{4r^3}e^{2r/b}\right)e^{-2r/b}\cos^2(\theta); \quad O_5(H) = O_4(H).$$

The operators of the square of the angular momentum are respectively:

$$O_1(\boldsymbol{L}^2) = \hat{\boldsymbol{L}}^2 - 2\hbar^2 b^2 \Delta + \frac{2\hbar^2 r^2}{3b^2} + 3\hbar^2, \tag{37}$$

$$O_2(\boldsymbol{L}^2) = \hat{\boldsymbol{L}}^2 - 28\hbar^2 b^2 \Delta + \frac{\hbar^2 r^2}{6b^2} + 3\hbar^2, \tag{38}$$

$$O_3(\boldsymbol{L}^2) = \hat{\boldsymbol{L}}^2 - 24\hbar^2 b^2 \Delta + \frac{\hbar^2 r^2}{10b^2}(2r^2 - z^2) - 5\hbar^2, \tag{39}$$

$$O_4(\boldsymbol{L}^2) = \hat{\boldsymbol{L}}^2 - 18\hbar^2 b^2 \Delta + \frac{\hbar^2 r^2}{20b^2}(3r^2 + z^2) - 5\hbar^2 - 3\hbar\hat{L}_z, \tag{40}$$

$$O_5(\boldsymbol{L}^2) = \hat{\boldsymbol{L}}^2 - 18\hbar^2 b^2 \Delta + \frac{\hbar^2 r^2}{20b^2}(3r^2 + z^2) - 5\hbar^2 + 3\hbar\hat{L}_z. \tag{41}$$

Finally, the operators of the third projection of the angular momentum are:

$$O_1(L_z) = O_2(L_z) = O_3(L_z) = \hat{L}_z; \tag{42}$$
$$O_4(L_z) = \hat{L}_z - \hbar; \quad O_5(L_z) = \hat{L}_z + \hbar.$$

We propose to solve the problem of finding states with minimum dispersion of three observables H, L_z, $\boldsymbol{L}^2$ numerically.

5 The Problem of Finding States with Minimum Dispersion of a Complete Set of Measured Observables

Let us formulate three functionals for finding minimum-dispersion states, not necessarily being eigenfunctions of the appropriate operators [15]:

$$F_1[a,b](H) \equiv \frac{\langle \Psi | O(H^2) | \Psi \rangle}{\langle \Psi | \Psi \rangle} - \left(\frac{\langle \Psi | O(H) | \Psi \rangle}{\langle \Psi | \Psi \rangle} \right)^2 \tag{43}$$

$$F_2[a,b](L_z) \equiv \frac{\langle \Psi | O(L_z^2) | \Psi \rangle}{\langle \Psi | \Psi \rangle} - \left(\frac{\langle \Psi | O(L_z) | \Psi \rangle}{\langle \Psi | \Psi \rangle} \right)^2 \tag{44}$$

$$F_3[a,b](\boldsymbol{L}^2) \equiv \frac{\langle \Psi | O((\boldsymbol{L}^2)^2) | \Psi \rangle}{\langle \Psi | \Psi \rangle} - \left(\frac{\langle \Psi | O(\boldsymbol{L}^2) | \Psi \rangle}{\langle \Psi | \Psi \rangle} \right)^2 \tag{45}$$

Each of functionals (43)–(45) is continuous with respect to a pair of variables $(a, b) \times \Psi$ and convex with respect to Ψ. The variable Ψ belongs to the (rigged) Hilbert space $L_2(\mathbb{R}^3)$. The variables (a, b) satisfy (due to $a_1^2+a_2^2+a_3^2+a_4^2+a_5^2 = 1$) the system of inequalities:

$$\left. \begin{array}{l} 0 \le a_1^2 \le 1; \quad 0 \le a_2^2 \le 1; \\ 0 \le a_3^2 \le 1; \quad 0 \le a_4^2 \le 1; \\ 0 \le a_1^2 + a_2^2 + a_3^2 + a_4^2(\equiv 1 - a_5^2) \le 1 \\ b_0 \le b \le_0^{-1} \text{ (in particular, } b_0 = 0.001) \end{array} \right\} \tag{46}$$

i.e., form the convex set $Q \subseteq (\mathbb{R}^5)$.

For the representation

$$\Psi(\boldsymbol{r}) = \sum_{n=1}^{n_{\max}} \sum_{l=1}^{n-1} \sum_{m=-l}^{l} C_{nlm} \Psi_{nlm}^{(0)}(\boldsymbol{r}). \tag{47}$$

Let us introduce partial sums with the notations:

$$\Psi_m(\boldsymbol{r}) \equiv \sum_{n=|m|+1}^{n_{\max}} \sum_{l=|m|}^{n-1} C_{nlm} \Psi_{nlm}^{(0)}(\boldsymbol{r}), \tag{48}$$

$$\Psi_l(\boldsymbol{r}) \equiv \sum_{n=l+1}^{n_{\max}} \sum_{m=-l}^{l} C_{nlm} \Psi_{nlm}^{(0)}(\boldsymbol{r}), \tag{49}$$

$$\Psi_n(\boldsymbol{r}) \equiv \sum_{l=0}^{n-1} \sum_{m=-l}^{l} C_{nlm} \Psi_{nlm}^{(0)}(\boldsymbol{r}). \tag{50}$$

now let us proceed to the formation of the minimum dispersion functionals for three operators $O(L_z)$, $O(\boldsymbol{L}^2)$, $O(H)$, respectively:

$$F_a(A) = \frac{\langle \Psi | O_a(A^2) | \Psi \rangle}{\langle \Psi | \Psi \rangle} - \left(\frac{\langle \Psi | O_a(A) | \Psi \rangle}{\langle \Psi | \Psi \rangle} \right)^2. \tag{51}$$

Then each of the three functionals

$$F_1(\Psi, a) \equiv \frac{\langle \Psi | O_a(L_z^2) | \Psi \rangle}{\langle \Psi | \Psi \rangle} - \left(\frac{\langle \Psi | O_a(L_z) | \Psi \rangle}{\langle \Psi | \Psi \rangle} \right)^2, \tag{52}$$

$$F_2(\Psi, a) \equiv \frac{\langle \Psi | O_a((L^2)) | \Psi \rangle}{\langle \Psi | \Psi \rangle} - \left(\frac{\langle \Psi | O_a(L^2) | \Psi \rangle}{\langle \Psi | \Psi \rangle} \right)^2, \tag{53}$$

$$F_3(\Psi, a) \equiv \frac{\langle \Psi | O_a(H^2) | \Psi \rangle}{\langle \Psi | \Psi \rangle} - \left(\frac{\langle \Psi | O_a(H) | \Psi \rangle}{\langle \Psi | \Psi \rangle} \right)^2 \tag{54}$$

can be formally presented as depending on three partial sums $\{\Psi_n\}$, $\{\Psi_l\}$, $\{\Psi_m\}$, and the problems of minimum determination will be considered relative to $\{\Psi_m\}$, $\{\Psi_l\}$, $\{\Psi_n\}$, respectively

$$F_1(a, \{\Psi_n\}, \{\Psi_l\}, \{\Psi_m\}) \xrightarrow[\{\Psi_m\}]{} \min \tag{55}$$

$$F_2(a, \{\Psi_n\}, \{\Psi_l\}, \{\Psi_m\}) \xrightarrow[\{\Psi_l\}]{} \min \tag{56}$$

$$F_3(a, \{\Psi_n\}, \{\Psi_l\}, \{\Psi_m\}) \xrightarrow[\{\Psi_n\}]{} \min \tag{57}$$

Note that in the QMK for the operators $\hat{H}$, $\hat{L}^2$, and $\hat{L}_z$ the relation is valid

$$\hat{H} = -\frac{Ze^2}{r} - \frac{\hbar^2}{2\mu r^2} \frac{\partial}{\partial r} \left(r^2 \frac{\partial}{\partial r} \right) + \frac{\hat{L}^2}{2\mu r^2} \tag{58}$$

and

$$\hat{L}^2 = -\hbar^2 \left\{ \frac{1}{\sin\theta} \frac{\partial}{\partial\theta} \left(\sin\theta \frac{\partial}{\partial\theta} \right) - \frac{\hat{L}_z^2}{\hbar^2 \sin^2\theta} \right\}. \tag{59}$$

Let us form two functionals

$$G_1 = \left[\langle \Psi | \left(O(H) + \frac{Ze^2}{r} + \frac{\hbar^2}{2\mu r^2} \frac{\partial}{\partial r} \left(r^2 \frac{\partial}{\partial r} \right) - \frac{O(L^2)}{2\mu r^2} \right) | \Psi \rangle \right] \Big/ \langle \Psi | \Psi \rangle, \tag{60}$$

$$G_2 = \left[\langle \Psi | \left(O(L^2) + \frac{\hbar}{\sin^2\theta} \frac{\partial}{\partial\theta} \left(\sin\theta \frac{\partial}{\partial\theta} \right) - \frac{O^2(L_z)}{\sin^2\theta} \right) | \Psi \rangle \right] \Big/ \langle \Psi | \Psi \rangle. \tag{61}$$

Analogously, it is first possible to form two operators

$$\hat{G}_1(a) \equiv O_a(H) + \frac{Ze^2}{r} + \frac{\hbar^2}{2\mu r^2} \frac{\partial}{\partial r} \left(r^2 \frac{\partial}{\partial r} \right) - \frac{O_a(L^2)}{2\mu r^2} \tag{62}$$

and

$$\hat{G}_2(a) \equiv O_a(H) + \frac{\hbar^2}{\sin^2\theta} \frac{\partial}{\partial\theta} \left(\sin\theta \frac{\partial}{\partial\theta} \right) - \frac{O_a(L_z)}{\sin^2\theta}. \tag{63}$$

Now for these operators one can formulate (a) the eigenvalue and eigenvector problem and (b) the problem of finding the minimum-dispersion states and the dispersion itself in these states. Since the three functionals F_1, F_2, F_3 relate to the dispersion of observables, we choose the following version for the minimum dispersion problem:

$$\hat{G}_1(a) = \left\| (\hat{G}_1 - \langle \Psi | \hat{G}_1 | \Psi \rangle) | \Psi \rangle \right\|^2, \hat{G}_2(a) = \left\| \left(\hat{G}_2 - \frac{\langle \Psi | \hat{G}_1 | \Psi \rangle}{\langle \Psi | \Psi \rangle} \right) | \Psi \rangle \right\|^2. \quad (64)$$

Now we group these two functionals into one

$$\hat{G}(a) = \hat{G}_1(a) + \frac{\hat{G}_2(a)}{(2\mu r^2)^2} \equiv \left\| \left(\hat{G}_1 - \frac{\langle \Psi | \hat{G}_1 | \Psi \rangle}{\langle \Psi | \Psi \rangle} \right) \Psi \right\|^2$$
$$+ \left(\frac{1}{2\mu r^2} \right)^2 \left\| \left(\hat{G}_2 - \frac{\langle \Psi | \hat{G}_2 | \Psi \rangle}{\langle \Psi | \Psi \rangle} \right) \Psi \right\|^2. \quad (65)$$

The extremum problem for this functional

$$G = G(a, \{\Psi_n\}, \{\Psi_l\}, \{\Psi_m\}) \xrightarrow[a]{} \min \quad (66)$$

closes the system F_1, F_2, F_3 of four functionals, which together form a problem of finding the Nash equilibrium point of functionals F_1, F_2, F_3, G with respect to variables $\{\Psi_m\}$, $\{\Psi_l\}$, $\{\Psi_n\}$, a [16].

6   Conclusion

Each quantum system is characterized by a set of statistical experiments, each of which consists of two independent stages: the preparation of the statistical state ρ and the measurement of the observable A. Real observables A are described (in orthodox quantum mechanics) by the Hermitian operators $\hat{A}$ with spectral decomposition $\hat{A} = \sum \int a \, E_a \, da$, where a is the eigenvalues of $\hat{A}$ and $\{E_a\}$ represents its spectral measure. The probability for A to be in a state ρ is equal to $Tr(\rho \hat{A})$.

In the process of measuring the observable A with a measuring instrument, the quantum part of which is in a state ρ_{ap} before the measurement procedure, the probability of detecting the observable A in a state ρ is equal to $Tr \, |_{H_{ap}} [(I \otimes E_a)(\rho_{ap} \otimes \rho)]$. Such a generalization of the Born-Von Neumann statistical quantum postulate from the case of orthodox quantum mechanics (without measurements) to the theory of quantum measurements, leads to a "broadening of spectral lines" and to the appearance of states with minimal dispersion that do not coincide with the "spectral proper states" of operator $O_{\rho_{ap}}$, which are measurable using ρ_{ap} of the observable A. Such "new" states in the model of quantum measurements of the valence electron in a hydrogen-like atom are of great interest, and this work is devoted to the study of their properties.

References

1. Zorin, A.V., Sevastianov, A.L., Sevastianov, L.A.: Application of the noncommutative theory of statistical decisions to the modeling of quantum communication channels. In: Conference: 2017 the 9th International Congress on Ultra Modern Telecommunications and Control Systems, pp. 26–31. IEEE (2017). https://doi.org/10.1109/ICUMT.2017.8255195
2. Zorin, A.V., Sevastianov, L.A., Tretyakov, N.P.: Application of the model of quantum measurments of valence electrons of alcali metals. In: International Congress on Ultra Modern Telecommunications and Control Systems and Workshops, pp. 437–442. IEEE Communications Society (2018)
3. Holevo, A.S.: On the mathematical theory of quantum communication channels. Probl. Inf. Transm. **8**(1), 47–54 (1972)
4. Zorin, A.V., Sevastianov, L.A.: Model of quantum measurements of Kuryshkin-Wodkiewicz. Vestnik RUDN. Matematika Informatika Fizika **3**, 99–104 (2010)
5. Zorin, A.V.: Moments of observables by the Kuryshkin quantization rule. Vestnik RUDN. Matematika Informatika Fizika **4**, 112–117 (2010)
6. Sevastyanov, L., Zorin, A., Gorbachev, A.: Pseudo-differential operators in an operational model of the quantum measurement of observables. In: Adam, G., Buša, J., Hnatič, M. (eds.) MMCP 2011. LNCS, vol. 7125, pp. 174–181. Springer, Heidelberg (2012). https://doi.org/10.1007/978-3-642-28212-6_17
7. Zorin, A.V.: Operational model of quantum measurements of Kuryshkin-Wodkiewicz. Vestnik RUDN. Matematika Informatika Fizika **2**, 43–55 (2012)
8. Sevastianov, L., Zorin, A., Gorbachev, A.: A quantum measurements model of Hydrogen-like atoms in maple. In: Gerdt, V.P., Koepf, W., Mayr, E.W., Vorozhtsov, E.V. (eds.) CASC 2013. LNCS, vol. 8136, pp. 369–380. Springer, Cham (2013). https://doi.org/10.1007/978-3-319-02297-0_30
9. Zorin, A.V.: Model of quantum measurements of a hydrogen-like atom in a rigged Hilbert space. Vestnik RUDN. Matematika Informatika Fizika **4**, 38–45 (2015)
10. Gorbachev, A.V., Zorin, A.V., Sevastianov, L.A.: The model of quantum measurements of Kuryshkin-Wodkiewicz for atoms and ions with one valence electron. Vestnik RUDN. Matematika Informatika Fizika **2**, 44–52 (2016)
11. Sevastianov, L.A., Zorin, A.V.: The computer-based model of quantum measurements. Phys. Atom. Nuclei **80**, 774–780 (2017)
12. Kurishkin, V.V.: Uncertainty principle and the problem of joint coordinate-momentum probability density in quantum mechanics. The Uncertainty Principle and Foundation of Quantum Mechanics, pp. 61–83. Wiley, London, New York (1977)
13. Zorin, A.V., Kurihshkin, V.V., Sevastianov, L.A.: Opisanie spectra vodorodopodobnogo atoma. Vestnik RUDN Fizika **6**(1), 62–66 (1998)
14. Zhidkov, E.P., Zorin, A.V.: Quantum theory with statistical interpretation: the hydrogen-like atom problem. J. Comput. Methods Sci. Eng. **2**(1–2), 293–307 (2002). Zhidkov, E.P., Zorin, A.V., Lovetsky, K.P., Tretyakov, N.P.: Matrix representation in quantum mechanics of nonnegative QDF by the example of a hydrogen-like atom. Dubna: JINR P11–2002-253, 28 p. (2002)
15. Zorin, A.V., Sevastianov, L.A.: States with minimal dispersion observed in Kuryshkin quantum mechanics. Vestnik RUDN Ser. Fizika **10**(1), 65–84 (2002)
16. Lovetsky, K.P., Sevastianov, L.A., Tretyakov, N.P.: An exact penalty function method for solving the full eigenvalue and eigenvector problem. J. Comput. Methods Sci. Eng. **2**(1–2), 189–194 (2002)

Multi-criteria Method for Calculating a SPTA Package for a Mobile Communications Vehicle

Bogdan Pankovsky$^{(\boxtimes)}$ and Sergey Polesskiy

National Research University "Higher School of Economics", Moscow, Russia
bepankovsky@gmail.com

Abstract. The disadvantages of the traditional method of calculating spare parts, tools and accessories package (SPTA) are considered. A multi-criteria methodology for calculating SPTA package has been developed for the simultaneous use of several criteria, such as weight, size or cost. Based on the convolution method, which allows to reduce the multicriterial task to a scalar one. To check the effectiveness of the developed method, SPTA package for a mobile communications vehicle was calculated using the current one-parameter and developed multi-parameter methods. The weight and volume of the required SPTA package was used as limitations. Based on the results obtained, an analysis of the developed method was carried out.

Keywords: SPTA package · Sufficiency · Reliability · Multi-criteria method

1 Introduction

In the age of high technology and almost complete cellular coverage, there are remote places with no cellular communication, or due to temporary circumstances cannot work correctly due to increased loads. At public events outside the city limits, mobile base stations are used to transmit cellular communications or live television broadcasting, as well as in emergency situations to maintain communication. In the field, fault tolerance in the operation of mobile telecommunications facilities is critical - this leads to excessive equipment reliability requirements in order to maintain the required level of reliability, a set of spare parts, tools and accessories (SPTA package) is included into the structure of the facility. The purpose of the usage of the spare parts kit is to restore the equipment, identify and eliminate the causes of malfunctions in the shortest period of time, during the entire period of usage at the site by the staff, taking into account the costs of maintenance, as well as the costs associated with forced idle time of the equipment. There are many approaches to calculating SPTA package, presented in [1,12], they all allow to carry out the calculation using one weight and size criterion; however, for mobile telecommunication objects, several parameters (such as weight and volume) can be equally critical. The existing

© Springer Nature Switzerland AG 2019
V. M. Vishnevskiy et al. (Eds.): DCCN 2019, LNCS 11965, pp. 520–533, 2019.
https://doi.org/10.1007/978-3-030-36614-8_40

methods for calculating the SPTA package use one-parameter techniques, which imposes a number of difficulties to comply with the operational requirements for the SPTA package considering the multi-criteria nature of the task. To maintain the required level of the reserve sufficiency measure or reliability index of the product, SPTA local package and a SPTA group package (warehouse or other product) can be used, «Central warehouse» can be used as the initial source of spare parts, which in turn provides the maintenance station with failing, but still repairable spare parts, if the spare part is not maintainable, it should be replaced.

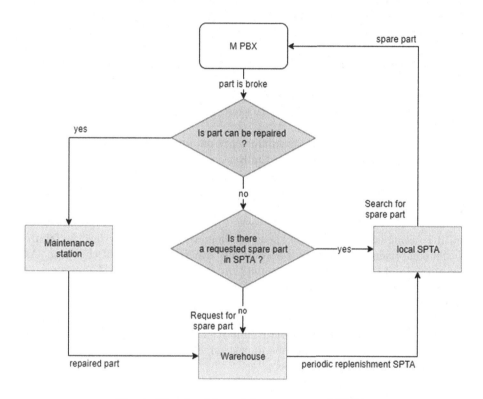

Fig. 1. The algorithm of the operation of SPTA

Additional repair levels are possible depending on the operating conditions of the facility, that is, a system is organized which is part of the facility maintenance and repair system.

2   Analysis of SPTA Methods

We will conduct a small comparative analysis of existing methods for calculating SPTA package. In the method presented by Ushakov [1], three basic schemes of spare parts and accessories are considered: SPTA local package, SPTA group package, and a repair kit. SPTA local package supports the performance of a single product. SPTA group package supports several SPTA local packages. SPTA package supports the performance of maintenance station. In this method, 6 replenishment strategies are considered: periodic replenishment; periodic replenishment with emergency deliveries; repair of failed items; unlimited repair of failed items; replenishment on the level of minimum stock; continuous stock replenishment; continuous replenishment at which the replenishment time is distributed exponentially; continuous replenishment at which the replenishment time is fixed. Solutions of problems are considered where elements of a product fail during idle time and the duration of repair of a failed element is fixed.

There are techniques presented in the textbook of Cherkesov [2], the addition of multi-level SPTA system in which an additional supply level is added to SPTA group package. The techniques presented in the tutorial allow to calculate SPTA package based on the criterion of reliability. One of the drawbacks is the complexity of the calculations and the lack of a universal method, since in this textbook, for each model of SPTA package the formulas and the course of the solution are given.

It is also worth noting the techniques presented in the American standards, which use a different structure from the Russian, they are represented in another textbook of Ushakov [7]. The techniques are based on the principle of prediction modeled on recovery tasks. They are empirical, since they were developed from the condition of consistency with known values of the recovery time for specific systems. The sample of restoration tasks is chosen based on the failure rate, and it is assumed that the time spent on diagnosing and correcting the failure of this component remains the same for any other component of this type. As operational data can show, this assumption is not always true. The system of warehouses is considered, which does not provide for the availability of SPTA package.

But in the course of using the calculation methods a critical drawback is revealed - the inability to use several limitations in the calculation. Only after calculating SPTA package using one limitation, check whether the other requirements of SPTA package meet is availble.

Below is an example of calculating a SPTA local package using the traditional calculation method [8], and a multi-criteria methodology for calculating a SPTA package will be presented. After which a comparative analysis of the obtained calculation results will be presented.

2.1   The Classical Method of Calculating SPTA

The classic objective function for calculating a SPTA is to calculate the reserve availability factor for a SPTA package:

$$K_{SPTA} \rightarrow \begin{cases} MIN_{1 \le i \le n}\{\sum_{i=1}^{n} C_i^{(1)}(x_i), \ldots, \sum_{i=1}^{n} C_i^{(M)}(x_i)\} \\ \prod_{i=1}^{n} R_i(x_i) \ge R_{req.} \end{cases} \tag{1}$$

where: R_i - Intermediate estimate for the SPTA of the i-th spare part, calculated depending on reserve replenishment strategy; C_i - cost of i-th spare part; n_i - amount of spare part of i-th type; $R_{req.}$ - required value of the calculated indicator.

Availability factor for SPTA calculated to be lower than 0,9:

$$K_{av.SPTA} = \prod_{i=1}^{n} K_i \geq 0{,}9 \tag{2}$$

where: K_i - Availability factor for i-th item.

Substitution rate of items typically do not greater than 0,1:

$$\Delta t_{SPTA} = \sum_{i=1}^{n} \Lambda_i \leq 0,1 \tag{3}$$

where: Λ_i - Substitution rate of i-th item.
In general, four types of replenishment strategies are used in SPTA packages:

- periodic replenishment (conditional index $\alpha_i = 1$);
- periodic replenishment with emergency deliveries ($\alpha_i = 2$);
- continuous replenishment ($\alpha_i = 3$);
- replenishment by the level of minimum stock ($\alpha_i = 4$).

In addition to the type (index α_i), each replenishment strategy is characterized by one (T_i) or two (T_i and β_i) numerical parameters that have values:

- at $\alpha_i = 1$ $T_i = T_{per.i}$ - the period of planned replenishment of the i-th stock, $\beta_i = 0$ - the parameter is not used;
- at $\alpha_i = 2$ $T_i = T_{per.i}$ - the period of planned replenishment of the stock, $\beta_i = T_{em.del.i}$ - the time of emergency delivery of the spare part i-th type;
- at $\alpha_i = 3$ $T_i = T_{del.i}(T_{rep.i})$ - delivery (repair) time of spare part i-th type, $\beta_i = 0$ - the parameter is not used;
- at $\alpha_i = 4$ $T_i = T_{del.i}$ $(T_{rep.i})$ - delivery time i-th spare part, $\beta_i = m_i$ - level of minimum stock-type.

Each individual stock in the SPTA kit can be replenished in the general case according to its own separate strategy, which differs from the others both in type (α_i) and in numerical parameters (T_i and β_i). In practical calculations, it is advisable to stock the spare part, which have approximately the same characteristics (replacement intensity, cost, dimensions, the possibility of recovery after a failure, etc.), to combine into groups with the same replenishment strategy.

When choosing the type and strategy of replenishment (if they are not specified in the technical documentation), the following recommendations should be considered:

A periodic replenishment strategy, which provides for replenishment of stocks with a planned periodicity of $T_{per.i}$, should be applied to stocks of non-recoverable spare parts with a relatively low intensity of demand for them and low costs (cost, weight). This strategy should be applied when replenishment of the SPTA kit in the time interval $(0, T_{per.i})$ is technically impossible or involves unreasonably high costs (for example, for SPTA equipment of remote and (or) hard-to-reach objects, SPTA aboard a long-voyage marine vessel, etc.).

The strategy of periodic replenishment with emergency deliveries provides that in addition to the planned replenishment of the stock with a frequency of $T_{per.i}$, in case of failure of the SPTA kit in stock, emergency replenishment can be carried out in time. At the same time, modifications to the strategy are possible.

The formation of the application may occur upon the failure of the SPTA kit or upon failure of the product for this type of parts. Both options coincide in non-reserved products and do not coincide in the presence of structural redundancy. This strategy is recommended to be used for those stocks that when the strategy $\alpha_i = 1$ are expensive (for example, in terms of overall dimensions or cost), that is, for stocks of large-sized expensive and relatively unreliable parts.

The number of spare parts requested in an emergency delivery application can vary from one spare part to a replenishment to an entry level. spare parts may be included in the application only of the type that led to the failure of the spare parts kit (or product failure), and in other cases the application may be multi-item.

It is advisable to apply this strategy when the additional financial costs for the implementation of emergency deliveries are acceptable (for example, at least two times lower than the cost of additional spare parts, which should be included in the SPTA with periodic replenishment).

A continuous replenishment strategy is used for stocks of renewable parts, which are either exchanged at a higher level SPTA or restored at a repair agency and returned to the SPTA kit from which they were removed.

This strategy should also be used in two-tier SPTA systems to replenish stocks in single sets from the group package of the SPTA system.

A characteristic of the continuous replenishment strategy is that a replenishment request is generated for each failed spare part separately, and the delivery (repair) time is then counted from the moment of removal from the SPTA kit (failure of the spare part in the product) and therefore can be significantly less than $T_{per.i}$ with periodic replenishment.

The replenishment strategy for the level of minimum stock provides that every time after the last call to the stock of this type there are spare parts left in it, it is replenished to the initial level over time. This strategy is recommended for stocks of relatively unreliable, but expensive spare parts in cases where long downtimes of the product due to their absence in the SPTA lead to a sharp decrease in efficiency or unacceptable technical and economic losses.

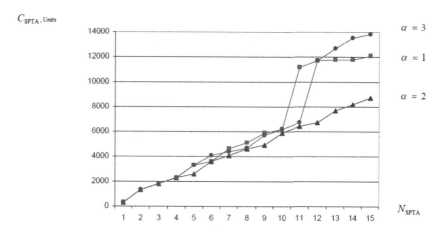

Fig. 2. Dependence of the total number of elements for various replenishment strategies on the cost of SPTA

The results obtained in this example give a complete picture of the formation of SPTA with various strategies for replenishing it. In fact, there may be other, intermediate strategies. But, obviously, any of them will be within the limits limited extreme strategies - periodic replenishment (worst case) and continuous replenishment (best case). Based on the data obtained, the engineer himself can decide on the rational composition of spare parts. If the forecasted situation is closer to the strategy of periodic filling (large remoteness from the warehouse, the inability to quickly deliver for one reason or another), then both factors - reliability requirements and cost limits will be critical factors affecting the required spare part composition. If the possibility of a relatively quick delivery of spare parts is predicted items from the warehouse (small distances, the availability of a cheap vehicle, etc.), then this situation is closer to the strategy of continuous replenishment. In this situation, the requirements for spare parts become less stringent, since due to the fast delivery a high level of reliability is ensured with a significantly smaller amount of spare parts.

In the methods of this standard, this replenishment strategy is used only in SPTA local. In evaluation and calculation models SPTA group package.

The average number of spare part requests received in the SPTA local kit for each type of stock for the replenishment period (delivery, repair time) is calculated by the formula:

$$a_i = k_i \lambda_i T_i \tag{4}$$

where: k_i - the number of i-type parts in the product; λ_i - the intensity of substitutions of an i-th part (or the intensity of demand for an i-type spare part), 1/h;

For each types of replenishment strategies are calculated intermediate estimate (R_i):

(a) periodic replenishment $(\alpha_i = 1)$;

$$R_i(a_i; n_i) = -\ln(\sum_{j=0}^{n_i} a_i^j/j! + 1/a_i[1 - e\sum_{j=0}^{n_i} a_i^j/j!]) \tag{5}$$

(b) periodic replenishment with emergency deliveries $(\alpha_i = 2)$;

$$R_i(a_i; n_i) = -\ln(1 - (T_{em.del.i}/T_i)\sum_{k=1}^{n_i} k[\sum_{j=k(n_i+1)}^{k(n_i+1)+n_i} (a^j/j!)e^{-n_i}]) \tag{6}$$

(c) continuous replenishment $(\alpha_i = 3)$;

$$R_i(a_i; n_i) = -\ln(1 - a_i^{n_i+1}/[(n_i + 1)!\sum_{j=0}^{n_i+1} a_i^j/j!]; \tag{7}$$

(d) replenishment by the level of minimum stock $(\alpha_i = 4)$.

$$R_i(a_i; n_i) = -\ln(1 - \frac{a_i^{m_i+2}}{a_i^{m_i+2} + (n_i - m_i)(1 + a_i)^{m_i+1}}), n_i \geq 2m_i + 2, \tag{8}$$

$$R_i(a_i; n_i) = -\ln(1 - \frac{a_i^{m_i+2}}{(a_i - m_i + n_i)(1 + a_i)^{m_i+1}}), n_i = 2m_i + 1. \tag{9}$$

In the general case, the algorithm for calculating a SPTA local package is as follows:

1. The initial number of spare parts is set based on the requirements, in general, it is set to 0
2. The calculated indicator Δ is calculated for each of the parts presented in the product range

$$\Delta_i = \frac{R_i(n_i; a_i) - R_i(n_i + 1; a_i)}{C_i} \tag{10}$$

Where: C_i - cost of i-th spare part.
3. Maximum of the calculated Δ is chosen;

4. To the spare part with the maximum indicator Δ, one spare part of this type is added;
5. Checks shall determine whether the resulting SPTA package set requirements: if the total costs resulting SPTA package set requirements and exceed the previous iteration satisfies them, then the calculation is over, otherwise we return to step 2.

2.2 Method of Multi-criteria Optimization

Adaptive multi-parameter optimization based on the criteria convolution method [12] allows using several criteria in the calculation at once. The reason for choosing this method, instead of artificial neural networks is the fact that the task of optimizing SPTA package is produced concerning a specific range of products and the use of learning sample is not possible, which directly affect the effectiveness of the methodology. The convolution method allow to reduce a multicriterial problem to a scalar form by introducing a generalized criterion (bringing several criteria to one) by normalizing, and then it is necessary to reduce it to one objective function. The objective function in this case will be as follows:

$$H(L_1..L_n) = \begin{cases} L_1..L_n \max R(L_1..L_n); \\ L_1..L_n \min \sum_{i=1}^{n} C_{iav} L_i \leq C_{avReq}; \end{cases} \tag{11}$$

$$C_{iav} = \frac{C_{1i}}{C_{1iav}} K + \frac{C_{2i}}{C_{2iav}} (K-1); 0 \leq K \leq 1; \tag{12}$$

$$C_{avReq} = \frac{C_{1req}}{C_{1iav}} K + \frac{C_{2req}}{C_{2iav}} (K-1); 0 \leq K \leq 1; \tag{13}$$

Where: R–intermediate calculated reliability indicator; L_i–reserve level of i-th type of SPTA local package (SPTA group package); $C_{1req.}$–cost measure used as a limit; C_{1i}–costs (cost, volume, weight, etc.) for one type of spare part of the i-th type in a set of SPTA local package (SPTA group package); C_{1iav}–averaged cost figure for the presented product range; K–"Weight" coefficient for the cost of spare part.

The criteria are normalized by calculating the average indicator for each of the criteria, based on the data of the product range. For example, to normalize the weight criterion, we find the average weight for the component of the entire product and then we find the ratio of the weight of the component to the average weight, thereby obtaining a normalized weight criterion. After the criteria are normalized, we set the weighting factor, at this stage of the research two criteria are used, the weighting factor is set respectively as a percentage relative to the degree of influence of the restriction on the final result of the calculation, for example, 70% of the first criterion/30% of the second criterion.

3   Object of Study

The object of the study was Mobile Communications Vehicles (MCV), the scheme of the basic configuration is shown in Fig. 3.

To calculate the SPTA package the following initial data nomenclature presented in Table 1 is used. The following limitations are used for SPTA package: weight 7,5 kg and volume 0,75 m^3, SPTA local package availability factor not less than 0.99, reserve replenishment strategy for the SPTA package is selected - continuous replenishment, as closest to the actual operating conditions.

It is worth noting that the required availability of the spare parts kit can be obtained using various spare parts, however, in this case the weight and volume of the final set can vary, the dependence of the volume weight and the availability factor of the spare parts kit is illustrated in the following graph (Fig. 4).

The graph (Fig. 4) was formed on the basis of the data presented above, by calculating intermediate coefficients for the MCV product range and based

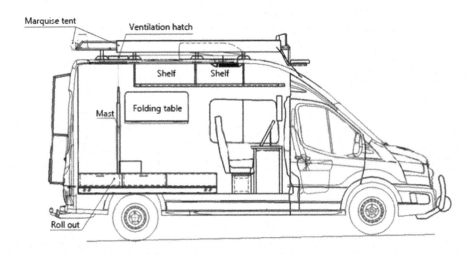

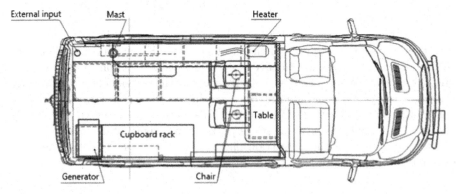

Fig. 3. The basic configuration of a MCV

Table 1. Nomenclature of MCV

No	Name	Amount of components	Replacement intensity, 10^{-6} 1/h	Volume, m^3	Mass of one component, kg	Period of scheduled replenishment
1	Signal conditioner	17	0,000186	1,400000	2,51	8760
2	Block of the ringing signal	1	0,000186	1	0,320000	8760
3	Universal time switch index	3	1,078811	2	15	8760
4	Converter of incoming signals	2	0,114155	0,6	1,500000	8760
5	Converter of outcoming signals	7	0,061764	0,900000	4,250000	8760
6	All-station generator	1	0,126839	0,550000	3,500000	8760
7	Multifrequency barreled receiver	1	0,236050	1,500000	10	8760
8	The central processor upgraded	1	0,236050	1,500000	10	8760
9	Multiplexer of the alarm signaling	1	0,336467	0,550000	3,580000	8760

Table 2. Results of the inverse optimization of SPTA local package

	Calculation criteria	
Number of parts of the nomenclature	weight	size
1	0	0
2	0	0
3	1	1
4	1	0
5	1	0
6	1	0
7	0	0
8	0	0
9	1	1
Size, m^3	1,150000	0,637500
Weight, kg	6,957500	4,645000
Amount of spare parts, pcs	5	2
Availability ratio, rel. units	0,995445	0,988621

on the presented parameters of weight and volume, a graph was formed that shows that, based on the product range, various sets of SPTA can be formed to meet the requirements sufficiency, while having different characteristics of weight and volume Based on the data presented in Table 1, SPTA calculations were performed using the classical method. Using a weight equal to 7,5 kg as a restriction in the first case, and in the second variant a restriction equal to 0,75 m^3 was used. The results of calculations, with the number of spare parts of each type, as well as the results of the characteristics of the sets, are presented in Table 2.

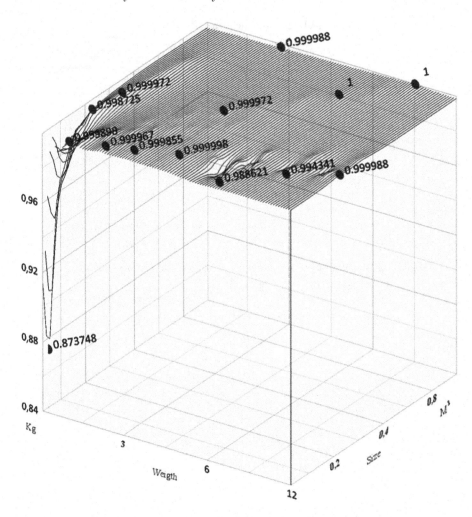

Fig. 4. Dependence of the intermediate calculation indicator on weight and volume

Since the methodology uses a single-criterion optimization method, the results obtained with its help can satisfy the requirements for weight, but not be suitable for the volume. At the same time, when calculating a SPTA package using volume as a limitation, we obtained a SPTA package that meets all the requirements set except the reserve availability factor, which is critical, such a result is possible when adding a part to the package will inevitably exceed the requirements, in this case volume. At the moment, in all the presented methods there is no consideration of an additional limitation, due to the limited mathematical apparatus, which makes it difficult to form a set that meets all the requirements. Therefore, there is a need for a multi-criteria optimization methodology that takes into account several limitations at once.

Table 3. The results of the calculation of the formation of SPTA package with the use of multi-criteria optimization techniques

	Weight coefficients in a weight/volume ratio		
No. of parts of the nomenclature	75%/25%	50%/50%	25%/75%
1	0	0	0
2	0	0	0
3	1	1	1
4	1	1	1
5	1	1	1
6	1	0	0
7	0	0	0
8	0	0	0
9	1	1	0
Size, m^3	1,150000	1,012500	0,875000
Weight, kg	6,957500	6,082500	5,187500
Amount of spare parts, pcs	5	4	3
Availability ratio, rel. units	0,995445	0,994341	0,991423

Below are summarized data (Table 3) using a multi-criteria optimization technique, using as a weight coefficient the percentage ratio of weight to volume in proportions: 75%/25%; 50%/50%; 25%/75%. The results are presented for each of the options of packages.

4   Comparative Analysis

The results obtained (Table 4) show that in some cases it is impossible to achieve the required values of SPTA packages that meets all the requirements, however, using a multiparameter calculation method allows the decision maker to obtain

Table 4. Summary table of the results of calculations for the formation of SPTA package

	Single criteria method		Multi-criteria method		
	Weight	Size	75%/25%	50%	25%/75%
Size, m^3	1,150000	0,637500	1,150000	1,012500	1,150000
Weight, kg	6,957500	4,645000	6,957500	6,082500	6,957500
Amount of spare parts, pcs	5	2	5	4	3
Availability ratio, rel. units	0,995445	0,988629	0,995445	0,994341	0,991423

SPTA package with minimum deviations from the stated requirements, based on the sample of SPTA packages. These manipulations are difficult and require more calculations using the one-parameter method, due to the need to re-calculate SPTA package for each of the criteria separately. Which in turn confirms the appropriateness of the use presented method.

5   Conclusion

The traditional method does not always meet all the requirements, due to which it is difficult to calculate a SPTA package. Therefore, there is a need to use a multi-criteria method that will allow you to get a spare parts kit with minimal deviations from the stated requirements, which can be adjusted by weighting factors in order to obtain the maximum permissible option. As a test of the developed method, a test calculation was made of a SPTA package for a MCV, the results of which were used to determine that this technique can be used for effective calculation, which has the following advantages:

- Reduces the time and complexity of calculations, due to the reduction in the number of calculations for a multicriteria task, through the use of a multicriteria method.
- This method allows you to visually analyze the final results of the calculation, without using additional sources or calculations.

Based on these results, we can conclude about the applicability of the concept of multicriteria method for calculating SPTA packages.

References

1. Ushakov, I.: Kurs teorii nadezhnosti sistem [Course in the theory of system reliability]. The Drofa, Moscow (2008)
2. Cherkesov, G.: Otsenka nadezhnosti s uchetom ZIP: ucheb. posobie [Evaluation of reliability with allowance for spare parts: Proc. allowance]. BHV-Petersburg, St. Petersburg (2012)
3. Hu, Q., Boylan, J.E., Chen, H., Labib, A.: OR in spare parts management: a review. Eur. J. Oper. Res. **266**, 395–414 (2018). https://doi.org/10.1016/j.ejor.2017.07.058
4. Hekimoğluab, M., Van der Laana, E., Dekker, R.: Markov - modulated analysis of a spare parts system with random lead times and disruption risks. Eur. J. Oper. Res. **269**, 909–922 (2018). https://doi.org/10.1016/j.ejor.2018.02.040
5. Feng, Y., Liu, Y., Xue, X., Lu, C.: Research on configuration optimization of civil aircraft spare parts with lateral transshipments and maintenance ratio. JNWPU **36**(6), 1059–1068 (2019). https://doi.org/10.1051/jnwpu/20183661059
6. Pan, G., Luo, Q., Li, X., Wang, Y.: Model of spare parts optimization based on GA for equipment. In: 2018 3rd International Conference on Modelling, Simulation and Applied Mathematics, pp. 44–47 (2018) https://doi.org/10.2991/msam-18.2018.10
7. Ushakov, I., Gnedenko, B.: Probabilistic Reliability Engineering. A Wiley-Interscience Publication, USA (1995). https://doi.org/10.1002/9780470172421

8. Zhadnov, V., Karapuzov, M., Kulygin V., Polesskiy, S.: Sravnenie lokalnykh vychis-litelnykh setey po kriteriyu trebovaniy k komplektam zapasnykh chastey [Compar-ison of local computer networks by the criterion of requirements to sets of spare parts]. Vestnik kompyuternykh i informatsionnykh tekhnologiy [Herald of com-puter and information technologies], no. 4, pp. 36–44 (2015)

9. Epstein, J., Ivry, O.: Spare parts supply chain shipment decision making in a deterministic environment. Mod. Manag. Sci. Eng. **5**(1), 10 (2017)

10. Mansik, H., Burcu, B., Charles, P.: End-of-life in-ventory control of aircraft spare parts under performance based logistics. Int. J. Prod. Econ. **204**, 186–203 (2018)

11. Savin, K., Khamkhanova, D.: Attestatsiya algoritmov opredeleniya vesovykh koef-fitsienty pokazateley kachestva [Attestation of algorithms of determination of weight quantity of quality indexes]. Sovremennye problemy nauki i obrazovaniya [Modern problems of science and education], no. 6 (2012)

12. Avdeev, D., Zhadnov, V.: Avtomatizatsiya proektirovaniya sistem ZIP [Automa-tion of design of spare parts systems]. Novye informatsionnye tekhnologii i menedzhment kachestva (NIT & QM). Materialy mezhdunarodnogo foruma. [New information technologies and quality management (NIT & QM). International forum], pp. 130–133. "Quality" Foundation, Moscow (2009)

13. Author, F.: Article title. Journal **2**(5), 99–110 (2016)

14. Author, F., Author, S.: Title of a proceedings paper. In: Editor, F., Editor, S. (eds.) CONFERENCE 2016, LNCS, vol. 9999, pp. 1–13. Springer, Heidelberg (2016). https://doi.org/10.10007/1234567890

Leaky Modes in Laser-Printed Integrated Optical Structures

A. A. Egorov[1], D. V. Divakov[2], K. P. Lovetskiy[2], A. L. Sevastianov[2],
and L. A. Sevastianov[2,3(✉)]

[1] A.M. Prokhorov General Physics Institute, Russian Academy of Sciences,
Moscow, Russia
yegorov@kapella.gpi.ru

[2] Peoples' Friendship University of Russia (RUDN University),
6 Miklukho-Maklaya Street, Moscow 117198, Russia
{divakov-dv,lovetskiy-kp,sevastianov-al,sevastianov-la}@rudn.ru

[3] Joint Institute for Nuclear Research (Dubna),
Joliot-Curie, 6, Dubna, Moscow Region 141980, Russia

Abstract. Some features of a promising technology for creating complex profiles of smoothly irregular and stepped integrated optical waveguide structures, namely, the technology of laser printing, are briefly discussed. The relevance and importance of this direction is due to the widespread and promising technology of femtosecond recording, as well as the active use of elements (chips) created in this way in integrated optics and nanophotonics. Three-dimensional photon schemes have a wide range of applications, from quantum information processing and miniature lasers to opto-mechanics and optical fluids.

When designing bulk integrated optical structures, complex problems arise, both of a theoretical nature and of numerical simulation of the waveguide propagation of optical radiation, because the guided waveguide modes experience radiation and leakage losses. In this regard, the creation of new methods for theoretical and numerical analysis of waveguide processes, in particular those related to the leakage of modes, is undoubtedly important and relevant in the development of the technology of "laser printing".

Our study shows that, first of all, the radiation outflow process should be considered as a wave process, and models should be built based on wave equations. Constructing a rigorous theory of waveguide leakage processes will make an important theoretical contribution to the theory of integrated optical waveguides and will contribute to improving the technology of "laser printing".

The publication has been prepared with the support of the "RUDN University Program 5-100"(Sevastianov L.A., mathematical model development). The reported study was funded by RFBR, project number 19-01-00645 (Egorov A.A., physical model development). The reported study was funded by RFBR, project number 18-07-00567 (Divakov D.V., numerical analysis).

V. M. Vishnevskiy et al. (Eds.): DCCN 2019, LNCS 11965, pp. 534–547, 2019.
https://doi.org/10.1007/978-3-030-36614-8_41

Keywords: Photonics · Laser-written waveguides · Dielectric
waveguides · Eigenvalues and eigenfunctions · Leaking modes ·
Sensors · Optofluidics · Computer simulation · Numerical modelling

1   Laser-Printed Integrated Optical Structures

In recent years, the technology of "laser printing" has been applied to create integrated optical waveguide structures having complex profiles (smoothly irregular and stepped) [1,2]. The method consists in forming optical waveguides by selectively affecting the volume of glass (or crystal) by a femtosecond laser, thereby creating a non-uniform waveguide core with an increased refractive index. Alternatively, when exposed to radiation, a waveguide cladding with a lower refractive index can be created. As a result, a femtosecond microprocessing forms a certain spatial profile of the refractive index of the glass, providing a waveguide regime. Similar structures can also be formed in multilayer waveguides. The relevance of this direction is due both to the wide spreading and prospects of femtosecond recording technology, and to the active use of similar structures in integrated optics and nanophotonics, including the creation of new promising elements (chips) of waveguide optical memory [3–6].

Optical waveguide structures with a complex architecture located on the chip surfaces suffer from signal crosstalk between intersecting channels [7]. New interconnect architectures, such as multilayer waveguides, are proposed as solutions to some of these problems.

Due to nonlinear processes occurring during ultrafast interaction of laser radiation and material [8,9], laser radiation can cause such a change in the refractive index in the volume of transparent materials, which allows waveguides to be made in glasses, polymers, lithium niobate and other crystals [10], with the prospects of important applications in integrated optics [10,11]. As a result, waveguiding channels are formed in the volume of the material, that can be narrowing, expanding, or look as shown in Figs. 1a and b.

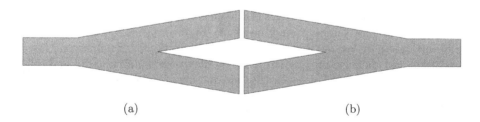

(a)   (b)

Fig. 1. Examples of waveguide channels printed by a femtosecond laser

In Ref. [1], a method for manufacturing optical waveguides in bulk materials using femtosecond laser burning out was demonstrated. This method allows real-time control of the parameters of the forming beam during recording, which facilitates the fabrication of more complex structures than using non-adaptive methods.

In Ref. [2], a three-dimensional conical mode-selective coupler in a photonic microcircuit was demonstrated. This ultra-wideband mode multiplexer based on a waveguide was manufactured applying the technology of direct femtosecond laser recording to a boroaluminosilicate glass chip. Linear cascades of such devices on a single photonic chip can be the dominant technology in the implementation of broadband multiplexing with mode separation to increase the throughput of optical fiber.

Special attention should be paid to the work of the Department of Physics of the University of Stockholm, which demonstrated the first experimental realization of the angular states created from classical light in three-dimensional photon structures recorded in glass samples using femtosecond laser technology [5]. In this paper, 3D optical waveguiding structures were fabricated using laser printing, which form optical fields with extremely specific and promising properties [5].

In the field of creating complex waveguide optoelectronic photonic devices, a femtosecond laser has recently been used to write irregular integrated optical structures from various transparent (dielectric) materials, such as glass, crystals and polymers, due to the unique ability to locally modify the refractive index near the focal spot. This made it possible to create three-dimensional photon schemes for a wide range of applications, from quantum information processing and miniature lasers to opto-mechanics and optical fluids [12,13].

Optofluidics is an active and rapidly expanding field of technology due to the many advantages of optofluidic devices over other photon structures.

(i) They can be easily integrated with other optical functional elements, allowing complete signal processing and creation of complete laboratory devices on a single chip.

(ii) By replacing the fluid the characteristics of the device can be easily changed.

(iii) They use the properties of liquids that have no analogue in solid-state systems.

To take advantage of these properties and to be able to make the required devices, a practically feasible configuration is necessary, for example, a microchannel with additional inlets that allow the efficient use of various liquids. Femtosecond laser irradiation and chemical etching (FLICE) is quickly becoming a useful and reliable micromachining technology for the manufacture of embedded microchannels in glass.

As a rule, a change in the refractive index of about $\sim 10^{-4}$-10^{-3} can be achieved by irradiating the inside of a glass or crystal sample with highly focused femtosecond laser pulses. This is sufficient to obtain single-mode waveguides with low propagation losses. An increase in the refractive index requires a stronger modification in the areas irradiated by a laser, which can be realized either at a higher energy flux density or by optimizing the composition of the substrate materials. The difficulty of forming large changes in the refractive index, which are smoothly distributed in the laser-irradiated regions, is a major obstacle to the creation of compact integrated photon devices.

These experimental works bring to the fore the need for a theoretical study of 3D integrated optical structures, providing an adequate three-dimensional vector description of waveguide fields, their transformation during propagation and interaction with the wave guiding structure (optical chip).

Big problems are facing the attempts of the theoretical description (including computer simulation) of the waveguide propagation of optical radiation in such integrated optical structures. Schematically, areas of maximum irregularity can be presented as shown in Fig. 1a, b. When encountering such irregular sections, guided waveguide modes (stationary before that) experience radiation losses and leakage. To account for such phenomena, in addition to guided modes, one should be able to describe radiative and leaky modes using the technologies of mathematical modelling and computer simulation.

2 Historical Review of Leaky Modes Studies

In the 70s of the last century, Snyder et al. published a series of papers on leaky modes and their use. Ogusu et al. described the "leaky" modes of a planar waveguide in Marcuse's notation. Basing of these results, Ogusu et al. presented a primary "operator" description of the properties of "leaky" waveguide modes [14]. The authors of Ref. [15] noted that the nature of the leaky modes is "equivalent" to that of tunneling in quantum mechanics. In recent decades, there has been a renewed interest to the study (theoretical and experimental) of leaky waves of dielectric waveguides and waveguides made from metamaterials. While foreign researchers pay more attention to experimental studies, Russian scientists are mainly engaged in theoretical research.

Oliner et al. [16–18] performed a thorough study of the complex roots of the dispersion equation that do not correspond to the guided modes. The study begins with the assumption that exponentially damped waves correspond to such roots, the experimental observation of which was reported in [19–21]. First, with the help of the ray technique, and then with the help of the mode analysis, the authors analyze the solutions corresponding to the four different roots of the fourth-order dispersion equation. Two of these roots correspond to solutions, exponentially decreasing in the direction of propagation and located symmetrically with respect to the coordinate axes and the origin. Many publications of that time were devoted to the analysis of the relative position of the variety of roots. Each of these roots specifies the solution running out at an angle θ (see below) to the guide surface. As the distance from the waveguide layer boundary increases, the corresponding solutions increase exponentially. There are an infinite number of such roots and the corresponding solutions (however, there is a finite number for each fixed angular outflow cone). Their complex "wave numbers" are arranged in pairs in the plane of "propagation" of outgoing waves.

As well as guided modes, the resulting waves are excited by a certain source of finite power, so that from an infinite number of admissible (from a mathematical point of view) solutions under specific physical conditions only a few "first" ones are implemented. These "slowly leaking waves" do not increase exponentially in the direction of "their exit" and can be supported by sources of a sufficiently large but finite power. The roots of the dispersion equation, which determine "outgoing waves", form "runaway" inhomogeneous waves at a distance (at an outflow angle) from the waveguide surface. This characteristic of the roots was taken for the definition of "outgoing waves" in the 80s of the last century and is currently practiced [27].

In quantum physics, such solutions of the stationary Schrödinger equation are called Gamow resonances [22,23] or Siegert pseudo-states [24–26].

3   Description of the Model of Leaky Modes Based on the Helmholtz Equation

Maxwell's equations in an isotropic dielectric medium with a refractive index $n : n^2 = \varepsilon_{rel}$ in the absence of external charges and currents are written in the form:

$$rot\,\boldsymbol{H} = \varepsilon_0 n^2 \frac{\partial \boldsymbol{E}}{\partial t}, \quad rot\,\boldsymbol{E} = -\mu_0 \frac{\partial \boldsymbol{H}}{\partial t}. \tag{1}$$

The waveguide structure consisting of three dielectrics is shown in Fig. 2

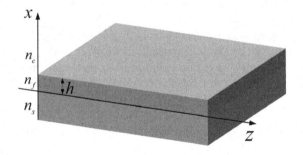

Fig. 2. Geometry of the dielectric waveguide structure

Waveguide polarized monochromatic radiation propagating along the axis Oz with frequency ω is sought in the form: $\boldsymbol{E}, \boldsymbol{H} \sim e^{i\omega t}e^{-i\beta z}$. Under the assumption $\partial/\partial y \equiv 0$, the system of Eq. (1) written in Cartesian coordinates is simplified and separated into two independent subsystems. The subsystem for the TE-modes takes the form:

$$-i\beta H_x - \frac{\partial H_z}{\partial x} = i\omega\varepsilon_0 n^2 E_y, \quad i\beta E_y = -i\omega\mu_0 H_x, \quad \frac{\partial E_y}{\partial x} = -i\omega\mu_0 H_z. \tag{2}$$

We look for the solution of Eq. (2) in the form:

$$E_y = \begin{cases} A\exp\{-\delta x\} & x \geq 0 \\ A\cos(\kappa x) + B\sin(\kappa x) & 0 \geq x \geq -d \\ \{A\cos(\kappa d) - B\sin(\kappa d)\}\exp\{\gamma(x+d)\} & x \leq -d \end{cases} \tag{3}$$

The boundary conditions at the points $x = 0$, $x = -d$ lead to the dispersion equation: $tg\kappa d = \kappa (\delta + \gamma) (\kappa^2 - \delta\gamma)$. After introducing concepts and notation for guided waveguide modes of a planar three-layer dielectric waveguide, Marcuse proceeds to the description of "leaky modes" basing on a single statement: everything is the same, but $\beta :$ $\operatorname{Re}\beta < n_s$ is a complex number, the real part of which is "unacceptable" for guided modes. Then in the substrate, the mode parameters can be written in the form:

$$
\begin{aligned}
\gamma &= \gamma_r + i\gamma_i = k_0 \operatorname{Re}\gamma_s + ik_0 \operatorname{Im}\gamma_s \\
\beta &= \beta_r + i\beta_i = k_0 \operatorname{Re}\beta + ik_0 \operatorname{Im}\beta
\end{aligned}
\tag{4}
$$

Using them, one can write the field E_y in the substrate in the form:

$$
E_y (x, z) = E_s \exp \{k_0 \operatorname{Re}\gamma_s x + k_0 \operatorname{Im}\beta z\} \exp \{i (k_0 \operatorname{Im}\gamma_s x - k_0 \operatorname{Re}\beta z)\}. \tag{5}
$$

Ogusu proposed the following choice of dispersion equation roots for β and square roots in expressions for γ

$$
\begin{aligned}
\operatorname{Re}\gamma_s &< 0, \ \operatorname{Im}\gamma_s > 0 \\
\operatorname{Re}\beta &> 0, \ \operatorname{Im}\beta < 0
\end{aligned}
\tag{6}
$$

Comment. Such a choice of roots is equivalent to setting the asymptotic conditions at the bottom of the substrate in the form of a runaway wave. The same can be done for the top of the waveguide layer, i.e., the asymptotic conditions can be set in the form of a wave running away from the top.

Comment. Ogusu claims that the power of such a waveguide eigenmode is radiated at an angle $\theta = arctg\,(-\operatorname{Im}\gamma_s/\operatorname{Re}\beta\,)$ to the axis Oz, but at this angle, the phase fronts of the escaping waveguide mode propagate. Moreover, Ogusu claims that $-\frac{\operatorname{Im}\gamma_s}{\operatorname{Re}\beta} = tg\theta = -\frac{\operatorname{Im}\beta}{\operatorname{Re}\gamma_s}$. So, the second equality means the orthogonality of the "planar" phase fronts to the lines of constant amplitude of the non-uniform runaway mode in the case of real n_s, n_f, n_c.

In Ref. [27], the evolution of ideas about outgoing waves in waveguide systems in the first decades of studying this phenomenon was described in sufficient detail and consistently. In our opinion, a more rigorous justification of the model of leaky waves of open wave guiding systems can be obtained by starting not from the Helmholtz equation, as is traditionally done, but from the wave equation, which more adequately reflects the wave nature of leaky waves (see [28,29]).

4   Statement of the Problem of Modelling Leaky Modes

In a symmetric three-layer planar waveguide consisting of a dielectric film of thickness h with a real refractive index n_f and surrounded by a covering layer with a real refractive index $n_c < n_f$, the propagation of radiation is described by the Maxwell equations. As the boundary conditions selecting the leaky modes, we consider [28,29] the conditions of "outgoing waves", corresponding to the Gamow-Siegert model [22–26].

The generally accepted model of the electromagnetic field in the planar (infinitely extended along the axis Oy) structure implies the fields independent of the variable y. In this case, Maxwell's equations are considerably simplified. Since $\partial E_\alpha / \partial y = \partial H_\alpha / \partial y \equiv 0$ for any $\alpha = x, y, z$, they are separated into two independent subsystems, corresponding to the so-called TE and TM modes. The subsystem for the TE mode can be presented as a single wave equation for the leading component E_y

$$\left(\frac{\partial^2}{\partial x^2} + \frac{\partial^2}{\partial z^2} - \frac{n^2(x)}{c^2} \frac{\partial^2}{\partial t^2} \right) E_y = 0, \tag{7}$$

where c is the electrodynamic constant. The subsystem for the TE-mode also includes two equations relating the components H_x, H_z with the leading component E_y.

In this case, the refractive index depends only on x, which allows separation of the variables in Eq. (7). As a result, we obtain solutions corresponding to the modes propagating in the positive direction of the axis z:

$$E_{yj}^+ \big|_{x>h} = A_{cj}^+ \cdot \exp \left\{ ik_0 \sqrt{n_c^2 - \beta_j^2} \, x + ik_0 \beta_j z - i\omega t \right\} \tag{8}$$

$$E_{yj}^+ \big|_{\substack{x<h \\ x>0}} = A_{fj}^+ \cdot \exp \left\{ ik_0 \sqrt{n_f^2 - \beta_j^2} \, x + ik_0 \beta_j z - i\omega t \right\}$$
$$+ B_{fj}^+ \cdot \exp \left\{ -ik_0 \sqrt{n_f^2 - \beta_j^2} \, x + ik_0 \beta_j z - i\omega t \right\} \tag{9}$$

$$E_{yj}^+ \big|_{x<0} = B_{cj}^+ \cdot \exp \left\{ -ik_0 \sqrt{n_c^2 - \beta_j^2} \, x + ik_0 \beta_j z - i\omega t \right\} \tag{10}$$

In each subdomain $x > h$, $0 < x < h$ and $x < 0$, the solution of the wave equation corresponding to the resulting modes can be presented as a wave with complex wave vector. In the case of a field corresponding to leaky modes running in the positive direction of the z-axis for $x > h$ and $x < 0$ due to the symmetry of the waveguide, the wave vector is determined as

$$\boldsymbol{k}_j^\pm = k_0 \begin{pmatrix} \pm\sqrt{n_c^2 - \beta_j^2} \\ \beta_j \end{pmatrix} \tag{11}$$

and in the waveguide layer ($0 < x < h$) there are two waves with the wave vectors

$$\boldsymbol{k}_{fj}^\pm = k_0 \begin{pmatrix} \pm\sqrt{n_f^2 - \beta_j^2} \\ \beta_j \end{pmatrix}, \tag{12}$$

the magnitudes of the wave vectors being equal to the corresponding wave numbers $\big| \boldsymbol{k}_j^\pm \big| = k_0 n_c$ and $\big| \boldsymbol{k}_{fj}^\pm \big| = k_0 n_f$. The problem of finding solutions corresponding to the following modes is formulated as an eigenvalue problem for a differential operator with non-self-adjoint boundary conditions.

5   Analysis of Leaky Modes in Terms of Inhomogeneous Waves

We will explicitly consider complex quantities $p_{cj} = \sqrt{n_c^2 - \beta_j^2} = p'_{cj} + ip''_{cj}$, $p_{fj} = \sqrt{n_f^2 - \beta_j^2} = p'_{fj} + ip''_{fj}$ and $\beta_j = \beta'_j + i\beta''_j$ in the solutions (8)–(10), and separate their real and imaginary parts, which will allow us to reformulate Eqs. (8)–(10) in terms of inhomogeneous waves, in which the amplitude is also a function of coordinates x and z:

$$E_{yj}^+\big|_{x>h} = A_{cj}^+ \cdot \exp\left\{-k_0\left|p''_{cj}x + \beta''_j z\right|\right\} \cdot \exp\left\{ik_0 p'_{cj}x + ik_0\beta'_j z - i\omega t\right\}, \quad (13)$$

$$E_{yj}^+\big|_{\substack{x<h \\ x>0}} = A_{fj}^+ \cdot \exp\left\{-k_0 p''_{fj}x - k_0\beta''_j z\right\} \cdot \exp\left\{ik_0 p'_{fj}x + ik_0\beta'_j z - i\omega t\right\}$$
$$+ B_{fj}^+ \cdot \exp\left\{k_0 p''_{fj}x - k_0\beta''_j z\right\} \cdot \exp\left\{-ik_0 p'_{fj}x + ik_0\beta'_j z - i\omega t\right\}, \quad (14)$$

$$E_{yj}^+\big|_{x<0} = A_{sj}^+ \cdot \exp\left\{k_0\left|p''_{cj}x - \beta''_j z\right|\right\} \cdot \exp\left\{-ik_0 p'_{cj}x + ik_0\beta'_j z - i\omega t\right\}. \quad (15)$$

In the obtained representation, the amplitudes of the inhomogeneous wave have the form:

$$E_{yj}^+\big|_{x>h} = A_{cj}^+ \cdot \exp\left\{-k_0\left|p''_{cj}x + \beta''_j z\right|\right\}, \quad (16)$$

$$E_{yj}^+\big|_{x<0} = A_{sj}^+ \cdot \exp\left\{-k_0\left|p''_{cj}x - \beta''_j z\right|\right\} \quad (17)$$

When $-k_0 p''_c x - k_0\beta'' z = 0$ and $p''_{cj}x - \beta''_j z = 0$, these expressions turn into constants A_c and A_{sj}^+, and within the cone $-k_0 p''_c x - k_0\beta'' z < 0$ and $p''_{cj}x - \beta''_j z \leq 0$ they decrease. Inside this cone, the traveling wave propagates along the direction given by the wave vectors $\boldsymbol{k}_c^+$, $\boldsymbol{k}_s^-$ with amplitudes decreasing in the vertical direction, at the boundary of the cone the wave propagates without decreasing and without increasing amplitude. The parameters of the cone $-k_0 p''_c x - k_0\beta'' z = 0$ and $p''_{cj}x - \beta''_j z = 0$ for the specific parameters of the waveguide will be investigated numerically.

Figure 3 shows the eigenvalues corresponding to the resultant modes of the considered symmetric waveguide in the region $n_c - 1 < \mathrm{Re}\,(\beta) < n_c$; the numerical values of the first 4 eigenvalues are given below:

$$\begin{aligned} \beta_1 &= 1.40442451178397 + 0.168764809837465i, \\ \beta_2 &= 1.15214250490115 + 0.495480866495216i, \\ \beta_3 &= 0.914455786974546 + 1.032045559681022i, \\ \beta_4 &= 0.826677810957949 + 1.63034755349560i. \end{aligned} \quad (18)$$

In the papers of the 70s, the fact of the increase in the amplitude of the leaky modes of a planar dielectric waveguide in the vertical direction and the asymptotic conditions contradicting it were discussed. In private conversations,

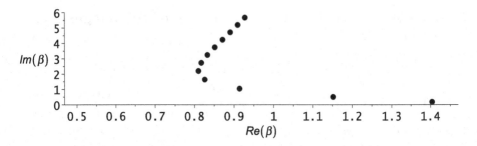

Fig. 3. The dots on the complex plane denote eigenvalues corresponding to the resulting modes of a symmetric waveguide with $n_c = 1.47$, $n_f = 1.565$, $\lambda = 0.55\mu$m, and $h = 1.1\lambda$

experimenters said that actually the flowing out energy density decreases due to the factors not taken into account in the model. The amplitude really decreases with distance from the waveguide layer after a certain distance. Some theorists believed that the use of geometric diffraction methods describing the region of the penumbra could allow a more rigorous description of this decrease. Perhaps the first were the works by Fock on the scattering theory [30]. Instead of a cumbersome analysis of the above factors, not knowing in advance that they correctly describe the experimentally observed behavior of the leaky mode amplitude in the vertical direction, we took into account the fact that it decreases "after a certain distance" in the model (13)–(15). This model dependence on the vertical coordinate causes the solution to decrease exponentially outside the guide cone in the vertical direction.

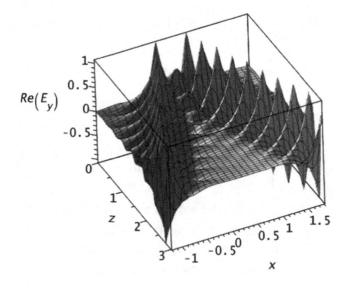

Fig. 4. The type of the instantaneous amplitude of the leaky mode corresponding to the first complex value $\beta_1 = 1.40442451178397 + 0.168764809837465i$. The distance along the axes x and z is given in microns. The value of the field component E_y is normalized - i.e. in dimensionless units.

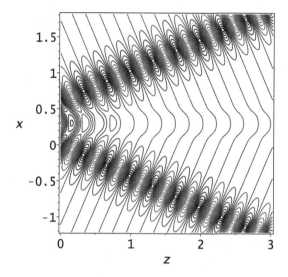

Fig. 5. Projection of the instantaneous amplitude value of Fig. 4 on the horizontal plane. The distance along the axes x and z is given in microns.

For clarity, in Fig. 4 we present the instantaneous values of the amplitude of the leaky mode, corresponding to the first complex value.

The projection of this amplitude onto the horizontal plane is shown in Fig. 5. The amplitude profile calculated according to the model (13)–(15) coincides qualitatively with the experimental data published in [31] and with the experimental observations of the first co-author of this work in smoothly irregular wave guiding liquid-crystal structures.

Such a solution of the problem (13)–(15) decreases in the domain, external with respect to the "guiding cone", mirror-symmetrically with respect to its inner region. In this case, the picture of local "maxima and minima" of the electric field strength coincides very well with those in Figs. 8, 10, 13 from Ref. [31].

6   Conclusion

In 1996, Hirao demonstrated a new method of processing materials by optical radiation for manufacturing of waveguides in glass using a femtosecond laser [47]. Since then, this technology has been actively developed, and in recent years, numerous photonic devices have been reported on various optical materials, including crystals. It is widely recognized that the rapid absorption of laser energy inside bulk glass and its subsequent scattering causes a change in material properties [48,49]. The degree of change in material properties is determined by various parameters, such as laser energy, laser pulse duration, repetition rate, light wavelength, volume of the material being processed, light polarization, etc. [50,51]. It is important to emphasize that the manufacture of photonic devices requires a laser beam with a controlled supply of energy inside the material volume being processed [48,52].

When waveguide propagation of electromagnetic radiation along a laser-printed integrated optical structure with smoothly irregular thicknesses and directions of waveguide channels, the regime of guided waveguide modes is sometimes changed for the regime of ingoing or outgoing waves, which later are transformed again into guided modes. To take into account the possibility of such transformations during the propagation of radiation in the simulation of laser-printed integrated optical structures, it is necessary to use the formalism of leaky modes. In this regard, such an analysis is important both from the point of view of solving the most significant problem radiation losses, including in the presence of interference (see, e.g., [32–38]), and from the point of view of using waveguides as sensors, including those incorporated in complex integrated optical fluid devices (see, e.g., [38–46]). This paper discusses the guided modes that exist when the waveguide thickness is below the critical one. Such modes are referred to as leaky modes. These waves occur when the thickness of the waveguide layer is below the critical value, and the waveguide properties cannot take place in the usual sense. The leaky waves propagate due to the effect of disturbed total internal reflection: during each act of disturbed total reflection at the interfaces of the media forming the waveguide, some of the power of this guided mode is radiated, i.e. "flows out" into the space surrounding the waveguide. The methods developed by us are undoubtedly useful for theoretical and numerical studies of dielectric and, in particular, optical waveguides supporting leaky modes, for example, when used as basic elements in the development of advanced sensors or various interface elements in integrated optical processors.

References

1. Salter, P.S., Jesacher, A., Spring, J.B., Metcalf, B.J., et al.: Adaptive slit beam shaping for direct laser written waveguides. Opt. Lett. **37**, 470–472 (2012)
2. Gross, S., Riesen, N., Love, J.D., Withford, M.J.: Three-dimensional ultra-broadband integrated tapered mode multiplexers. Laser Photonics Rev. **8**(5), L81–L85 (2014)
3. Heilmann, R., Greganti, C., Gräfe, M., Nolte, S., Walther, P., Szameit, A.: Tapering of fs Laser-written Waveguides (2017). arXiv:1707.02941
4. Corrielli, G., Seri, A., Mazzera, M., Osellame, R., de Riedmatten, H.: An Integrated Optical Memory based on Laser Written Waveguides (2015). arXiv:1512.09288v1
5. El Hassan, A., Kunst, F.K., Moritz, A., Andler, G., Bergholtz, E.J., Bourennane, M.: Corner states of light in photonic waveguides (2018). arXiv:1812.08185v1
6. Pavlov, I., Tokelet, O., Pavlova, S., et al.: Femtosecond laser written waveguides deep inside silicon. Opt. Lett. **42**, 3028–3031 (2017)
7. Sherwood-Droz, N., Lipson, M.: Scalable 3D dense integration of photonics on bulk silicon Opt. Express **19**(18), 17758–17765 (2011)
8. Stuart, B.C., Feit, M.D., Herman, S., Rubenchik, A.M., Shore, B.W., Perry, M.D.: Nanosecond-to-femtosecond laser-induced breakdown in dielectrics. Phys. Rev. B **53**(4), 1749–1761 (1996)

9. Carr, C.W., Radousky, H.B., Rubenchik, A.M., Feit, M.D., Demos, S.G.: Localized dynamics during laser-induced damage in optical materials. Phys. Rev. Lett. **92**(8), 087401 (2004)

10. Stone, A., et al.: Direct laser-writing of ferroelectric single-crystal waveguide architectures in glass for 3D integrated optics. Sci. Rep. **5**, 10391 (2015)

11. Corrielli, G., et al.: Rotated waveplates in integrated waveguide optics. Nat. Commun. **5**, 4249 (2014)

12. Bellouard, Y., Said, A.A., Bado, P.: Integrating optics and micro-mechanics in a single substrate: a step toward monolithic integration in fused silica. Opt. Express **13**, 6635 (2005)

13. Osellame, R., Hoekstra, H.J.W.M., Cerullo, G., Pollnau, M.: Femtosecond laser microstructuring: an enabling tool for optofluidic lab-on-chips. Laser Photon. Rev. **5**, 442 (2011)

14. Ogusu, K., Miyag, M., Nishida, S.: Leaky TE modes in an asymmetic three-layered slab waveguide. J. Opt. Soc. Am. **70**, 68–72 (1980)

15. Marcuvitz, N.: On field representations in terms of leaky modes or eigenmodes. IRE Trans. Antennas Propag. **4**(3), 192–194 (1956)

16. Goldstone, L.O., Oliner, A.A.: Leaky-wave antennas I: rectangular waveguides. IRE Trans. Antennas Propag. AP **7**, 307–319 (1959)

17. Tamir, T., Oliner, A.A.: Guided complex waves, part I: fields at an interface. Proc. inst. Elec. Eng. **110**, 310–324 (1963)

18. Tamir, T., Oliner, A.A.: Guided complex waves, part II: relation to radiation patterns. Proc. Inst. Elec. Eng. **110**, 325–334 (1963)

19. Barone, S.: Leaky wave contributions to the field of a line source above a dielectric slab. Report R-532-546, PIB-462, Microwave Research Institute, Polytechnic Institute of Brooklyn, 26 November 1956

20. Barone, S., Hessel, A.: Leaky wave contributions to the field of a line source above a dielectric slab-part II. Report R-698-58, PIB-626, Microwave Research Institute, Polytechnic Institute of Brooklyn, December 1958

21. Cassedy, E.S., Cohn, M.: On the existence of leaky waves due to a line source above a grounded dielectric slab. IRE Trans. Microwave Theor. Tech. **9**, 243–247 (1961)

22. Gamow, G.: Zur quantentheorie de atomkernes. Z. Phys. **51**, 204–212 (1928)

23. Bohm, A., Gadella, M., Mainland, B.: Gamow vectors and decaying states. Am. J. Phys. **57**, 1103–1108 (1989)

24. Siegert, A.F.J.: On the derivation of the dispersion formula for nuclear reactions. Phys. Rev. **56**, 750–752 (1939)

25. Tolstikhin, O.I., Ostrovsky, V.N., Nakamura, H.: Siegert pseudo-states as a universal tool: resonances, S matrix, green function. Phys. Rev. Lett. **79**, 2026 (1997)

26. Tolstikhin, O.I., Ostrovsky, V.N., Nakamura, H.: Siegert pseudostate formulation of scattering theory: one-channel case. Phys. Rev. A **58**, 2077 (1998)

27. Monticone, F., Alu, A.: Leaky-wave theory, techniques, and applications: from microwaves to visible frequencies. Proc. IEEE **103**(5), 793–821 (2015)

28. Divakov, D., Drevitskiy, A., Egorov, A., Sevastianov, L.: Numerical modeling of leaky electromagnetic waves in planar dielectric waveguides. Proc. SPIE **11066**, 110660R (2019)

29. Divakov, D., Tiutiunnik, A., Sevastianov, A.: Symbolic-numeric computation of the eigenvalues and eigenfunctions of the leaky modes in a regular homogeneous open waveguide. MATEC Web Conf. **186**, 4 p., Article ID 01009 (2018)

30. Fock, V.A.: Electromagnetic Diffraction and Propagation Problems. Pergamon Press, London (1965)

31. Martínez-Ros, A.J., Gómez-Tornero, J.L., Clemente-Fernández, F.J., Monzó-Cabrera, J.: Microwave near-field focusing properties of width-tapered microstrip leaky-wave antenna. IEEE Trans. Antennas Propag. **61**(6), 2981–2990 (2013)

32. Egorov, A.A.: Theory of laser radiation scattering in integrated optical waveguide with 3D-irregularities in presence of noise: vector consideration. Laser Phys. Lett. **1**(12), 579–585 (2004). https://doi.org/10.1002/lapl.200410140

33. Egorov, A.A.: Use of waveguide light scattering for precision measurements of the statistic parameters of irregularities of integrated optical waveguide materials. Opt. Eng. **44**(1), 014601-1-10 (2005). https://doi.org/10.1117/1.1828469

34. Egorov, A.A.: Inverse problem of theory of the laser irradiation scattering in two-dimensional irregular integrated optical waveguide in the presence of statistic noise. Laser Phys. Lett. **2**(2), 77–83 (2005). https://doi.org/10.1002/lapl.200410129

35. Egorov, A.A.: Theoretical, experimental and numerical methods for investigating the characteristics of laser radiation scattered in the integrated-optical waveguide with three-dimensional irregularities. Quant. Electron. **41**(7), 644–649 (2011). https://doi.org/10.1070/QE2011v041n07ABEH014560

36. Egorov, A.A.: Study of bifurcation processes in a multimode waveguide with statistical irregularities. Quant. Electron. **41**(10), 911–916 (2011). https://doi.org/10.1070/QE2011v041n10ABEH014683

37. Egorov, A.A.: Theoretical and numerical analysis of propagation and scattering of eigen- and non-eigenmodes of an irregular integrated-optical waveguide. Quant. Electron. **42**(4), 337–344 (2012). https://doi.org/10.1070/QE2012v042n04ABEH014809

38. Egorov, A.A., Shigorin, V.D., Ayriyan, A.S., Ayryan, E.A.: Study of the effect of pulsed-periodic electric field and linearly polarized laser radiation on the properties of liquid-crystal waveguide. Phys. Wave Phenom. **26**(2), 116–123 (2018). https://doi.org/10.3103/S1541308X18020012

39. Egorov, A.A., Egorov, M.A., Tsareva, Y.I., Chekhlova, T.K.: Study of the integrated-optical concentration sensor for gaseous substances. Laser Phys. **17**(1), 50–53 (2007). https://doi.org/10.1134/S1054660X07010100

40. Egorov, A.A., Egorov, M.A., Chekhlova, T.K., Timakin, A.G.: Study of a computer-controlled integrated optical gas-concentration sensor. Quant. Electron. **38**(8), 787–790 (2008). https://doi.org/10.1070/QE2008v038n08ABEH013589

41. Egorov, A.A.: Theory of absorption integrated optical sensor of gaseous materials. Opt. Spectrosc. **109**(4), 625–634 (2010). https://doi.org/10.1134/S0030400X1010022X

42. Egorov, A.A., Andler, G., Sevastyanov, A.L., Sevastyanov, L.A.: On some properties of smoothly irregular waveguide structures critical for information optical systems. Commun. Comput. Inf. Sci. **919**, 387–398 (2018). https://doi.org/10.1007/978-3-319-99447-5

43. Egorov, A., Sevastianov, L., Shigorin, V., Andler, G., Ayriyan, A., Ayriyan, E.: Experimental and numerical study of properties of nematic liquid crystal waveguide structures. IEEE Xplore **8631282**, 448–452 (2019). https://doi.org/10.1109/ICUMT.2018.8631282

44. Wang, Y., H., Li, Zhao, L., Liu, Y., Liu, S., Yang, J.: Tapered optical fiber waveguide coupling to whispering gallery modes of liquid crystal microdroplet for thermal sensing application. Opt. Express **25**(2), 918–926 (2017). https://doi.org/10.1364/OE.25.000918

45. Liu, J.-M.: Photonic Devices. University Press, Cambridge (2005)

46. Rigneault, H., Lourtioz, J.-M., Delalande, C., Levenson, A. (eds.): Nanophotonics. ISTE Ltd. (2006)

47. Davis, K.M., Miura, K., Sugimoto, N., Hirao, K.: Writing waveguides in glass with a femtosecond laser. Opt. Lett. **21**(21), 1729–1731 (1996)
48. Ams, M., et al.: Investigation of ultrafast laser-photonic material interactions: challenges for directly written glass photonics. IEEE Sel. Top. Quant. Electron. **14**(5), 1370–1381 (2008)
49. Nandi, P., et al.: Femtosecond laser written channel waveguides in tellurite glass. Opt. Express **14**(25), 12145–12150 (2006)
50. Nejadmalayeri, A.H., Herman, P.R.: Ultrafast laser waveguide writing: lithium niobate and the role of circular polarization and picosecond pulse width. Opt. Lett. **31**, 2987–2989 (2006)
51. Yang, W., Kazansky, P.G., Svirko, Y.P.: Non-reciprocal ultrafast laser writing. Nat. Photonics **2**, 99–104 (2008)
52. Eaton, S.M., Chen, W., Zhang, L., Iyer, R., Aitchison, J.S., Herman, P.R.: Telecomband directional coupler written with femtosecond fiber laser. IEEE Photonics Technol. Lett. **18**(20), 2174–2176 (2006)

Information Flow Control on the Basis of Meta Data

Alexander Grusho$^{(\boxtimes)}$, Nick Grusho, and Elena Timonina

Institute of Informatics Problems of Federal Research Center "Computer Science and Control" of the Russian Academy of Sciences, Vavilova 44-2, 119333 Moscow, Russia
grusho@yandex.ru, info@itake.ru, eltimon@yandex.ru

Abstract. Usually security of information flows is supported by Firewalls, Proxy servers, Intrusion Detection Systems. These mechanisms work when there can be malicious information flows. The paper presents the new security mechanism which can be used instead of traditional measures. This is due to the strong limitation of the existence of non needed information flows.

The main objective of our method is the prevention of interactions which are initiated by a malicious code. For the solution of this problem it is offered to define legal communications by meta data.

The review of the results connected with the usage of the method of meta data for creation of an authorization system of connections in a distributed information system and, in particular, in information systems on the basis of SDN is provided in the paper. It is shown that the found vulnerabilities can be compensated by means of an expansion of the functionality of controlling tasks $\mathcal{N}$ and $\mathcal{M}$, and by cryptography.

Using the fact that meta data reflect a reduced model of relationships of cause and effect in information technologies it is possible to solve an inverse problem connected with localization of failures and errors in data.

Keywords: Information security · Flow control · Meta data · Distributed information system

1 Introduction

The main objective of our research is the prevention of interactions which are initiated by a malicious code [1,2]. The concept of permitted interactions is defined, i.e. such interactions are necessary for the fulfilment of the information technology (IT). Any other interactions are considered as forbidden.

For such security policy it is necessary to develop special means of its realization. For the solution of this problem it is offered to define legal communications by meta data.

To realize this control the started task appeals to meta data for the permission of the interaction with the other task, necessary for its decision. On the

Partially supported by Russian Foundation for Basic Research (project 18-07-00274).

V. M. Vishnevskiy et al. (Eds.): DCCN 2019, LNCS 11965, pp. 548–562, 2019.
https://doi.org/10.1007/978-3-030-36614-8_42

basis of meta data the need of such interaction is identified and the permission which cannot be forged or bypassed is given. Really interactions are implemented through a network by means of sessions.

Usually security of information flows is supported by Firewalls, Proxy servers, Intrusion Detection Systems. These mechanisms work when there can be malicious information flows. The paper presents the new security mechanism which can be used instead of traditional measures. This is due to the strong limitation of the existence of non needed information flows.

The two-level hierarchical model of a distributed information system (DIS) consists of the level of tasks and the level of network. The network consists of hosts. The connection equipment allows to connect each host with any other host without interactions with intermediate hosts. Communication is implemented only by means of sessions, for example, under the TCP protocol. In further considerations the network equipment will not be considered.

Hosts as computer systems contain computing resources, information resources and software for the solution of various tasks.

2   Modelling of Tasks

For formation of meta data it is necessary to make several steps. The first step is to construct the sequence of tasks from models of business processes.

The basis of these models is the concept of a task. The simplest definition of a task is a transformation of the input data into the output data. Any transformation consists in a performance of some functions. It means that the task has to realize some functionality. For the task A we will designate the functionality of the transformation by $\pi(A)$. For the performance of the transformation of the task A it is necessary to provide $\pi(A)$ with the input data and the values of configuration parameters of the functionality. These data should be obtained from the outside. Besides, it is necessary to distribute results of the transformation or to keep these data in the memory. The union of these functions of the organization of the solution of the task A we will designate through $\varphi(A)$. Thus, the task A is defined as the couple $(\pi(A), \varphi(A))$.

Further we will generalize this definition, when the task A consists of several tasks $\{B_1, B_2, ..., B_s\}$.

According to the above definition
$$A = (\pi(A), \varphi(A)), \text{ and } B_i = (\pi(B_i), \varphi(B_i)), \ i = 1, ..., s.$$

Definition 1. It is possible to identify the compound task A with the set of the tasks $\{B_1, B_2, ..., B_s\}$ in the only case when the following conditions are satisfied:

1. the functionality of the set $\{\pi(B_i), \ i = 1, ..., s\}$ covers the functionality of $\pi(A)$;
2. the object $\varphi(A)$ has a possibility of interaction with each of the objects $\{\varphi(B_i), \ i = 1, ..., s\}$ for:
   - receiving and transfers of all input data to each task $B_i, \ i = 1, ..., s$;

- distributions of all output data of the tasks B_i, $i = 1, ..., s$, which have to be defined in the task A;
- definitions of the order of the performance of the tasks B_i, $i = 1, ..., s$, for the solution of the task A;
- communications of the tasks B_i, $i = 1, ..., s$, among themselves.

This definition is similar to the concept of client-server architecture in which clients B_i organize their work under the control of server $\varphi(A)$.

The Definition 1 is connected with hierarchical decomposition of tasks as sets of subtasks. Development of this view on a task is the understanding that subtasks of the initial task can be represented as a set of tasks. Let the compound task $A = \{B_1, B_2, ..., B_s\}$ be defined. In the same time every B_i may be also the compound task $B_i = \{C_{i,j}\}$. Then there is the hierarchical decomposition of the task A on subtasks. Such hierarchical decomposition in the theory of artificial intelligence is called the reduction [3]. In [3] algorithms of creation of a reduction are given. At the same time it is necessary that all tasks were unambiguously identified.

Let's note that $\pi(C_{i,j})$ unambiguously is defined by the name of the task $C_{i,j}$, and the object $\varphi(C_{i,j})$ is connected with the object $\varphi(B_i)$. In this regard we use the following designation: $\varphi_X(C)$ is the object of the organization of the solution of the task C as the subtask of the task X. Tasks in a hierarchical decomposition can repeat. At the same time if the values of the objects $\pi(\cdot)$ for all repetitions of one task are identical, then the values of the objects $\varphi(\cdot)$ for all repetitions are various. Therefore the repeating tasks unambiguously are identified.

Reuse of tasks can be realized in three ways.

At first assume that the task be the shared object. Then a parallel address to this task from its various repetitions is possible. For example, DBMS can process in parallel several requests from various sources. Then it is possible that such task A is characterized by the address to it by means of various objects $\varphi_X(A)$.

The second way is a creation of queue of the usage of the task A. This queue will consist of the chain of objects $\varphi_X(A)$ for various X.

The third way is connected with a creation of copies of the task A.

Thus, each compound task A presented in the form of a set of subtasks forms the tree of subtasks $L(A)$ with the root in A. This tree bijectively determines the tree $L(\varphi(A))$. Nodes of $L(\varphi(A))$ are objects $\varphi_X(C)$.

In the tree $L(\varphi(A))$ there is all information about interactions of tasks provided that all initial data for these tasks are collected in the object $\varphi(A)$. In principle, each such tree can be considered as the maximum meta data, which defines the order of the solution of the task A. Therefore this tree can be used as permission information for connections in network. However these data are complex for usage in real time.

The model of DIS can be presented as the compound task including all tasks which can be solved by means of this system. It is possible to consider that the object $\varphi(DIS)$ is capable to allocate and configure various sets of tasks. Then the IT model is the subset of tasks of DIS model united with model of business process and the concrete configurations of these tasks. The object $\varphi(\text{IT})$ also determines the queue of the transformation of the initial data for the realization of this IT.

3   Model of Meta Data

Different tasks can be solved on different hosts of a network. Then the network allows to collect initial data for tasks and to distribute results of processing. Therefore, there is a binary relation $H(A)$ between the host H and the concrete copy of the task A presented by the software and information resources.

Security policy of DIS demands the control of interactions of hosts in the network which comes down to monitoring of interactions and management of connections. Control of interactions of hosts in the network allows to reduce threats of infiltration and distribution of a malicious code via network equipment and communication channels. Management of interactions of hosts in the network should be realized by means of meta data.

Let's designate the set of tasks realized in DIS through Ω and we will consider on this set the binary relation (A_i, A_j), called generation, where A_i and A_j are tasks, and A_i defines the start or participates in start of the task A_j. Generations in which the initial task is on one host and generated task is on another host, we will designate through $\mathcal{B}$ and we will call it by *meta data* (simplified meta data).

For the tasks entering relation $\mathcal{B}$ we will define three additional tasks $\mathcal{M}, \mathcal{N}, \mathcal{R}$, which manage interactions in network on the basis of meta data $\mathcal{B}$. The task $\mathcal{M}$ distributes applications for the solution of tasks between hosts (for simplicity we will speak about distribution of tasks on hosts). The task $\mathcal{M}$ defines the binary relation $H(A)$, meaning that on the host H the task A can be calculated.

Results of the task $\mathcal{M}$ are used by the task $\mathcal{N}$. The task $\mathcal{N}$ can interact with each host and is responsible for permission and providing to hosts the information on the request for interactions of tasks on different hosts. Permission is based on meta data $\mathcal{B}$.

The task $\mathcal{R}$ builds the main and reserve routes on the instructions of the task $\mathcal{N}$. For example, $\mathcal{R}$ is in the controller of SDN (Software Defined Network).

If we support an isolation of tasks on a host then it allows to exclude non admissible transit of information flows. Basic element of providing information security is the task $\mathcal{N}$ on the host $H(\mathcal{N})$. $\mathcal{N}$ is also the task containing meta information about other tasks.

For its execution the task A on the host $H(A)$ has to address an immediate task A_1 as it follows from scheme of the hierarchical decomposition. For this purpose the task A generates the request for the possibility of the appeal of the task A to the task A_1, and addresses the managing program $\mathcal{N}(H(A))$. Each host H in DIS has in the managing task $\mathcal{N}(H)$ cryptographic facilities and the unique key $k(H)$ for the communication with the managing task $\mathcal{N}$ on the host $H(\mathcal{N})$. All communications of any host are realized through this managing task agent $\mathcal{N}(H)$.

The task $\mathcal{N}(H(A))$ forms the encoded message on the key $k(H(A))$ with the request to allow the interaction of the task A with the task A_1. The task $\mathcal{N}$ checks need of the appeal to A_1 with the help of available for it meta data for the solution of the task A. At the positive decision $\mathcal{N}$ forms the encoded message for the task $\mathcal{N}(H(A))$ in which there is the address of the host $H(A_1)$, the number of the port for communication with the task $\mathcal{N}(H(A_1))$ and the session key $k(A, A_1)$. The similar message is also formed for the host $H(A_1)$. The address of the host $H(A)$, the port number of the task $\mathcal{N}(H(A))$, the permission for starting of the task A_1 for the benefit of the task A, and the common key for their communication $k(A, A_1)$ are specified in this message. After obtaining this information the host $H(A)$ initiates the session of the encoded communication with the host $H(A_1)$.

Completion of the session happens standardly. If there is a failure, then it comes to light by means of identification codes MAC (Message Identification Code). In case of need there is a restart of the protocol. If we have agents $\mathcal{N}(H)$ in every host, then there is no need to use all other protocols to control parameters of the network. All necessary control information can be gathered by secure interaction of available hosts with $H(\mathcal{N})$. It can be done by several sets of meta data. One of them may be network control meta data. It helps to forbid service flows.

Cryptographical part of the model resembles well known protocol Kerberos, but it supports different functionality.

4   Formation of Meta Data

Questions of the organization of meta data are connected with possible examples of interrelation of tasks. The protocol forbids any connection which is not reflected as a legal in $\mathcal{N}$. Therefore, additional questions of completeness, of consistency, of possibility of modification, and of scalability are connected with the organization of meta data.

Without applying for completeness, we will give several examples.

Example 1. Addressing of a Task A to Database
Let T be the parameter of the task A for DBMS. The protocol will organize the session with the host containing the DBMS, but the access is possible only if the parameter T is legal.

Let's show how it is supported by the functionality of Oracle DBMS. The addressing to the task A_1 with the parameter T (the appeal to the DBMS with a request from the task A) is equivalent to creation of the user process on the host $H(A_1)$. In Oracle DBMS the user process generates the server process. Isolation of server processes by the mechanisms of discretionary and mandatory security policies are used [4]. In this regard restrictions of the process with the parameter T is implemented by the standard policies of access control which are built in the DBMS. The trust to these functions is determined by the certification documents.

Example 2. Formation of Meta Data by Means of Models of Business Processes

Formation of meta data about interactions of tasks can be made on the basis of business process modelling methods. A set of advanced methods is developed for business process modeling: IDEF, ARIS, UML, BPMN, etc. [5].

For creation of models an automation software, for example, of BPwin and ERwin [5] are created. For creation of sequential diagrams it is possible to use methodology of IDEF3 [5]. The methodology of IDEF3 is supported by software of the Computer Associates companies, etc.

The methodology of ARIS assumes several abstraction layers and serves for the complex description of activities of the enterprizes. In this methodology the set of models for the adequate description of system and its processes is created. Software tools are created to support of ARIS which are added by modeling languages of UML, BPMN and etc.

Example 3. Meta Data for Tasks with Uncertainty

It is the most difficult to apply the offered approach of support of information security in DIS to tasks with uncertainties. An example of uncertainty is the question to the task $\mathcal{N}$: "Whether the task A can be solved by means of the task A_1?" For the consideration of such questions it is possible to use semantic methodology. Semantic methods are based on the description of ontologies. The ontology is understood as [5] hierarchical data structure containing meanings of information and their communications. The formal language of descriptions of ontologies is the standard of Web Ontology Language [5].

Example 4. The Description of Admissible Communications of Tasks by Means of Data Mining

This method of formation of an authorization system is connected with search of such tasks which can be the useful in the analysis of the task A. This method has common features with the example 3. However complication of search in comparison with an example 3 consists that there is no accurate description of the required ontologies. Therefore for creation of an algorithm of the decision about admissible communications for this class of tasks more thin methods of data mining are necessary [6].

Example 5. Statistical Method of Formation of Meta Data

Let DIS work in the free mode in some time in the supposition that all it services are legal. Meta data are created by results of observation over interactions of tasks in DIS in the free mode. In some time point all interactions of tasks in each of information technologies are fixed. The received mold of interactions defines meta data for monitoring of further interactions.

In case of such method of formation of meta data next errors are possible:

- some operation modes of information technologies can demand further additional requests for information resources or the software. I.e. in this method we get an increased number of non admissible interactions, than it is necessary;
- in the free mode some interactions could be excessive, and can generate information flows, which are dangerous to functioning of information technologies.

5 Threats and Security

The two-level organization of information protection by means of meta data also has vulnerabilities. A considerable part of these vulnerabilities was investigated for SDN supporting DIS. In this regard it is necessary to give the short review of these networks.

Usually SDN is presented in the form of three planes. The plane of data (Data Flow Level) consists of hosts and switches. Each switch has FT (Flow Table). In this table there are rules for the switch for forwarding data. The number of such rules can exceed 1000.

The packet arriving on the switch is being processed as follows. The switch looks for rules in FT, after finding the rule the rule counter increases its value, and fulfills the action intended for such packet. If the rule isn't found in FT, then the packet header comes to the controller or the packet is discarded at all.

Rules are created by the controller. The plane of the controller is described as follows. The host of the controller is connected to switches either via the common channel, or via the special channel. The controller creates routes for connection of hosts.

On the third plane there are applications which support the functions of the controller.

If hosts organize connections in a network it may be dangerous from the point of view of information security. So, the malicious code can be the initiator of the connection. Therefore, it is necessary to use fire-walls, IDS and other security features. However, if interactions of hosts are defined only by necessary interactions of legal tasks, then problems with information security become less. The controller has the task of formation of routes for interactions of legal tasks, then the filtering of flows by means of fire-walls and IDS isn't necessary. The main thing, that the necessity of the appeal of switches to the controller disappears.

6 Analysis of Security of SDN Ruled by Meta Data

The problems of SDN security discuss in many scientific papers [7,8]. The main differences between the known vulnerabilities and the present analysis consists on the fact that switches cannot transmit their data to the controller, but the controller can transmit its data to switches. It helps to reduce vulnerabilities which have the traditional protocol OpenFlow. The following network attacks should be considered as most important.

Man-in-the-Middle Attack
If one or several communication links are listened, then an adversary can transfer the intercepted information through its own channels outside of the network. However, if the adversary has no channels outside of the network, then transmission of the intercepted information is impossible because in the switch there is no allowed route to the hostile host with the legally solvable task. But man-in-the-middle attack can lead to distortion of information. This attack is blocked by the cryptography and the unique identifier of executed IT.

Capture of a Switch

The problem of the capture of a switch was discussed in [9, 10].

This attack allows to solve many problems for an adversary. In traditional SDN there is a possibility of duplicating of a traffic. However, in our model of transmission through a network of a duplicated traffic is possible only in the case the captured switch is directly connected to a hostile host. In remaining cases the transmission is impossible. The captured switch gives more opportunities for a distortion of the transit data. It is equivalent that a part of sessions is locked. Therefore, hosts easily detect such attacks. The attack with distortion of data can be considered as a failure, and for the restoration of the session it is necessary to use reserve routes.

However, there is a complex task of localization of the captured switch.

The failure detection of some non-localized switch in SDN can be done in case of violation of legal connections between hosts. In case of detection of the forced rupture of the connection the host transfers the message about it to $\mathcal{N}$. However, bypassing of the failure can be done only after localization of the faulty switch. We will suppose that there can be only one faulty switch in SDN.

Consider the problem of localization of the faulty switch after detection of the forced rupture of the connection. For this purpose it is necessary to determine some features of the network topology.

Let the network contain n of identical hosts and two selected hosts. The network should allow creation of the connection of any host with any host.

We will consider the simplest network topology that consists of the set of identical switches and allows to connect every host with other hosts. Such network represents a k-tree L, where k is the number of ports of each of identical switches. For the simplification we will propose that $k = 3$ for identical switches.

Among hosts there are two hosts allowing to provide the management of the network. These are the controller and the host $H(\mathcal{N})$. We assume that the tree of switches L is a root tree and symmetric with regard to its root. Therefore, $n = 2^{h_0}$, where h_0 is the tree height, and unlike other switches the root switch has 4 ports:

- two ports on the descending branches of the tree;
- one port on the SDN controller;
- one port on the host $H(\mathcal{N})$.

The host $H(\mathcal{N})$ is the only one at the network which can communicate with any other host without permissions, or respond to any request of any other host. The connection $H(\mathcal{N})$ with the controller is carried out on the constant basis outside of SDN.

Let D be a switch located on the route from X to Y. If at this time there is an active route passing through D, then D cannot be the reason of the rupture of the connection between X and Y.

Choose the host U and construct two new routes from U to hosts X and Y, which pass through a switch on the faulty route. Remember the assumption that a failure can be only in one switch. Then one of new constructed routes should function normally while the second route should be faulty. Along the new constructed faulty route, it is necessary to look for the faulty switch from the intersection switch to the ending hosts X or Y. If both new routes are faulty, then the selected switch on the route between X and Y is the faulty one. Repeat the previous procedure with a random switch on the new faulty route. Repeating these steps, we'll come to the faulty switch.

Capture of a Controller

Capture of a switch in traditional SDN can initiate an attack on the controller. For this purpose in traditional SDN an overflowing of input buffers of the switch is imitated. In this case together with heads of packets the whole packets are sent to the controller. Thus, it is possible to increase the traffic in the channel between the controller and the switch. Thus, denial-of-service attack on the controller is built. In our model such attack is impossible since all routes are fixed and the appeal to the controller from switches is forbidden.

Direct Capture of the Controller or $H(\mathcal{N})$

Physical capture of the controller or the host $H(\mathcal{N})$ completely compromises network functioning and information security at the level of tasks.

Capture of Hosts $H(A)$ or $H(A_1)$

Capture of hosts $H(A)$ or $H(A_1)$ can create a possibility of distribution of a malicious code or a harmful influence into network hosts, and also can attack the controller or $H(\mathcal{N})$. This attack is based on the implementation in the captured host of a malicious code and it means the possibility of its implementation into other hosts.

It is easy to find some strategy of the information security violator at the level of meta data. Having obtained the information on the list of tasks with which the adversary can set connection, but without knowing the hidden copy of the identifier of the real IT, which should be executed on the captured host, the adversary can request the connection at random. In this case mechanisms of management of connections at the level of meta data (task $\mathcal{N}$, see [11]) identify the captured host at once.

In [12] more sophisticated attack which creates delays in execution of all acting IT is considered. However at the correct distribution of tasks of hosts (the task $\mathcal{M}$, see [11]) this attack is compensated. The most complex problems arise if the adversary begins accurately to change the output data of tasks which are legally being solved on this host. The simplest variant of this problem is considered in [12]. If the adversary changes the output data of the solvable task onto already used old data of this task, then the creation of databases of hash function values of exits of all tasks is possible. It can be the centralized or the local databases (DB) [12]. The comparison of the next values of hash functions of the output data with the data stored in DB allow to reveal use of the old data and to reveal the captured host. The probability of repetition of values in DB is low. Emergence of single repetition of variable value and absence of repetitions in other variables indicates the attack.

Let's define conditions when the identification of a substitution of data by the adversary is impossible.

The detailed analysis of the existing tools of the description of business process (BP) models [5] showed that they do not suppose usage of specific values of data in these models and the related with them IT. Variables and descriptions of domains of these variables are used everywhere in these models, but values of data are not used anywhere. It is easy to understand that all BP and IT suppose repeated usage of variables. Therefore, the data values cannot enter the description of transformations realized by BP and IT except as in the form of variables, definition ranges and areas of values.

From here it is possible to draw the following conclusion. Any mappings of transformations entering the BP models or the IT models do not depend on values of data. Results of transformations of BP and IT are also described by variables, therefore the description of meta data does not include concrete values of variables.

From here it follows.

Assertion. *Meta data cannot control deliberate changes of data in tasks, except as with the help of an exit of values of variables into the field of prohibited values (bans).*

Proof. Change of values of variables by the adversary corresponds to change of values of parameters determined by some variables in the descriptions of BP and IT. Meta data are functions from the transformations participating in the descriptions of BP and IT. Therefore all variables which are found in meta data do not contain specific values, except as in the descriptions of domains or bans.

Corollary. *The adversary who captured a host and selected the strategy of deliberate change of data cannot be detected without additional information about domains of values of variables or bans obtained by means of observations of monitoring data.*

It means that for the identification of the adversary who selected the strategy of change of data, it is necessary to know bans and to hope that the changes of data will lead to emergence of bans.

It is possible to offer several ways of the description of domains and bans at the different strategies of change of data by the adversary. The most general case is the definition of search of prohibited values of the executed functions when the emergence of bans means the emergence of incorrect data in some tasks.

Explain this method on the following example. Let A be the task obtaining initial data from tasks B and C. Let the host $H(B)$ be captured and $H(B)$ changes data, and the host $H(C)$ is not captured. Then the task A obtains from the task B the distorted initial data x, and from the task C obtains the correct initial data y.

Execution of the task A means the calculation of some function f_A. Let's put $f_A(x, y) = z$. The values x and y are often connected among themselves by initial conditions. Therefore the changed data x can break these connections and it affects the value z. If $z \notin D_A$, where D_A is the set of permissible values of function f_A, then this fact can be interpreted the failure of the initial data. From this it follows that it is necessary to investigate hosts $H(B)$ and $H(C)$ on the possibility of the capture.

The research of sets of values of all functions used in IT is the complex problem. Usually there is a preliminary testing of the most often found initial data. Then any task A generates the set $D'_A \subseteq D_A$, where D'_A is the empirical estimation of the set D_A. The usage of the set D'_A instead of the set D_A generates false alarms, when the received initial data differ from often found data. Therefore this technique is usually applied to the aggregated tasks, and their extreme values have been researched.

"Honey Pot" IT differs from "traps" used against a malicious code. The following idea of the identification of an adversary changing data is offered.

At the stage of the trial operation all trajectories of IT execution, i.e. the input and output data of all tasks are stored (the set U). Elements of the set U are considered as "correct" ones. In the case of the change of data at least for one of tasks anomalies appear. According to some (random) schedule instead of the next copy of IT the copy $u \in U$ of IT is started. If the adversary constantly replaces data on the captured host, then this host comes to light on start of test technology u, because on the trajectory u the output data of all tasks is completely defined. If the adversary applies the "Byzantine" strategy, i.e. changes the correct data not on each copy of IT, but on randomly selected data, then the possibility of the identification of the captured host can be defined on the basis of the following probabilistic model.

Let p be the probability of start of the controlled copy of IT in the sequence of started IT $(n = 1, 2, ...)$. Let q be the probability of the change of data in the considered copy of IT. Then the probability of the identification of the change of data in the next copy of IT is equal to pq. According to the model of the geometrical distribution the probability of the identification of the change of data at n steps is equal to $1 - (1 - pq)^n$. Mathematical expectation of the number of steps before the identification of the change of data is equal to $\frac{1}{pq}$.

In the considered method it is supposed that there are no repetitions of controlled trajectories, otherwise the identification of the control by the adversary is possible, since the adversary remembers values of parameters, which pass through the host captured by him.

The implementation of the method of "traps" demands the special architecture of DIS. It can be the client-server architecture or variants close to it [13]. However in the client-server architecture a need of the direct interaction of hosts among themselves disappears.

In the concept of the network protection on the basis of meta data defined in the works ([11–14]), there is no mentioning about time response characteristics. Let's remind that meta data contain the information on validity of the connection of the hosts $H(B)$ and $H(A)$ on the basis of the need of the interaction of tasks B and A for current IT.

At the same time in the information system several IT which have the right to address to the task A can function at the same time. Then the emergence of a queue for the usage of the task A is possible.

Assume that during the execution of technologies the queue of tasks $B_1, B_2, ..., B_s$ be formed. The distribution of tasks through hosts is carried out at the level of BP models by means of the task $\mathcal{M}$, and this distribution of tasks can be considered as random. The start of several IT and the solution of the sequence of tasks implementing these technologies can be considered as random process.

If to consider the fixed task A and its host $H(A)$, then during the execution of several IT it is possible to suppose that the appeals to the task A are independent random events. As meta data do not control the extension of the queue, it is possible to violate the information security due to an exceeding of the opportunity of the host $H(A)$ to connect with hosts $H(B_1), H(B_2), ..., H(B_s)$ of tasks $B_1, B_2, ..., B_s$, expecting the appeal to the task A.

This violation of security is similar to the DDoS attack. The distinction consists in that inadmissible queue can randomly arise in different hosts of the network.

The following solution is proposed in [15]. The task $\mathcal{N}$ receives from the task A through the agent in the host $H(A)$ the message that A is busy. The agent in $H(\mathcal{N})$ reports about this to all hosts $H(B_1), H(B_2), ..., H(B_s)$, which submitted requests to the task $\mathcal{N}$ about the connection with the task A. In this case the task $\mathcal{N}$ can create a queue, consistently reporting to hosts $H(B_i)$, $i = 1, ... s$, about their possibility of the connection with the host $H(A)$. I.e. the host $H(A)$ should send to the host $H(\mathcal{N})$ the message that the task A is free for the next usage.

Such approach provides the resolution of the conflict between information technologies and does not lead to the DDoS failure. However the waiting time of the fulfillment of IT becomes indefinite and random. Besides, this method considerably increases the load of the task $\mathcal{N}$. Therefore this method should be "strengthened" by a possibility of a reconfiguration of the network and an upgrade of meta data.

Summarizing the results of the short analysis of the security of the elementary protocol, it is possible to claim that the considerable part of threats connected to a redirection and a duplicating of information flows is liquidated. The possibility of the start of illegal tasks is reduced, and the possibility of an interaction of illegal tasks is restricted. Even the DDoS attack can be prevented.

7   Extension of Methods of Meta Data Usage

It was noted earlier meta data is not adapted for the search of violations in values of data. The attacks with data modification at the captured host do not exhaust the problem of a treatment of data. If during the solution of the task there is a need to take an optional data from earlier solved tasks, then it can be not permitted by meta data. Therefore, such request to the task $\mathcal{N}$ will not be processed. Also the additional request for data retrieval in DB will not be processed if this task is not permitted by meta data.

However, these tasks can naturally arise, and can sometimes contradict the security policy of handling of information resources. Therefore, it is necessary to enter the additional changes into the system of work with meta data. Let's consider several ways of solving the task of collecting optional data.

1. Let's assume that the task $\mathcal{N}$ remembers the list of tasks, the order of these tasks and the identifiers of hosts on which they were solved. Designate this list of tasks through $\{B\}$. Then the current task A can address the task $\mathcal{N}$ for receiving the necessary optional data from the task B which was solved earlier. The task $\mathcal{N}$ allows the connection of $H(A)$ with $H(B)$ with the identifier of the current technology for the normal protocol of connections, if such returnable connection does not contradict the security policy. During the session the task A obtains the optional data and specifications, and after that the task A discards the connection.

2. Let the task $\mathcal{N}$ store the list of tasks $\{B\}$ and the list of hosts on which they were solved in the current copy of IT. In this case the task $\mathcal{N}$ acts as a broker, i.e. the task A requests the task $\mathcal{N}$ using the identifier of the current copy of IT for receiving the optional data from the task B which was solved earlier. The task $\mathcal{N}$ checks the possibility of the access to the required data at the task B. In case of the positive decision the host $H(\mathcal{N})$ connects to the host $H(B)$ with the identifier of current IT, and with the request to the task B for providing the necessary data to the task A. The task B transfers data, necessary for the task A, to the task $\mathcal{N}$ through the secure channel between $H(B)$ and $H(\mathcal{N})$. The task $\mathcal{N}$ transfers the required data to the task A through the secure channel between $H(\mathcal{N})$ and $H(A)$. Note that the specified protocol also protects the integrity of the transmitted data.

   In this case, if the task B does not remember the required data, then the restart of the task B with the subsequent data transmission to the task A through the task $\mathcal{N}$ is possible.

   The high complexity of this algorithm is the need of the storage of old data. Not to allow an expansion of the search the requests stated above should be followed by the identifier of the copy of IT which at the solution of the task A initiated the search of the optional data.

3. The search an optional data can demand an exit out of limits of the current copy of IT. Then the task $\mathcal{N}$ sequently addresses higher vertices of the tree $L(A)$ for the permission of the security policy for obtaining the required data. When receiving the optional data their transfer to the task A is carried out by the rules of the security policy through the located below vertices of the tree $L(A)$.

   Let's note once again that the task $\mathcal{N}$ stores the graph of the sequence of solvable tasks, and can organize the poll of these tasks in the inverse order. The allowing method by means of meta data is based on relationships of cause and effect of these tasks. From there is a possibility of an usage of this graph for the solution of the problem of monitoring of technical failures and errors in data. As it was noted in the paper [16], many problems of the search of a failure demand the complex analysis of the possible cause of a failure. This analysis is based on relationships of cause and effect of the processes arising at the solution of tasks. Then the graph of the solved tasks allows to assume the place of the cause of the arisen failure. The selected section of the alleged cause can be detailed, using hierarchical approach until the failure causes are established.

Note that the method of the return to the data of already solved tasks adds the approach to the identification of distortions of data on the host captured by an adversary. If on the captured host in the memory there are no distorted data of already solved tasks, then new distortions identify the adversary. For this purpose, it is necessary that the task $\mathcal{N}$ remembered all data transferred from a task to a task. If the captured host remembers distortions, then the method of "traps" allows to detect the adversary.

8 Conclusion

The review of the results connected with the usage of the method of meta data is presented. The creation of the authorization system for connections in DIS and, in particular, in information systems on the basis of SDN is provided in the paper. It is shown that the found vulnerabilities can be compensated by means of the expansion of the functionality of tasks $\mathcal{N}$ and $\mathcal{M}$, and by means of the cryptography.

Using the fact that meta data reflects the reduced model of relationships of cause and effect in IT it is possible to solve the inverse problem connected with localization of failures and errors in data. This approach differs from the observation of logs of processes of IT implementations since it is the aggregation of these processes. The method gives the time stamp of an error and the list of possible sources of failures and errors. Thus, meta data and data storage in the task $\mathcal{N}$ represent the new mechanism of monitoring in DIS. The possibility of such usage of meta data was checked experimentally with participation of the DIS system administrator.

References

1. Rieck, K., Stewin, P., Seifert, J.-P. (eds.): DIMVA 2013. LNCS, vol. 7967, 207 p. Springer, Heidelberg (2013). https://doi.org/10.1007/978-3-642-39235-1
2. Skorobogatov, S., Woods, C.: Breakthrough silicon scanning discovers backdoor in military chip. In: Prouff, E., Schaumont, P. (eds.) CHES 2012. LNCS, vol. 7428, pp. 23–40. Springer, Heidelberg (2012). https://doi.org/10.1007/978-3-642-33027-8_2
3. Nilsson, N.J.: Problem-Solving Methods in Artificial Intelligence. McGraw-Hill Pub. Co., New York (1971)
4. TCSEC: Department of Defense Trusted Computer System Evaluation Criteria. DoD (1985)

5. Samuylov, K.E., Chukarin, A.V., Yarkina, N.V.: Business Processes and Information Technologies in Management of the Telecommunication Companies. Alpina Pablisherz, Moscow (2009)
6. Finn, V.K. (ed.): Automatic Hypotheses Generation in Intelligent Systems. KD "LIBROKOM", Moscow (2009)
7. Tran, C.N., Danciu, V.: A general approach to conflict detection in software-defined networks. SN Comput. Sci. **1**, 9 (2020). https://doi.org/10.1007/s142979-019-0009-9
8. Shu, Z., Wan, J., Li, D., Lin, J., Vasilakos, A.V., Imran, M.: Security in software-dened networking: threats and countermeasures. J. Mob. Netw. Appl. **21**(5), 764–776 (2016). https://doi.org/10.1007/s11036-016-0676-x
9. Petrov, I.S.: Systems of detection of compromised switches in software-defined networks. J. Inf. Technol. **25**(3), 131–142 (2019)
10. Petrov, I.S.: A problem of detection of compromised switchs in SDN. J. REDS: Telecommun. Devices Syst. **7**(4), 515–518 (2017)
11. Grusho, A.A., Timonina, E.E., Shorgin, S.Y.: Modelling for ensuring information security of the distributed information systems. In: 31th European Conference on Modelling and Simulation Proceedings, pp. 656–660. Digitaldruck Pirrot GmbHP Dudweiler, Germany (2017)
12. Grusho, A., Timonina, E., Shorgin, S.: Security models based on stochastic meta data. In: Rykov, V.V., Singpurwalla, N.D., Zubkov, A.M. (eds.) ACMPT 2017. LNCS, vol. 10684, pp. 388–400. Springer, Cham (2017). https://doi.org/10.1007/978-3-319-71504-9_32
13. Grusho, A.A., Timonina, E.E., Shorgin, S.Y.: Hierarchical method of meta data generation for control of network connections. J. Inf. Appl. **12**(2), 44–49 (2018)
14. Grusho, A., Grusho, N., Zabezhailo, M., Zatsarinny, A., Timonina, E.: Information security of SDN on the basis of meta data. In: Rak, J., Bay, J., Kotenko, I., Popyack, L., Skormin, V., Szczypiorski, K. (eds.) MMM-ACNS 2017. LNCS, vol. 10446, pp. 339–347. Springer, Cham (2017). https://doi.org/10.1007/978-3-319-65127-9_27
15. Grusho, A.A., Timonina, E.E., Shorgin, S.Y.: Overcoming of protection of the network where connections are controled by meta data. J. Syst. Means Inf. **28**(4), 22–30 (2018)
16. Grusho, A.A., Zabezhailo, M.I., Zatsarinnyy, A.A., Nikolaev, A.V., Piskovski, V.O., Timonina, E.E.: Erroneous states classification in distributed computing systems and sources of their occurrence. J. Syst. Means Inf. **27**(2), 30–41 (2017)

Application of Machine Learning Algorithms to Handle Missing Values in Precipitation Data

Andrey Gorshenin[1,2]([envelope]) [iD], Mariia Lebedeva[2], Svetlana Lukina[2], and Alina Yakovleva[2]

[1] Federal Research Center "Computer Science and Control" of the Russian Academy of Sciences, Moscow, Russia
agorshenin@frccsc.ru
[2] Faculty of Computational Mathematics and Cybernetics, Lomonosov Moscow State University, Moscow, Russia
mash.lebedeva2010@yandex.ru, svetl.luckina2016@yandex.ru, alyna_yakovleva@mail.ru

Abstract. The paper presents two approaches to filling gaps in precipitation based on classification (Support-Vector Machines) and regression (EM, Random Forests, k-Nearest Neighbors) machine learning algorithms as well as the pattern-driven methodology. These methods are among of the most powerful tools for data mining in a wide range of research areas including meteorology and climatology due to the presence of a large amount of temporal and spatial observations. When collecting observations from weather stations, there are a lot of missing records. Data processing algorithms are often very sensitive to the presence of incomplete data, so missing values should be firstly imputed and only after that the complete samples can be analyzed. The possibility of a correct filling data even for high missing levels based on suggested methods is demonstrated. The observations in Potsdam and Elista for about 60 years were used. Also, comparison of various algorithms for data imputation taking into account different missing levels is presented. The proposed methodology can be successfully used for real-time data processing of information flows.

Keywords: Precipitation · Missing values · Support-Vector Machines · XGBoost · Patterns · EM algorithm · Random Forests

1 Introduction

Methods of time series analysis are one of the most powerful tools for data mining in a wide range of research areas. Different approaches based on vari-

All ideas for imputation methodology based on patterns and machine learning techniques were suggested by Andrey Gorshenin whose work was supported by the **Russian Science Foundation** (project **18-71-00156**). Software tools were written in Python and tested on real data by MSc students (M. Lebedeva, S. Lukina, A. Yakovleva).

V. M. Vishnevskiy et al. (Eds.): DCCN 2019, LNCS 11965, pp. 563–577, 2019.
https://doi.org/10.1007/978-3-030-36614-8_43

ous autoregressive models as well as a family of machine learning algorithms, including artificial neural networks, are traditionally used. These methods are very effective for meteorological problems due to the presence of a large amount of temporal and spatial data and, as a consequence, significant problems of their processing. In this paper, we focus on precipitation, but similar tools and approaches can equally well be used for another types of observations.

Precipitation is one of the most difficultly analyzed atmospheric parameters due to the nature of its variability. When collecting data from weather stations, there are a lot of missing records by various reasons. Data processing algorithms are often very sensitive to the presence of missing values, since they can seriously distort the results of analysis [20]. In particular, statistical tests and thresholds for extreme precipitation events [13] based on incomplete data can be incorrect.

Under the circumstances, in practice researchers should firstly fill in gaps, and only after that the complete set of observations can be analyzed and used as parameters of hydrologic models [16]. It is also worth noting that one can use an approach based on exclusion from the processing of subsamples with gaps until the time moment when the data becomes complete. However, in meteorological data a missing value may appear at any random time moment. Another approach may be based on reanalysis, but such data often exhibits variations compared with observations obtained by other techniques.

A well-known filling approach is inverse distance weighting method [24], including its extensions based on artificial neural networks [3]. Precipitation is usually weakly correlated with other meteorological parameters collected by weather stations. Moreover, the neighboring stations, which observations could be used to fill gaps correctly, are not in all geographic locations. It is worth noting, there are some suitable processing methods for case of a cluster of stations [21]. Having summarized, filling methods that use only initial data themselves without invoking any additional features are of particular interest.

This paper uses such well-known [27] machine learning algorithms as k-Nearest Neighbors (k-NN) [1], Expectation-Maximization (EM) [18], Support-Vector Machines (SVM) [7], and Random Forests (RFs) [4]. They are still actively used to solve applied scientific problems in various fields. For example, researches related to the software [28] and big data problems [19,26] can be mentioned. The theoretical aspects are also improved, and results are published in the worlds leading high-ranking journals, see [2,8]. The novelty of our paper is based on a joint use of the pattern-based methodology [11] and machine learning algorithms to fill a significant amount (up to 40%) of missing values in data. The effectiveness of our approach using the precipitation data in Potsdam and Elista for about 60 years [30] is also demonstrated.

The paper is organized as follows. Section 2 introduces the approach to imputation missing values based on classification. It implies that each observation is replaced by "D" (no rain) or "W" (rain), and the imputation does not take into account the exact precipitation daily volume. SVM is used as a classification machine learning (ML) algorithm. Section 3 describes the results of regression ML methods to handle missing values in precipitation data, i.e., "continuous" forecasts are implemented. The k-NN, EM algorithm and Random Forests are involved. Section 4 is devoted to discussion of the obtained results.

2 Approach Based on Classification

In this section, a classification approach to handling the missing values is introduced. The initial non-negative precipitation data should be modified as follows: if any positive value is observed in the current day, it is replaced by "W" (i. e., wet day), otherwise the symbol "D" (i. e., dry day) is used. Thus, the continuous time-series become discrete. Subsequences of some length of such modified sample are data patterns. The so-called wet and dry periods are defined as follows:

$$\ldots - D - \underbrace{W - W - W - \ldots - W}_{\text{wet period}} - D - \ldots$$

$$\ldots - W - \underbrace{D - D - D - \ldots - D}_{\text{dry period}} - W - \ldots$$

2.1 Simulation of Incomplete Data

There are two complete precipitation data sets for the period from 1950 to 2009 for analysis. Therefore, the missing values should be artificially inserted into the samples. Then, we can jointly use patterns and machine learning algorithms for imputations and subsequent comparison of our results with true values (in terms of "D-W" classification). It is possible to determine the frequencies of appearance of each pattern as the ratio of the number of such sets of fixed length N to the total number of possible chains (obviously, 2^N). Patterns with size $N = 5$ will be used throughout this work. The detailed description of the corresponding numerical characteristics for this case is given in [12].

In this section, the cases of 1, 2 and 3 consecutive missing values (MVs) are considered, and their total number varies from 5% to 40% of the sample size. That are so called missing levels (for example, see paper [17] where the threshold methodology is used to filling gaps at levels 5%, 10%, and 15–18%). The procedure of inserting the missing values is described using pseudocode, see Algorithm 1. Within this approach, the upper bounds of missing levels are as follows:

- up to 20% for only one MV in the window which size equals 5;
- up to 33% for two consecutive MVs;
- up to 43% for three consecutive MVs.

It explains the different limits for the curves on Figs. 1 and 2 in Sects. 2.2 and 2.3.

Algorithm 1. Simulation of incomplete data

1: INPUT(MV);   //*Number of consecutive MVs*
2: $\mathcal{I}_0=$ DROPOUT(Sample, MissLvl); //*Array of indices for possible insertion of MVs*
3: **if** $\mathcal{I}_0$(end)$\geqslant$LENGTH(Sample)-N **then**
4:     $\mathcal{I}_0$(end)=[];   //*MVs cannot be inserted into the end of data*
5: **end if**
6: k=0;
7: **for** i=1:length($\mathcal{I}_0$)-1 **do**
8:     **if** $\mathcal{I}_0$(i+1)-$\mathcal{I}_0$(i)$\geqslant$ N+MV **then**
9:         k=k+1;
10:        **for** j=0:MV-1 **do**
11:            $\mathcal{I}$(k+j)=$\mathcal{I}_0$(i)+j;   //*Array of indices of MVs*
12:        **end for**
13:    **end if**
14: **end for**

2.2 Imputation Methodology Based on Binary Patterns

In this section, an algorithm of "pure probabilistic" pattern-based filling is given.

1. Let us consider the subsample with a missing value (further it is denoted by symbol $\mathcal{X}$):

$$\ldots - D - W - D - D - D - \boxed{\mathcal{X}} - W - D - D - W - \ldots$$

2. All subsamples of pre-selected length that contain this gap are chosen (the value $N = 5$ is still used):
   (a) $\ldots - D - \boxed{W - D - D - D - \mathcal{X}} - W - D - D - W - \ldots$
   (b) $\ldots - D - W - \boxed{D - D - D - \mathcal{X} - W} - D - D - W - \ldots$
   (c) $\ldots - D - W - D - \boxed{D - D - \mathcal{X} - W - D} - D - W - \ldots$
   (d) $\ldots - D - W - D - D - \boxed{D - \mathcal{X} - W - D - D} - W - \ldots$
   (e) $\ldots - D - W - D - D - D - \boxed{\mathcal{X} - W - D - D - W} - \ldots$

3. There are only two possible "D–W" patterns in each situation. For example, the subsample from item 2a can be only as follows:

$$W - D - D - D - \boxed{W} \quad \text{or} \quad W - D - D - D - \boxed{D}$$

4. To fill the missing value, the pattern with maximum possible frequency should be chosen. The corresponding element in this pattern is considered as a decision:

$$\ldots - D - \boxed{W - D - D - D - \mathcal{X}} - W - D - D - W - \ldots \Rightarrow W - D - D - D - \boxed{D}$$

$$\ldots - D - W - \boxed{D - D - D - \mathcal{X} - W} - D - D - W - \ldots \Rightarrow D - D - D - \boxed{W} - W$$

$$\ldots - D - W - D - \boxed{D - D - \mathcal{X} - W - D} - D - W - \ldots \Rightarrow D - D - \boxed{D} - W - D$$

$$\ldots - D - W - D - D - \boxed{D - \mathcal{X} - W - D - D} - W - \ldots \Rightarrow D - \boxed{D} - W - D - D$$

$$\ldots - D - W - D - D - D - \boxed{\mathcal{X} - W - D - D - W} - \ldots \Rightarrow \boxed{W} - W - D - D - W$$

5. Then, from the set of such decisions, the most frequent element ("D" or "W") should be selected, and then it is used to fill in data:

$$\ldots - D - W - D - D - D - \boxed{D} - W - D - D - W - \ldots$$

Table 1. Accuracy of pattern-based data imputation, Potsdam.

	Missing levels								
	1%	5%	10%	15%	20%	25%	30%	35%	40%
One MV	72.4%	71.8%	71.5%	71.08%	70.41%	–	–	–	–
Two MVs	68.73%	68.12%	67.98%	66.7%	66.5%	66.32%	66.1%	–	–
Three MVs	54.11%	48.94%	48.63%	47.2%	46.8%	45.87%	45.7%	45.1%	44.7%

On test data, it was found that presence of missing values do not significantly change the frequency for the patterns, so this method can be successfully applied to real incomplete data. Figure 1 and Table 1 demonstrate examples for various missing levels.

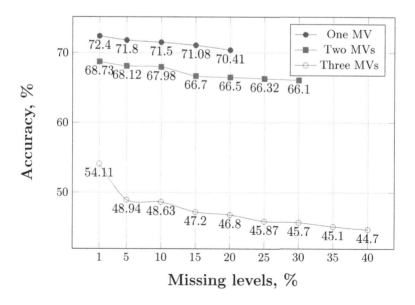

Fig. 1. Accuracy of pattern-based data imputation, Potsdam.

The described algorithm is a very simple and natural to use, however, the obtained values of the filling accuracy, especially for the case of several consecutive gaps, should be increased.

We demonstrate the case of only one consecutive missing value. In the remaining situations, one can act similarly with the corresponding modifications.

2.3   SVM for Handling the Missing Values

In this section, we use SVM, a supervised learning model with associated algorithms, to fill in the gaps in data based on "D–W" classification. The following features have been used: min, max and mean temperatures, mean dew point, mean wind speed. The corresponding results will be compared with decisions based on a probabilistic approach described in Sect. 2.2. Figure 2 and Table 2 demonstrate accuracy of SVM-based data imputation for 1, 2 and 3 consecutive missing values.

The x-axis presents missing levels from 1% to 40%, and the y-axis corresponds to a percentage of elements correctly filled by "D" or "W".

Table 2. Accuracy of SVM-based data imputation for patterns, Potsdam.

| | Missing levels | | | | | | | | |
	1%	5%	10%	15%	20%	25%	30%	35%	40%
One MV	80.17%	79.28%	78.41%	78.3%	77.52%	–	–	–	–
Two MVs	77.91%	76.34%	76.05%	75.12%	74.87%	74.31%	74.17%	–	–
Three MVs	75.68%	75.17%	73.69%	73.02%	73.32%	72.02%	71.87%	71.26%	70.19%

With increasing the total number of missing values, accuracy is reduced in all cases for both methods. But the values for a "pure probabilistic" approach are unsuitable for practical use even for 1% level especially if three consecutive MVs are allowed (see Fig. 1). The SVM accuracy even for the 40% missing level and the same number of consecutive MVs does not fall below 70%. Errors for the upper and lower levels differ by no more than 5.5%, that should also be considered as a good result for practice. It is also important to note that the SVM accuracies in the classification problem are close to each other for various numbers of consecutive MVs.

Table 3 presents the comparison of average prediction accuracies of SVM and pure probabilistic pattern-based filling.

For the case of one consecutive missing value, the difference is about 7%, for two and three consecutive MVs is about 8% and 25%, respectively. That is, when increasing the number of MVs, the accuracy of the probabilistic approach dramatically decreases, while SVM is less sensitive to this situation. Thus, SVM accuracy is higher than "pure probabilistic" one. So, these results indicate the possibility of practical applications in real problems. Some examples of using SVM-based classification for observations obtained in another climatic zone are also given in paper [23].

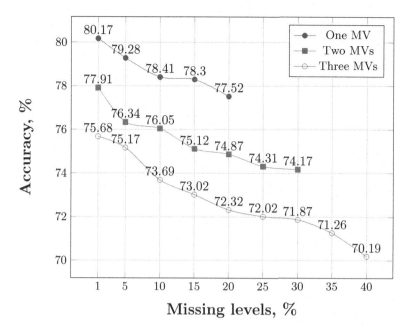

Fig. 2. Accuracy of SVM-based data imputation for patterns, Potsdam.

Table 3. Comparison of average prediction accuracies for "pure probabilistic" approach and SVM (Potsdam).

	Patterns	SVM
One MV	71.44%	78.74%
Two consecutive MVs	67.21%	75.54%
Three consecutive MVs	47.45%	72.8%

3 Approach Based on Regression

In this section, machine learning algorithms will be used to fill in the exact values of the MVs, that is, the regression problem will be solved. In this case, the patterns discussed in Sect. 2 will also be used as an auxiliary tool to improve the quality of the methods.

3.1 Simulation of Incomplete Data

In this section, we use a slightly different method of randomly selecting the positions for insertion of missing values.

1. The sample are divided into K equal parts, where the length of each is 5:

$$\boxed{W - D - D - D - W} - \boxed{W - D - D - W - W} - \ldots - \boxed{D - D - D - D - W}$$

2. Depending on the required percentage of the missing values, L subsamples are randomly selected from K ones mentioned above to replace with an unknown value:

$$W - D - D - D - W - \boxed{W - D - D - W - W} - \ldots - \boxed{D - D - D - D - W}$$

3. Then, in each of the L subsamples, the position to replace is randomly selected:

$$W - D - D - D - W - \boxed{W - \boxed{x} - D - W - W} - \ldots - \boxed{\boxed{x} - D - W - D - W}$$

This algorithm is simple to implement, and gaps are filled in a random but a controlled way. It is worth noting that this method also allows us to prevent the consecutive missing values. Using this algorithms, test samples for 1%, 5%, 10%, 15% and 20% levels were simulated using initial complete observations from Potsdam and Elista for about 60 years.

3.2   Algorithms for Handling the Missing Values

In this section, we briefly describe the regression machine learning algorithms that will be used to fill missing values.

Mean imputation, *Means*, is the simplest method of filling the missing values. All missing elements are replaced by an arithmetic mean of a selected subsample or the full time-series. The corresponding drawbacks are obvious. For example, this method can be not suitable well for the non-stationary time-series and data with outliers.

EM algorithm is one of the most popular regression algorithms that allows us to work effectively with large data volumes. It is assumed that the distribution of analyzed data can be approximated by a linear combination of multidimensional normal distributions.

k-Nearest Neighbors algorithm is one of the most used non-parametric method for data prediction. The missed observation should be filled by mean of values of k nearest neighbors. It is worth noting that k-NN is one of the simplest machine learning algorithms.

Random Forests are the ensemble learning methods for regression based on decision trees [15]. RFs present mean prediction as output in this problem. An ensemble training is used to obtain better results than could be obtained from any of the constituent methods alone. It is an attractive method for imputing missing data, wherein such data can be analyzed even without filling. Possible modification for better performance is discussed, for example, in [22].

Extreme gradient boosting **XGBoost** [6] is widely used to solve classification and regression problems in a wide range of applications (see, e.g., [5,25,29]).

Below, we consider the application of these algorithms to real precipitation data and compare the imputation accuracy for different missing levels. The volume of each test sample is more than 20000 observations.

3.3 Data Imputation with Pattern-Based Classification

Let us suppose that the filling procedure was initially carried out using the pattern-based classification. Thus, the "D–W" sample does not contain gaps. However, the exact values of daily volumes are still unknown. Thus, they need to be imputed using the regression approach.

First, it should be checked whether the missing value belongs to the dry or wet period. Indeed, if the MV is located in a dry period, then the best filling is realized by Means due to the zeros are correctly placed into the corresponding positions. In this case, other algorithms can forecast small but nonzero values, and thus it leads to increase the total error due to the test data contains a lot of zero values.

Table 4. Accuracy of data imputation for dry periods in Potsdam and Elista.

City	Missing levels				
	1%	5%	10%	15%	20%
Potsdam	89.4%	87.7%	84.2%	83.2%	77.3%
Elista	89.6%	87.9%	86.2%	85.1%	83.3%

Table 4 and Fig. 3 demonstrate the accuracies of data imputation for dry periods in Potsdam and Elista.

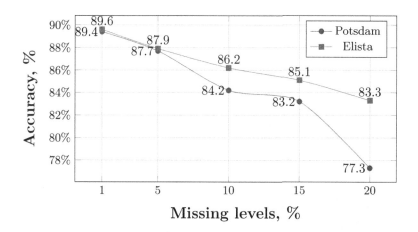

Fig. 3. Accuracy of data imputation for dry periods in Potsdam and Elista.

As a disadvantage, as mentioned earlier, it is worth noting that a pattern-based accuracy strongly decreases with increasing missing levels. The accuracy can be undoubtedly increased using SVM (see Sect. 2).

To evaluate the imputation accuracy for wet periods, the following metric is used: $\varepsilon_m = V_m^{-1} RMSE_m$, where $RMSE_m$ is a Root Mean Square Error corresponding to a m-th wet period and V_m is a total precipitation volume of the same period. Note that values V_m are determined using complete data. In fact, this is the normalization of observations over the wet period. It makes possible to accurately compare errors ε_m for different wet periods, and also to compute their mean value for all intervals with gaps.

A row vector $\varepsilon = \{\varepsilon_m\}_{m=\overline{1,L}}$ corresponds to the consecutive errors for all wet periods containing missing values. Then, the total error $Err = L^{-1}\varepsilon 1_{L\times 1}$, where $1_{L\times 1}$ is a column vector consisting of L ones.

Table 5. Accuracy of data imputation for wet periods in Potsdam.

Method	Missing levels				
	1%	5%	10%	15%	20%
Means	83.2%	82.4%	80.1%	78.8%	77.3%
k-NN	83.5%	82.3%	80.5%	79.1%	78.4%
EM algorithm	84.1%	83.6%	82.9%	80.4%	78.8%
Random forest	83.4%	82.3%	81.5%	79.7%	77.6%

Tables 5, 6 and Figs. 4, 5 present comparison of various ML algorithms for data imputation in Potsdam and Elista taking into account different missing levels. ML methods were implemented using *scikit-learn* library for the Python programming language.

Table 6. Accuracy of data imputation for wet periods in Elista.

Method	Missing levels				
	1%	5%	10%	15%	20%
Means	83.6%	82.4%	82.1%	80.2%	78.3%
k-NN	83.8%	82.9%	81.7%	80.3%	78.4%
EM algorithm	84.3%	83.7%	83.1%	81.2%	78.9%
Random forest	83.7%	82.5%	81.5%	80.3%	78.8%

In most cases, the Means accuracy is minimal among all compared methods. The exception is demonstrated for Elista on the 10% missing level (see Fig. 5). The differences for Potsdam on the 5% missing level (see Fig. 4) can be explained by computational errors.

The best results for all situations are given by EM algorithm. On test data, it turned out that EM and Means set the upper and lower bounds of accuracies. For Potsdam data on 20% missing level, EM accuracy is 1.5% more compared to

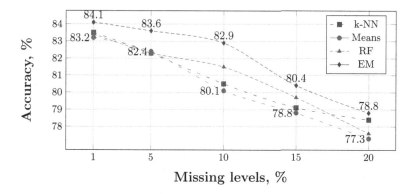

Fig. 4. Accuracy of data imputation for wet periods in Potsdam.

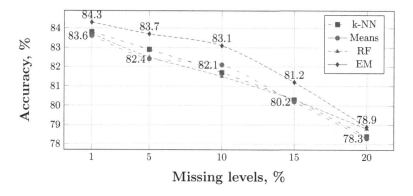

Fig. 5. Accuracy of data imputation for wet periods in Elista.

Means one (see Table 5). This is a significant advantage for real data. For Elista, this difference is about 1% (see Table 6). With the increasing missing levels, the RF results tends to EM values.

3.4 Data Imputation with SVM Classification

Let us suppose that the filling procedure was initially carried out using improved machine learning classification based on SVM (see Sect. 2.3). Tables 7 and 8 present improved results for various regression algorithms for one and two consecutive MV's.

Comparing Tables 5 and 7, it is easy to see that extreme gradient boosting provides a five percent advantage over the best results of Sect. 3.3. This is most clearly shown in Fig. 6.

The obvious advantage of this approach allows maintaining the accuracy of forecasting continuous data at a level of more than 80%. The results for two consecutive passes are presented in Table 8. It can be seen that the accuracy decreases significantly with increasing missing levels, but for SVM with XGBoost

Table 7. Improved accuracy of classification-regression data imputation: one MV, Potsdam.

Missing levels	Method			
	SVM+XGBoost	SVM+EM	SVM+RF	SVM+k-NN
1%	89.74%	84.27%	85.79%	86.20%
5%	88.94%	81.94%	84.49%	84.76%
10%	87.52%	81.01%	81.70%	79.33%
15%	84.51%	80.87%	79.83%	77.39%
20%	82.89%	78.81%	76.79%	75.06%

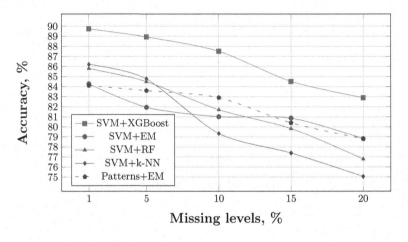

Fig. 6. Improved accuracy of data imputation in Potsdam.

Table 8. Improved accuracy of classification-regression data imputation: two MV's, Potsdam.

Missing levels	Method			
	SVM+XGBoost	SVM+EM	SVM+RF	SVM+k-NN
1%	82.63%	79.51%	71.96%	83.75%
5%	76.37%	75.15%	71.56%	74.87%
10%	75.92%	73.32%	72.04%	70.11%
15%	74.27%	72.67%	68.22%	67.14%
20%	73.74%	71.08%	66.22%	67.95%
25%	73.19%	69.88%	68.33%	70.59%
30%	71.39%	68.79%	65.99%	66.53%

Table 9. The learning rates in seconds.

Missing levels	Method			
	SVM+XGBoost	SVM+EM	SVM+RF	SVM+k-NN
1%	2.03	1.71	1.98	1.82
5%	1.99	1.93	1.88	1.74
10%	1.89	2.01	1.76	1.60
15%	1.64	2.55	1.54	1.39
20%	1.53	2.89	1.29	1.13
25%	1.18	3.05	0.98	0.84
30%	1.00	3.13	1.01	0.77

it remains above 70%. Moreover, it follows from Table 9 that the learning rate for this method is quite comparable with others, therefore, it can be used for solving real-time problems.

4   Conclusion

The paper presents two approaches to filling gaps in precipitation based on classification and regression machine learning algorithms as well as the pattern-driven methodology. The possibility of correct imputation even for high missing levels is demonstrated. The Python implementations lead to the possibility of using the high-performance computing, in particular, to solve the problem of the effective selection of various hyperparameters. The obtained results are quite suitable for the analysis of real incomplete data. Therefore, the analysis of data sets from distributed weather stations in Russia and neighboring countries, Europe, Asia, etc., should be mentioned as a direction for further researches. In addition, these methods can be applied to data of a different nature, in particular, to various information systems.

For the observed data, the optimal algorithm is based on SVM classification and XGBoost regression. However, the MVs problem does not have a universal solution, and there is no the only method that would be superior in quality to all others for all situations. Each ensemble requires an individual approach taking into account the physical nature of data. The further research in this area can also be focused on involving neural networks in classification, regression or both stages, because an accuracy of about 97% was obtained [14] for one-step pattern-based forecasts.

It is worth noting that the suggested methodology can be successfully used for real-time data processing of information flows [9,10]. For example, it can be useful for telecommunication loads or traffic, where different states exist due to the packet loss or hacker attacks.

Acknowledgments. The authors are grateful to **Professor V. Yu. Korolev** for useful discussions and joint researches. The authors would like to thank **Professor**

K. E. Samuylov for careful reading of material and the valuable comments that helped us to improve the manuscript.

References

1. Altman, N.: An introduction to kernel and nearest-neighbor nonparametric regression. Am. Stat. **46**(3), 175–185 (1992). https://doi.org/10.1080/00031305.1992.10475879
2. Athey, S., Tibshirani, J., Wager, S.: Generalized random forests. Ann. Stat. **47**(2), 1148–1178 (2019). https://doi.org/10.1214/18-AOS1709
3. Barrios, A., Trincado, G., Garreaud, R.: Alternative approaches for estimating missing climate data: application to monthly precipitation records in South-Central Chile. For. Ecosyst. **5**, 28 (2018). https://doi.org/10.1186/s40663-018-0147-x
4. Breiman, L.: Random forests. Mach. Learn. **45**, 5–32 (2001). https://doi.org/10.1023/A:1010933404324
5. Chatzis, S., Siakoulis, V., Petropoulos, A., Stavroulakis, E., Vlachogiannakis, N.: Forecasting stock market crisis events using deep and statistical machine learning techniques. Expert Syst. Appl. **112**, 353–371 (2018). https://doi.org/10.1016/j.eswa.2018.06.032
6. Chen, T., Guestrin, C.: XGBoost: a scalable tree boosting system. In: Proceedings of the 22nd ACM SIGKDD International Conference on Knowledge Discovery and Data Mining, pp. 785–794 (2016). https://doi.org/10.1145/2939672.2939785
7. Cortes, C., Vapnik, V.N.: Support-vector networks. Mach. Learn. **20**(3), 273–297 (1995). https://doi.org/10.1007/BF00994018
8. Fernandez-Gonzalez, P., Bielza, C., Larranaga, P.: Random forests for regression as a weighted sum of k-potential nearest neighbors. IEEE Access **7**, 25660–25672 (2019). https://doi.org/10.1109/ACCESS.2019.2900755
9. Gorshenin, A., Kuzmin, V.: Online system for the construction of structural models of information flows. In: Proceedings of the 7th International Congress on Ultra Modern Telecommunications and Control Systems and Workshops, pp. 216–219 (2015). https://doi.org/10.1109/ICUMT.2015.7382430
10. Gorshenin, A., Kuzmin, V.: On an interface of the online system for a stochastic analysis of the varied information flows. AIP Conf. Proc. **1738**(220009) (2016). https://doi.org/10.1063/1.4952008
11. Gorshenin, A.: Pattern-based analysis of probabilistic and statistical characteristics of precipitations. Informatika i ee Primeneniya **11**(4), 38–46 (2017). https://doi.org/10.14357/19922264170405
12. Gorshenin, A.: Investigation of parameters of meteorological models based on patterns. In: CEUR Workshop Proceedings, vol. 2177, pp. 4–10 (2018). http://ceur-ws.org/Vol-2177/paper-01-a005.pdf
13. Gorshenin, A., Korolev, V.: Determining the extremes of precipitation volumes based on a modified "Peaks over Threshold". Informatika i ee Primeneniya **12**(4), 16–24 (2018). https://doi.org/10.14357/19922264180403
14. Gorshenin, A., Kuzmin, V.: Neural network forecasting of precipitation volumes using patterns. Pattern Recognit. Image Anal. **28**(3), 450–461 (2018). https://doi.org/10.1134/S1054661818030069
15. Ho, T.: The random subspace method for constructing decision forests. IEEE Trans. Pattern Anal. Mach. Intell. **20**(8), 832–844 (1998). https://doi.org/10.1109/34.709601

16. Kalteh, A., Hjorth, P.: Imputation of missing values in a precipitation-runoff process database. Hydrol. Res. **40**(4), 420–432 (2009). https://doi.org/10.2166/nh.2009.001
17. Kim, J., Ryu, J.: Quantifying a threshold of missing values for gap filling processes in daily precipitation series. Water Resour. Manag. **29**(11), 4173–4184 (2015). https://doi.org/10.1007/s11269-015-1052-5
18. Korolev, V.Y.: Probabilistic and Statistical Methods of Decomposition of Volatility of Chaotic Processes. Moscow University Publishing House, Moscow (2011)
19. Lulli, A., Oneto, L., Anguita, D.: Mining big data with random forests. Cogn. Comput. **11**(2), 294–316 (2019). https://doi.org/10.1007/s12559-018-9615-4
20. Sattari, M., Rezazadeh-Joudi, A., Kusiak, A.: Assessment of different methods for estimation of missing data in precipitation studies. Hydrol. Res. **48**(4), 1032–1044 (2017). https://doi.org/10.2166/nh.2016.364
21. Simolo, C., Brunetti, M., Maugeri, M., Nanni, T.: Improving estimation of missing values in daily precipitation series by a probability density function-preserving approach. Int. J. Climatol. **30**(10), 1564–1576 (2010). https://doi.org/10.1002/joc.1992
22. Tang, F., Ishwaran, H.: Random forest missing data algorithms. Stat. Anal. Data Min. **10**(6), 363–377 (2017). https://doi.org/10.1002/sam.11348
23. Teegavarapu, R., Aly, A., Pathak, C., Ahlquist, J., Fuelberg, H., Hood, J.: Infilling missing precipitation records using variants of spatial interpolation and data-driven methods: use of optimal weighting parameters and nearest neighbour-based corrections. Int. J. Climatol. **38**(12), 776–793 (2018). https://doi.org/10.1002/joc.5209
24. Teegavarapu, R., Chandramouli, V.: Improved weighting methods, deterministic and stochastic data-driven models for estimation of missing precipitation records. J. Hydrol. **312**(1–4), 191–206 (2005). https://doi.org/10.1016/j.jhydrol.2005.02.015
25. Torres-Barran, A., Alonso, A., Dorronsoro, J.: Regression tree ensembles for wind energy and solar radiation prediction. Neurocomputing **326**, 151–160 (2019). https://doi.org/10.1016/j.neucom.2017.05.104
26. Wang, W., Du, X., Wang, N.: Building a cloud IDS using an efficient feature selection method and SVM. IEEE Access **7**, 1345–1354 (2019). https://doi.org/10.1109/ACCESS.2018.2883142
27. Wu, X., et al.: Top 10 algorithms in data mining. Knowl. Inf. Syst. **14**(1), 1–37 (2008). https://doi.org/10.1007/s10115-007-0114-2
28. Yang, N., Wang, Y.: Identify silent data corruption vulnerable instructions using SVM. IEEE Access **7**, 40210–40219 (2019). https://doi.org/10.1109/ACCESS.2019.2905842
29. Zhang, D., Qian, L., Mao, B., Huang, C., Huang, B., Si, Y.: A data-driven design for fault detection of wind turbines using Random Forests and XGboost. IEEE Access **6**, 21020–21031 (2018). https://doi.org/10.1109/ACCESS.2018.2818678
30. Zolina, O., Simmer, C., Belyaev, K., Kapala, A., Gulev, S.: Improving estimates of heavy and extreme precipitation using daily records from European rain gauges. J. Hydrometeorol. **10**, 701–716 (2009). https://doi.org/10.1175/2008JHM1055.1

The Bans in Finite Probability Spaces and the Problem of Small Samples

Alexander Grusho$^{(\boxtimes)}$ ⓘ, Nick Grusho, and Elena Timonina ⓘ

Institute of Informatics Problems of Federal Research Center,
"Computer Science and Control" of the Russian Academy of Sciences,
Vavilova 44-2, 119333 Moscow, Russia
{grusho,eltimon}@yandex.ru, info@itake.ru

Abstract. Traditional statistical methods of analysis of small samples produce many "false alarms" and a possibility to miss searched information with a high probability. The paper presents formal arguments about the cause of this problem. When finite probability spaces are considered, it becomes possible to use theory of bans and to solve the problem of small samples.

In the model of this theory all probabilities of "false alarms" are equal to zero. At the same time in many cases it is possible to prove either the consistency of the sequence of criteria, or the correct acceptance of the alternative with probability 1 on a finite step.

Bans of probability measures in finite probability spaces considerably simplify the identification of samples with the probability distributions which differ from supposed distributions.

The simplest models of small samples in finite probability spaces are considered. Partially these models were used in control systems for search of a malicious insider in a big communities of users.

The generalization of this research is the introduction of multiple information spaces. This paper describes the usage of bans in the model of small samples from two information spaces.

The finiteness of probability spaces allows to expand a set of possible models. The models considered in the paper far do not exhaust a class of possible generalizations.

Keywords: Small samples · Bans in finite probability space · Statistical tests

1 Introduction

The problem of small samples begun to draw an attention of experts in the field of mathematical statistics long ago [1–3].

Let's assume that the set of small samples is given, a part from which is received according to some finite set of distributions $\{P_\theta,\ \theta \in \Theta\}$, and another

Partially supported by Russian Foundation for Basic Research (project 18-29-03081).

V. M. Vishnevskiy et al. (Eds.): DCCN 2019, LNCS 11965, pp. 578–590, 2019.
https://doi.org/10.1007/978-3-030-36614-8_44

part of them is received according to other set of distributions $\{Q_\omega, \omega \in \Omega\}$. One of wordings of the problem consists in check of hypothesis that in the set of samples there are no samples received according to the second set of distributions against alternative that such samples are present. In this paper we will be interested in this wording of the problem.

The model discussed in the paper is closely related to modern methods of machine learning in intelligent data analysis. In the review [4] the big list of articles in the field of Small Sample Learning (SSL) is provided. In particular, our formulation of the problem of small samples corresponds to the case of Long Tail Distribution. In a classification task where many frequently encountered classes are so dominant that rarely encountered classes are often skipped and give rise to classification errors. The proposed bans of probability measures correspond to the emergence of a new meaning in Concept Learning in SSL. Concept Learning is based on a variety of contents and rules to correlate the received sample to some existing content. In our case, content is described by a set of distributions in different information spaces, and compliance rules by statistical criteria. Bans can appear in any of the information spaces, hence the classifier can learn new content and recognize known content.

The applied nature of the model considered in the paper is the next. In the description of a subject to classification several information spaces are used, each of which has finite power. In finite information spaces there are contents which are not present at often found classes. Often contents are expressed in appearance of samples with positive probabilities. New contents have in such distributions probability 0 and are bans [5].

Such situations can be interpreted depending on applied tasks, for example

– emergence of failure in an information technology,
– violation of security rules at execution of tasks of an information technology.

An example of identification of the harmful malicious insider with use of many information spaces was reviewed in the paper [6].

In [6] we introduced the method of several information spaces for analyzing of different kinds of behavior data for search of a malicious insider. The idea of the method consists in gathering heterogeneous information from different processes and combining of this information for getting a common result. Using [6] we consider vectors of small samples which synchronously appear for analyzing.

The idea to introduce a binary sequence γ which regulates the appearance of bans also is connected with the problem of search of a malicious insider. The clever insider will not use his capacity to get additional information without of some aim. Mostly the reason to break rules comes when some extremely interesting information appears for the malicious insider. For example, huge sums of money are transferred. Such events can be known to security officers. This information can be used to increase the attention on users' behaviors. When the correlation of such events and of search for additional information is established then security officers get more arguments to accuse the insider.

In [7] the theory of bans was used in the problem of search of malicious insider. A finite number of analysts get tasks to analyze parts of data from database. The database is organized as rows of a table of big data. Columns of the table are named attributes. Every task demands for analyst to look in a bounded number of attributes. These numbers are much less than the total number of attributes. During the work at a task it is forbidden to increase the number of attributes that are needed for the current analyzing task. The main idea of [7] is that a malicious insider tries to increase a quantity of attributes which are necessary to solve his current task. As every analyst gets short time for his current task, but is supposed to solve many tasks he deals with the set of small samples of attributes.

In [7] bans are defined by appearance of additional attributes in samples which exceed the permitted sets of attributes. Due to dividing of samples into two classes (permitted and unpermitted samples) it became possible to construct invariants for differing samples. It helped to unify small samples and to gather enough data for every analyst to describe his behavior.

So the theory of bans was used for search of a malicious insider among big set of analysts.

In fact many applications can be interpreted as statistical decision in spaces of small samples. Analyze of heterogeneous data with usage of the theory of bans helps to exclude different cases of false alarms.

However there are some other wordings of the problem when some integrated characteristics of a set of distributions of small samples are investigated. In [1] the approach of creation of some invariant statistics, which allow to reduce the set of small samples until one array of statistical data was considered.

In articles with two classes of distributions [3, 8, 9] various assumptions, concerning proximity of classes of distributions which allow to test hypotheses, are done.

Considerable part of papers (see, for example, [10]) has experimental character. Attempts to find measures of proximity, allowing to estimate characteristics of a set of small samples are done in these articles.

In our paper the attempt to apply the theory of bans [11] for tasks with many small samples on finite spaces is done. All probabilities of "false alarms" are equal to zero in this theory. At the same time in many cases it is possible to prove either the consistency of the sequence of criteria, or the correct acceptance of an alternative with probability 1 on a finite step. The simplest models of small samples on finite probabilistic spaces are considered in the paper.

The paper is organized as follows. Section 2 describes problems in analysis of a set of small samples. Section 3 describes the small samples model. Section 4 deals with model of insertions. Section 5 shows ways of generalizations. In the Conclusion we summarize results and shortly describe future researches.

2   Problems of Analysis of a Set of Small Samples

The complexity of the problem of analysis of a set of small samples consists in the fact that, having constructed criterion for a small sample with an error of the "false alarm" limited to $\alpha > 0$, and errors to miss a presence of samples with alternative probability distributions $\beta(\omega) > 0$, we will receive the following picture [12,13]. If we reduce the value α, then β begins to grow so, that the probability to miss required samples becomes big. Therefore in many cases traditional methods of mathematical statistics do not work.

However in [12,13] the proof of this fact is carried out by numerical examples. That is why in this paper the other model is offered allowing to prove the specified property in wide class of distributions.

Let X be a finite set of small samples. Assume small samples of this set be chosen independently of each other from the same measurable space, but with different probability measures. As a matter of convenience we will renumber samples of the set X, i.e. $X = \{x_1, ..., x_N\}$. It is known that among these samples only one sample is received according to the probability distribution P_1, and other samples are received according to the probability distribution P_0. The problem consists in identification of the only sample received according to the probability distribution P_1.

Let T be a test with a critical set S which for each sample x_i makes the decision $T(x_i) = 1$ if it is possible to consider that sample x_i is received according to the probability distribution P_1, and $T(x_i) = 0$, if it is possible to consider that the sample x_i is received according to the probability distribution P_0. As a matter of convenience we write $T(x_i) = T_i$, $i = 1, ..., N$.

Also we enter a random variable $f(x_i) = f_i$, which is equal to 1, if the sample received according to the probability distribution P_1 is located on i-th place, and $f_i = 0$, if the sample received according to the probability distribution P_0 is located on the place i.

Let's consider a joint probability distribution of random variables f_i, T_i, $i = 1, ..., N$. Denote $P(f_i = 1, T_i = 0)$ be the probability that the sample received according to the probability distribution P_1 is located on i-th place, but the test defines it is the sample received according to the probability distribution P_0. Thus it expresses the probability of loss of required sample on i-th place.

Expression $P(f_i = 0, T_i = 1)$ is the probability of "false alarm", when the sample received according to the probability distribution P_0 is located on i-th place, but it is accepted as the sample received according to the probability distribution P_1.

Expression $P(f_i = 0, T_i = 0)$ is the probability of the correct decision if on place i the sample is received according to the probability distribution P_0, and expression $P(f_i = 1, T_i = 1)$ is the probability of the correct decision, if on place i the sample is received according to the probability distribution P_1.

The specified probabilities can be expressed as follows:

$$P(f_i = 1, T_i = 0) = \frac{1}{N}P(T_i = 0| f_i = 1) = \frac{1}{N}P_1(T_i = 0), \qquad (1)$$

and

$$P(f_i = 0, T_i = 1) = \frac{N-1}{N}P(T_i = 1| f_i = 0) = \frac{N-1}{N}P_0(T_i = 1). \qquad (2)$$

Other probabilities are similarly expressed.

Let's note that all probabilities defined above are identical at $i = 1, ..., N$. Therefore in the formulas, which will be considered further, we will miss the index i.

We use (2) for calculation of the mathematical expectation of number of "false alarms":

$$N \frac{N-1}{N} P_0(T = 1) = (N-1)P_0(T = 1). \tag{3}$$

Remember that in the set X there is only one sample received by the probability distribution P_1. Then by the choice of function T it is possible to receive the following relation.

$$(N-1)P_0(T = 1) = C_1 > 0.$$

From here

$$P_0(T = 1) = \frac{C_1}{N-1},$$

that for large N can be considered as an asymptotic estimation of the probability $P_0(T = 1)$ in the task of the analysis of the set of small samples.

Let's consider the asymptotical behaviour of the probability of the loss of the required sample received according to the probability distribution P_1. As it was shown earlier, the probability of the correct decision that in the fixed place there is the sample from the probability distribution P_1, is equal to

$$\frac{1}{N} P_1(T = 1).$$

Then it follows, that the probability of the wrong decision that in this place there is the sample received according the probability distribution P_1, is equal to

$$1 - \frac{1}{N} P_1(T = 1). \tag{4}$$

Probability that the sample received according the probability distribution P_1 "is lost" on this place doesn't depend on the same checking procedure for any other place. Therefore the probability of loss of the sample received according to the probability distribution P_1 after testing of each sample of the set X is equal to

$$(1 - \frac{1}{N} P_1(T = 1))^N.$$

Let $N \to \infty$. Then the probability of loss of the sample received according to the probability distribution P_1 is asymptotically equal to

$$\exp(-P_1(T = 1))(1 + O(\frac{1}{N})).$$

Note that when $P_1(T = 1)$ tends to 1, the probability of loss of this sample is big ($\sim \frac{1}{e}$). It is possible because the required case can be considered as "false alarm".

This simple example confirms complexity of the problem of small samples. If there are a few samples received according to the probability distribution P_1, but the ratio of the number of such samples to the total number of samples tends to 0, when $N \to \infty$, then the essence the picture remains the same. But similar calculations in this case are complicated, because it is necessary to consider joint probability distributions for the crossed subsets of possible appearance of samples received according to the probability distribution P_1.

It is interesting to compare probabilities of loss of required samples, which are received in various works. In [13] it is supposed, that there are two attempts per a day to fulfill intrusions into a computer system at the total number of records of audit per a day, equals to 10^6. The factor of "false alarm" demands decrease of probability of this factor to 2×10^{-5} (i.e. 20 "false alarms"). The Bayesian method of identification of intrusions in these conditions gives the estimation of the probability of loss of the records identifying intrusions, as 0.62, that approaches the estimates received in this work.

Similar results are received in [8] by consideration of the problem of prediction of wrong data (samples according to the probability distribution P_1) on the basis of the Weibull model.

3   Model of Small Samples

For simplicity we will consider only two information spaces to support SSL. Then there are two types of small samples out of the first set of probability distributions. Let two finite alphabets A_1 and A_2 be given. Let's assume that all samples have length N. Let's designate through $X_1 = A_1^N$ and $X_2 = A_2^N$ the sets of words of length N in alphabets A_1 and A_2 respectively. Let $X = X_1 \times X_2$. From here we receive the sequence of couples of small samples $x^{(n)} = (x_1, x_2)^{(n)}$, $n = 1, 2, ...$, where $x_1 \in X_1$, $x_2 \in X_2$.

Let's define two probability distributions P_1 on X_1 and P_2 on X_2. Their supports are
$$D_1 = \{x_1 \in X_1, P_1(x_1) > 0\}$$
and
$$D_2 = \{x_2 \in X_2, P_2(x_2) > 0\}$$
At the same time we suppose that $D_1 \neq X_1$ and $D_2 \neq X_2$. It means that in the set X_1 there is $x_1 \in X_1$, that $P_1(x_1) = 0$. Similarly, in the set X_2 there is $x_2 \in X_2$, that $P_2(x_2) = 0$.

Let's designate through S_1 and S_2 the sets $S_1 = X_1 \setminus D_1$ and $S_2 = X_2 \setminus D_2$ respectively. We will call elements of sets S_1 and S_2 by bans of the probability measures P_1 and P_2 respectively.

Let's define the probability measure $P^{(1)}$ on the set $X = X_1 \times X_2$ as follows:
$$P^{(1)}(x) = P_1(x_1) \cdot P_2(x_2), \ x = (x_1, x_2) \in X.$$

The probability measure on the set of small samples X^n is defined as follows

$$P^{(n)}(\overline{x}^{(n)}) = \prod_{i=1}^{n} P^{(1)}(x^{(i)}),$$

where $\overline{x}^{(n)} = (x^{(1)}, ..., x^{(n)}) \in X^n$.

It is clear, that it is the sequence of consistent measures. Then on space of the infinite sequences X^∞ with σ-algebra $\mathcal{A}$, generated by cylindrical sets, the single probability measure P is defined, for which probability measures $P^{(n)}$ are projections on the first n coordinates [14], i.e.

$$P(\overline{x}^{(n)} \times X^\infty) = P^{(n)}(\overline{x}^{(n)}).$$

Also σ-algebra $\mathcal{A}$ is Borel σ - algebra in Tychonoff product X^∞ with the discrete topology in X [14,15]. Further for X^∞ let's designate $x|_n$ be the vector $\overline{x}^{(n)}$, which consists of the first n coordinates of the sequence x.

The support of the measure $P^{(1)}$ is the set $D = D_1 \times D_2$. Then support of measure $P^{(n)}$ is the set D^n, and the set of bans of the probability measure $P^{(n)}$ is the set $X^n \setminus D^n$, i.e. if

$$\overline{x}^{(n)} \in X^n \setminus D^n,$$

then the equality $P^{(n)}(\overline{x}^{(n)}) = 0$ is fulfilled.

The support of the measure P is defined by the following limit [11]:

$$\Delta(P) = \bigcap_{n=1}^{\infty} (D^n \times X^\infty).$$

Let probability measures Q_1 and Q_2 on sets X_1 and X_2 be defined. Assume that their supports D_1' and D_2' fulfill the following conditions:

$$D_1 \subset D_1', \ D_2 \subset D_2'.$$

Then there exists such $\varepsilon > 0$, that

$$Q_1(D_1) < 1 - \varepsilon, \ Q_2(D_2) < 1 - \varepsilon.$$

Define the probability measure $Q^{(1)}$ on X by the next condition

$$Q^{(1)}(x_1, x_2) = Q_1(x_1) \cdot Q_2(x_2), \ x_1 \in X_1, \ x_2 \in X_2.$$

From here we have the following estimation:

$$Q^{(1)}(D_1 \times D_2) < (1 - \varepsilon)^2 = \delta < 1.$$

Support of measure $Q^{(1)}$ is defined from the expression $D' = D_1' \times D_2'$. As before

$$Q^{(n)}(\overline{x}^{(n)}) = \prod_{i=1}^{n} Q^{(1)}(x^{(i)}),$$

where $\overline{x}^{(n)} = (x^{(1)}, ..., x^{(n)}) \in X^n$.

These probability measures are consistent, therefore on $(X^\infty, \mathcal{A})$ there exists the single probability measure Q, for which $Q^{(n)}$ is the projection of Q on the first n coordinates.

Let $\Delta(Q)$ be the support of the probability measure Q. For every $\overline{x}^{(n)} \in X^n$ the next equality is carried out:

$$Q(\overline{x}^{(n)} \times X^\infty) = Q^{(n)}(\overline{x}^{(n)}).$$

Let's consider the sequence of hypotheses

$$H_{0n} : P^{(n)},$$

$$H_{1n} : Q^{(n)}.$$

We need to define the infinite sequence of events $B_n = D^n$, $n = 1, 2, ...$ Each event consists of elements of D^n. It follows from definition of the probability measure Q that

$$Q^{(n)}(B_n) < \delta^n.$$

From here it follows that

$$\sum_{n=1}^\infty Q(B_n \times X^\infty) = \sum_{n=1}^\infty Q(D^{(n)} \times X^\infty) \leq \sum_{n=1}^\infty \delta^n = \frac{1}{1-\delta}.$$

Then Borel-Cantelli lemma [16] says that in the sequence defined above a finite number of events happens with probability 1, i.e. identification of Q happens with probability 1 for a finite number of steps. Then $\Delta P \subset \Delta(Q)$, $Q(\Delta(P)) = 0$, and it means that there is N such, that for every $\forall n \geq N$ the next equations are fulfilled

$$Q(X^\infty \setminus \Delta(P)) = 1.$$

It means that there is N such, that for every $n \geq N$, if the set $X^n \setminus D^{(n)}(P)$ can be taken as the critical set, then $\alpha = 0$, and the power function of the criterion for any n is estimated by $1 - \delta^n$.

However, such definition of the critical set is inconvenient, because the complexity of calculation of belonging to the critical set is rather high.

Let's define the critical set of criterion with less complicated function of belonging. Let

$$S = (S_1 \times D_2) \cup (S_2 \times D_1) \cup (S_1 \times S_2).$$

For n samples from X^n the critical set for all hypothesis H_{0n} against alternative H_{1n} is defined as S^n. It is clear, that it can be a serial function evaluation of belonging to the critical set, sequentially considering $S^1, S^2, ..., S^n$. If the first couple of samples out of X doesn't belong to S^1, then we analyze the second couple. If the second couple doesn't belong to S^1, then the first two couples don't belong to S^2. Thus, belonging to S^n is calculated sequentially by means of calculation of belonging of each couple of $\overline{x}^{(i)}$, $i = 1, ..., n$, to the set S if all previous couples of $\overline{x}^{(j)}$, $j = 1, ..., i$, fulfill the condition $\overline{x}^{(j)} \notin S^j$, $j = 1, ..., i$.

The constructed critical set S^n satisfies the condition

$$\overline{x}^{(n)} \in S^n \Rightarrow P^{(n)}(\overline{x}^{(n)}) = 0,$$

i.e. the probability of "false alarm" is equal to zero.

Estimation of power function of criterion is still equal to $1 - \delta^n$, as the constructed critical set S^n coincides with the set $X^n \setminus D^{(n)}(P)$.

4   Model of Insertions

Let the binary sequence γ be fixed. It begins with 0 and contains the infinite number of 1 and 0. Let's define the bans $a \in X_1$ and $b \in X_2$ such that $a \notin D_1$, $b \notin D_2$. The random process P is defined as it was made before. The set $\Delta(P)$ is the support of the probability measure P in the space $(X^\infty, \mathcal{A})$.

Assume that any sequence from $\Delta(P)$ can be changed by means of the sequence γ as follows. If in the i-th place in the sequence γ the element equals to 1, then on this place in any sequence $x \in \Delta(P)$ the element (a, b) is inserted. Further the new sequence x is generated in such a way that the element $x^{(i)}$ becomes the element $x^{(i+1)}$ of the new sequence. Thus, the sequence of bans of the probability measure P is inserted in all sequences $x \in \Delta(P)$ by means of the sequence γ.

Let Δ be a set of sequences from X^∞, which are received by an insertion of one-type bans (a, b) into all sequences of $\Delta(P)$. Let's consider the support $D^{(n)}$ of the measure $P^{(n)}$, and let's limit the sequence γ by its first n of coordinates. Denote limited vector as $\gamma|_n$. Let the number of 1 in the vector $\gamma|_n$ equals to m. Define the probability measures

$$Q^{(n+m)}(\overline{x}^{(n+m)}) = P^{(n)}(\overline{x}^{(n)}),$$

where $\overline{x}^{(n)} \in D^{(n)}$ and all $\overline{x}^{(n+m)}$ are generated from $\overline{x}^{(n)} \in D^{(n)}$ by inserting of m bans according to 1 in the sequence $\gamma|_n$.

Let's designate the received set through $\widetilde{D}^{(n+m)}$. As inserting to the same places is carried out for all $\overline{x}^{(n)} \in D^{(n)}$, then $\widetilde{D}^{(n+m)}$ contains expansions of all vectors from $D^{(n)}$, and only them. Then

$$\sum_{\overline{x}^{(n+m)} \in \widetilde{D}^{(n+m)}} Q^{(n+m)}(\overline{x}^{(n+m)}) = \sum_{\overline{x}^{(n)} \in D^{(n)}} P^{(n)}(\overline{x}^{(n)}) = 1.$$

The probability measures $Q^{(n)}$ are defined for all n. Really, let $Q^{(n+m)}$ be defined as it was described above. Then $Q^{(n+m-1)}$ is defined by:

1. the probability measure $P^{(n-1)}$ for $D^{(n-1)}$, when the number of units in $\gamma|_{n-1}$ is still equaled to m,
2. the probability measure $P^{(n)}$ for $D^{(n)}$, when the vector $\gamma|_n$ has 1 in n-th place.

In the first case probability measures $Q^{(n+m)}$ and $Q^{(n+m-1)}$ are consistent, because for any $\overline{x}^{(n-1)} \in D^{(n-1)}$ the next equality is carried out:

$$P^{(n)}(\overline{x}^{(n-1)}, X) = P^{(n-1)}(\overline{x}^{(n-1)}).$$

In the second case, throwing out the ban, we will receive:

$$Q^{(n+m)}(\overline{x}^{(n+m)}) = Q^{(n+m-1)}(\overline{x}^{(n+m-1)}),$$

since $P^{(n)}(\overline{x}^{(n)})$ determines both probabilities.

The consistency of finite-dimensional probabilistic distributions follows from conditions:

$$Q^{(n+m)}(\overline{x}^{(n+m-1)}X) = Q^{(n+m-1)}(\overline{x}^{(n+m-1)}).$$

The consistency of the probability measures $Q^{(n)}$ means that there is the only probability measure Q in $(X^\infty, \mathcal{A})$. As $\widetilde{D}^{(n)}$ are supports of the probability measure Q^n, it is obvious that the sequence of cylindrical sets $\widetilde{D}^{(n)} \times X^\infty$ does not increase (though some of these cylinders coincide). Then [11]

$$\Delta(Q) = \bigcap_{n=1}^{\infty} (\widetilde{D}^{(n)} \times X^\infty)$$

is the support of the probability measure Q. At the same time all sequences from $\Delta(Q)$ have bans according to the sequence γ, and any sequence from $\Delta(P)$ has no bans.

Theorem 1. *The sequence of hypotheses H_{0n} and H_{1n} allows to receive the power of criterion equaled to 1 on a finite step.*

Proof. It is clear, that

$$\Delta(P) \bigcap \Delta(Q) = \emptyset.$$

Then according to the theorem in [12] the sequence of hypotheses H_{0n} and H_{1n} allows to receive the power of criterion equaled to 1 on a finite step.   □

This result is clear without such complex model because it is supposed, that we know that (a, b) is the ban. But this model becomes important when we do not know that (a, b) is the ban for undefined probability distribution P. In this case the traditional statistical methods should be used to prove that the regularity of (a, b) appearances differs from regularities of appearances of other pairs of small samples.

Introduction of insertions of bans helps to prove that there is no intersection between the supports of the probability measures. From here it follows that the sequence of criteria satisfies the property of consistency of tests.

5   Discussion

1. If at the last section instead of a binary sequence γ we use 4-valued sequence γ, then it is possible to refuse of the limitation on bans by the type (a, b), and to consider bans on any component of a vector $\overline{x}^{(1)} \in X$. For this case the same results about the sequences of tests can be proved. I.e. the probabilities of "false alarm" are equal to 0, and that the power of any criterion reaches 1 in a finite step is evident.

2. The next generalization is simultaneous insertions of different bans by means of several sequences $\gamma_1, ..., \gamma_k$, $k < \infty$. Cases of coincidence of places of the bans don't lead to contradictions in the constructed reasoning. Every γ_i, $i = \overline{1; k}$ generates the probability measure $Q(i)$, and supports of these probability measures don't intersected with $\Delta(P)$.

3. The case of a two-dimensional design of receiving small samples was considered above. The dimension of X can be increased to any finite natural number. All proofs remain in force, but a complexity of their statement increases.

4. It is possible to consider the one-dimensional sequence of small samples receiving which will demand an additional mechanism of management of their emergence. However, such mechanism isn't connected with emergence of bans in each sample. Therefore, for independent emergence of samples of different type the same results of identification of bans are true.

6   Conclusion

In the paper the problem of testing hypotheses about a probability distribution of data given in a set of small samples is considered. Usage of bans of probability measures on finite probability spaces considerably simplifies identification of existence of samples with probability distributions which differ from supposed probability distributions.

As a matter of convenience probabilistic spaces of vectors on a finite alphabet are considered. At the same time spaces are divided into two parts:

– the area of the vectors appearing with positive probability;
– the area of vectors, everyone of them has zero probability, i.e. is a ban.

Advantages of bans in finite probabilistic spaces do not mean that the theory of bans has no problems. Namely, we will denote two problems.

The first problem consists in finding of bans belonging to a set of bans. According to the finiteness of probabilistic spaces there is always a possibility of exhaustive search of vectors of this space and estimation of function of belonging to the set of bans.

However with growth of dimensions of vectors there is a problem of complexity of an estimation algorithm of function of belonging to the set of bans. Some decrease in complexity can be received by the introduction of the concept of the smallest ban [11], and by the opportunity to express any ban through the smallest ban. However such method of decrease in complexity should be improved for practical applications.

The second problem consists in fast calculation of emergence of a ban in the observed random sequence. In the scheme of many small samples the problem of search of a ban in appeared samples has a limited complexity. However with growth of number N of samples the calculation of the criterion based on bans becomes a labour-intensive task.

The interesting approach to the task of the description of bans and expeditious calculation of existence of a ban in small sample is given in the paper [7]. The main idea of this approach consists in presenting each event of the probabilistic space in the form of a set of characteristics from a some initial set, i.e. to carry out a detailing of each elementary event in the probabilistic space. Then the small sample is presented in the form of a set of characteristics, and the detailing is chosen in such a way that bans are characterized by the appearance of characteristics which are not belonging to usual events. So the ban is characterized by the appearance of characteristics which are not peculiar to samples with the initial probability distribution. Then the function of belonging of the sample to the set of bans is defined by the existence of an "excess" characteristic that considerably simplifies estimation of function of belonging.

Such simplification allowed to describe bans in the paper [7] by a set of small samples connected with work of many analysts. At the same time invariant statistics of relatively standard sets of characteristics of separate events is constructed. Usage of invariant statistics allowed to unite the analysis of the set of small samples and to carry out the analysis by standard statistical methods.

The finiteness of probability spaces allows to considerably expand a set of possible models. The models considered in the paper far do not exhaust a class of possible generalizations.

Introduction of sequences of insertions of bans does not reduce a generality since these sequences are unknown to an analyst (unknown parameter). However, a finite set of such sequences allow to prove the nonintersecting of supports of appropriate measures.

References

1. Linnik, Yu.V.: Statistical Tasks with Disturbing Parameters. Science, Moscow (1966)
2. Volodin, I.N.: Testing of statistical hypotheses on the type of distribution by small samples. Kazan. Gos. Univ. Učen. Zap. **125**(6), 3–23 (1966). (Russian)
3. Petrov, A.A.: Verification of statistical hypotheses on the type of a distribution based on small samples. Theory Probab. Appl. **1**(2), 223–245 (1956)
4. Shu, J., Xu, Z., Meng D.: Small sample learning in big data era. arXiv:1808.04572v3 [cs.LG], 22 August 2018
5. Grusho, A., Timonina, E.: Prohibitions in discrete probabilistic statistical problems. J. Discret. Math. Appl. **21**(3), 275–281 (2011)
6. Grusho, A.A., Zabeshailo, M.I., Smirnov, D.V., Timonina, E.E.: The model of the set of information spaces in the problem of insider detection. J. Inform. Appl. **11**(4), 65–69 (2017)
7. Martyanov, E.A.: Bans of probability measures in the problem of insider detection. J. Syst. Means Inform. **27**(4), 144–149 (2017)

8. Wang, H., Yang, G., Bai, L., Juanyin, Li, Q.: Small sample fault data prediction study based on Weibull model. In: Proceedings of International Conference on Computer Science and Mechanical Automation (CSMA), pp. 9–14 (2015)

9. Liu, Q., Li, Y., Boyett, J.M.: Controlling false positive rates in prognostic factor analyses with small samples. Stat. Med. **16**(18), 2095–2101 (1997)

10. Yu, C., Pan, Q., Cheng, Y., Zhang, H.: Small sample size problem of fault diagnosis for process industry. In: IEEE ICCA 2010, pp. 1721–1725 (2010)

11. Grusho, A., Grusho, N., Timonina, E.: Quality of tests defined by bans. In: Proceedings of the 16th Applied Stochastic Models and Data Analysis International Conference (ASMDA2015) with Demographics 2015 Workshop, Piraeus, Greece, pp. 289–295 (2015)

12. Axelsson, S.: The base-rate fallacy and its implications for the difficulty of intrusion detection. In: Proceedings of the 6th Conference on Computer and Communications Security, pp. 1–7 (1999)

13. Axelsson, S.: The Base-Rate Fallacy and the Difficulty of Intrusion Detection. ACM Trans. Inf. Syst. Secur. **3**(3), 186–205 (2000)

14. Prokhorov, Yu.V., Rozanov, Yu.A.: Theory of probabilities. Science, Moscow (1993)

15. Bourbaki, N.: Topologie G'en'erale. Russian translation. Science, Moscow (1968)

16. Shiryaev, A.N.: Probability. GTM, vol. 95. Springer, New York (1995). https://doi.org/10.1007/978-1-4757-2539-1

Reliability Evaluation of a Distributed Communication Network of Weather Stations

Dmitry Aminev[1,3], Evgeny Golovinov[3], Dmitry Kozyrev[1,2(✉)] [ID],
Andrey Larionov[1] [ID], and Alexander Sokolov[1]

[1] V. A. Trapeznikov Institute of Control Sciences of Russian Academy of Sciences,
65, Profsoyuznaya Street, Moscow 117997, Russia
aminev.d.a@ya.ru, larioandr@gmail.com, aleksandr.sokolov@phystech.edu
[2] Department of Applied Probability and Informatics,
Peoples' Friendship University of Russia (RUDN University),
6 Miklukho-Maklaya Street, Moscow 117198, Russian Federation
kozyrev-dv@rudn.ru
[3] Federal State Budgetary Scientific Institution All-Russian Research Institute
for Hydraulic Engineering and Land Reclamation (VNIIGiM), Moscow, Russia
evgeny@golovinov.info

Abstract. Distributed automatic weather stations networks (AWSN) play an important role in modern weather forecasting and digitalization of agriculture. These networks allow to monitor various environment parameters and transmit actual data in real time to data centers, which makes possible to dramatically increase the efficiency of technical processes controlling farming. Since these networks may cover large areas, weather stations may use other stations to connect to gateways via multihop routes. The nodes are deployed outdoors and they are subject to failures due to harsh environment, deterioration of equipment, battery discharge, etc. If a station is used as a relay for other nodes, after its failure other stations may also become unavailable. To study the network reliability in this paper we propose a general methodology, consisting of six consolidated procedures, and apply the apparatus of the multidimensional alternating stochastic processes. We demonstrate the application of this analytical method for a special case of the minimal topology AWSN. General topology cases are studied with the use of the simulation approach. To study the numerical results, a discrete-event simulation model in Python language was developed. The paper presents numerical reliability analysis for three types of topologies: a simple network with three stations, a forest of ternary trees and random multihop networks with one or more gateways. In all scenarios we estimate reliability for the cases of static and dynamic routing. Different ways to enhance the distributed network reliability are discussed.

The reported study was funded by RFBR, projects number 19-29-06043 and 17-07-00142.

V. M. Vishnevskiy et al. (Eds.): DCCN 2019, LNCS 11965, pp. 591–606, 2019.
https://doi.org/10.1007/978-3-030-36614-8_45

Keywords: Automated weather stations · Distributed network · Wireless communications · Access point · Meteorological measurements · Agrometeo-parameters · WiFi · Operational failure rate · Reliability

1 Introduction

Modern agriculture and civilization require an increase in food production with the rapid growth of the world population. In agriculture, new technologies and solutions are being introduced to provide the best alternative for collecting and processing information while improving net productivity. At the same time, alarming climate changes, the growing water crisis, and natural disasters require the modernization of agriculture utilizing the latest technologies available on the market, and improved methodologies for the modern era of agriculture [1].

In the current trend of digitization of agriculture, one of the tasks is to equip agricultural fields with the means of automatic monitoring of agrometeorological parameters, such as temperature, surface, and atmospheric humidity, precipitation, soil moisture at various levels, etc. [2,3].

Such means is a distributed automated weather stations network (AWSN), which is a multi-radial topology network (Fig. 1), distributed on the ground and consisting of access points and automated weather stations (AWS), remote from each other at distances of several kilometers and connected to the nearest access points via communication channels [4,5]. The equipment of weather stations and the features of its operation are considered in [6–8] and other papers. An automatic weather station (AWS) is defined as a means that automatically transmits or records the obtained agrometeorological observations using measuring instruments. In the AMS, measurements of meteorological elements are converted into electrical signals using sensors. Then the signals are processed and converted into meteorological data. The received data is finally transmitted by wire or radio or is automatically stored on the recording medium [6]. In [7], a micro-meteorological station is presented, which can measure temperature, relative humidity, pressure, and wind speed, is portable and has high accuracy. The AWS consists of a micro-sensor chip, an anemometer, a measurement system, an indication system, and a power management system. In [8], an IoT weather station is presented as a tool or device that provides weather information in a neighboring environment. For example, the AWS can provide detailed information about ambient temperature, atmospheric pressure, humidity, etc. Therefore, this device mainly measures temperature, pressure, humidity, light intensity, and the amount of rain. In the prototype there are various types of sensors with which all of the above parameters can be measured.

Automated weather station networks are increasingly being used to collect meteorological data for agricultural and other bio-climatic purposes. The use of AWS networks has grown rapidly during the 1980s due to improvements in battery-powered data recording systems and computer communications [9]. Despite a significant amount of scientific work in this area [10–12], the problem

Fig. 1. Distributed communication network of weather stations

of determining the reliability of a distributed network of weather stations is not sufficiently developed. Therefore, the urgent task is to search and study existing and develop new concepts, models, algorithms and methods for constructing means for reliability evaluation of a distributed network of weather stations.

To solve this problem, it is necessary to create a reliability model for which, at the initial stage, it is necessary to analyze the composition and components of the AWSN, formalize the statement of the problem of ensuring reliability, and develop a methodology for reliability evaluation and analysis. Then it is necessary to develop a structural diagram of reliability, and, based on an expert assessment of the operational failure rate of component elements, conduct modeling using both analytical and simulation approaches, and compare the calculation results. Based on the simulation results, recommendations should be developed on the reservation of components and the composition of spare parts.

2   AWSN Architecture and Composition

The block diagram of the interaction between the AWSs and the gateways to the mobile network is presented in Fig. 2.

The core of the weather station is a micro-controller (CPU), which receives and processes data from meteorological sensors and GPS receivers, controls data

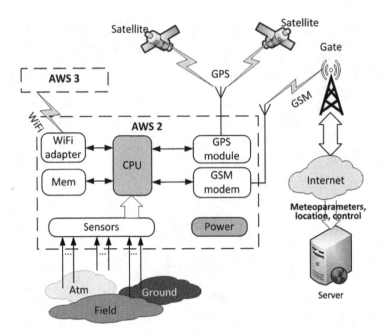

Fig. 2. Block diagram showing interaction between the AWSs and the gates

transmission via GSM modem and WiFi module. Telemetry and location data are transmitted via a GSM modem through the Internet to a monitoring server for further processing, analysis and presentation to the operator. One can monitor the location of the AWS and read meteorological parameters at any time. The power supply is a typical battery.

3   Problem Setting

The initial data for determining the reliability of the AWSN are its topology, the structure of weather stations, the type of communication channels, information about the operational failure rate of all the print nodes and communication links, as well as clear criteria for the system's operability, time schedule, and operating conditions. The performance criteria of the AWSN at the upper level of its hierarchy are determined by the requirements for the coverage area, that is, the territory on which the AWSN can monitor the agrometeorological parameters. The coverage area of the AWSN is conditionally divided into areas relative to access points (Fig. 3).

In the general case, weather stations (denoted by U in the diagram, $U*$—intermediate AWSs equipped with a repeater) are located near each gateway. At that, the number of AWSs for each access point may be different—$m_1, m_2, \ldots m_n$. Since the AWSN is spatially distributed, it is reasonable to distinguish two states of functioning—working state and failure state. In the

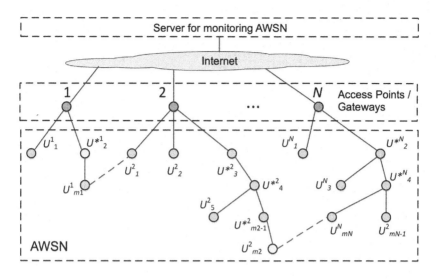

Fig. 3. AWSN coverage area

working state, all AWSN nodes are fully functioning without malfunction. The conditions for a failure state are the inoperability of at least two AWSs at each gateway. Mathematically, these criteria can be expressed as follows:

$$Q_0 = \overline{\{U^1_{1...m_1} \& U^2_{1...m_2} \& ... \& U^n_{1...m_n}\}} \times \Delta t_i,$$

where Q_0—the failure function, Δt_i—time during which a failure occurs.

The mean time to failure of the AWSN should be at least 6 months (4500 h) under round-the-clock service. As communication channels, there can be used telegraph, telephone, radio relay, tropospheric, fiber optic, cellular and WiFi communication links. The performance recovery of the AWSN should be ensured by the cold redundancy (by replacing the failed components from the spare parts kit), and the recovery time should not exceed 24 h.

4   Methodology for Reliability Evaluation

We propose a methodology, consisting of six consolidated procedures, for determining the reliability of the AWSN. The methodology allows a comprehensive assessment of reliability indicators with the analysis of the results obtained, the search for the most critical nodes and making recommendations. The IDEF0 diagram of the proposed methodology is shown in Fig. 4.

The methodology includes the following functions:

- A1 – the formation of the structural reliability scheme (SRS) of the AWSN and its component parts based on the analysis of limitations of the technical requirements, failure criteria, requirements for reliability indicators;

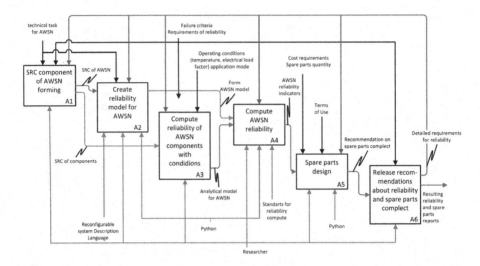

Fig. 4. IDEF0 diagram for the reliability evaluation methodology

- A2 – creation of a formalized reliability model - an analytical model based on the logical-probabilistic method and a simulation model based on the discrete-event approach using Python;
- A3 – downward and upward calculation of the reliability characteristics of components first, then of electronic modules and components from the AWSN, taking into account the temperature conditions, load factors and other parameters affecting the reliability characteristics. The average daily cycles of the application modes of the AWSN elements are also taken into account;
- A4 – a comprehensive evaluation of the reliability of the AWSN, based on a formalized reliability model that describes the behavior of the AWSN in various states and reliability indicators of the individual components obtained when performing functions A2 and A3, respectively;
- A5 – calculation and optimization of sets of spare parts, tools and accessories (SPTA) relevant to the current configuration and operating model of the AWSN taking into account the selected replenishment strategy;
- A6 – a comprehensive analysis of the results of the calculation of the AWSN, including the reliability evaluation of the item and the elements included at all levels, searching for critical elements by identifying those that make the greatest contribution to reducing the reliability of the entire system, comparing the data with the data in the technical specifications, and analyzing the SPTA acquisition.

The necessary data for the formation of SRS at A1 are: the topology of the AWSN, a list of elements (specification). The formalized model at A2 is built on the basis of the SRS and the reliability indicators data from the technical requirements. The main task of designing spare parts at A5 is to ensure maintainability indicators of the AWSN. To calculate and optimize the spare parts sets, it is recommended to use the Python system. According to the results of A6, reporting documentation is issued that contains data on the reliability of the AWSN and recommendations for SPTA acquisition.

5   Analytical Reliability Model of AWSN

To evaluate and analyze the reliability of the distributed AWSN we apply the apparatus of the multidimensional alternating stochastic processes. To demonstrate the application of this method we consider the minimal configuration AWSN (Fig. 5).

We construct a mathematical model of the considered AWSN as a heterogeneous redundant repairable system consisting of $n = 3$ different components working in a hot standby:

1 — AWS 1,
2 — AWS 2,
3 — AWS 3.

Failure of any 2 out of 3 AWSs is sufficient for the failure of the entire minimal configuration system. At that, failure of AWS 2 leads to the unavailability of the adjacent AWS 3 and, thus, to the failure of the system.

Fig. 5. Minimal topology AWSN

The states of this system are described by the vector $\mathbf{j} = (j_1, j_2, j_3)$ with binary components j_k $(k = 1, 2, 3)$,

$$j_k = \begin{cases} 0, & \text{if } k\text{-th AWS is operable,} \\ 1, & \text{if } k\text{-th AWS is out of service.} \end{cases}$$

We denote the set of states of the considered system by $\mathcal{E} = \{\mathbf{j} = (j_1, j_2, j_3), j_k = \{0, 1\}\ (k = 1, 2, 3)\}$ with a finite number $N = 2^3 = 8$ of states, and the sets of working and failure states are denoted by $\mathcal{E}_0$ and $\mathcal{E}_1$ respectively.

Consider a stochastic process with binary components, defined over the state space $\mathcal{E}$ by the relation:

$$\mathbf{X}(t) = \mathbf{j}, \quad \text{if at time } t \text{ the process is in the state } \mathbf{j}.$$

Assuming that the uptime and repair time of the AWSs are exponentially distributed with parameters α_k, β_k for the k-th AWS, then this stochastic process is a multidimensional Markov alternating process with transition rates

$$\lambda(\mathbf{i},\mathbf{j}) = \begin{cases} \alpha_k & \text{for } \mathbf{i}: i_k = 0, \ \mathbf{j} = \mathbf{i} + \mathbf{e}_k, \\ \beta_k & \text{for } \mathbf{i}: i_k = 1, \ \mathbf{j} = \mathbf{i} - \mathbf{e}_k, \end{cases}$$

where $\mathbf{e}_k$ is a vector of zeros with a one in the k-th position, while the transitions from the state $\mathbf{j}$ are possible only to the "neighboring" states $\mathbf{j} + \mathbf{e}_k \ \mathbf{j} - \mathbf{e}_k$.

For the sake of convenience of representation of the transition rate matrix of the process and its transition graph, we renumber the states according to the relation $j = j_1 + 2j_2 + 4j_3$, i.e. as follows:

$$(0,0,0) = 0, \ (1,0,0) = 1, \ (0,1,0) = 2, \ (1,1,0) = 3,$$
$$(0,0,1) = 4, \ (1,0,1) = 5, \ (0,1,1) = 6, \ (1,1,1) = 7$$

and simultaneously with the vector process $\mathbf{X}(t)$ we consider the corresponding scalar process $X(t)$.

If we arrange the states according to the number of failed AWSs (and the transition graph) in the order $E = \{0, 1, 2, 4, 3, 5, 6, 7\}$, then the transition rate matrix $\boldsymbol{\Lambda} = [\lambda_{i,j}]$ can be represented in a block diagonal form with blocks, formed by the states $\{0\}$; $\{1, 2, 4\}$; $\{3, 5, 6\}$; $\{7\}$,

$$\boldsymbol{\Lambda} = \left(\begin{array}{cccc|ccc|c} -\lambda_{0,0} & \alpha_1 & \alpha_2 & \alpha_3 & 0 & 0 & 0 & 0 \\ \hline \beta_1 & -\lambda_{1,1} & 0 & 0 & \alpha_2 & \alpha_3 & 0 & 0 \\ \beta_2 & 0 & -\lambda_{2,2} & 0 & \alpha_1 & 0 & \alpha_3 & 0 \\ \beta_3 & 0 & 0 & -\lambda_{4,4} & 0 & \alpha_1 & \alpha_2 & 0 \\ \hline 0 & \beta_2 & \beta_1 & 0 & -\lambda_{3,3} & 0 & 0 & \alpha_3 \\ 0 & \beta_3 & 0 & \beta_1 & 0 & -\lambda_{5,5} & 0 & \alpha_2 \\ 0 & 0 & \beta_3 & \beta_2 & 0 & 0 & -\lambda_{6,6} & \alpha_1 \\ \hline 0 & 0 & 0 & 0 & \beta_3 & \beta_2 & \beta_1 & -\lambda_{7,7} \end{array} \right), \qquad (1)$$

where the diagonal elements of $\lambda_{j,j}$ are the rates of leaving the corresponding states. Notice, that taking in account that the failure of AWS 2 leads to the unavailability of the adjacent AWS 3, the sets of working and failure states are equal to $\mathcal{E}_0 = \{0, 1, 4\}$ and $\mathcal{E}_1 = \{2, 3, 5, 6, 7\}$ respectively.

At that, the transition graph of the process $X(t)$ for the chosen state numbering has the form shown in Fig. 6.

We set the problem of calculating the stationary distribution of the system states and the stationary probability of failure-free operation (stationary reliability) of the system.

The system of differential equations with the initial condition for the state probabilities $p_{\mathbf{j}}(t) = \mathbf{P}\{\mathbf{J}(t) = \mathbf{j}\} \ (\mathbf{j} \in E)$ of the process $X(t)$ in matrix form is given by (2),

$$\dot{\boldsymbol{p}}'(t) = \boldsymbol{p}'(t)\boldsymbol{\Lambda}, \qquad \boldsymbol{p}'(0) = \boldsymbol{e}_0', \qquad (2)$$

where vector $\boldsymbol{e}_0$ is a vector of zeros with a 1 in the first position.

The existence of stationary probabilities $p_j = \lim_{t \to \infty} p_j(t)$ is obvious since the phase space of the considered process is finite and represents a single class of communicating recurrent states.

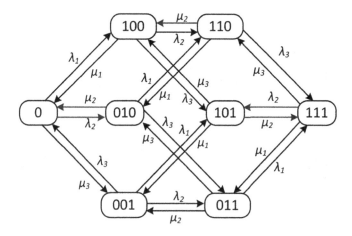

Fig. 6. Transition graph of the process $X(t)$

The stationary states probabilities $p_\mathbf{j}$ satisfy the system of equilibrium equations in coordinate form:

$$\sum_{k:\,j_k=1} \alpha_k p_{\mathbf{j}-\mathbf{e}_k} - \left(\sum_{k:\,j_k=0} \alpha_k + \sum_{k:\,j_k=1} \beta_k \right) p_\mathbf{j} + \sum_{k:\,j_k=0} \beta_k p_{\mathbf{j}+\mathbf{e}_k} = 0 \quad (\mathbf{j} \in E) \quad (3)$$

with the additional normalization condition $\sum_{\mathbf{j}\in E} p_\mathbf{j} = 1$.

By direct substitution, it is easy to verify that the solution of the last system of equations has the form of a multiplicative representation

$$p_\mathbf{j} = \prod_{1 \le i \le n} \frac{\rho_i^{j_i}}{1+\rho_i} \quad \text{where} \quad \rho_i = \frac{\alpha_i}{\beta_i}, \tag{4}$$

For numerical calculations, real and expert data was used—the average values of uptime and repair time of AWSs for particular equipment. This data is as follows:

- AWS 1 mean time to failure $m_{1,0} = 3500$ (hours);
- AWS 2 mean time to failure $m_{2,0} = 4000$ (hours);
- AWS 3 mean time to failure $m_{3,0} = 4500$ (hours);
- AWSs mean repair time $m_{1,1} = 24$ (hours);

These data were used to calculate the corresponding transition rates of the modeling process $\alpha_k = m_{k,0}^{-1}$, $\beta_k = m_{k,1}^{-1}$.

Numerical calculations using the developed software in the MATLAB environment allowed to obtain the following results:

- stationary state probabilities vector:

$$\boldsymbol{p}' = (0.982,\, 0.0067,\, 0.0059,\, 4 \cdot 10^{-5},\, 0.0052,\, 3.6 \cdot 10^{-5},\, 3.1 \cdot 10^{-5},\, 2.15 \cdot 10^{-7}).$$

– system failure probability $p_{\text{failure}} = p_2 + p_3 + p_5 + p_6 + p_7 = 0.006000129$;
– system failure-free probability (stationary reliability) $p_{\text{failure-free}} = 1 - p_{\text{failure}} = 0.9939998711$;

For the analysis of the dependence of the AWSN stationary reliability on the ratio between the average uptime and the average repair time of the AWSs, a plot of the stationary failure-free probability was constructed (Fig. 7).

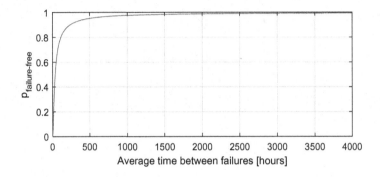

Fig. 7. Stationary reliability of the minimal topology AWSN.

6   Simulation Model and Numerical Analysis

In the numerical experiments we use a discrete event simulation model, implemented in Python language. Source code of the model is available at GitHub[1].

To measure the network reliability under various scenarios we consider three types of topologies:

1. Minimal topology with three stations;
2. Forest of ternary trees;
3. Random connected networks.

AWS mean uptime equal to 4000 h is very large compared to the typical 24-hours repair interval, so to simplify computations we assume by default that the mean uptime is equal to 720 h (1 month). In other experiments we vary failure rate in such a way that mean uptime varies from a week to 4000 h (approx. 5 and half months). The network is simulated over 10 years time interval.

For larger topologies we introduce one more assumption regarding repair time: we assume that the first repair interval (which takes places after a critical number of stations became unavailable due to failure or loosing connection) is 24 h, while all successive repair intervals are 2 h in average. This assumption is

---
[1] Experiment source code: https://github.com/larioandr/2019-dccn-sensors.

advocated by the fact that a repair team need much more time to visit a stock, take the equipment and get to the first station (especially in case of difficult terrains) and much less time to reach other failed stations after repairing the first one.

We also assume by default that the network is considered unavailable when 2 or more AWS are offline, but for larger topologies with 20–30 stations we extend the critical number of failed AWS to 4. It will be shown, that the effect of dynamic routing is especially clear in this case.

In all experiments we consider both static and dynamic routing. In the first case, the routes are predefined, and if an intermediate relay station becomes unavailable all other nodes connected to the gateway via this station become offline. In the second case, if the station became offline due to broken path after intermediate station failure, it tries to build an alternative route to any of the gateways. As it will be shown below, the dynamic routing utilization slightly decreases the average number of offline nodes and increases the network uptime probability.

In all experiments we assume that both repair and uptime intervals are distributed exponentially. All gateways are considered absolutely reliable.

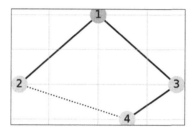

Fig. 8. Minimal topology with an alternative connection from node 4 to node 2. Node 1 is a gateway, all other nodes are AWS.

6.1   Minimal Topology

First of all, we run a numerical experiment on a basic minimal topology, shown in Fig. 8. We assume that while the leaf station 4 is connected to a relay station 3, it can also reconnect to station 2 if station 3 becomes unavailable and dynamic routing is used.

The reliability estimation under various AWS mean uptime for this basic topology is shown in Fig. 9. It can be seen that the system-level reliability reaches 0.96 when the dynamic routing is used and each node becomes unavailable after 936 h (39 days) in average. If stations are less reliable, the dynamic routing strongly outperforms the static routing. On the other hand, as the stations become more reliable, estimations for static and dynamic routing become closer to each other.

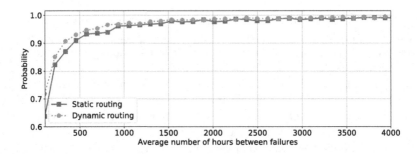

Fig. 9. Reliability of the minimal topology AWSN.

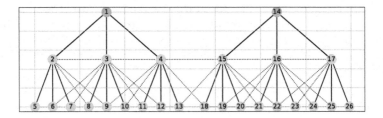

Fig. 10. A ternary forest topology. Solid and dashed lines represent static and backup connections respectively. Nodes 1 and 14 are gateways, all other nodes are AWS.

6.2   Forest of Ternary Trees

In the second experiment we generalized the basic minimal topology described above by considering a forest, each tree having a gateway at the root. We assume that the forest contains two ternary trees, both having depth equal to 2 (see Fig. 10).

In this experiment we also consider fast successive repairs as described above: when a critical number of stations become offline and the network becomes unavailable, the first AWS is repaired in 24 h in average and each another broken AWS is repaired in 2 h. Since the number of stations in this topology is much higher than in the basic network, we also consider the case when 4 AWS are required to be broken to consider the network as unavailable. The reliability estimation results are shown in Fig. 11.

Figures 12 and 13 show the estimations of the expected number of failed and offline nodes, and the functions of the number of failed and offline nodes regarding time for a given failure rate, respectively. While the number of failed nodes for static and dynamic routing is the same, dynamic routing allows to effectively decrease the number of offline nodes due to utilization of alternative paths. Dynamic routing efficiency increases along with the minimum number of unavailable nodes needed to consider the network unavailable, as shown in Fig. 11.

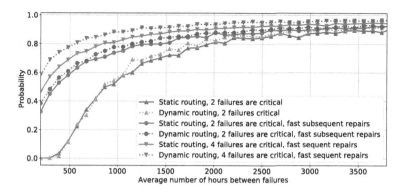

Fig. 11. Reliability of the forest with two ternary trees.

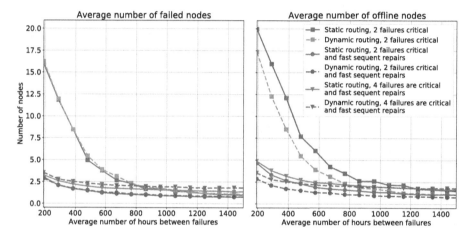

Fig. 12. Stationary number of failed and offline stations for a forest of ternary trees network.

6.3 Random Topologies

In the last experiment we analyze random network topologies. We assume that the network contains 20 stations (AWSs) and 1 or more gateways. An example of the generated topology is shown in Fig. 14. For each experiment we generated four random networks and estimated reliability and the expected number of offline nodes by averaging results over these networks.

The topology generation algorithm finds places for gateways by putting them within distances $(\underline{R}, \overline{R})$ from each other. Then it iteratively finds a random place for AWS S_i, such that the distance $\rho(S_i, N_j) \geq r_0$ for any other node (AWS or gateway) N_j, and $\rho(S_i, N_j) \leq r$ for some another node (AWS or gateway). Here $\underline{R}$ and $\overline{R}$ are the minimal and maximal distances between gateways, r_0 denotes the minimal distance between any network nodes and r is the effective range of the station radio coverage. Nodes N_i, N_j can be connected if $\rho(N_i, N_j) \leq r$. It

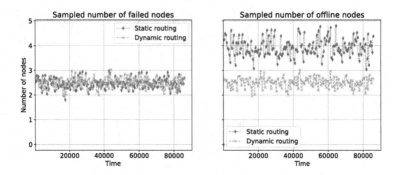

Fig. 13. Estimated number of failed and offline stations for a forest of ternary trees network depending on the time. Mean interval between node failures is 1 month.

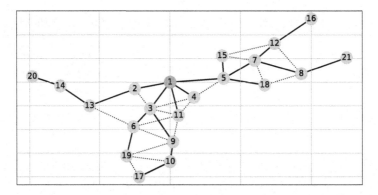

Fig. 14. Random topology example. Solid and dashed lines represent static and backup connections respectively. Node 1 is a gateway.

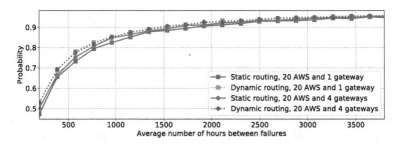

Fig. 15. Reliability of the random topologies with 1 and 4 gateways and fast successive repairs. Network is considered unavailable when 4 or more stations are offline.

can be seen that any network generated by this algorithm will have at least one spanning forest with at least one gateway in each tree.

Figures 15 and 16 show the estimations of the network reliability and the number of offline nodes. As before, we analyzed both static and dynamic routing.

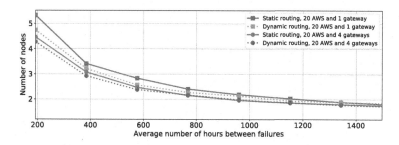

Fig. 16. Number of offline AWSs in the random topologies with 1 and 4 gateways and "fast" successive repairs. Network is considered unavailable when 4 or more stations are offline.

We also assumed that network becomes unavailable when 4 or more AWS become offline, and considered fast successive repairs as in the forest scenario: the first repair takes 24 h in average, while each successive repair (until all stations are repaired) takes 2 h in average. Besides, we considered a case when additional gateways are deployed to increase the network reliability.

7   Conclusion

The numerical results for the forest topology show, that the AWS network reliability can be strongly increased even with unreliable stations if dynamic routing is used along with higher tolerance on the number of unavailable stations.

Summing up and analysing the obtained results, we can conclude that the dynamic routing increases the network reliability and decreases the number of offline nodes. However, in case of random topologies the effect of dynamic routing is smaller, then in case of the ternary forest: random networks may have less alternative paths compared to the forest. We should also note, that actual numerical results strongly depend on the parameters of the network generator, and the results should be re-estimated more precisely if more information about the expected topology is known.

The results in Fig. 15 show, that deploying additional gateways leads to increased network reliability. However, the increase is not very high, and if the gateways are expensive, other ways to enhance the reliability (like reduction of repair time) may provide better results.

References

1. Suciu, G., Ijaz, H., Zatreanu, I., Drăgulinescu, A.-M.: Real time analysis of weather parameters and smart agriculture using IoT. In: Poulkov, V. (ed.) FABULOUS 2019. LNICST, vol. 283, pp. 181–194. Springer, Cham (2019). https://doi.org/10.1007/978-3-030-23976-3_18

2. Moummadi, K., Abidar, R., Medromi, H.: Generic model based on constraint programming and multi-agent system for M2M services and agricultural decision support. In: 2011 International Conference on Multimedia Computing and Systems (ICMCS), pp. 1–6 (2011)

3. Lee, M., Hwang, J., Yoe, H.: Agricultural production system based on IoT. In: 16th IEEE International Conference on Computational Science and Engineering, Sydney, NSW, pp. 833–837 (2013)

4. Fridzon, M.B., Ermoshenk, Yu.M: Development of the specialized automatic meteorological observational network based on the cell phone towers and aimed to enhance feasibility and reliability of the dangerous weather phenomena forecasts. Russ. Meteorol. Hydrol. **34**(2), 128–132 (2009)

5. Sarkar, I., Pal, B., Datta, A., Roy, S.: Wi-Fi-based portable weather station for monitoring temperature, relative humidity, pressure, precipitation, wind speed, and direction. In: Tuba, M., Akashe, S., Joshi, A. (eds.) Information and Communication Technology for Sustainable Development. AISC, vol. 933, pp. 399–404. Springer, Singapore (2020). https://doi.org/10.1007/978-981-13-7166-0_39

6. Ahmad, L., Habib Kanth, R., Parvaze, S., Sheraz Mahdi, S.: Automatic weather station. Experimental Agrometeorology: A Practical Manual, pp. 83–87. Springer, Cham (2017). https://doi.org/10.1007/978-3-319-69185-5_12

7. Fang, Z., Zhao, Z., Du, L., Zhang, J., Pang, C., Geng, D.: A new portable micro weather station. In: 2010 IEEE 5th International Conference on Nano/Micro Engineered and Molecular Systems, Xiamen, pp. 379–382 (2010). https://doi.org/10.1109/NEMS.2010.5592239

8. Kodali, R.K., Mandal, S.: IoT based weather station. In: 2016 International Conference on Control, Instrumentation, Communication and Computational Technologies (ICCICCT), Kumaracoil, pp. 680–683 (2016). https://doi.org/10.1109/ICCICCT.2016.7988038

9. Snyder, R.L., Brown, P.W., Hubbard, K.G., Meyer, S.J.: A guide to automated weather station networks in North America. In: Stanhill, G. (ed.) Advances in Bioclimatology. Advances in Bioclimatology, vol. 4, pp. 1–61. Springer, Heidelberg (1996). https://doi.org/10.1007/978-3-642-61132-2_1

10. Aminev, D., Zhurkov, A., Polesskiy, S., Kulygin, V., Kozyrev, D.: Comparative analysis of reliability prediction models for a distributed radio direction finding telecommunication system. In: Vishnevskiy, V.M., Samouylov, K.E., Kozyrev, D.V. (eds.) DCCN 2016. CCIS, vol. 678, pp. 194–209. Springer, Cham (2016). https://doi.org/10.1007/978-3-319-51917-3_18

11. Rykov, V.V., Kozyrev, D.V.: Reliability model for hierarchical systems: regenerative approach. Autom. Remote Control **71**(7), 1325–1336 (2010). https://doi.org/10.1134/S0005117910070064

12. Rykov, V.V., Kozyrev, D.V.: Analysis of renewable reliability systems by markovization method. In: Rykov, V.V., Singpurwalla, N.D., Zubkov, A.M. (eds.) ACMPT 2017. LNCS, vol. 10684, pp. 210–220. Springer, Cham (2017). https://doi.org/10.1007/978-3-319-71504-9_19

Large-Scale Centralized Scheduling of Short-Range Wireless Links

Alexander Pyattaev[1(✉)] and Mikhail Gerasimenko[2]

[1] Peoples' Friendship University of Russia (RUDN University),
6 Miklukho-Maklaya Street, Moscow 117198, Russian Federation
alex.pyattaev@gmail.com
[2] Tampere University, Korkeakoulunkatu 1, 33720 Tampere, Finland
mikhail.gerasimenko@tuni.fi

Abstract. In 5G networks we expect femtocells, mmWave and D2D communications to take over the more typical long-range cellular architectures with pre-planned radio resources. However, as the connection length between the nodes become shorter, locating feasible, non-interfering combinations of the links becomes more and more difficult. In this paper a new approach to this problem is presented. In particular, through guided heuristic search, it is possible to locate non-interfering combinations of wireless connections in a highly effective manner. The approach enables operators to deploy centralized scheduling solutions for emerging technologies such as network-assisted WiFi-Direct and LTE Direct, and others, especially those which lack efficient medium arbitration mechanisms.

1 Introduction

The D2D communications gain support by mobile devices with every year, with such bright examples as Bluetooth and WiFi Direct [8,11]. Lately, LTE Direct also joined the race towards consumer markets [13], and mmWave systems like IEEE 802.11ad and 5G NR are also ramping up [1]. Finally, latest versions of 802.11ax standard imply the usage of scheduling for individual WiFi cells to be performed by their access points, yet no indication on how exactly they should coordinate is given. In this work we argue that we must improve the medium access control mechanisms controlling the multiplexing of all these links. In what follows we will mostly talk about WiFi, yet the major conclusions apply equally to other systems.

It is important to note that having 100 WiFi stations connected to an access point (AP) is very different from having 100 independent P2P links over the same area, or, equivalently, having 100 APs with 1 client station each. In particular, in the centralized case, the AP will most likely be the major source of traffic, while stations will listen, causing minimal contention for resources in either downlink

The publication has been prepared with the support of the "RUDN University Program 5-100".

or uplink direction. In the P2P case, however, all devices will try to access the channel. With CSMA-based MAC (such as in WiFi and LTE Direct), this results in poor performance for all but few lucky users, as shown e.g. in [7]. Similar conditions are present when multiple legacy WiFi access points share the channel. In a scheduled system such as 802.11ax or LTE direct one needs to solve a similar coordination issue, as otherwise interference can not be controlled and may result in some links becoming unusable.

Further, when multiple stations are deployed in a way that their access patterns depend on each other (but do not form just one collision domain) the approaches normally used to analyze and optimize WiFi no longer apply. In what follows, we will show that WiFi's default medium access, when applied over randomly deployed links, tends to elect just a few of them that will consume all of the resources, while the rest have no chance to get on the channel. The resulting resource allocation is also surprisingly stable (as in, it will only change when the "lucky" nodes stop transmitting, and even then the change is likely to be localized). Clearly, for P2P applications such resource sharing may be undesirable [9].

Based on this observation, we proceed to propose a control algorithm that constructs sets of non-intersecting links in a centralized fashion, and then assigns them to the end-users using a different radio technology (e.g. LTE). By using a separate radio for control signaling, we are following the ideas of LTE phantom cells, and enable a central controller to assign resources (or transmission opportunities) to the devices in controlled area. Similar management has been used successfully for other cases [2,3].

2   Analyzing WiFi Behavior in a Multi-cluster Environment

The well-studied part of 802.11 WiFi protocol and its many derivatives typically ends with the words "hidden" or "exposed" terminal. By some convention, the scenarios where such terminals appear, are mysteriously considered to be largely irrelevant, and thus left for further studies, or out of scope entirely. Unfortunately, it is exactly the kind of deployments that are most critical in our daily lives, with such typical examples like airports, conference facilities, exhibition halls, etc. It must come at no surprise then, that most of such settings are not handled by the WiFi MAC particularly well.

The key to understanding what exactly is wrong there is in the way WiFi (or, for that matter, nearly any other random-access) MAC operates. Essentially, the chances of any given device to obtain channel access are maintained inversely proportional to the number of *active* (i.e, looking for transmit opportunity) terminals around it. Apparently, a design assumption was made that as users get their transmit opportunities they will stop asking for more and thus change the set of active terminals, eventually making the system *stochastically fair*. Unfortunately, the random processes that are supposed to make it happen operate on timescales of multiple seconds, which is too slow for any QoS-demanding

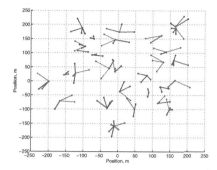

Fig. 1. Multi-cluster WiFi network example resource distribution (Color figure online)

application. As a result, observing a WiFi cluster for short periods of time (1 s in our case), allows to capture the unfair distribution of the resources in the network that is a result of this assumption.

In order to properly illustrate the issue, as well as in further study, we employ a simulation model for WiFi implemented as part of WiterSim framework [10],[?]. The simulation uses standard parameters for 802.11n EDCF protocol with full buffer traffic. The access points are deployed uniformly, and their associated stations are dropped uniformly within a circle of 50 m radius around the access points.

Figure 1 illustrates an example of a WiFi network deployed over isotropic environment by showing the percent of time spent on the channel (with pure red corresponding to 100%, and blue corresponding to 0%). For the sake of example all traffic is uplink.

As one can clearly see, the network does not end up being fair at all. In particular, as many as half of the users barely got any service whatsoever, while others got a large share all for themselves. Importantly, this has nothing to do with the link length and RF conditions, as those are kept sufficiently high for any link to achieve maximal rate. The CDF plot (Fig. 2) illustrates this quite conclusively, with under 5% of the users getting throughput above 15 Mbps, while most of them barely get 1 Mbps rate. Of some importance is additional observation that some cells behave in a significantly more fair manner than others, indicating that sometimes the timescale of stochastic fairness can be very short, but that is not always the case.

To illustrate such distribution of resources in practical terms, under 5% of the users in the example given could watch an HD video stream, while 40% of the users would experience noticeable delays while getting an email downloaded. Further, some of the users would be under false impression that there is absolutely nothing wrong with random access procedure, as their immediate neighbors all get good performance.

However, none of the users can even expect the resource allocation to remain static over a longer time window. In a matter of seconds, the bandwidth distribution across users would normally change, thus denying the high-bitrate users, and providing excess bitrate to others. Naturally, such behavior is wasteful and

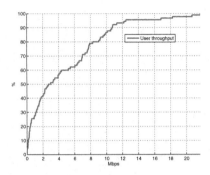

Fig. 2. CDF of the resource distribution

unstable, and an alternative, more predictable, solution is desirable for many applications. Unfortunately, fixing it within legacy WiFi MAC may prove to be quite a challenge, since by its nature global fairness requires global knowledge. In what follows, we will consider a way to deliver such global knowledge via external (e.g. cellular) network assistance, or inter-AP collaboration channel.

3   Proposed Spatial Multiplexing Approach

Before we proceed to design the procedure to manage the WiFi links, let us make several simplification assumptions:

1. *All radio links are running at constant power, and their power can not be adjusted.* Most WiFi devices are unable to perform dynamic power control due to protocol limitations and RF hardware costs associated with it.
2. *At a given time, at most one link can use the channel.* Both endpoints must be able to transmit and receive without interfering with any other devices.
3. *Observed over a short period of time (e.g. 100 us), any link can be assumed to be either active or inactive.* The activation of a given link is considered to be a MAC decision, and we simply observe the two binary states. For the purposes of further discussion we thus ignore RTS-CTS handshakes that do not result in a data transmissions.

To allocate the radio resource to a an irregular deployment of links, we need to guarantee that none of the active links interferes with any other active link, while the maximal possible number of links is active. In any regular deployment it is fairly easy to find out which links can run at once and how many of them could be running at once for a given area. However, in a random deployment it is much more difficult to find which combination of active links is optimal. Indeed, any link activated generally prevents activation of some other links. As a result, activation side-effects for any one link can "propagate" over the entire network.

In such conditions, it is easy to show that network planning is an NP-hard problem known as "maximal matching problem" [4], itself a case of knapsack packing problem. Counter-intuitively, the existing distributed coordination algorithms can operate in such conditions reasonably well, with such examples as CSMA/CA and Bluetooth's FHSS being particularly effective. However, the distributed algorithms mentioned above lack the capability to find optimal solutions. Even if a reasonable solution is found, this solution can never be fair, as the set of activated links will not evolve in a fair manner over time.

We propose that the network would dynamically plan which links would be active in a centralized manner, as to maintain the fairness and efficiency, on a level unachievable with distributed coordination. To achieve that, the devices would need to report an up-to-date list of their neighbors to the network coordinator, such that it can base its decisions on up-to-date information. Based on these reports, the network can construct a graph of all desired connections and potential interferers.

Such level of awareness about the topology of the network is typical only for cellular systems, but is also anticipated in the forthcoming 5G networks. Furthermore, latest 802.11 specs (past 802.11-rev MC) recommend support for "measurement report" packets for the purposes of reporting interference conditions.

It is also possible for an oracle entity to be available to implement any needed access coordination. For instance, in Network-assisted D2D scenarios, a special Proximity Service entity called D2D server is installed in the core of the cellular network. Such entity is expected to coordinate all sorts of D2D connectivity and discovery, and therefore would have a global view of all processes happening with WiFi and other D2D technologies, as was illustrated in trial activities [11]. For WiFi-only systems such oracle could be part of a WiFi controller.

4   Proposed Algorithm and Evaluation

Utilizing the above model for the D2D links, as well as the availability of a central control server, let us try to solve the global network planing problem. Generally, we need to find multiple of sets of non-interfering links, such that the size of the set is maximized and fairness is ensured. Unfortunately, there is no algorithm that could reliably find such solution sets. Moreover, complete search of all possible combinations is also infeasible for a network with more than about 30 nodes. Knowing this to be the case, we will not search for an optimal solution, but instead will try to find one that exceeds baseline CSMA/CA performance by a reasonable margin.

The proposed network planning algorithm works in cycles, with the length of the cycle chosen short enough such that dynamic fairness can be provided. Each cycle consists of 4 operations:

- **Update the information about the heard interfering links.** Each device periodically uploads the list of heard MAC addresses (in hashed form). The list is then compressed and sent to the planning server
- **Each device indicates to planning server which link it wishes to use.**
- **Planning server constructs the graph of all links.** Links are one of two categories: desired and interfering.

- **On the graph, the link selection algorithm is run.** The algorithm first assigns activation priorities to all links. If a link to be activated is conflicting with already activated one, it is skipped. Links with same priority are activated in random order relative to each other.
- **After the ranking and selection is completed, the decisions are disseminated to all clients.**

The presented algorithm is entirely feasible due to its very low complexity (linear after activation priority is assigned), allowing it to run on a local eNB, especially with the recent promotion of small data services and MTC, that allow to efficiently transport tiny packets over LTE RAN [6,12].

The most tricky part of the algorithm is the link ranking. The requirements to the link ranking algorithm are very tight, as it has to operate extremely quickly on large number of links. Further we discuss several heuristics and their relative performances.

4.1   Network Planning Heuristics

The heuristics presented next essentially define the sorting order for all available links when the access list is constructed at the planning server.

Shortest Links First. We could prioritize the shortest links first (based on time of flight of RX power), which would be close to what LTE normally does with its proportional fair scheduler. While a valid strategy to maximize throughput, it does not take into account fairness.

Least Connected First. This policy tries to activate the links that are least likely to block other links. In this case, the activation priority is inversely proportional to the number of neighboring links that the one under consideration would have blocked if activated. This strategy favoring the links that have least contention.

Most Clustered First. The most clustered (links that have most common neighbors for endpoints) first policy attempts to promote segregation of the links into independent clusters, such that the start and endpoint of the link would have mostly identical sets of neighbors. This approach maximizes the number of independent clusters of links, where activation of any other link within the cluster would immediately block all other links in that cluster. Furthermore, of all links in cluster, the one is chosen that has least external neighbors.

In terms of resulting performance, this policy is very similar to least connected first, but allows more flexibility in the choice of links, as we will see further on the fairness comparisons.

Random. Finally, random policy is simply a random combination of non-interfering links. Clearly, this should provide the most fair distribution of resources, as no link is favored.

Figure 3 shows all of the presented schemes at work, making a decision at a given time instant (blue stands for idle and red for active, gray lines indicate possible links). One can easily see that on many decisions they agree, and in some cases there are variations. What is common, however, is the fact that no interfering pairs are ever scheduled at the same time.

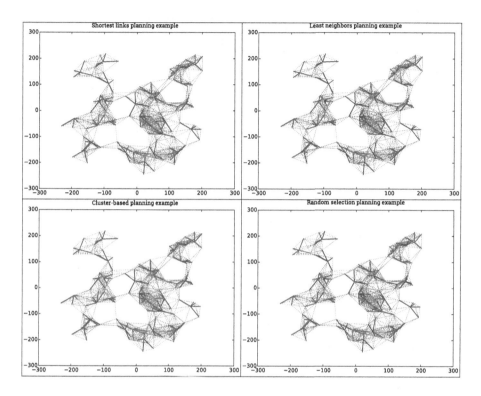

Fig. 3. Examples of different scheduling schemes at work (Color figure online)

4.2   Fairness and Throughput Conflict

Naturally, system-wide fairness and throughput are conflicting targets [5], and thus in the coordination scheme we also employ a link mixer. The role of the mixer is to disallow the same links to be scheduled ahead of others that have similar priority and have not been recently scheduled. In particular, every link under system's control is given a counter, which tells how many cycles ago it was active. All links with the same counter value are grouped into scheduling slices. The planning begins with the largest counter value slice, and once a feasible set of links is chosen from there, all other slices are considered according to descending counter values. Once the planning cycle is done, all counters are updated. Such mixing algorithm essentially turns the system into an advanced round-robin scheduler. As a result the system is forced to be fair no matter which heuristic is used.

Let us conclude the comparisons of all the proposed schemes between each other and also with CSMA. All systems are evaluated in full-buffer setting over the same deployment, WiFi MAC is used for all of them with 802.11n timings, and no MIMO. The packet size is 300 bytes, and the aggregation is disabled. As a result of this, any contention really penalizes the system, and the inefficiency of resource allocation is more clearly visible.

The Table 1 clearly shows that any network-assistance based scheme clearly outperforms conventional WiFi by a factor of two. The reason for it is lack of collisions, which are no longer necessary as all links run orthogonal to each other.

Table 1. Performance comparison of different management schemes

Scheme	System throughput, Mbps (mixed/unmixed)	Links active (mixed/unmixed)	Jain's index (mixed/unmixed)
WiFi (baseline)	−/157.83	−/150.0	−/0.24
Shortest links	283.2 /111.5	13.10/16	0.941/0.082
Least connected	284.7 /406.0	13.10/16	0.944/0.088
Cluster-based	287.1/370.4	12.95/14	0.949/0.078
Random	285.0/285.8	13.20/13.15	0.942/0.405

Finally, it is worth noting that the random selection policy is not far behind of any of the heuristics. The reason for this is that the mixer essentially randomizes most of the processes after link sorting. The unmixed experiments clearly indicate that it is the case, and also show why exactly the mixer is absolutely necessary for all but random heuristic. Even though the heuristics are able to squeeze as many as 16 orthogonal links into the network, the system does not evolve and thus has poor fairness. On the other hand, mixing process essentially equalizes everyone, but at the same time brings the throughput almost to the level of random selection. Overall, probably the best scheme in a practical case is cluster-based, because it responds best to the mixing of the links, which is absolutely necessary to provide decent fairness.

5   Conclusions

Expanding the concept of network-assisted D2D, we have proposed a way to perform network-wide radio network planning and resource allocation for as many P2P links as necessary for practical purposes. Proposed planning process in the simulation takes typically only a few microseconds for a thousand links, and its complexity is logarithmic with network size, with opportunities for multithreading. The planning approach allows one to improve upon WiFi's area throughput, as well as provide the much wanted fairness in the system.

Further, we have shown that any network-assisted scheme (that does not schedule conflicting links) outperforms CSMA by a huge margin. On top of that, one can trade fairness for resource efficiency to achieve as much as 2.5

times increase in spectral efficiency. Obviously, there are costs associated with reporting the neighbor lists to the network, delays, and other effects we could not anticipate within this study. However, we believe that it is useful to think outside of the boundaries of conventional distributed scheduling and proceed to promote planning of massive networks in a semi-centralized fashion instead.

Acknowledgments. The publication has been prepared with the support of the "RUDN University Program 5-100" (recipient Aleaxander Pyattaev, simulation model development).

References

1. Ericsson Mobility Report, June 2019
2. Andreev, S., Galinina, O., Pyattaev, A., Johnsson, K., Koucheryavy, Y.: Network-assisted offloading of cellular data sessions onto device-to-device connections. IEEE J. Sel. Areas Commun. (2014)
3. Andreev, S., Pyattaev, A., Johnsson, K., Galinina, O., Koucheryavy, Y.: Cellular traffic offloading onto network-assisted device-to-device connections. IEEE Commun. Mag. **52**, 20–31 (2014). https://doi.org/10.1109/MCOM.2014.6807943
4. Borbash, S., Ephremides, A.: The feasibility of matchings in a wireless network. IEEE Trans. Inf. Theory **52**, 2749–2755 (2006). https://doi.org/10.1109/TIT.2005.860471
5. Gerasimenko, M., et al.: Adaptive resource management strategy in practical multi-radio heterogeneous networks. IEEE Access **5**, 219–235 (2016)
6. Gudkova, I., et al.: Analyzing impacts of coexistence between M2M and H2H communication on 3GPP LTE system. In: Mellouk, A., Fowler, S., Hoceini, S., Daachi, B. (eds.) WWIC 2014. LNCS, vol. 8458, pp. 162–174. Springer, Cham (2014). https://doi.org/10.1007/978-3-319-13174-0_13
7. Ometov, A.: Fairness characterization in contemporary IEEE 802.11 deployments with saturated traffic load. In: Proceedings of 15th Conference of Open Innovations Association FRUCT, St. Petersburg, Russia, 21–25 April 2014, pp. 99–104 (2014). https://doi.org/10.1109/FRUCT.2014.6872427
8. Ometov, A., et al.: Toward trusted, social-aware D2D connectivity: bridging across the technology and sociality realms. IEEE Wirel. Commun. **23**, 103–111 (2016)
9. Pyattaev, A., Galinina, O., Andreev, S., Katz, M., Koucheryavy, Y.: Understanding practical limitations of network coding for assisted proximate communication. IEEE J. Sel. Areas Commun. **33**(2) (2015). https://doi.org/10.1109/JSAC.2014.2384232
10. Pyattaev, A., Andreev, S., Vinel, A., Sokolov, B.: Client relay simulation model for centralized wireless networks. In: Proceedings of the 7th EUROSIM Congress on Modelling and Simulation, pp. 672–677. Czech Technical University in Prague (2010)
11. Pyattaev, A., et al.: 3GPP LTE-assisted Wi-Fi-Direct: trial implementation of live D2D technology. ETRI J. **37**, 877–887 (2015)
12. Third Generation Partnership Project: TR 36.888: Machine-Type Communications (MTC) User Equipments (UEs) based on LTE. Technical report (2013)
13. Çabuk, U., Kanakis, G., Dalkılıç, F.: LTE direct as a device-to-device network technology: use cases and security. Int. J. Adv. Res. Comput. Commun. Eng. (IJAR-CCE) **5**, 401 (2016)

Author Index

Printed in the United States
By Bookmasters